T0320602

Nonlinear Optical Systems

Guiding graduate students and researchers through the complex world of laser physics and nonlinear optics, this book provides an in-depth exploration of the dynamics of lasers and other relevant optical systems, under the umbrella of a unitary spatio-temporal vision.

Adopting a balanced approach, the book covers traditional as well as special topics in laser physics, quantum electronics and nonlinear optics, treating them from the viewpoint of nonlinear dynamical systems. These include laser emission, frequency generation, solitons, optically bistable systems, pulsations and chaos and optical pattern formation. It also provides a coherent and up-to-date treatment of the hierarchy of nonlinear optical models and of the rich variety of phenomena they describe, helping readers to understand the limits of validity of each model and the connections among the phenomena. It is ideal for graduate students and researchers in nonlinear optics, quantum electronics, laser physics and photonics.

Luigi Lugiato is a Professor Emeritus at the Università dell'Insubria, Como, Italy. He has received numerous honors as a result of his many pioneering contributions in nonlinear optics and quantum optics. These include the Albert A. Michelson Medal of the Franklin Institute, the Quantum Electronics Prize of the European Physical Society, the Max Born Award of the Optical Society of America and the Fermi Prize and Medal of the Italian Physical Society. He is a member of the Academia Europaea and a Fellow of the Optical Society of America, the American Physical Society and the European Physical Society.

Franco Prati is an Associate Professor at the Università dell'Insubria, Como, Italy. He has worked in laser physics and nonlinear optics since 1987. His main contributions concern the study of temporal and spatio-temporal instabilities in nonlinear optical systems. He is the co-author of more than 80 papers in peer-reviewed journals, and he is an Outstanding Referee of the American Physical Society and a member of the Optical Society of America.

Massimo Brambilla is an Associate Professor at the Politecnico di Bari, Italy. He has worked in nonlinear and quantum optics since 1986, participating in the early studies of optical pattern formation and dynamics and contributing to more than 100 journal papers and numerous conference presentations. He is active in theoretical research on nonlinear optical systems, with a special focus on the spatio-temporal dynamics of coherent field and solitonic phenomena. He is a member of the Optical Society of America.

Nonlinear Optical Systems

LUIGI LUGIATO

Università degli Studi dell'Insubria, Italy

FRANCO PRATI

Università degli Studi dell'Insubria, Italy

MASSIMO BRAMBILLA

Politecnico di Bari, Italy

CAMBRIDGE
UNIVERSITY PRESS

CAMBRIDGE
UNIVERSITY PRESS

University Printing House, Cambridge CB2 8BS, United Kingdom

One Liberty Plaza, 20th Floor, New York, NY 10006, USA

477 Williamstown Road, Port Melbourne, VIC 3207, Australia

314-321, 3rd Floor, Plot 3, Splendor Forum, Jasola District Centre, New Delhi - 110025, India

79 Anson Road, #06-04/06, Singapore 079906

Cambridge University Press is part of the University of Cambridge.

It furthers the University's mission by disseminating knowledge in the pursuit of education, learning and research at the highest international levels of excellence.

www.cambridge.org
Information on this title: www.cambridge.org/9781107062672

© L. Lugiato, F. Prati & M. Brambilla 2015

First published 2015

A catalogue record for this publication is available from the British Library

ISBN 978-1-107-06267-2 Hardback

To Vilma, Roberta, Monica

Contents

Preface

The aim of our book is to provide a unified and compact vision of the tree of nonlinear optical models and of the wealth of phenomena that can be described by them. In doing that, we adopt the viewpoint of the general field of nonlinear dynamical systems, even if we keep the treatment at a certain level of simplicity, performing an in-depth analysis but avoiding all of the technicalities which are not strictly necessary.

The discussion encompasses static aspects, temporal phenomena and spatial effects, including both those arising in the longitudinal direction in which the light beam propagates and those which occur in the transverse directions. The selected material gathers and organizes a wealth of knowledge scattered in a vast literature from the sixties of the past century to our days.

The volume is subdivided into three parts of decreasing extent. The first seventeen chapters derive from the fundamental laws which govern electromagnetic radiation and matter, a variety of models that describe the radiation–matter interaction both in free propagation and in optical cavities, and discuss mainly the stationary solutions of such models. Most space is devoted to two-level systems, but attention is paid also to parametric systems and to the effects of atomic coherence in multilevel systems. Part II (Chapters 18–25) illustrates the dynamical aspects of lasers and other amplifying or absorbing systems and, in particular, the onset of instabilities that lead to phenomena of spontaneous pulsations and chaos. Part III (Chapters 26–30) deals with the phenomena which arise in the transverse section of light beams, such as Gaussian modes, spontaneous spatial pattern formation and cavity solitons.

The book combines topics that are usually considered in courses on laser physics/quantum electronics and nonlinear optics. The natural attention to the standard laser is extended to other kinds of laser, such as lasers with saturable absorber or injected signal, to other light sources as the optical parametric oscillator and to passive systems that exhibit optical bistability. The description of classic laser phenomena such us, for example, relaxation oscillations, giant pulses and mode locking, is naturally extended to the spontaneous un-damped pulsations which emerge from temporal instabilities, providing a modern vision of multimodal phenomena in lasers. The same holds for the extension of the classic analysis of Gaussian beams to the spatial instabilities which produce patterns in the beam cross section. These features allow one to connect in a natural way to the vast field of the phenomena which arise in nonlinear dynamical systems in general.

The volume is addressed to students and teachers of graduate courses and to researchers in the areas of nonlinear optics, laser physics/quantum electronics/photonics and dynamics of nonlinear optical systems.

The treatment is consistently limited to the semiclassical theory but, whenever possible, we avail ourselves of the description of the radiation field in terms of photons, as happens, for example, in the description of quadratic and cubic nonlinearities.

All models treated in this book are derived, of course, from the fundamental set of Maxwell equations and the Schrödinger equation. We have taken special care in constructing a solid, coherent and logically compact building of models, with clear interconnections among them and a very linear and economical set of derivations. This point is especially important, because in the literature one meets models of very different kinds, for example models that include explicitly the field propagation, models expressed in terms of modal amplitudes, or models that include the transverse diffraction effects, and it is necessary to have a clear global picture of how all such models branch from the tree of the Maxwell–Schrödinger equations. Particular attention has been devoted to avoiding all phenomenological steps that it is possible to avoid, to describing the physical aspects of all steps with precision and to making quite clear the limits of validity of each model.

The vision provided aims at being comprehensive, but the length of the volume is kept reasonable thanks to a strict selection of the topics discussed and to the fact that for related topics we refer the reader to excellent textbooks, reviews or research articles in the literature. Unavoidably, the selection of the topics presented here is affected by our personal preferences and by the limits of our knowledge.

In this connection, it is necessary to spell out some aspects that in this book are treated only marginally or not treated at all. First, the polarization of light is included only in its simplest configuration of linear polarization, and polarization effects are not discussed. The same is true for fluctuations, not only of quantum but also of classical origin. Noise is considered only when necessary, and its effects are not discussed.

Insofar as the atoms are concerned, they are described only in terms of their internal degrees of freedom, whereas their external degrees, i.e. their motion, are taken into account only to include the collisional broadening of the atomic line or to describe inhomogeneously broadened lasers. This implies that here we do not consider the case of cold atomic systems, i.e. we assume that the atoms are sufficiently hot that their momenta are very large compared with the momentum transfer from the ponderomotive force, hence the atomic motion is unaffected and the atomic density does not change appreciably in time.

We would like to mention also not only that the material presented in this book is better organized than in the literature, but also that a few of the results are even new, being included to improve the balance and the impact of the discussion. Examples can be found in Sections 14.3, 20.3, 21.2, 21.3 and 23.1.

We are grateful to E. Arimondo, S. Coen, W. Firth, A. Gatti, D. Gomila, G.-L. Oppo, M. Saffman, M. San Miguel and G. Tissoni for their precious advice and help.

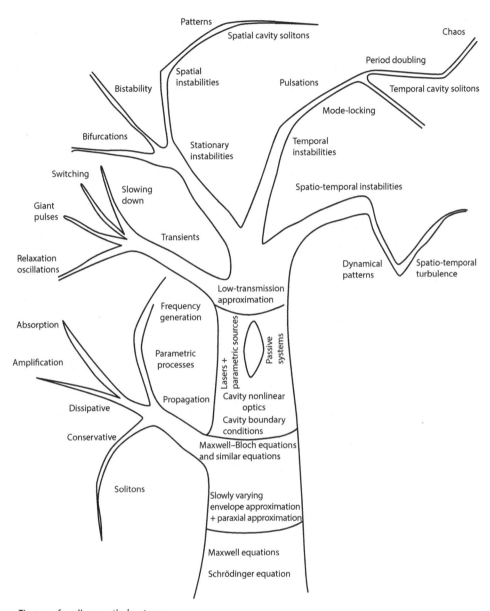

The tree of nonlinear optical systems

PART I

MODELS, PROPAGATION, STATIONARY PHENOMENA

Introduction to Part I

Three elements, introduced in Part I of the book, play a key role in the book itself.

The first is the derivation of the Maxwell–Bloch equations, which is carried out in Chapters 2–4. In particular the derivation of the field-envelope equation, discussed in Section 3.1, is explained with special care. The derivation is based, of course, on the slowly varying envelope approximation and on the paraxial approximation. In textbooks these two approximations are usually discussed separately and rather briefly. We show that the two approximations are intimately connected, and only in this way it is possible to demonstrate that, while the second-order derivatives with respect to time and to the longitudinal coordinate z can be neglected, the second-order derivatives with respect to the transverse coordinates x and y must be kept. It is also possible to demonstrate that the term $\nabla(\nabla \cdot \mathbf{E})$ in the Maxwell wave equation for the electric field can indeed be neglected. An additional bonus of this approach is that, once one has shown in detail the derivation of the envelope equation in the simplest context, it becomes straightforward to generalize it to the more complex cases of frequency dispersion in the background refractive index (Section 5.4) and of quadratic and cubic nonlinearities (Chapters 6 and 7).

The second key element is the use of a multiple limit in the parameters, which in the literature is called the mean-field limit or uniform-field limit, whereas in this book we prefer to name it more precisely the *low-transmission limit*. This plays a central role when we derive modal equations and single-mode models from sets of equations with propagation. In this way one avoids, for example, steps like the phenomenological introduction of a damping term in the field equation(s) to describe the escape of photons from the cavity, and derives the damping from the boundary conditions of the cavity in a transparent way (Chapters 12 and 14). In the book, the low-transmission limit is introduced in Section 9.2.

The last key element is the adiabatic elimination principle, which allows one to notably simplify the dynamical equations and obtain simpler pictures. This is discussed in Chapter 10.

The first chapter of the book illustrates the classic rate-equation model for the laser. This is the only case in which we discuss a phenomenological model. We do that because of the historical importance of this model and for pedagogical reasons. After Chapter 1 we turn to the semiclassical theory, and in Chapter 19 in Part II we derive this model in a rigorous way, with a clear specification of its limits of validity, in the case of class-B lasers.

Chapters 2–7 discuss the radiation–matter interaction in free propagation, whereas the following chapters of the book mainly consider the case in which the atomic medium is contained in an optical cavity. For the sake of simplicity, we consider mainly ring cavities, but in Chapter 14 and Section 22.7 we turn our attention to Fabry–Perot cavities.

The standard laser configuration is considered in Chapters 9, 15 (in this case, with inhomogeneous broadening) and 16. Chapter 16 focusses on the case of semiconductor lasers for two reasons, in addition to the technological importance of such lasers. One reason is that this gives us the chance of showing that, despite the complexity of the physics which governs the system, after a number of steps one arrives at a relatively simple model which, as shown in Section 19.3 in Part II, is identical to that which describes class-B lasers. The other reason is that in Section 30.3 of Part III we describe extensively the topic of cavity solitons in semiconductor lasers.

On the other hand, the case of passive, optically bistable systems is discussed in detail in Chapter 11, while Chapter 13 deals with kinds of light sources that represent variants of the laser, such as the laser with injected signal (Section 13.1) and the laser with saturable absorber (Section 13.2), or other sources of light such as the optical parametric oscillator (Section 13.4).

The final Chapter 17 extends the vision to systems that involve multilevel atoms, with the aim of illustrating some striking phenomena that arise from atomic coherence and quantum interference.

1   The rate-equation model for the laser

The aim of this chapter is to introduce some basic concepts about the light–matter interaction such as absorption, stimulated emission and spontaneous emission, and some basic concepts about the laser, which not only is the most important nonlinear optical system, but also lies at the very root of nonlinear optics, because optical nonlinearities can become manifest only thanks to the power of laser light. We will derive the commonly used *rate equations* for the laser on the basis of phenomenological arguments, and will utilize this model to describe the stationary behavior of the laser. The phenomenological derivation uses the photon concept but, indeed, the quantization of the electromagnetic field is not necessary to formulate the rate-equation model. In the following chapters of this book we will pursue a first-principles treatment, the semiclassical theory, in which the electric field is treated as a classical c-number, and much later, in Chapter 19, we will show how the rate equations for the laser can be obtained from the semiclassical equations. Apart from the use of the semiclassical approximation, the semiclassical approach is complete and allows one to appreciate the exact limits of validity of each model that is derived in its framework. The semiclassical equations allow one to describe the phenomena which arise from the coherence of the light–matter interaction and are not accessible within the rate-equation approximation.

1.1 Absorption, stimulated emission and spontaneous emission

In 1917 Albert Einstein wrote an article [1] in which he was able to provide a physical explanation for the spectral energy density (energy per unit volume and unit frequency) of blackbody radiation [2],

$$u(v, T) = \frac{8\pi h v^3}{c^3} \frac{1}{e^{hv/(k_B T)} - 1},$$
(1.1)

where c is the light velocity in vacuum, T is the absolute temperature and k_B is the Boltzmann constant. He considered a situation of thermal equilibrium between a gas and the electromagnetic radiation, and analyzed the possible mechanisms which induce exchanges of energy between the radiation and the atoms of the gas. For a mode of the radiation field of frequency v equal or close to the atomic Bohr transition frequency $v_a = \Delta E / h$, where ΔE is the energy difference between the two levels involved in the transition and h is Planck's constant, the possible interaction processes between the mode and the atoms are as follows (see Fig. 1.1):

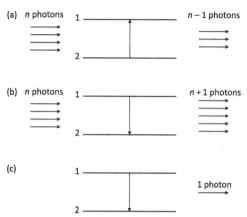

Figure 1.1 The processes of (a) absorption, (b) stimulated emission and (c) spontaneous emission.

(i) Absorption, in which the atom is initially in the lower energy level and goes to the upper state by absorbing a photon of energy $h\nu$.

(ii) Stimulated emission, in which the atom is initially in the upper state, and the interaction with the radiation mode causes the transition to the lower level with the simultaneous emission of an additional photon identical to the mode photons, i.e. with the same energy, momentum and polarization.

(iii) Spontaneous emission, in which the atom is initially in the upper state and the radiation mode has no photons at all (the vacuum state). The atom decays spontaneously to the lower level, with the simultaneous emission of a photon of energy $h\nu_a$. The decay law is exponential as in the case of radioactive decay, i.e. the probability that at time t the atom is still in the upper level is given by $p(t) = \exp(-At)$, where A is the inverse of the lifetime of the upper state. The photon is emitted in a random direction, whereas in stimulated emission the direction is that of the incoming photons.

Note the substantial difference between the first two processes and the third. In the cases of absorption and stimulated emission the transition is induced by the interaction with an electromagnetic field of nonvanishing energy, so these processes have a classical analogue represented by a forced harmonic oscillator that exchanges energy with an external system. In contrast, in spontaneous emission the electromagnetic field has initially zero energy and a fully quantum treatment is necessary to describe this process, i.e. not only the atom, but also the radiation field, must be quantized. In 1917 the theory for this quantization had not been worked out yet, therefore Einstein introduced spontaneous emission in a phenomenological way, showing that it is only by taking this process into account that one can justify Planck's law.

The transition rates per unit time in the processes of absorption and stimulated emission are proportional to the numbers of atoms N_2 and N_1 which are in the lower and in the upper state, respectively, and to the number of photons or, in the case of blackbody radiation, to the energy spectral density of the electromagnetic field. The spontaneous-emission rate is instead proportional only to the number of atoms in the upper state and the proportionality coefficient is A. If we indicate the upper level with subscript 1 and the lower level with subscript 2, we can define the three rates at which these processes take place, in the following way:

$$R_{\text{st.em.}} = B_{1\to 2}N_1 u(\nu), \quad R_{\text{abs}} = B_{2\to 1}N_2 u(\nu), \quad R_{\text{sp.em.}} = AN_1, \tag{1.2}$$

where $B_{2\to 1}$ and $B_{1\to 2}$ are the proportionality coefficients for stimulated emission and absorption, respectively, which a priori are different from each other. At this point we can write the rate equations which govern the time evolution of the populations N_1 and N_2 of the two levels, by simply equating the time derivative of each variable to the sum of the rates, each with the appropriate sign, of the transition processes involved:

$$\dot{N}_1 = R_{\text{abs}} - R_{\text{st.em.}} - R_{\text{sp.em.}} = [B_{2\to 1}N_2 - B_{1\to 2}N_1]\,u(\nu) - AN_1, \tag{1.3}$$

$$\dot{N}_2 = -R_{\text{abs}} + R_{\text{st.em.}} + R_{\text{sp.em.}} = [-B_{2\to 1}N_2 + B_{1\to 2}N_1]\,u(\nu) + AN_1, \tag{1.4}$$

where we have used the standard notation $\dot{N}_i$ to indicate the derivative of N_i with respect to time. The fact that $\dot{N}_1 + \dot{N}_2 = 0$ is the obvious consequence of the circumstance that the total number of atoms $N = N_1 + N_2$ is constant in time. At thermal equilibrium with temperature T the time derivatives vanish, so

$$u(\nu, T) = \frac{A}{B_{2\to 1}\dfrac{N_2}{N_1} - B_{1\to 2}}. \tag{1.5}$$

On the other hand, Boltzmann's thermal distribution prescribes that

$$\frac{N_2}{N_1} = \frac{e^{-E_2/(k_{\text{B}}T)}}{e^{-E_1/(k_{\text{B}}T)}} = e^{\Delta E/(k_{\text{B}}T)} = e^{h\nu/(k_{\text{B}}T)}, \tag{1.6}$$

where $\Delta E = E_1 - E_2 = h\nu$ is the energy difference between the two levels. Hence

$$u(\nu, T) = \frac{A}{B_{2\to 1}e^{h\nu/(k_{\text{B}}T)} - B_{1\to 2}}. \tag{1.7}$$

This formula reduces to Planck's law (1.1), and becomes identical to it if one assumes that $B_{1\to 2} = B_{2\to 1} \equiv B$ and that

$$\frac{A}{B} = \frac{8\pi h\nu^3}{c^3} = \frac{2\hbar\omega^3}{\pi c^3}, \tag{1.8}$$

where $\hbar = h/(2\pi)$. This is the conclusion of Einstein's work, which leaves the two co-efficients A and B undetermined. To calculate the B coefficient it is sufficient to use a semiclassical description, in which the atom is described quantum mechanically whereas the radiation field is described classically. To calculate the coefficient A it is necessary to quantize the field, and the result satisfies Eq. (1.8). In the following we calculate B and, assuming the validity of Eq. (1.8), we obtain A.

1.2 Calculation of the B coefficient

In order to calculate the B coefficient we will utilize the standard perturbative treatment of the radiation–matter interaction [3–5]. Let us consider a harmonic perturbation of the free-atom Hamiltonian

$$H_1 = T_- e^{-i\omega t} + T_+ e^{i\omega t}, \tag{1.9}$$

with $T_+ = T_-^\dagger$, and ω is close to resonance with the two-level transition frequency $\omega \approx \omega_a = \Delta E / \hbar$. Hence the transition probability at time t from the lower to the upper level is given by

$$P_{2\to1}(t) = \frac{4}{\hbar^2} |(T_-)_{12}|^2 \frac{\sin^2[(\omega_a - \omega)t/2]}{(\omega_a - \omega)^2}, \tag{1.10}$$

with $(T_-)_{12} = \langle 1|T_-|2\rangle$, where we use Dirac's notation and $|i\rangle$ ($i = 1, 2$) are the states of the two-level atom. In the dipole approximation (see the next chapter) and considering a monochromatic plane wave electric field $\mathbf{E} = E_0 \cos(kz - \omega t)\hat{\mathbf{e}}$, where $\hat{\mathbf{e}}$ is a unit vector, orthogonal to the z axis, which indicates the field polarization, we have

$$(T_-)_{12} = -\frac{1}{2}\mathbf{d}\cdot\hat{\mathbf{e}}\, E_0 e^{ikz}, \tag{1.11}$$

where $\mathbf{d} = \langle 1|e\mathbf{r}|2\rangle$ is the electric-dipole matrix element between levels 1 and 2 and e is the electron charge. Hence the transition probability (1.10) becomes

$$P_{2\to1}(t) = \frac{(\mathbf{d}\cdot\hat{\mathbf{e}})^2 E_0^2}{4\hbar^2} \frac{\sin^2[\pi(\nu_a - \nu)t]}{\pi^2(\nu_a - \nu)^2}, \tag{1.12}$$

with $\omega = 2\pi\nu$ and $\omega_a = 2\pi\nu_a$. The energy density associated with the wave is

$$w = \frac{1}{2}(\epsilon_0 \mathbf{E}\cdot\mathbf{E} + \mu_0 \mathbf{H}\cdot\mathbf{H}) = \epsilon_0 \mathbf{E}\cdot\mathbf{E} = \epsilon_0 E_0^2 \cos^2(kz - \omega t), \tag{1.13}$$

where $\mathbf{H}$ is the magnetic field and ϵ_0 and μ_0 are the vacuum dielectric constant and magnetic permittivity, respectively. Hence the time-averaged energy amounts to $\overline{w} = \epsilon_0 E_0^2/2$, so the transition probability can be written as

$$P_{2\to1}(t) = \frac{1}{2\epsilon_0}\left(\frac{\mathbf{d}\cdot\hat{\mathbf{e}}}{\hbar}\right)^2 \overline{w} \frac{\sin^2[\pi(\nu_a - \nu)t]}{\pi^2(\nu_a - \nu)^2}. \tag{1.14}$$

This equation holds for a monochromatic field. When, as in the case of a blackbody, the radiation has a broad spectrum, with energy spectral density $u(\nu)$, we can set $\overline{w} = u(\nu)d\nu$ and, by integrating over the spectrum, we obtain

$$P_{2\to1}(t) = \frac{1}{2\epsilon_0}\left(\frac{\mathbf{d}\cdot\hat{\mathbf{e}}}{\hbar}\right)^2 t^2 \int_0^\infty \frac{\sin^2[\pi(\nu_a - \nu)t]}{[\pi(\nu_a - \nu)t]^2} u(\nu)d\nu. \tag{1.15}$$

The integral involves the square of a function sinc $x = \sin x/x$, with $x = \pi(\nu_a - \nu)t$, peaked at ν_a, multiplied by $u(\nu)$. The width of the sinc^2 function is determined by the first two minima around the central maximum, which lie at $\nu_a \pm 1/t$. Therefore, if $1/t$ is much smaller than the frequency band of the field we can calculate the integral by evaluating u in ν_a and taking it out of the integral, obtaining

$$P_{2\to1}(t) = \frac{1}{2\epsilon_0}\left(\frac{\mathbf{d}\cdot\hat{\mathbf{e}}}{\hbar}\right)^2 \frac{t}{\pi} u(\nu_a) \int_{-\infty}^\infty \text{sinc}^2 x \, dx = \frac{1}{2\epsilon_0}\left(\frac{\mathbf{d}\cdot\hat{\mathbf{e}}}{\hbar}\right)^2 u(\nu_a)t, \tag{1.16}$$

where we have extended the lower limit of the integration domain to $-\infty$ because this changes the result in an irrelevant way. This amounts to treating the function sinc$^2 x/\pi$ as a Dirac delta function. As a consequence, the transition probability no longer oscillates in time but grows linearly with time. We must also take into account that the polarization direction of the field assumes all possible values, so that we must integrate over the solid angle Ω

$$\overline{P_{2\to 1}} = \frac{1}{2\epsilon_0 \hbar^2} u(\nu_a) t \frac{1}{4\pi} \int d\Omega\, d^2 \cos^2\theta = \frac{d^2}{6\epsilon_0 \hbar^2} u(\nu_a) t, \tag{1.17}$$

with d being the modulus of $\mathbf{d}$ and θ being the angle between the vectors $\mathbf{d}$ and $\hat{\mathbf{e}}$, taking into account that $\int d\Omega \cos^2\theta = 4\pi/3$. In conclusion the transition probability per unit time, i.e. the rate R_{abs} for the absorption process, is given by

$$\frac{R_{abs}}{N_2} = \frac{d\,\overline{P_{2\to 1}}}{dt} = \frac{d^2}{6\epsilon_0 \hbar^2} u(\nu_a). \tag{1.18}$$

In the same way one verifies that the stimulated emission rate per atom $R_{st.em.}/N_1$ has the same expression as R_{abs}/N_2. All of this implies that

$$B = \frac{d^2}{6\epsilon_0 \hbar^2} \quad \to \quad A = \frac{d^2 \omega_a^3}{3\pi \epsilon_0 \hbar c^3}, \tag{1.19}$$

where we have taken Eq. (1.8) into account.

1.3 The laser

From the equality of the B coefficients for absorption and stimulated emission, it follows that the latter process will prevail over the first if $N_1 > N_2$. Under such conditions the atoms are out of the thermal equilibrium state and the radiation will be amplified as long as it propagates along the medium. Furthermore, since the photons emitted by stimulated emission are in phase with those which generate them, the output radiation will be coherent. This is the basic laser principle; the acronym LASER means light amplification by stimulated emission of radiation. In order to construct a laser, it is therefore necessary to realize a situation of *population inversion* in the atomic medium with respect to equilibrium. This configuration is obtained by appropriate *pump mechanisms*, which are contrasted by the spontaneous decay of the atom, that occurs with rate A. As we see from Eq. (1.19), A scales as the third power of the frequency, and this feature limits the frequency at which a laser can operate. On the other hand, spontaneous emission has also a fundamental positive role in the laser operation, because it originates the first photon which starts the emission process (the radiation field has initially no photons), activating the amplification of light by stimulated emission. The spontaneous decay rate of the atoms is enhanced by processes involving inelastic collisions among the atoms (see Chapter 4).

Another essential ingredient in the laser is the optical cavity (Fig. 1.2) which confines most of the radiation inside a finite volume, forcing the photons to travel several times through the atomic medium, thereby multiplying the processes of stimulated emission. The confinement time of the photons in the cavity, however, must be finite because at

A Fabry–Perot cavity with spherical mirrors. The rectangle represents the active medium. The mirrors are partially reflecting in order to let the radiation escape from the cavity.

least one mirror of the cavity must be partially transmitting in order to allow the radiation to escape from the cavity. Summarizing, the laser is a device that generates coherent light by exploiting the phenomenon of stimulated emission. It is formed essentially by an *active* atomic medium, i.e. a medium in which, via a pump mechanism, one realizes the population inversion between the two atomic levels in which the laser transition occurs, and by an optical cavity that provides the necessary feedback mechanism.

On the basis of these considerations it is not difficult to construct a phenomenological dynamical model for the laser. We must, first of all, calculate the probabilities of absorption and stimulated emission in this case, which is different from that of blackbody radiation because in the laser case we are dealing with essentially monochromatic radiation with frequency ν_0. We must observe, on the other hand, that the atom has a finite linewidth, which we call $\gamma_\perp/(2\pi)$, linked to the circumstance that it decays to the lower level. Here we must take into account not only the radiative decay due to spontaneous emission, but also nonradiative processes that arise from the atomic elastic and inelastic collisions and increase the atomic linewidth (see Chapter 4). As a consequence of the finite linewidth, the atoms no longer see a monochromatic wave [6], but instead experience a field with a spectral energy density that, as a consequence of the exponential decay of the atom, turns out to be a Lorentzian,

$$u(\nu) = \frac{\epsilon_0 E_0^2}{2} \frac{\gamma_\perp}{2\pi^2} \frac{1}{[\gamma_\perp/(2\pi)]^2 + (\nu - \nu_0)^2} = \frac{\gamma_\perp \epsilon_0 E_0^2}{\gamma_\perp^2 + 4\pi^2(\nu - \nu_0)^2}, \qquad (1.20)$$

where ν_0 is the center field frequency. Therefore the transition probability is given by Eq. (1.16) with $u(\nu)$ defined by Eq. (1.20), i.e. by

$$P_{2\to 1} = \frac{\gamma_\perp}{2} \left(\frac{dE_0}{\hbar}\right)^2 \frac{1}{\gamma_\perp^2 + 4\pi^2(\nu_a - \nu_0)^2} t, \qquad (1.21)$$

where we have assumed that the field is linearly polarized and that the direction of the dipole moment coincides with that of the electric field, so that $\mathbf{d} \cdot \hat{\mathbf{e}} = d$. If, instead, the atomic dipoles are randomly oriented, we must integrate over the solid angle exactly as in Eq. (1.17) and the result must be therefore multiplied by $1/3$. In the remainder of this book we will assume that $\mathbf{d} \cdot \hat{\mathbf{e}} = d$. By introducing the *atomic detuning parameter*

$$\Delta = \frac{\omega_a - \omega_0}{\gamma_\perp}, \qquad (1.22)$$

which measures the off-resonance between the field and the atoms since $\omega_a = 2\pi \nu_a$ and $\omega_0 = 2\pi \nu_0$, we can write

$$P_{2\to 1} = \frac{1}{2\gamma_\perp} \left(\frac{dE_0}{\hbar}\right)^2 \frac{1}{1+\Delta^2} t. \tag{1.23}$$

On the other hand, if we introduce the Rabi frequency (not to be confused with the solid angle in Eq. (1.17))

$$\Omega = \frac{dE_0}{\hbar}, \tag{1.24}$$

we can write

$$P_{2\to 1} = \frac{1}{1+\Delta^2} \frac{\Omega^2 t}{2\gamma_\perp}. \tag{1.25}$$

Hence, the rates of absorption and stimulated emission read $R_{abs} = RN_2$ and $R_{st.em.} = RN_1$, with

$$R = \frac{d}{dt} P_{2\to 1} = \frac{1}{1+\Delta^2} \frac{\Omega^2}{2\gamma_\perp}. \tag{1.26}$$

The rate equations for the populations N_1 and N_2 for a laser can be therefore written as Eqs. (1.3) and (1.4) by adding a term that describes the presence of the pump. Let us set

$$R_{pump} = \gamma_\uparrow N_2, \tag{1.27}$$

i.e. let us assume that the pumping, which brings the atoms from the lower to the upper level, is proportional to the number of atoms in the lower level, with a proportionality coefficient $\gamma_\uparrow$ that depends on the pump intensity, and is a controllable parameter, in contrast with A and B. Hence the rate equations for N_1 and N_2 read

$$\dot{N}_1 = R(N_2 - N_1) - \gamma_\downarrow N_1 + \gamma_\uparrow N_2, \tag{1.28}$$
$$\dot{N}_2 = R(N_1 - N_2) + \gamma_\downarrow N_1 - \gamma_\uparrow N_2, \tag{1.29}$$

where the downwards transition rate $\gamma_\downarrow$ is the sum of the spontaneous-emission decay rate A and the nonradiative decay rate arising from inelastic collisions.

Because the total number of atoms $\mathcal{N} = N_1 + N_2$ is constant, these two equations are redundant. It suffices to write one equation for the population inversion $\Delta N = N_1 - N_2 = 2N_1 - \mathcal{N} = \mathcal{N} - 2N_2$,

$$\dot{\Delta N} = 2\dot{N}_1 = -2R\,\Delta N - (\gamma_\uparrow + \gamma_\downarrow)\Delta N + (\gamma_\uparrow - \gamma_\downarrow)\mathcal{N}. \tag{1.30}$$

If we define the parameters

$$\gamma_\parallel = \gamma_\uparrow + \gamma_\downarrow, \qquad \sigma = \frac{\gamma_\uparrow - \gamma_\downarrow}{\gamma_\uparrow + \gamma_\downarrow}, \tag{1.31}$$

we obtain

$$\dot{\Delta N} = -2R\,\Delta N - \gamma_\parallel(\Delta N - \mathcal{N}\sigma). \tag{1.32}$$

This equation shows that in the absence of an electric field ($R = 0$) the population inversion tends to the value $\mathcal{N}\sigma$ given by the balance of the pump and the decay processes. One has population inversion if $\sigma > 0$, i.e. $\gamma_\uparrow > \gamma_\downarrow$, which means that the pump prevails.

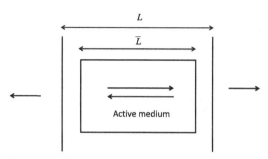

Figure 1.3 A Fabry–Perot cavity with planar mirrors.

A further simplification can be obtained by introducing the new variables D and $\tilde{I}$, which denote the normalized population difference and a normalized intensity, respectively:

$$D = \frac{\Delta N}{\mathcal{N}\sigma}, \qquad \tilde{I} = \frac{2R}{\gamma_\parallel} = \frac{\Omega^2}{\gamma_\perp \gamma_\parallel (1 + \Delta^2)} = \frac{d^2 E_0^2}{\hbar^2 \gamma_\perp \gamma_\parallel (1 + \Delta^2)}. \tag{1.33}$$

Hence the time-evolution equation for D reads

$$\dot{D} = -\gamma_\parallel [D(1 + \tilde{I}) - 1]. \tag{1.34}$$

Up to now we have treated the field intensity as a constant. Actually, it varies in time due to stimulated emission and absorption. Furthermore, the photon number undergoes a loss process because the photons escape from the cavity at a rate that, as we will show in Chapter 12, is given by

$$2\kappa = 2\frac{cT}{2L}, \tag{1.35}$$

where κ is given by the product of the inverse of the roundtrip transit time $2L/c$ in the cavity (Fig. 1.3), where L is the cavity length, and the transmissivity coefficient T (not to be confused with temperature) of the mirrors of the cavity. Let us now write the time-evolution equation for the number of photons in the cavity, m:

$$\dot{m} = -2\kappa m + RN_1 - RN_2 = -2\kappa m + R\mathcal{N}\sigma D = 2\kappa \left(\frac{R\mathcal{N}\sigma L}{cT} D - m \right), \tag{1.36}$$

where we have neglected spontaneous emission. Note that m denotes the number of photons traveling in the direction of the cavity axis and that spontaneous-emission photons are emitted in all directions. We neglect the contribution of photons spontaneously emitted along the cavity axis, i.e. the photons which start the laser emission process. This contribution can be described only in a full quantum treatment, and this neglect is consistent with the semiclassical approximation used in this book.

The first term of Eq. (1.36) which appears within the bracket is proportional to the field intensity and therefore to the photon number. As a matter of fact, the photon number m is equal to the energy of the electromagnetic field divided by the single-photon energy $\hbar\omega_0$, $\omega_0 = 2\pi\nu_0$. The energy of the electromagnetic field is equal to the product of the energy density per unit volume and the volume of the cavity. If for the sake of theoretical simplicity we consider a Fabry–Perot cavity with planar mirrors (Fig. 1.3), where V is the volume of the active medium and $\bar{L}$ is its length, then, assuming that the cavity and the active region have the same section, the cavity volume is given by $VL/\bar{L}$ and therefore the photon number corresponds to

$$m = \frac{VL\bar{w}}{\bar{L}\hbar\omega_0} = \frac{VL}{\bar{L}\hbar\omega_0}\frac{\epsilon_0 E_0^2}{2}. \tag{1.37}$$

Therefore, using Eqs. (1.26) and (1.37), R can be written as

$$R = \frac{1}{2}\frac{d^2 E_0^2}{\hbar^2 \gamma_\perp (1 + \Delta^2)} = \frac{d^2 \omega_0 \bar{L}}{\epsilon_0 \hbar \gamma_\perp (1 + \Delta^2) VL}m, \tag{1.38}$$

Hence we can set $R \mathcal{N} \sigma L/(cT) = am$ with

$$a = \frac{d^2 \omega_0 N \sigma \bar{L}}{\epsilon_0 \hbar \gamma_\perp (1 + \Delta^2) V c T}, \tag{1.39}$$

and the equation for m becomes

$$\dot{m} = 2\kappa(aD - 1)m. \tag{1.40}$$

Since m is proportional to $\tilde{I}$ we obtain immediately from this equation the corresponding equation for $\tilde{I}$,

$$\dot{\tilde{I}} = 2\kappa(aD - 1)\tilde{I}, \tag{1.41}$$

which can be coupled to Eq. (1.34). Finally it is convenient to introduce the normalized decay rate

$$\gamma = \frac{\gamma_\|}{2\kappa}, \tag{1.42}$$

and to indicate with a prime the derivative with respect to the normalized time $2\kappa t$, so that Eqs. (1.41) and (1.34) take the final forms

$$\tilde{I}' = (aD - 1)\tilde{I}, \tag{1.43}$$
$$D' = \gamma\left[1 - D(1 + \tilde{I})\right]. \tag{1.44}$$

Let us now discuss the stationary solutions of these dynamical equations, which coincide with the asymptotic solutions that the system approaches in the long time limit. If we set the time derivatives equal to zero in Eqs. (1.43) and (1.44) we get two stationary solutions,

$$\tilde{I}_s = 0, \qquad D_s = 1 \tag{1.45}$$

and

$$\tilde{I}_s = a - 1, \qquad D_s = \frac{1}{a}, \tag{1.46}$$

where the solution (1.46) exists only for $a > 1$, because $\tilde{I}$ is positive by definition. The value of the parameter a determines two regions: the region $a < 1$, where the laser is below threshold and the emitted intensity is zero if spontaneous emission is neglected; and the region $a > 1$, where the solution (1.45) is unstable and therefore unphysical, and the laser emits coherent radiation [6]. The value $a = 1$ marks the laser threshold. The diagram of $\tilde{I}_s$

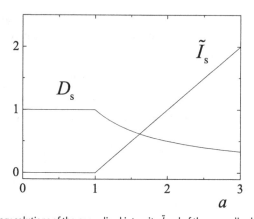

Figure 1.4   A plot of the stable stationary solutions of the normalized intensity $\tilde{I}$ and of the normalized population difference D as functions of the parameter a.

and D_s as functions of the parameter a is shown in Fig. 1.4. Note that in correspondence to the threshold there is a discontinuity in the derivative both in $\tilde{I}_s$ and in D_s.

Three remarks are in order. The first is that the stability of the stationary solutions will be analyzed in Part II of this book. The second is that, in a fully quantum theory in which one takes into account the effect of spontaneous emission in the direction of the cavity, the average stationary emitted intensity is not zero below threshold, and the emitted light is fully incoherent when the system is well below threshold and fully coherent when it is well above threshold. The last remark is that, using the semiclassical approach, one can calculate also the frequency of the field emitted by the laser, which is not possible on the basis of the treatment of this chapter. A partial list of textbooks on laser physics is [6–21].

2 The interaction of a system of two-level atoms with the electromagnetic field

The concept of a *two-level atom* is clearly a schematization. However, it amounts to an excellent approximation of reality in all those cases in which the electromagnetic field is resonant or quasi-resonant with the transition between the two levels in play. In this chapter we derive the *optical Bloch equations* which describe in the *dipole* and *rotating-wave approximations* the interaction of a two-level atom with the radiation field. They are a set of time-reversible equations exactly like the Schrödinger (or Liouville–Von Neumann) equation from which they are derived. The dissipative processes necessary to describe real physical systems will be introduced in Chapter 4.

The formalism for a two-level atom is mathematically equivalent to that for a spin 1/2 that interacts with a magnetic field that is periodically oscillating in time. In 1937 Rabi studied this problem and demonstrated that the spin exhibits a nutation around the magnetic field, passing periodically from the ↑ state to the ↓ state and vice versa. These are called *Rabi oscillations* and are the basics of nuclear magnetic resonance.

As a consequence of its analogy with the spin, also a two-level atom exhibits *optical* Rabi oscillations under the action of a monochromatic electromagnetic field. In the perfectly resonant configuration the analogy is complete. In this case the atom oscillates between the two levels at a frequency equal to the Rabi frequency. The exact treatment coincides with the description provided by the time-dependent perturbation theory in the limit in which the Rabi frequency is much smaller than the modulus of the atomic detuning.

2.1 The interaction Hamiltonian in the dipole approximation

Let us consider an effective one-electron atom such as an alkali atom. The interaction of the electron with the electromagnetic field is described by the Hamiltonian [3, 22]

$$H = \frac{1}{2m}[\mathbf{p} - e\mathbf{A}(\mathbf{x}, t)]^2 + eV(\mathbf{x}, t) + U(r), \tag{2.1}$$

where m and e are the mass and the charge of the electron and $\mathbf{x} = \mathbf{r}_0 + \mathbf{r}$ is the position vector of the electron in a given reference frame, with $\mathbf{r}$ its position vector with respect to the nucleus and $\mathbf{r}_0$ the position vector of the nucleus. In our treatment we do not quantize the external degrees of freedom of the atom and therefore only $\mathbf{r}$, which is an internal coordinate, is treated as an operator, whereas $\mathbf{r}_0$ is a c-number vector. The symbol $\mathbf{p}$ denotes the canonical momentum operator conjugate to $\mathbf{r}$. $\mathbf{A}(\mathbf{x}, t)$ and $V(\mathbf{x}, t)$ are the vector and scalar potentials of the external electromagnetic field and $U(r)$ is the spherically symmetric potential which describes how the electron is bound to the atom.

The electric and magnetic fields are expressed in terms of the vector and scalar potentials as follows:

$$\mathbf{E} = -\nabla V - \frac{\partial \mathbf{A}}{\partial t}, \tag{2.2}$$

$$\mathbf{B} = \nabla \times \mathbf{A}. \tag{2.3}$$

We use the Coulomb gauge in which we assume that $\nabla \cdot \mathbf{A} = 0$. We assume also that $V = 0$, so that the electromagnetic field is fully described by the vector potential $\mathbf{A}(\mathbf{r}_0 + \mathbf{r}, t)$. We also suppose that the radiation is a monochromatic plane wave with wavelength λ.

In the dipole approximation, one assumes that the atomic size is much smaller than the wavelength, i.e. $|\mathbf{k} \cdot \mathbf{r}| \ll 1$, where $\mathbf{k}$ is the wave vector, which is linked to the wavelength by the relation $k = 2\pi/\lambda$. In this limit, which holds very well at optical frequencies, we can approximate the vector potential in the following way:

$$\begin{aligned}
\mathbf{A}(\mathbf{r}_0 + \mathbf{r}, t) &= \tilde{\mathbf{A}}(t)\exp[i\mathbf{k} \cdot (\mathbf{r}_0 + \mathbf{r})] + \text{c.c.} \\
&= \tilde{\mathbf{A}}(t)\exp(i\mathbf{k} \cdot \mathbf{r}_0)(1 + i\mathbf{k} \cdot \mathbf{r} + \cdots) + \text{c.c.} \\
&\approx \tilde{\mathbf{A}}(t)\exp(i\mathbf{k} \cdot \mathbf{r}_0) + \text{c.c.} = \mathbf{A}(\mathbf{r}_0, t).
\end{aligned} \tag{2.4}$$

Since we quantize neither the external degrees of freedom of the atom nor the electromagnetic field, the vector potential reduces to a c-number vector, because any dependence on the position operator $\mathbf{r}$ has disappeared. The Schrödinger equation for the wave vector ψ of the electron, in the Coulomb gauge and in the dipole approximation, therefore reads

$$H\psi(\mathbf{r}, t) = \left\{ \frac{1}{2m}[\mathbf{p} - e\mathbf{A}(\mathbf{r}_0, t)]^2 + U(r) \right\} \psi(\mathbf{r}, t) = i\hbar \frac{\partial \psi(\mathbf{r}, t)}{\partial t}. \tag{2.5}$$

The Hamiltonian can be rephrased in a form similar to the classical one, in which the interaction energy between the electron and the field is given by $-\mathbf{d} \cdot \mathbf{E}$, by introducing a gauge transformation defined by the unitary operator

$$T(\mathbf{r}, t) = \exp\left[-i\frac{e}{\hbar}\mathbf{r} \cdot \mathbf{A}(\mathbf{r}_0, t) \right]. \tag{2.6}$$

By applying this transformation to the Schrödinger equation we obtain

$$THT^{\dagger}T\psi = i\hbar T\frac{\partial \psi}{\partial t}. \tag{2.7}$$

Now $TUT^{\dagger} = U$, where the dagger indicates the adjoint operator, because the potential depends only on the position vector $\mathbf{r}$. Furthermore, we obtain

$$T[\mathbf{p} - e\mathbf{A}(\mathbf{r}_0, t)]^2 T^{\dagger} = \{T[\mathbf{p} - e\mathbf{A}(\mathbf{r}_0, t)]T^{\dagger}\}^2 = [T\mathbf{p}T^{\dagger} - e\mathbf{A}(\mathbf{r}_0, t)]^2. \tag{2.8}$$

Since $\mathbf{p} = -i\hbar\,\nabla$ and $\nabla(\mathbf{r}\cdot\mathbf{A}) = \mathbf{A}$ for any vector $\mathbf{A}$ independently of $\mathbf{r}$, we get

$$
\begin{aligned}
T\mathbf{p}T^{\dagger}\psi &= T(-i\hbar\,\nabla)\left\{\exp\left[i\frac{e}{\hbar}\mathbf{r}\cdot\mathbf{A}(\mathbf{r}_0,t)\right]\psi\right\} \\
&= T\left\{\exp\left[i\frac{e}{\hbar}\mathbf{r}\cdot\mathbf{A}(\mathbf{r}_0,t)\right][-i\hbar\,\nabla\psi + e\mathbf{A}(\mathbf{r}_0,t)\psi]\right\} \\
&= [\mathbf{p} + e\mathbf{A}(\mathbf{r}_0,t)]\psi,
\end{aligned}
\tag{2.9}
$$

which implies that

$$
THT^{\dagger} = \frac{\mathbf{p}^2}{2m} + U(r) = H_0,
\tag{2.10}
$$

where H_0 is the free-atom Hamiltonian. Furthermore

$$
T\frac{\partial\psi}{\partial t} = \frac{\partial(T\psi)}{\partial t} - \frac{\partial T}{\partial t}\psi,
\tag{2.11}
$$

and

$$
\frac{\partial T}{\partial t} = -i\frac{e}{\hbar}\mathbf{r}\cdot\frac{\partial\mathbf{A}(\mathbf{r}_0,t)}{\partial t}T = \frac{i}{\hbar}e\mathbf{r}\cdot\mathbf{E}T.
\tag{2.12}
$$

Hence by defining the new wave function $\psi' = T\psi$ the Schrödinger equation is rewritten in the form

$$
H_0\psi' = i\hbar\left[\frac{\partial\psi'}{\partial t} - \frac{i}{\hbar}e\mathbf{r}\cdot\mathbf{E}\psi'\right],
\tag{2.13}
$$

i.e., by omitting the apex,

$$
(H_0 + H_1)\psi = i\hbar\frac{\partial\psi}{\partial t}, \qquad H_1 = -e\mathbf{r}\cdot\mathbf{E}.
\tag{2.14}
$$

The gauge transformation introduces a local dependence in the phase of the wave function and, with respect to the potentials of the radiation field, induces the transition from the Coulomb gauge $(\mathbf{A}, 0)$ to the gauge $(0, -\mathbf{r}\cdot\mathbf{E})$. From now on, in all equations, we replace the symbol $\mathbf{r}_0$ by the symbol $\mathbf{x}$.

2.2 The two-level atom and its analogy with spin 1/2

Among the infinitely many energy levels of the atom, let us consider two levels with energies E_1 and E_2, with $E_1 > E_2$, such that the Bohr transition frequency

$$
\omega_a = \frac{E_1 - E_2}{\hbar}
\tag{2.15}
$$

is close to resonance with the frequency of the monochromatic radiation field.

A two-level atom [23–25] is formally described by a two-dimensional Hilbert space in which we select the orthonormal basis formed by the two eigenstates $|1\rangle$ (upper level) and $|2\rangle$ (lower level) of the free-atom Hamiltonian H_0. If we use the matrix notation we can write the two basis vectors as

$$|1\rangle = \begin{pmatrix} 1 \\ 0 \end{pmatrix}, \qquad |2\rangle = \begin{pmatrix} 0 \\ 1 \end{pmatrix}, \tag{2.16}$$

with $H_0|1\rangle = E_1|1\rangle$ and $H_0|2\rangle = E_2|2\rangle$. The generic state of the atom can be described as

$$|\psi\rangle = a|1\rangle + b|2\rangle = \begin{pmatrix} a \\ b \end{pmatrix}, \tag{2.17}$$

whereas the generic observable O is described by the Hermitian matrix

$$O = \begin{pmatrix} O_{11} & O_{12} \\ O_{21} & O_{22} \end{pmatrix}, \qquad O_{ij} = \langle i|O|j\rangle, \tag{2.18}$$

with O_{11} and O_{22} real and $O_{21} = O_{12}^*$, where * indicates complex conjugation, since $O = O^\dagger$. The analogy with the case of a spin 1/2 is evident, in particular the two levels $|1\rangle$ and $|2\rangle$ of the two-level atom correspond to the states $|\uparrow\rangle$ and $|\downarrow\rangle$ of the spin, respectively.

Let us now consider the observables which are most relevant for the two-level atom. If we fix the energy reference level halfway between the two levels, so that $E_1 = \hbar\omega_a/2$ and $E_2 = -\hbar\omega_a/2$, the free-atom Hamiltonian is given by

$$H_0 = \begin{pmatrix} \hbar\omega_a/2 & 0 \\ 0 & -\hbar\omega_a/2 \end{pmatrix} = \frac{\hbar\omega_a}{2}\sigma_z, \tag{2.19}$$

where

$$\sigma_z = \begin{pmatrix} 1 & 0 \\ 0 & -1 \end{pmatrix}$$

is the third Pauli matrix. The dipole moment of the electron er corresponds to the matrix

$$er = \begin{pmatrix} 0 & \mathbf{d} \\ \mathbf{d} & 0 \end{pmatrix} = \mathbf{d}\sigma_x, \tag{2.20}$$

where

$$\sigma_x = \begin{pmatrix} 0 & 1 \\ 1 & 0 \end{pmatrix}$$

is the first Pauli matrix and we have assumed that the two states $|1\rangle$ and $|2\rangle$ have well-defined and opposite parity so that

$$\mathbf{d}_{11} = \mathbf{d}_{22} = 0, \qquad \mathbf{d}_{12} = \mathbf{d}_{21}^* = \langle 1|er|2\rangle = \mathbf{d}, \tag{2.21}$$

and the same value for the magnetic quantum number m, i.e. in the transition one has $\Delta m = 0$, so that $\mathbf{d}$ is real [23].

Therefore, in the dipole approximation the Hamiltonian which includes the interaction of the electron with an external electric field $\mathbf{E}$ reads

$$H = H_0 - er \cdot \mathbf{E} = \frac{\hbar}{2}\begin{pmatrix} \omega_a & -\Omega_R \\ -\Omega_R & -\omega_a \end{pmatrix}, \tag{2.22}$$

where we have introduced the *Rabi frequency*

$$\Omega_R = \frac{2\mathbf{d} \cdot \mathbf{E}}{\hbar}. \tag{2.23}$$

2.3 The rotating-wave approximation. Optical Bloch equations

Since we are interested in a statistical description of the atom under the action of the radiation field, let us consider the temporal evolution of the statistical operator (or density operator) [26] $\rho(\mathbf{x}, t)$ of a two-level atom in the position $\mathbf{x}$, which provides all information about the state of the atom, allowing one to calculate at every instant the mean value of any observable O since it follows

$$\langle O \rangle(\mathbf{x}, t) = \text{Tr}[\rho(\mathbf{x}, t)O], \tag{2.24}$$

where Tr indicates the trace operation.

The statistical operator is characterized by the following properties: (i) it is Hermitian $\rho^\dagger = \rho$; (ii) it is positive definite $\rho \geq 0$, i.e. $\langle \psi | \rho | \psi \rangle \geq 0$ for any vector $|\psi\rangle$; and (iii) it has unit trace $\text{Tr}\, \rho = 1$. Furthermore, if the system is in a pure state so that $\rho = |\psi\rangle\langle\psi|$, one has that also $\text{Tr}\, \rho^2 = 1$.

In matrix notation one introduces the density matrix which, for a two-dimensional Hilbert space, has the form

$$\rho = \begin{pmatrix} \rho_{11} & \rho_{12} \\ \rho_{21} & \rho_{22} \end{pmatrix}, \qquad \rho_{ij} = \langle i | \rho | j \rangle = \rho_{ji}^*. \tag{2.25}$$

Next, let us consider the projection operators P_1 and P_2 onto the states $|1\rangle$ and $|2\rangle$, respectively:

$$P_1|\psi\rangle = \begin{pmatrix} 1 & 0 \\ 0 & 0 \end{pmatrix} \begin{pmatrix} a \\ b \end{pmatrix} = \begin{pmatrix} a \\ 0 \end{pmatrix}, \qquad P_2|\psi\rangle = \begin{pmatrix} 0 & 0 \\ 0 & 1 \end{pmatrix} \begin{pmatrix} a \\ b \end{pmatrix} = \begin{pmatrix} 0 \\ b \end{pmatrix}. \tag{2.26}$$

The diagonal elements ρ_{11} and ρ_{22} coincide with the mean values of P_1 and P_2, respectively,

$$\langle P_1 \rangle(\mathbf{x}, t) = \text{Tr}[\rho(\mathbf{x}, t)P_1] = \rho_{11}(\mathbf{x}, t), \tag{2.27}$$

$$\langle P_2 \rangle(\mathbf{x}, t) = \text{Tr}[\rho(\mathbf{x}, t)P_2] = \rho_{22}(\mathbf{x}, t). \tag{2.28}$$

Therefore ρ_{11} and ρ_{22} represent the probabilities of finding the atom in the states $|1\rangle$ and $|2\rangle$, respectively, and the $\text{Tr}\, \rho = 1$ property amounts to the condition of the total probability $\rho_{11} + \rho_{22} = 1$. Insofar as the off-diagonal matrix elements are concerned, we remark that

$$\langle e\mathbf{r} \rangle(\mathbf{x}, t) = \text{Tr}[\rho(\mathbf{x}, t)e\mathbf{r}] = \mathbf{d}[\rho_{12}(\mathbf{x}, t) + \rho_{12}(\mathbf{x}, t)^*]. \tag{2.29}$$

Hence these elements are linked to the electric dipole moment induced by the electric field, i.e. to the microscopic polarization of the material. In the presence of N atoms per unit volume, all oriented along the same direction defined by the unit vector $\hat{\mathbf{u}}$, the macroscopic atomic polarization per unit volume (atomic density) is given by

$$\mathbf{P}(\mathbf{x}, t) = N\langle er\rangle(\mathbf{x}, t) = Nd[\rho_{12}(\mathbf{x}, t) + \rho_{12}^*(\mathbf{x}, t)]\hat{\mathbf{u}}. \tag{2.30}$$

The temporal evolution of the statistical operator is governed by the Liouville–von Neumann equation [26]

$$\frac{\partial \rho}{\partial t} = -\frac{i}{\hbar}[H, \rho]. \tag{2.31}$$

The dynamical equations for the matrix elements are

$$\frac{\partial \rho_{ij}}{\partial t} = -\frac{i}{\hbar}\sum_k \left[H_{ik}\rho_{kj}(\mathbf{x}, t) - \rho_{ik}(\mathbf{x}, t)H_{kj}\right]. \tag{2.32}$$

Since $\rho_{21} = \rho_{12}^*$ and $\rho_{22} = 1 - \rho_{11}$, we can limit ourselves to considering only the dynamical equations for the matrix elements ρ_{12} and ρ_{11},

$$\frac{\partial \rho_{12}}{\partial t} = -\frac{i}{\hbar}\{(H_{11} - H_{22})\rho_{12}(\mathbf{x}, t) + H_{12}[\rho_{22}(\mathbf{x}, t) - \rho_{11}(\mathbf{x}, t)]\}, \tag{2.33}$$

$$\frac{\partial \rho_{11}}{\partial t} = -\frac{i}{\hbar}[H_{12}\rho_{21}(\mathbf{x}, t) - H_{21}\rho_{12}(\mathbf{x}, t)]. \tag{2.34}$$

By using Eq. (2.22) for the Hamiltonian of the two-level atom we obtain

$$\frac{\partial \rho_{12}}{\partial t} = -i\omega_a\rho_{12}(\mathbf{x}, t) + \frac{i}{2}\Omega_R(\mathbf{x}, t)[1 - 2\rho_{11}(\mathbf{x}, t)], \tag{2.35}$$

$$\frac{\partial \rho_{11}}{\partial t} = \frac{i}{2}\Omega_R(\mathbf{x}, t)[\rho_{12}(\mathbf{x}, t)^* - \rho_{12}(\mathbf{x}, t)]. \tag{2.36}$$

The dependence of the matrix elements ρ_{12} and ρ_{11} on $\mathbf{x}$ arises from the spatial dependence of Ω_R. Let us now assume that the electric field has the configuration of a plane wave traveling in the direction of the z axis and polarized along the direction of the unit vector $\hat{\mathbf{e}}$ orthogonal to the z axis, and that it is monochromatic with frequency ω_0 equal or close to the atomic transition frequency ω_a.

$$\mathbf{E}(z, t) = \mathcal{E}(z, t)\hat{\mathbf{e}}, \qquad \mathcal{E}(z, t) = \frac{E_0(z, t)}{2}e^{i(k_0 z - \omega_0 t)} + \text{c.c.}, \tag{2.37}$$

where c.c. indicates complex conjugate and k_0 will be defined in Section 3.1. E_0 is an envelope that varies in space and time much more slowly than the carrier wave $e^{i(k_0 z - \omega_0 t)}$. As a consequence of Eq. (2.37), the Rabi frequency (2.23) can be written as

$$\Omega_R(z, t) = \Omega e^{i(k_0 z - \omega_0 t)} + \text{c.c.}, \qquad \Omega = \frac{\mathbf{d} \cdot \hat{\mathbf{e}}\, E_0}{\hbar}. \tag{2.38}$$

Since the polarization of the medium is induced by the electric field $\mathbf{E}$, we can assume that the elementary dipoles $\mathbf{d}$ are oriented like the electric field, i.e. that $\hat{\mathbf{u}} = \hat{\mathbf{e}}$, so that

$$\Omega = \frac{d E_0}{\hbar}. \tag{2.39}$$

Furthermore, the spatial configuration of Ω_R produces the same spatial configuration in the macroscopic polarization via the interaction described by Eq. (2.22), so we can write

$$\rho_{12} = r(z, t)e^{i(k_0 z - \omega_0 t)}, \tag{2.40}$$

where $r(z, t)$ is an envelope function. Correspondingly, using Eq. (2.30) we have

$$\mathbf{P}(z, t) = \mathcal{P}(z, t)\hat{\mathbf{e}}, \qquad \mathcal{P}(z, t) = \frac{P_0(z, t)}{2}e^{i(k_0 z - \omega_0 t)} + \text{c.c.} \qquad (2.41)$$

with

$$P_0(z, t) = 2Ndr(z, t). \qquad (2.42)$$

Furthermore, let us introduce the variable

$$r_3(z, t) = \rho_{11}(z, t) - \rho_{22}(z, t) = 2\rho_{11}(z, t) - 1, \qquad (2.43)$$

so that the quantity Nr_3 has the meaning of population inversion per unit volume, because it is positive when the population in the medium is inverted with respect to thermal equilibrium, i.e. the population of the upper level is larger than that of the lower level. Next, if we insert Eqs. (2.38), (2.40) and (2.43) into (2.35) and (2.36) we obtain

$$\frac{\partial r}{\partial t} = i(\omega_0 - \omega_a)r - \frac{i}{2}\left[\Omega + \Omega^* e^{-2i(k_0 z - \omega_0 t)}\right]r_3, \qquad (2.44)$$

$$\frac{\partial r_3}{\partial t} = i\left[\Omega r^* - \Omega^* r - \Omega r e^{2i(k_0 z - \omega_0 t)} + \Omega^* r^* e^{-2i(k_0 z - \omega_0 t)}\right]. \qquad (2.45)$$

On the right-hand side of each of these equations there are terms that vary slowly in time and space because they depend only on envelopes, and terms that have a fast variation, i.e. they oscillate at a frequency twice the optical frequency. The *rotating-wave approximation* (RWA) amounts to neglecting the fast-varying terms, thereby considering the slow temporal evolution which characterizes the envelopes. In such a way we get

$$\frac{\partial r}{\partial t} = -i\delta r - \frac{i}{2}\Omega r_3, \qquad (2.46)$$

$$\frac{\partial r_3}{\partial t} = i(\Omega r^* - \Omega^* r), \qquad (2.47)$$

where we have introduced the atomic detuning parameter, which measures how much the electric field is off-resonance with respect to the atomic frequency,

$$\delta = \omega_a - \omega_0. \qquad (2.48)$$

Equations (2.46) and (2.47) are called *optical Bloch equations* for a system of two-level atoms.

2.4 The Bloch vector and its nutation

In this section we solve the optical Bloch equations [23, 26] for a given initial condition and we show that the dynamical evolution of a two-level atom under the action of a constant electric-field envelope can be visualized as the nutation of an appropriate vector, called the *Bloch vector*, in an appropriate three-dimensional space associated with the atom. We will assume that the atom is in a pure state $\rho = |\psi\rangle\langle\psi|$, so that the statistical operator obeys the condition $\text{Tr}\,\rho^2 = 1$.

Let us consider the Bloch equations (2.46) and (2.47). The first is a complex equation whereas the second is real. We want, first of all, to rephrase them into three real equations for the variables r_1, r_2 and r_3, with r_1 and r_2 being the two quadrature components of the dipole moment, which multiply the functions $\cos(k_0 z - \omega_0 t)$ and $\sin(k_0 z - \omega_0 t)$. From Eq. (2.29) and using Eq. (2.40) we can write

$$
\begin{aligned}
\langle e\mathbf{r} \rangle &= \mathbf{d} \left[r e^{i(k_0 z - \omega_0 t)} + r^* e^{-i(k_0 z - \omega_0 t)} \right] \\
&= \mathbf{d} \left[r_1 \cos(k_0 z - \omega_0 t) + r_2 \sin(k_0 z - \omega_0 t) \right],
\end{aligned} \tag{2.49}
$$

from which we see that r_1 and r_2 are proportional to the real and to the imaginary part of r, respectively,

$$
r_1 = r + r^* = 2 \operatorname{Re} r, \tag{2.50}
$$

$$
r_2 = i(r - r^*) = -2 \operatorname{Im} r. \tag{2.51}
$$

Let us now assume that the electric-field envelope E_0, and therefore also the Rabi frequency Ω, are real so that

$$
\Omega_R = 2\Omega \cos(k_0 z - \omega_0 t). \tag{2.52}
$$

Hence r_1 and r_2 acquire the meaning of quadrature components of the atomic polarization in phase and in quadrature with the electric field, respectively. Furthermore, let us assume that E_0 and Ω are stationary in time and uniform in space, i.e. constant, so that any spatial dependence disappears from the dynamical equations (2.46) and (2.47) and we can write total derivatives with respect to time instead of partial derivatives. These equations hold also in the case of a single two-level atom instead of a system of two-level atoms. From Eqs. (2.46) and (2.47) we obtain

$$
\dot{r}_1 = -\delta r_2, \tag{2.53}
$$

$$
\dot{r}_2 = \delta r_1 + \Omega r_3, \tag{2.54}
$$

$$
\dot{r}_3 = -\Omega r_2, \tag{2.55}
$$

where we indicate the time derivative by a dot. Next, we can consider r_1, r_2 and r_3 as the components of a vector in a three-dimensional space, and we call it the *Bloch vector*. Its length is given by

$$
\begin{aligned}
r_1^2 + r_2^2 + r_3^2 &= 4 \left[(\operatorname{Re} r)^2 + (\operatorname{Im} r)^2 \right] + r_3^2 = 4|r|^2 + r_3^2 \\
&= 4|\rho_{12}|^2 + (\rho_{11} - \rho_{22})^2.
\end{aligned} \tag{2.56}
$$

Since the atom is initially in a pure state we have that $\operatorname{Tr} \rho^2 = 2|\rho_{12}|^2 + \rho_{11}^2 + \rho_{22}^2 = 1$, hence

$$
r_1^2 + r_2^2 + r_3^2 = 2 - 2\rho_{11}^2 - 2\rho_{22}^2 + \rho_{11}^2 + \rho_{22}^2 - 2\rho_{11}\rho_{22} = 2 - (\rho_{11} + \rho_{22})^2 = 1. \tag{2.57}
$$

Because this equality is satisfied throughout the time evolution, the length of the Bloch vector is constant (and fixed to unity):

$$\frac{d}{dt}(r_1^2 + r_2^2 + r_3^2) = 2(r_1\dot{r}_1 + r_2\dot{r}_2 + r_3\dot{r}_3)$$

$$= 2(-\delta r_1 r_2 + \delta r_2 r_1 + \Omega r_2 r_3 - \Omega r_3 r_2) = 0. \tag{2.58}$$

Therefore the tip of the Bloch vector lies on a sphere of unit radius, which is called the *Bloch sphere*. For $\delta = 0$ Eqs. (2.53)–(2.55) are identical to the original Bloch equations [27, 28] which describe the evolution of a spin 1/2 under the action of a magnetic field directed along the x axis. This is because the atomic dipole moment is proportional to the σ_x Pauli matrix. In the spin case r_1, r_2 and r_3 correspond to the three components of the spin along the axes x, y and z, respectively.

If we introduce the vectorial notation (note, however, that the three components of ρ are not the spatial components of a vector)

$$\rho = \begin{pmatrix} r_1 \\ r_2 \\ r_3 \end{pmatrix}, \qquad \Omega = \begin{pmatrix} -\Omega \\ 0 \\ \delta \end{pmatrix}, \tag{2.59}$$

the dynamical equations for the Bloch vector components can be rewritten in the compact form [23]

$$\dot{\rho} = \Omega \times \rho, \tag{2.60}$$

which is an equation completely identical to that which describes the nutation of a magnetic dipole around the magnetic field. The analogy with the Larmor nutation is reinforced by the circumstance that the Rabi and Larmor frequencies are defined in a closely similar way. We can therefore say that the Bloch vector ρ performs a precession around the vector Ω in such a way that

(a) the angle ζ between ρ and Ω is constant and given by

$$\cos\zeta = \frac{\rho \cdot \Omega}{|\rho||\Omega|}, \tag{2.61}$$

(b) the angular velocity of the nutation is given by the generalized Rabi frequency

$$|\Omega| = \sqrt{\Omega^2 + \delta^2} \equiv \Omega_\delta. \tag{2.62}$$

By definition the vector Ω lies in the 1–3 plane. These remarks allow us to get a qualitative idea of the motion of the Bloch vector and hence of the dynamical evolution of the atom.

We calculate the solution of the Bloch equations corresponding to the initial condition such that the atom is initially in the lower state ($r_3(0) = -1$) or in the upper state ($r_3(0) = 1$). Since the Bloch vector has unit length this implies that $r_1(0) = r_2(0) = 0$, meaning that the initial dipole moment vanishes. Therefore the constant of motion $\rho \cdot \Omega$ amounts to

$$\rho \cdot \Omega = -\Omega r_1(t) + \delta r_3(t) = -\Omega r_1(0) + \delta r_3(0) = \delta r_3(0) \tag{2.63}$$

for all $t \geq 0$, and the angle ζ is given by

$$\cos \zeta = \frac{\delta r_3(0)}{\Omega_\delta}. \tag{2.64}$$

Let us now focus on the population inversion r_3. By deriving Eq. (2.55) with respect to time and taking Eqs. (2.54) and (2.63) into account, we arrive at the following closed equation for r_3:

$$\ddot{r}_3 = -\Omega \dot{r}_2 = -\Omega \delta r_1 - \Omega^2 r_3 = \delta^2 [r_3(0) - r_3] - \Omega^2 r_3, \tag{2.65}$$

i.e.

$$\ddot{r}_3 + \Omega_\delta^2 r_3 = \delta^2 r_3(0), \tag{2.66}$$

which corresponds to the equation of a forced harmonic oscillator with frequency equal to the generalized Rabi frequency. The solution is the sum of the general solution of the associated homogeneous equation and of the stationary solution ($\ddot{r}_3 = 0$),

$$r_3 = A \cos(\Omega_\delta t + \varphi) + \frac{\delta^2}{\Omega_\delta^2} r_3(0). \tag{2.67}$$

By imposing the initial conditions $r_3(0)$ and $\dot{r}_3(0) = -\Omega r_2(0) = 0$ (see Eq. (2.55)) we obtain

$$r_3 = \frac{r_3(0)}{\Omega_\delta^2} \left[\Omega^2 \cos(\Omega_\delta t) + \delta^2 \right]. \tag{2.68}$$

The population inversion periodically oscillates between the values $r_3(0)$ and $-r_3(0) \times (\Omega^2 - \delta^2)/(\Omega^2 + \delta^2)$, at a frequency equal to the generalized Rabi frequency. Insofar as the first two components of the Bloch vector are concerned, we have from Eq. (2.63)

$$r_1 = \frac{\delta}{\Omega}[r_3 - r_3(0)] = r_3(0) \frac{\delta \Omega}{\Omega_\delta^2} [\cos(\Omega_\delta t) - 1] \tag{2.69}$$

and from Eq. (2.55)

$$r_2 = -\frac{\dot{r}_3}{\Omega} = r_3(0) \frac{\Omega}{\Omega_\delta} \sin(\Omega_\delta t). \tag{2.70}$$

Hence, on taking Eq. (2.49) into account, the interaction with the electric field induces an oscillating dipole moment.

Let us now focus on the special case where the atom is initially in the lower state ($r_3(0) = -1$). The oscillations of the three components of the Bloch vector are shown in Fig. 2.1(a). Since $r_3 = p_1 - p_2$ and $1 = p_1 + p_2$ it follows that the transition probability to the upper state, which coincides in this case with the occupation probability p_1 of the upper state, is linked to r_3 by the relation

$$p_1 = \frac{1}{2}(1 + r_3) \tag{2.71}$$

and therefore

$$p_1(t) = \frac{1}{2} \frac{\Omega^2}{\Omega_\delta^2} [1 - \cos(\Omega_\delta t)] = \frac{\Omega^2}{\Omega_\delta^2} \sin^2 \left(\frac{\Omega_\delta t}{2} \right). \tag{2.72}$$

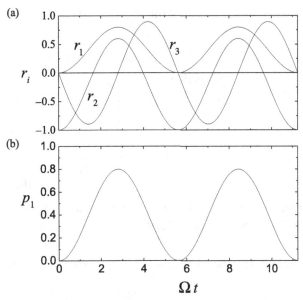

Figure 2.1 The time evolution of the components of the Bloch vector (a) and of the occupation probability of the excited state (b) when the atom is initially in the ground state and $\Omega = 2\delta$.

Hence the probability of the upper state oscillates between the values 0 and $\Omega^2/(\Omega^2 + \delta^2)$, as shown in Fig. 2.1(b).

2.4.1 The resonant case

When the electric field is in exact resonance with the atom ($\delta = 0$) the vector $\boldsymbol{\Omega}$ lies along the axis 1 and the Bloch vector is orthogonal to it, since $\cos\zeta = 0$. The generalized Rabi frequency coincides with the Rabi frequency and the solution of the Bloch equations reads

$$r_1 = 0, \tag{2.73}$$

$$r_2 = r_3(0)\sin(\Omega t), \tag{2.74}$$

$$r_3 = r_3(0)\cos(\Omega t). \tag{2.75}$$

The motion of the Bloch vector occurs in the 2–3 plane, and its tip describes the circle with unit radius. Only the polarization component in quadrature with the field is excited in this case. The inversion oscillates between -1 and 1, i.e. the atom passes from the lower to the upper state and vice versa, at the Rabi frequency. In this case there is a perfect analogy with a spin-1/2 particle [27, 28] and therefore these are called Rabi oscillations. If the atom starts from the lower state the occupation probability of this state is given by

$$p_1(t) = \sin^2\left(\frac{\Omega t}{2}\right). \tag{2.76}$$

In the resonant case, the solution of the Bloch equations can be directly obtained in a very straightforward manner. Let us assume that the atom is initially in the lower state, and

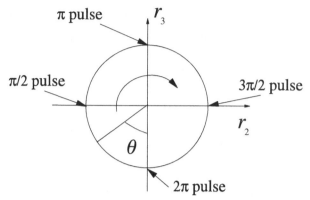

Figure 2.2 Illustration of the motion of the Bloch vector in the resonant case when an atom initially in the ground state interacts with a pulse of area equal to the indicated values.

let us introduce the angle θ (see Fig. 2.2), called the *tipping angle*, such that

$$r_2(t) = -\sin\theta(t), \qquad r_3(t) = -\cos\theta(t). \tag{2.77}$$

By replacing these expressions for r_2 and r_3 in Eq. (2.55), we obtain

$$\dot{\theta} = \Omega, \tag{2.78}$$

from which $\theta(t) = \Omega t$, so that we recover the expressions (2.74) and (2.75), for r_2 and r_3.

In connection with an electric-field pulse with envelope $E_0(z, t)$, one defines in general an "area" in the following way

$$A(z, t) = \int_{-\infty}^{t} \Omega(z, t')dt'. \tag{2.79}$$

For an envelope (pulse) such that $\Omega(z, t) = 0$ for $t < 0$ and $t > \bar{t}$, and $\Omega(z, t) = \Omega$ constant for $0 < t < \bar{t}$, the area for $t = \bar{t}$ is given by $\Omega\bar{t}$, which coincides with the angle $\theta(\bar{t})$. A pulse of area $\pi/2$ (mod 2π) brings the atom to the state $r_2 = -1$, $r_3 = 0$, a pulse of area π (mod 2π) brings the atom to the upper state, a pulse of area $3\pi/2$ (mod 2π) brings the atom to the state $r_2 = 1$, $r_3 = 0$, and finally a pulse of area $2n\pi$ brings the atom back to the lower state. The latter case also applies for the solitons we treat in Section 3.3, as we shall see in the next chapter. Solitons are not, however, pulses of constant amplitude.

2.4.2 The strongly off-resonance case

This is the opposite limit with respect to the resonant case, namely the situation in which the Rabi frequency is much smaller than the detuning, $\Omega \ll |\delta|$. In this case the vector $\boldsymbol{\Omega}$ has direction very close to that of the 3 axis and the Bloch vector is close to being antiparallel to it, since $\cos\zeta \approx -1$. On assuming $\delta > 0$ we have that $\Omega_\delta \approx \delta$ and the motion of the

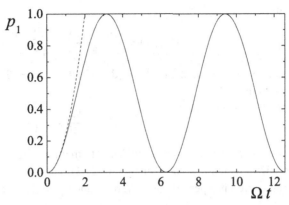

Figure 2.3 The occupation probability of the exact theory (solid line) is compared with that of the perturbative theory (dashed line) in the resonant case.

Bloch vector is described by the equations

$$r_1 \approx r_3(0)\frac{\Omega}{\delta}[\cos(\delta t) - 1], \tag{2.80}$$

$$r_2 \approx r_3(0)\frac{\Omega}{\delta}\sin(\delta t), \tag{2.81}$$

$$r_3 \approx r_3(0)\left[1 - 2\frac{\Omega^2}{\delta^2}\sin^2\left(\frac{\delta t}{2}\right)\right], \tag{2.82}$$

which hold up to corrections of third order with respect to Ω/δ. The population inversion r_3 remains always very close to -1, i.e. the atom substantially remains in the lower state, while the polarization components r_1 and r_2 execute small oscillations around their time-averaged values $\langle r_1 \rangle = -r_3(0)\Omega/\delta$ and $\langle r_2 \rangle = 0$, respectively. Hence also the polarization induced by the field is very small, and only the quadrature in phase with the electric field has a time average different from zero. The occupation probability of the upper state, when the atom starts from the lower state, remains always very small,

$$p_1(t) \approx \frac{\Omega^2}{\delta^2}\sin^2\left(\frac{\delta t}{2}\right), \tag{2.83}$$

since its maximum value is $\Omega^2/\delta^2 \ll 1$.

2.4.3 Comparison with perturbation theory

When the atom is initially in the lower state, the quantity $p_1(t)$ represents also the transition probability from the lower to the upper state. The expressions that we have found for $p_1(t)$ can be therefore compared with those obtained using perturbation theory, which provides the transition probability between the two levels in the case of a time-dependent perturbation of the form given by Eqs. (2.14) and (2.37). With the help of Eqs. (1.10) and (1.11) we obtain

$$P_{2\to1}(t) = \left(\frac{\mathbf{d}\cdot\hat{\mathbf{e}}E_0}{\hbar}\right)^2 \frac{\sin^2[\pi(\nu_a - \nu_0)t]}{4\pi^2(\nu_a - \nu_0)^2}. \tag{2.84}$$

By taking into account that $\Omega = (\mathbf{d}\cdot\hat{\mathbf{e}})E_0/\hbar$ and $\delta = 2\pi(\nu_a - \nu)$ this equation can be rewritten in the form

$$P_{2\to1}(t) = \frac{\Omega^2}{\delta^2}\sin^2\left(\frac{\delta t}{2}\right). \tag{2.85}$$

The expression for $P_{2\to1}(t)$ coincides with that for $p_1(t)$ only in the strong detuning limit, $\Omega \ll |\delta|$, in which $p_1(t)$ is given by Eq. (2.83). This condition defines the situation in which the interaction is a perturbation. At resonance ($\delta = 0$) we have

$$P_{2\to1}(t) = \frac{\Omega^2 t^2}{4}, \tag{2.86}$$

which coincides with the expression for $p_1(t)$ given by Eq. (2.76) only for short times, namely for times much smaller than Ω^{-1} (see Fig. 2.3).

3

The Maxwell–Bloch equations

The optical Bloch equations describe how a given electric field drives a system of two-level atoms. In this chapter we derive, first of all, the equation which governs the converse action, i.e. how a given atomic polarization generates the radiation field. This equation is obtained by starting from the Maxwell equations and applying a rigorous formulation of the paraxial and slowly varying envelope approximations. The next step is to couple such a wave equation with the optical Bloch equations, thereby obtaining the basic set of equations which describes the radiation–matter interaction. They are called *Maxwell–Bloch equations* in the literature, and a feature of paramount importance is that they constitute a nonlinear system of equations. Finally, we apply them the to the description of two intriguing phenomena, self-induced transparency and superradiance/superfluorescence.

3.1 The Maxwell equations. Paraxial and slowly varying envelope approximations

Let us consider the Maxwell equations in the international system of units MKSA. If we set the free electric charge density ρ and current density $\mathbf{J}$ equal to zero and neglect the magnetization of the medium by setting $\mathbf{B} = \mu_0 \mathbf{H}$ we have

$$\nabla \times \mathbf{E} = -\mu_0 \frac{\partial \mathbf{H}}{\partial t}, \tag{3.1}$$

$$\nabla \cdot \mathbf{H} = 0, \tag{3.2}$$

$$\nabla \cdot \mathbf{D} = 0, \tag{3.3}$$

$$\nabla \times \mathbf{H} = \frac{\partial \mathbf{D}}{\partial t}. \tag{3.4}$$

The $\mathbf{D}$ and $\mathbf{E}$ fields are linked by the constitutive relation

$$\mathbf{D} = \epsilon_0 \mathbf{E} + \mathbf{P}, \tag{3.5}$$

where $\mathbf{P}$ is the polarization density of the medium. By applying the $\nabla \times$ operation to both sides of Eq. (3.1) and taking into account Eqs. (3.4) and (3.5), the identity $\nabla \times (\nabla \times \mathbf{E}) = \nabla(\nabla \cdot \mathbf{E}) - \nabla^2 \mathbf{E}$ and the fact that $\epsilon_0 \mu_0 = 1/c^2$ we obtain

$$\nabla^2 \mathbf{E} - \nabla(\nabla \cdot \mathbf{E}) - \frac{1}{c^2} \frac{\partial^2 \mathbf{E}}{\partial t^2} = \frac{1}{\epsilon_0 c^2} \frac{\partial^2 \mathbf{P}}{\partial t^2}. \tag{3.6}$$

In general, the atomic polarization has two components [29]; one is the polarization of the two-level atoms that we have considered up to this point and the other is the polarization $\mathbf{P}_B$ of a background material that hosts the two-level atoms. We assume that $\mathbf{P}_B$ depends linearly on the electric field, i.e. $\mathbf{P}_B = \epsilon_0 \chi_B \mathbf{E}$, where χ_B is a real constant, i.e. we neglect the linear absorption due to the background medium, which will be considered later in Section 5.4.2. Therefore, in Eqs. (3.5) and (3.6) we must replace $\mathbf{P}$ by $\mathbf{P} + \mathbf{P}_B$, and, defining the background refractive index n_B as $n_B^2 = 1 + \chi_B$, we obtain

$$\nabla^2 \mathbf{E} - \nabla(\nabla \cdot \mathbf{E}) - \frac{n_B^2}{c^2} \frac{\partial^2 \mathbf{E}}{\partial t^2} = \frac{1}{\epsilon_0 c^2} \frac{\partial^2 \mathbf{P}}{\partial t^2}, \tag{3.7}$$

while Eq. (3.3) reads

$$\epsilon_0 n_B^2 \nabla \cdot \mathbf{E} = -\nabla \cdot \mathbf{P}. \tag{3.8}$$

Let us assume that the electric field propagates in the positive-z direction (*unidirectional propagation*),

$$\mathbf{E}(x, y, z, t) = \frac{1}{2} \left[\mathbf{E}_0(x, y, z, t) e^{-i(\omega_0 t - k_0 z)} + \text{c.c.} \right], \tag{3.9}$$

where ω_0 is the central frequency of the field, $k_0 = \omega_0 n_B / c$, and $\mathbf{E}_0$ is the field envelope, which varies in space and time much more slowly than the carrier factor $e^{-i(\omega_0 t - k_0 z)}$. Similarly, we will write

$$\mathbf{P}(x, y, z, t) = \frac{1}{2} \left[\mathbf{P}_0(x, y, z, t) e^{-i(\omega_0 t - k_0 z)} + \text{c.c.} \right]. \tag{3.10}$$

We can expand $\mathbf{E}$ on the plane-wave basis

$$\mathbf{e}_{\mathbf{k}}^{(i)} e^{i \mathbf{k} \cdot \mathbf{x}}, \qquad \mathbf{k} \equiv (k_x, k_y, k_z), \qquad i = 1, 2, 3, \tag{3.11}$$

where $\mathbf{e}_{\mathbf{k}}^{(i)}$ are the three orthogonal unit vectors with $\mathbf{e}_{\mathbf{k}}^{(3)} = \mathbf{k}/k$. The *paraxial approximation* consists in assuming that

$$|k_x|, |k_y| \ll k_z \sim k_0. \tag{3.12}$$

In the following we show that, even if the condition $\nabla \cdot \mathbf{E} = 0$ does not hold exactly, it becomes valid in the limit of the paraxial approximation and of the slowly varying envelope approximation that we define below. Hence we discard the unit vector $\mathbf{e}_{\mathbf{k}}^{(3)}$ and keep only $\mathbf{e}_{\mathbf{k}}^{(i)}, i = 1, 2$, which are orthogonal to $\mathbf{k}$. In this approximation the component of $\mathbf{E}$ in the z direction is much smaller than the orthogonal components because the condition $\nabla \cdot \mathbf{E} = 0$ applied to the plane wave (3.11) with $i = 1, 2$ gives

$$\left| \left(\mathbf{e}_{\mathbf{k}}^{(i)} \right)_z \right| = \left| \frac{k_x}{k_z} \left(\mathbf{e}_{\mathbf{k}}^{(i)} \right)_x + \frac{k_y}{k_z} \left(\mathbf{e}_{\mathbf{k}}^{(i)} \right)_y \right| \ll 1. \tag{3.13}$$

We can write

$$\mathbf{E}(x, y, z, t) = \frac{1}{2} [\hat{\mathbf{e}} E_{0T}(x, y, z, t) + \hat{\mathbf{e}}_z E_{0L}(x, y, z, t)] e^{-i(\omega_0 t - k_0 z)} + \text{c.c.}, \tag{3.14}$$

$$\mathbf{P}(x, y, z, t) = \frac{1}{2} [\hat{\mathbf{e}} P_{0T}(x, y, z, t) + \hat{\mathbf{e}}_z P_{0L}(x, y, z, t)] e^{-i(\omega_0 t - k_0 z)} + \text{c.c.}, \tag{3.15}$$

where $\hat{\mathbf{e}}_z$ is the unit vector of the z axis and $\hat{\mathbf{e}}$ is a unit vector orthogonal to it, T indicates transverse and L longitudinal.

In this section we will follow a simplified version of the approach of Lax [30, 31]. The fields are characterized by three spatial scales: the wavelength in the background material $\lambda = 2\pi / k_0$ and the lengths d_0 and L over which the envelopes vary appreciably in the transverse and longitudinal directions, respectively. If we take into account that k_x, $k_y \sim d_0^{-1}$ the paraxial approximation (3.12) amounts to the condition

$$\frac{\lambda}{d_0} \ll 1. \tag{3.16}$$

On the other hand, the so-called *slowly varying envelope approximation* prescribes that the field amplitude varies in z and t much more slowly than the carrier wave, i.e. that

$$\left| \frac{\partial \mathbf{E}_0}{\partial z} \right| \ll k_0 |\mathbf{E}_0|, \qquad \left| \frac{\partial \mathbf{E}_0}{\partial t} \right| \ll \omega_0 |\mathbf{E}_0|, \tag{3.17}$$

hence we have

$$\frac{\lambda}{L} \ll 1. \tag{3.18}$$

More quantitative conditions can be obtained if, as is reasonable, we assume that L and d_0 are linked by the same (apart from a factor of 2, see Eq. (26.6)) relation as the diffraction length and the beam width in the paraxial approximation, i.e. that

$$L = k_0 d_0^2 \quad \Rightarrow \quad d_0 = \left(\frac{\lambda L}{2\pi} \right)^{1/2}. \tag{3.19}$$

We can then introduce a smallness parameter ϵ such that

$$\epsilon = \frac{1}{k_0 d_0} = \frac{d_0}{L} \quad \Rightarrow \quad \frac{1}{k_0 L} = \epsilon^2, \tag{3.20}$$

which agrees with, but is more precise than, inequality (3.18).

If we now assume that $P_{0T}/(\epsilon_0 E_{0T})$, $P_{0L}/(\epsilon_0 E_{0L}) \sim \epsilon^2$, as will be confirmed by our subsequent analysis, then from Eq. (3.8) it follows that the condition $\nabla \cdot \mathbf{E} = 0$ (which holds exactly for a linear medium) is approached in the limit $\epsilon \to 0$. Moreover, in the limit $\epsilon \ll 1$ a simple relation that links E_{0L} with E_{0T} can be found because if we divide the ∇ operator into a transverse part and a longitudinal part,

$$\nabla = \nabla_\perp + \hat{\mathbf{e}}_z \frac{\partial}{\partial z}, \qquad \nabla_\perp = \hat{\mathbf{e}}_x \frac{\partial}{\partial x} + \hat{\mathbf{e}}_y \frac{\partial}{\partial y}, \tag{3.21}$$

and apply the slowly varying envelope approximation, so that $\partial E_z / \partial z \propto i k_0 E_{0L}$, the condition $\nabla \cdot \mathbf{E} = 0$ is equivalent to

$$E_{0L} = \frac{i}{k_0} \nabla_\perp \cdot (\hat{\mathbf{e}} E_{0T}), \tag{3.22}$$

which implies that $|E_{0L}/E_{0T}| \sim (k_0 d_0)^{-1} = \epsilon \ll 1$.

Even though we have used the condition $\nabla \cdot \mathbf{E} = 0$ to determine Eq. (3.22) for E_{0L}, we insert into Eq. (3.7) the exact relation (3.8) in order to be able to evaluate exactly the order of magnitude of all terms for $\epsilon \ll 1$. Thus, we obtain the equation

$$\nabla^2 \mathbf{E} - \frac{n_B^2}{c^2} \frac{\partial^2 \mathbf{E}}{\partial t^2} = \frac{1}{\epsilon_0 c^2} \frac{\partial^2 \mathbf{P}}{\partial t^2} - \frac{1}{\epsilon_0 n_B^2} \nabla(\nabla \cdot \mathbf{P}). \tag{3.23}$$

By inserting Eqs. (3.14) and (3.15), equating separately the terms containing $e^{-i(\omega_0 t - k_0 z)}$ and dividing the result by the exponential factor, we obtain the following equation for the transverse parts directed along $\hat{\mathbf{e}}$:

$$\nabla_\perp^2 E_{0T} + 2ik_0 \frac{\partial E_{0T}}{\partial z} + \frac{\partial^2 E_{0T}}{\partial z^2} + 2i \frac{k_0 n_B}{c} \frac{\partial E_{0T}}{\partial t} - \frac{n_B^2}{c^2} \frac{\partial^2 E_{0T}}{\partial t^2}$$
$$= -\frac{1}{\epsilon_0 n_B^2} \left\{ k_0^2 P_{0T} + 2i \frac{k_0 n_B}{c} \frac{\partial P_{0T}}{\partial t} - \frac{n_B^2}{c^2} \frac{\partial^2 P_{0T}}{\partial t^2} \right.$$
$$\left. + \nabla_\perp \left[\nabla_\perp \cdot (\hat{\mathbf{e}} P_{0T}) + \left(ik_0 + \frac{\partial}{\partial z} \right) P_{0L} \right] \right\}, \tag{3.24}$$

where we have taken into account that $\omega_0 = ck_0/n_B$ and the transverse Laplacian $\nabla_\perp^2$ is defined as

$$\nabla_\perp^2 = \frac{\partial^2}{\partial x^2} + \frac{\partial^2}{\partial y^2}. \tag{3.25}$$

If we now divide both terms of Eq. (3.24) by $2ik_0$, and introduce the scaled dimensionless variables

$$\bar{x} = \frac{x}{d_0}, \quad \bar{y} = \frac{y}{d_0}, \quad \bar{z} = \frac{z}{L}, \quad \bar{t} = \frac{ct}{n_B L}, \tag{3.26}$$

we can recast Eq. (3.24) as follows:

$$\frac{1}{2i} \bar{\nabla}_\perp^2 E_{0T} + \frac{\partial E_{0T}}{\partial \bar{z}} + \frac{\partial E_{0T}}{\partial \bar{t}} + \frac{\epsilon^2}{2i} \left(\frac{\partial^2 E_{0T}}{\partial \bar{z}^2} - \frac{\partial^2 E_{0T}}{\partial \bar{t}^2} \right)$$
$$= \frac{1}{2\epsilon_0 n_B^2} \left\{ \frac{i}{\epsilon^2} P_{0T} - 2 \frac{\partial P_{0T}}{\partial \bar{t}} - i\epsilon^2 \frac{\partial^2 P_{0T}}{\partial \bar{t}^2} \right.$$
$$\left. + i \bar{\nabla}_\perp \left[\bar{\nabla}_\perp \cdot (\hat{\mathbf{e}} P_{0T}) + \frac{1}{2} \left(1 + \epsilon^2 \frac{\partial}{\partial \bar{z}} \right) P_{0L} \right] \right\}. \tag{3.27}$$

From this we see that, consistently with our initial assumption, the ratio $P_{0T}/(\epsilon_0 E_{0T})$ is of order ϵ^2. Similarly, from the equation for the longitudinal part (which we do not write explicitly here) we obtain also that $P_{0L}/(\epsilon_0 E_{0L}) \sim \epsilon^2$, and since $E_{0L}/E_{0T} \sim \epsilon$ it follows that $P_{0L}/P_{0T} \sim \epsilon$ and $P_{0L}/\epsilon_0 E_{0T} \sim \epsilon^3$. Hence, the fields coincide with their transverse parts to leading order and we write

$$\mathbf{E}(x, y, z, t) = \frac{\hat{\mathbf{e}}}{2} E_0(x, y, z, t) e^{-i(\omega_0 t - k_0 z)} + \text{c.c.}, \tag{3.28}$$

$$\mathbf{P}(x, y, z, t) = \frac{\hat{\mathbf{e}}}{2} P_0(x, y, z, t) e^{-i(\omega_0 t - k_0 z)} + \text{c.c.} \tag{3.29}$$

The final equation for E_0 and P_0 is obtained from Eq. (3.27) by dropping the terms proportional to ϵ^2 and ϵ^4 and the suffix T, coming back to the original variables x, y, z and

t, and taking into account that $L = k_0 d_0^2$:

$$\frac{1}{2ik_0} \nabla_\perp^2 E_0 + \frac{\partial E_0}{\partial z} + \frac{n_{\rm B}}{c} \frac{\partial E_0}{\partial t} = i \frac{k_0}{2\epsilon_0 n_{\rm B}^2} P_0. \tag{3.30}$$

We note in particular that no term originating from $\nabla(\nabla \cdot \mathbf{E})$ survives in Eq. (3.30).

3.2 The Maxwell–Bloch equations. The plane-wave approximation

At this point we can combine the wave equation (3.30) with the optical Bloch equations (2.46) and (2.47) in a consistent way. In the wave equation we note the slowly varying envelopes of the field E_0 and of the macroscopic polarization density P_0; in the Bloch equations we note the Rabi frequency Ω and the slowly varying envelope of the microscopic polarization r. Since, in accord with Eqs. (2.39) and (2.42) one has $\Omega = dE_0/\hbar$ and $P_0 = 2Ndr$, where N is the atomic density, we obtain the closed set of equations

$$\frac{1}{2ik_0} \nabla_\perp^2 \Omega + \frac{\partial \Omega}{\partial z} + \frac{n_{\rm B}}{c} \frac{\partial \Omega}{\partial t} = i \frac{k_0}{\epsilon_0 n_{\rm B}^2} N \frac{d^2}{\hbar} r, \tag{3.31}$$

$$\frac{\partial r}{\partial t} = -i\delta r - \frac{i}{2}\Omega r_3, \tag{3.32}$$

$$\frac{\partial r_3}{\partial t} = i(\Omega r^* - \Omega^* r), \tag{3.33}$$

where Ω, r and r_3 depend on the variables x, y, z and t. These are the *Maxwell–Bloch equations* [32, 33] which govern the interaction of a system of two-level atoms with the electric field. Such equations hold in the (i) dipole, (ii) rotating wave, (iii) paraxial and (iv) slowly varying envelope approximations.

It is of fundamental importance to observe that, when the Rabi frequency is not a given function of time and space but rather is a variable coupled to the atomic variables by the Maxwell–Bloch equations, the dynamics is nonlinear because Eqs. (3.32) and (3.33) contain products of variables.

In the remainder of this chapter we will also use the *plane-wave approximation*, which consists in assuming that the functions Ω, r and r_3 depend only on the longitudinal variable z and on time t and are independent of the transverse variables x and y. Thus, the field equation reduces to

$$\frac{\partial \Omega}{\partial z} + \frac{n_{\rm B}}{c} \frac{\partial \Omega}{\partial t} = i \frac{k_0}{\epsilon_0 n_{\rm B}^2} N \frac{d^2}{\hbar} r. \tag{3.34}$$

We note that the combination of the paraxial plus slowly varying envelope approximation and the plane-wave approximation introduces a drastic simplification because it leads to a dynamical model in which all the equations contain only first-order derivatives.

3.3 Self-induced transparency, the sine–Gordon equation and solitons

There are conditions such that a light pulse propagates along a medium without modifying its shape. Therefore the medium turns out to be transparent to the pulse, and one calls this phenomenon, which was predicted and experimentally demonstrated by McCall and Hahn [34], *self-induced transparency*. In order that this may happen it is necessary that the pulse area defined by Eq. (2.79) in the limit $t \to +\infty$ becomes equal to 2π. In this way the passage of the pulse causes a 2π rotation of the Bloch vector and leaves the medium in the same state as it was in before the pulse.

Let us consider[1] the Maxwell–Bloch equations (3.34), (3.32) and (3.33) in the resonant case and let us re-express the Bloch equations in terms of the quantities r_1 and r_2 defined by Eqs. (2.50) and (2.51). Since $\delta = 0$, r_1 is constant in time, and, if we assume that initially the atoms are in the lower state for all z, then $r_1 = 0$ for all values of time and space and therefore we can ignore it. Hence the Maxwell–Bloch equations read (setting $n_B = 1$ for the sake of simplicity)

$$c\frac{\partial \Omega}{\partial z} + \frac{\partial \Omega}{\partial t} = \beta^2 r_2, \tag{3.35}$$

$$\frac{\partial r_2}{\partial t} = \Omega r_3, \tag{3.36}$$

$$\frac{\partial r_3}{\partial t} = -\Omega r_2, \tag{3.37}$$

where

$$\beta^2 = \omega_0 N \frac{d^2}{2\epsilon_0 \hbar}, \tag{3.38}$$

is a parameter with dimensions equal to those of a squared frequency. The Rabi-frequency envelope remains real if it is real initially, and the motion of the Bloch vector occurs over the circle of unit radius in the (r_2, r_3) plane. We now express the variables r_2 and r_3 as a function of the tipping angle θ in Fig. 2.2 as in Eq. (2.77), so that we obtain an equation essentially identical to Eq. (2.78),

$$\frac{\partial \theta}{\partial t} = \Omega(z, t), \tag{3.39}$$

where we take $t = -\infty$ as the initial time, so that $\theta(-\infty) = 0$. If we insert the equation $r_2 = -\sin\theta$ into Eq. (3.35) we obtain

$$c\frac{\partial \Omega}{\partial z} + \frac{\partial \Omega}{\partial t} = -\beta^2 \sin\theta. \tag{3.40}$$

---

[1] The following treatment in this section is inspired by lectures delivered by Rodolfo Bonifacio.

Next, taking Eq. (3.39) into account, we arrive at a closed equation for the tipping angle θ:

$$c\frac{\partial^2\theta}{\partial z\,\partial t} + \frac{\partial^2\theta}{\partial t^2} = -\beta^2\sin\theta. \tag{3.41}$$

If we transform to a retarded time, i.e. we set

$$\tau_1 = t - \frac{z}{c}, \quad z_1 = z, \tag{3.42}$$

Eq. (3.41) takes the form

$$c\frac{\partial^2\theta}{\partial z_1\,\partial\tau_1} = -\beta^2\sin\theta, \tag{3.43}$$

which is called the *sine–Gordon equation*.[2] Let us now look for a solution of Eq. (3.41) with the form of a pulse that propagates along z with arbitrary velocity v, i.e. $\theta(t - z/v)$. If we define $\tau = t - z/v$ we obtain from Eq. (3.41) or from Eq. (3.43) the following equation:

$$\frac{d^2\theta}{d\tau^2} = \beta'^2\sin\theta, \quad \beta'^2 = \frac{\beta^2}{c/v - 1}, \tag{3.44}$$

where we have assumed that $v < c$. This is a pendulum equation, if θ is measured from the unstable equilibrium point. Let us now observe that, if θ is a solution of the equation

$$\frac{d\theta}{d\tau} = 2\beta'\sin\left(\frac{\theta}{2}\right), \tag{3.45}$$

then it is also a solution of Eq. (3.44), since if we derive Eq. (3.45) we obtain

$$\frac{d^2\theta}{d\tau^2} = \beta'\cos\left(\frac{\theta}{2}\right)\frac{d\theta}{d\tau} = 2\beta'^2\cos\left(\frac{\theta}{2}\right)\sin\left(\frac{\theta}{2}\right) = \beta'^2\sin\theta. \tag{3.46}$$

The fact that θ is a solution of Eq. (3.45) implies that with the initial condition $\theta(-\infty) = 0$ there is associated also the other initial condition $\dot{\theta}(-\infty) = 0$. In order to find the solution of Eq. (3.45) let us introduce the substitution $\theta = 4\arctan x$, from which it follows that

$$\frac{d\theta}{d\tau} = \frac{4}{1 + x^2}\frac{dx}{d\tau}. \tag{3.47}$$

---

[2] This name originates from the equation

$$\frac{\partial^2\theta}{\partial t^2} - c^2\frac{\partial^2\theta}{\partial z^2} = -\beta^2\sin\theta,$$

which, if one replaces the sin θ term by the linear term θ, reduces to the one-dimensional Klein–Gordon equation. If, in such an equation, one introduces the transformation

$$\tau_1 = t - \frac{z}{c}, \quad z_1 = z + ct,$$

the equation takes the form

$$4c\frac{\partial^2\theta}{\partial z_1\,\partial\tau_1} = -\beta^2\sin\theta,$$

which, apart from a numerical factor, coincides with Eq. (3.43).

Furthermore, since $\sin(2\alpha) = 2\tan\alpha/(1 + \tan^2\alpha)$ we have also that

$$\sin\left(\frac{\theta}{2}\right) = \sin(2\arctan x) = \frac{2x}{1 + x^2}, \tag{3.48}$$

hence the equation for x is simply $dx/d\tau = \beta'x$ from which

$$x = Ae^{\beta'\tau}. \tag{3.49}$$

The circumstance that $x(-\infty) = 0$ is in accord with the fact that $x = \tan(\theta/4)$ and $\theta(-\infty) = 0$. We have also that $x(+\infty) = +\infty$, and this implies that $\theta(+\infty) = 2\pi$, i.e. the pulse area, is equal to 2π.[3] Finally, if we assume that the pulse is symmetrical in τ and therefore $\theta(0) = \pi$ from which $x(0) = 1$, we have that $A = 1$. Therefore, using Eq. (3.48), the solution of Eq. (3.44) has the form

$$\sin\left(\frac{\theta}{2}\right) = \frac{2x}{1 + x^2} = \frac{2e^{\beta't}}{1 + e^{2\beta't}} = \frac{1}{\cosh(\beta'\tau)} = \mathrm{sech}(\beta'\tau). \tag{3.50}$$

From Eq. (3.45) it follows that

$$\Omega(z, t) = \frac{\partial\theta}{\partial t} = \frac{d\theta}{d\tau} = 2\beta'\,\mathrm{sech}(\beta'\tau) = 2\beta'\,\mathrm{sech}\left[\beta'\left(t - \frac{z}{v}\right)\right]. \tag{3.51}$$

This equation describes a pulse that propagates with constant velocity while conserving its shape. Such pulses are called *self-induced transparency solitons*. If we call $\tau_0 = 1/\beta'$ the pulse width, we have

$$\Omega(z, t) = \frac{2}{\tau_0}\,\mathrm{sech}\left(\frac{\tau}{\tau_0}\right) = \frac{4}{\tau_0}\frac{1}{e^{\tau/\tau_0} + e^{-\tau/\tau_0}}, \tag{3.52}$$

which shows that the pulse is higher the shorter it is, in accord with the fact that the area must be equal to 2π. The propagation velocity is given by the relation $\beta'^2 = \beta^2/(c/v - 1)$, from which

$$v = \frac{c}{1 + \beta^2\tau_0^2}, \tag{3.53}$$

in accord with the condition $v < c$. The pulse is faster the shorter it is. One can show that the solitons can collide, because a faster soliton can overtake a slower one, but at the end of the collision they assume their original shapes again. Figure 3.1 shows the soliton for various values of τ_0 and also the corresponding profile for the inversion r_3 which, taking Eqs. (2.77) and (3.50) into account, is given by

$$r_3 = \frac{6 - e^{-2\beta'\tau} - e^{2\beta'\tau}}{2 + e^{-2\beta'\tau} + e^{2\beta'\tau}}. \tag{3.54}$$

The energy density, and therefore the radiation intensity, is proportional to the square of the electric field and, taking Eq. (3.9) into account and averaging over time, is proportional to $|E_0|^2$ and therefore to $|\Omega|^2$. Hence in Fig. 3.1 we plot Ω^2 (Ω is real on resonance).

---

[3] In principle the pulse area might be equal to any integer multiple of 2π, but one can demonstrate that pulses of area $2n\pi$ with $n > 1$ are unstable and decay into n pulses of area 2π.

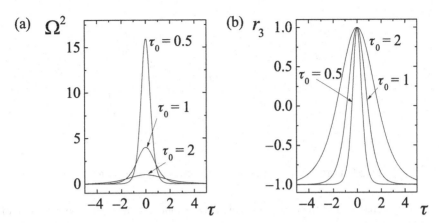

Figure 3.1 (a) The profile of the soliton intensity for some values of τ_0. (b) The profile of the inversion for the same values of τ_0.

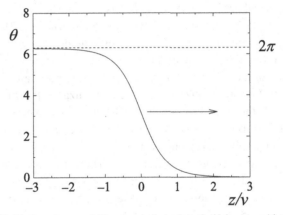

Figure 3.2 A plot of the function $\theta(\tau)$ for $t = 0$, $\tau_0 = 1$. The arrow indicates how the kink moves with time.

Since from Eq. (3.48, 3.49)

$$\theta(\tau) = 4 \arctan x(\tau) = 4 \arctan e^{\beta'(t-z/v)}, \tag{3.55}$$

the graph of θ as a function of z for a fixed value of t has the kink structure shown in Fig. 3.2.

3.4 Superradiance and superfluorescence

In 1954 Dicke [35] (see also [36]) predicted that a system of many two-level atoms, contained in a small region of volume smaller than the wavelength to the third power and prepared in such a way that the atomic system has a macroscopic dipole moment, emits a radiation pulse with intensity proportional to the square of the number of atoms, instead of the standard emission with intensity proportional to the number of atoms, and he called this phenomenon *superradiance*. This remarkable behavior arises from the cooperative

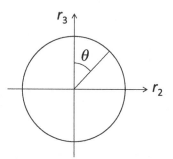

Figure 3.3   The definition of the tipping angle θ used for superradiance/superfluorescence.

character of the emission, which is the coherent emission from a system of aligned dipoles. Subsequent investigations considered a system of atoms that, instead of being point-like, occupies a pencil-shaped region, studying the propagation of the emitted light along the sample and predicting that also in this case one has superradiant emission.

In the 1970s Bonifacio, Schwendimann and Haake [37] considered the case in which the atomic system has initially no macroscopic dipole, but is in the upper state, and showed that also in this case there can be the emission of a pulse with intensity proportional to the square of the number of atoms. The atomic dipoles are initially in a disordered state, but nonetheless they realize a sort of self-organization that produces the *cooperative emission*. While in the case of superradiance the phenomenon is classical because the atomic system emits just like a classical antenna, in the case of initially excited atoms the emission starts as spontaneous emission and therefore needs a quantum description. Since in standard fluorescence (spontaneous emission) the intensity is simply proportional to the number of atoms, Bonifacio introduced later the name *superfluorescence* to designate the cooperative spontaneous emission with peak intensity proportional to the square of the number of atoms. It is good to keep in mind that in the literature the name superfluorescence is used also for kinds of amplified spontaneous emission in which the proportionality to the square of the number of atoms is missing, but here we will stick to the definition of superfluorescence given above.

Again, let us assume exact resonance between the radiation field and the atoms ($\delta = 0$) and let us start from the Maxwell–Bloch equations (3.35)–(3.37) for a pencil-shaped sample of length L. In the case of superradiance/superfluorescence it is convenient to consider a tipping angle θ measured from the North pole of the Bloch circle as in Fig. 3.3, so that

$$r_2(z, t) = \sin \theta(z, t), \qquad r_3(z, t) = \cos \theta(z, t), \tag{3.56}$$

and Eq. (3.39) remains unaltered, whereas Eq. (3.40) becomes

$$c\frac{\partial \Omega}{\partial z} + \frac{\partial \Omega}{\partial t} = +\beta^2 \sin \theta, \tag{3.57}$$

where, again, we have set $n_B = 1$. At this stage, we introduce an approximation that is rather crude but allows one to perform an analytical treatment and to capture the essential features of the phenomena in play, leading to results in qualitative agreement with the experimental findings. The propagation term $c\, \partial\Omega/\partial z$ represents a loss because the radiation escapes from the edge $z = L$ of the pencil-shaped sample. The approximation consists in replacing this term by a damping term $k\Omega$, with $k = c/L$, and in neglecting the spatial variation by considering directly the field emitted at $z = L$, so that Eq. (3.57) becomes

$$\dot{\Omega} + k\Omega = \beta^2 \sin\theta, \tag{3.58}$$

where we have written $\dot{\Omega}$ because now the problem is in time only. As we will see in Chapter 12, this step becomes rigorous when the atomic system is contained in an optical cavity of high quality. Using Eq. (3.39) we obtain the following damped pendulum equation for the tipping angle:

$$\ddot{\theta} + k\dot{\theta} = \frac{1}{\tau_c^2} \sin\theta, \tag{3.59}$$

where we have introduced the parameter $\tau_c = 1/\beta$, which in the literature on superradiance/superfluorescence is called the cooperation time. It was introduced by Arecchi and Courtens [38].

An initial condition for Eq. (3.59) is $\dot{\theta} = 0$ because the electric field is zero at the beginning of the emission. The initial value $\theta(0) = \theta_0$ of the tipping angle is different for superradiance and superfluorescence. In the case of superradiance one has $r_2(0) = 1$, $r_3(0) = 0$ and therefore $\theta_0 = \pi/2$; in the case of superfluorescence one has $r_2(0) = 0$, $r_3(0) = 1$ and therefore $\theta_0 = 0$. However, the state $\theta(0) = 0$, $\dot{\theta}(0) = 0$ corresponds to a pendulum starting from the unstable equilibrium state with zero velocity and therefore the solution of Eq. (3.59) would be the (unstable) stationary solution $\theta(t) = 0$, i.e. the emission would not start. This is due to the fact that superfluorescent emission is initiated by spontaneous emission in the direction of the z axis, which is missing in the semiclassical picture. Superfluorescence needs a quantum picture to describe its initiation. The quantum-mechanical analysis of [37] shows that, in the semiclassical picture, the quantum initiation can be simulated by taking for the tipping angle the initial value $\theta(0) = \sqrt{2/\mathcal{N}}$, where $\mathcal{N}$ is the number of atoms in the pencil-shaped sample. This small initial deviation from the unstable equilibrium point allows the emission to start. Hence the initial conditions are

$$\theta(0) = \theta_0 = \frac{\pi}{2}, \quad \dot{\theta}(0) = 0 \tag{3.60}$$

for superradiance and

$$\theta(0) = \theta_0 = \sqrt{\frac{2}{\mathcal{N}}}, \quad \dot{\theta}(0) = 0 \tag{3.61}$$

for superfluorescence. The most typical configuration for superradiance/superfluorescence is that defined by the condition $k\tau_c \gg 1$ (*pure superradiance/superfluorescence*). In this limit the pendulum described by Eq. (3.59) is overdamped, i.e. approaches the stable equilibrium state $\theta = \pi$ without oscillations, and the inertial term $\ddot{\theta}$ can be dropped, so the pendulum equation reduces to

$$\dot{\theta} = \frac{2}{\tau_R} \sin\left(\frac{\tilde{\theta}}{2}\right), \tag{3.62}$$

where we have set $\tilde{\theta} = 2\theta$ and introduced the superradiance/superfluorescence time τ_R defined as

$$\tau_R = k\tau_c^2 \longrightarrow k\tau_R = (k\tau_c)^2. \tag{3.63}$$

Note that the condition $k\tau_c \gg 1$ implies that $k\tau_R \gg 1$. In addition, if we introduce the cooperation length $L_c = c\tau_c$, the condition $k\tau_c \gg 1$, since $k = c/L$, is equivalent also to $L \ll L_c$.

A very remarkable fact is that Eq. (3.62) is identical to Eq. (3.45) for self-induced transparency if in the latter we replace θ by $\tilde{\theta}$ and β' by $1/\tau_R$. Hence to solve Eq. (3.62) we follow the same steps as for Eq. (3.45). In the case of Eq. (3.62) the solution expressed in terms of the variable x reads

$$x(t) = Ae^{t/\tau_R}, \tag{3.64}$$

and, by imposing the initial condition $\theta(0) = \theta_0$, which implies that $x(0) = \tan(\theta(0)/4) = \tan\theta_0/2$, we obtain

$$x(t) = e^{(t-t_D)/\tau_R}, \qquad t_D = -\tau_R \ln\left(\tan\frac{\theta_0}{2}\right), \tag{3.65}$$

where we have introduced the time t_D, which corresponds to the delay of the emission peak with respect to the initial time $t = 0$. Next, exactly as in Eq. (3.50), we arrive at the result

$$\sin\left(\frac{\tilde{\theta}}{2}\right) = \sin\theta = \text{sech}\left(\frac{t - t_D}{\tau_R}\right), \tag{3.66}$$

from which we conclude that the quantity $\Omega^2(t)$, which is proportional to the emitted intensity, is given by

$$\Omega^2 = \dot{\theta}^2 = \left(\frac{\dot{\tilde{\theta}}}{2}\right)^2 = \frac{1}{\tau_R^2}\left[\text{sech}\left(\frac{t - t_D}{\tau_R}\right)\right]^2. \tag{3.67}$$

In the case of superradiance $\theta_0 = \pi/2$ and $t_D = 0$, so the emission is maximum at the start of the process, whereas in the case of superfluorescence there is initially a lethargic stage and the emission peak is located with a delay $t_D = -\tau_R \ln(\tan((1/2)\sqrt{2/\mathcal{N}})) \approx -\tau_R \ln(1/\sqrt{2\mathcal{N}}) = (\tau_R/2)\ln(2\mathcal{N})$ (see Fig. 3.4(a)). The most remarkable feature that emerges from Eq. (3.67) is that the peak height is proportional to τ_R^{-2}, which is proportional to the square of the atomic density N, as follows from Eq. (3.63), owing to the fact that $\tau_c = \beta^{-1}$ and Eq. (3.38). Correspondingly, the width of the emission peak is on the order of τ_R and therefore is inversely proportional to the atomic density. We must also note that, even if Eq. (3.62) for pure superradiance/superfluorescence is identical in form to Eq. (3.45) for self-induced transparency, there are basic differences between the two cases. The first main difference is that in the case of self-induced transparency, see Eq. (3.51), Ω^2 is proportional to β'^2 and therefore to β^2, hence on the basis of Eq. (3.38) the peak is simply proportional to the atomic density N and not to the square of the atomic density as in superradiance/superfluorescence. The second difference is that in the case of superradiance/superfluorescence, as a consequence of the fact that $r_3 = \cos\theta$ and of Eq. (3.66), the

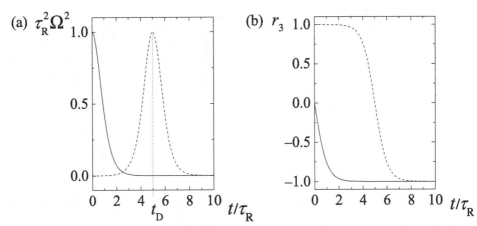

Figure 3.4   Pure superradiance/superfluorescence ($k\tau_c \gg 1$). (a) The time evolution of the emitted intensity in the case of superradiance ($\theta_0 = \pi/2$) and in the case of superfluorescence ($\theta_0 = 2e^{-5}$). (b) The corresponding time evolution of the population inversion. The solid curves correspond to superradiance, the broken curves to superfluorescence.

time evolution of the population inversion is given by (see Fig. 3.4(b))

$$r_3 = -\tanh\left(\frac{t - t_D}{\tau_R}\right),\tag{3.68}$$

which is quite different from the expression (3.54) for self-induced transparency. The difference arises from the fact that Eq. (3.62) is for $\tilde{\theta}$ whereas Eq. (3.45) is for θ. Hence in the superradiant/superfluorescent emission the atoms decay cooperatively in a time τ_R that is inversely proportional to the number of atoms. Pure superfluorescence was experimentally observed for the first time by Vrehen and Gibbs [39].

Let us now reduce the product $k\tau_c$ to values of order unity, which means also $k\tau_R$ of order unity, and is equivalent to the condition $L \approx L_c$. In this situation the pendulum is no longer overdamped and approaches the stable equilibrium state $\theta = \pi$ with some oscillations (*oscillatory superradiance/superfluorescence* [40]). Correspondingly, the emission intensity exhibits a number of peaks. After each peak, except the last one, the light is partially reabsorbed by the atoms, so the emission becomes partially reversible.

The time scale is still τ_R and this implies that the emitted intensity, which is proportional to $\Omega^2 = \dot{\theta}^2$, scales as the square of the atomic density. The height of the highest peak is lower than that of the pure superradiant/superfluorescent peak, and decreases when $k\tau_c$ is decreased. If we introduce the normalized time $t_1 = t/\tau_R$, Eq. (3.59) reads

$$\frac{d^2\theta}{dt_1^2} = (k\tau_c)^2\left(-\frac{d\theta}{dt_1} + \sin\theta\right).\tag{3.69}$$

Figures 3.5(a) and (b) show the time evolution of Ω^2 and of r_3, respectively, for $k\tau_c = \sqrt{0.1}$. The first experimental observation of oscillatory superfluorescence was reported by Feld and his collaborators [41].

Finally, let us further reduce the product $k\tau_c$ to values much smaller than unity, which is equivalent to assuming that $L \gg L_c$. In this configuration the radiation is emitted in a

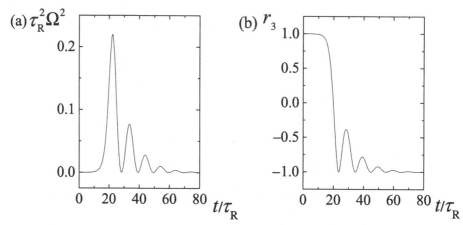

Figure 3.5   Oscillatory superfluorescence for $k\tau_c = \sqrt{0.1}$. (a) The emitted intensity. (b) Population inversion. The value of θ_0 is $\theta_0 = 2e^{-5}$.

very slowly damped sequence of peaks. The time scale is no longer τ_R but τ_c and, as a consequence, Ω^2 scales as the atomic density and no longer as the square of the atomic density, hence the emission is no longer superradiant/superfluorescent. If we introduce the normalized time $t_2 = t/\tau_c$, Eq. (3.59) takes the form

$$\frac{d^2\theta}{dt_2^2} + (k\tau_c)\frac{d\theta}{dt_2} = \sin\theta. \tag{3.70}$$

In the limit $k\tau_c = 0$, i.e. when L/L_c approaches infinity [42], this equation reduces to a pure pendulum equation. Its solutions correspond to an elliptic function with a period on the order of τ_c. The radiation is periodically emitted and reabsorbed in a reversible way. The quantum model, of which Eq. (3.70) in the limit $\tau_c = 0$ represents the semiclassical approximation, was formulated by Tavis and Cummings [43]. It describes the reversible dynamics of $\mathcal{N}$ atoms interacting with a cavity mode.

In the case of superfluorescence quantum noise is important not only in the initiation stage of the emission, but also throughtout the time evolution, and gives rise to anomalously large fluctuations, so that only a fully quantum picture is adequate.

The paradigm for the atom–field interaction at the quantum level is given by the Jaynes–Cummings model [44], which governs the reversible exchange of energy between a two-level atom and a cavity mode in the absence of any irreversible process. The development of the field of *cavity quantum electrodynamics* during the 1980s, with the realization of microcavities with strong coupling between a single atom and a cavity mode and an extremely small rate of escape of photons from the cavity, led to the realization of physical conditions corresponding to the Jaynes–Cummings model [45, 46].

4

Inclusion of the irreversible processes in the atomic equations

The Maxwell–Bloch equations discussed in the previous chapter describe a Hamiltonian dynamics that neglects irreversible processes. In this chapter we introduce in a phenomenological manner the dissipative decay and excitation processes that affect the atoms and therefore must be included in the Bloch equations. In the study of phenomena in optical cavities we will take into account that also the radiation field undergoes a loss process, due to the escape of photons from the cavity. This feature will be discussed in Chapter 12.

4.1 Irreversible transition processes between the two levels

In the first two sections of this chapter (Sections 4.1 and 4.2) we neglect the propagating radiation field, i.e. we set $\Omega = 0$, and consider only the irreversible processes. The decay of the two-level atom from the upper to the lower state is due to two distinct processes [47].

(i) Spontaneous emission: in the transition the atom emits a photon in a random direction, in general different from the propagation direction of the radiation field.
(ii) Inelastic collisions among atoms: in this case the transition from the upper to the lower level is not accompanied by photon emission.

The only rigorous treatment of spontaneous emission is provided by the quantum theory. Here we follow a simple phenomenological approach. The atomic decay is described by a phenomenological term in the equation for the occupation probability of the upper level,

$$\dot{p}_1 = -\gamma_\downarrow p_1, \tag{4.1}$$

with

$$\gamma_\downarrow = A + \gamma_a, \tag{4.2}$$

where the Einstein coefficient A describes spontaneous emission and γ_a is the contribution of inelastic collisions. Since $p_1 = 1 - p_2$, the equation for p_2 is

$$\dot{p}_2 = -\dot{p}_1 = \gamma_\downarrow p_1 = -\gamma_\downarrow(p_2 - 1). \tag{4.3}$$

The solutions of Eqs. (4.1) and (4.3) are, respectively,

$$p_1(t) = p_1(0)e^{-\gamma_\downarrow t} \tag{4.4}$$

and

$$p_2(t) = 1 + (p_2(0) - 1)e^{-\gamma_\downarrow t}, \tag{4.5}$$

from which one sees that, as a consequence of the decay processes, the atoms approach the state in which the electron is in the lower state $p_2 = 1$, $p_1 = 0$.

On the other hand, in opposition to the decay processes there are also two kinds of excitation processes.

(iii) Thermal effects: at absolute temperature $T > 0$ there is a non-vanishing probability that, as a consequence of thermal fluctuations, the upper level is occupied. Using Boltzmann's statistics, the thermal equilibrium probabilities are such that

$$\frac{p_1}{p_2} = \frac{e^{-E_1/(k_B T)}}{e^{-E_2/(k_B T)}} = e^{-(E_1 - E_2)/(k_B T)} = e^{-\hbar\omega_a/(k_B T)}. \tag{4.6}$$

This ratio is, however, very small at optical frequencies and room temperature: By assuming $\hbar\omega_a \approx 1$ eV and $k_B T \approx 0.025$ eV, one obtains $p_1 \approx e^{-40} p_2 \approx 10^{-18} p_2$. We will call a material in which the thermal effect is the only excitation mechanism a *passive medium*.

(iv) Pumping mechanisms: with this name we designate all the external procedures by which one can realize the population-inversion configuration, i.e. one provides the atoms with energy such that the probability of the upper level is larger than the probability of the lower level, $p_1 > p_2$. The pumping can be realized by several different processes, the most common are an electrical discharge or electrical current and the excitation provided by an external optical field. We will call a material in which there is population inversion an *active medium*.

In order to describe the excitation processes of the atom, one introduces a phenomenological decay term in the equation for the probability p_2 of the lower level,

$$\dot{p}_2 = -\gamma_\uparrow p_2, \tag{4.7}$$

and a corresponding term in the equation for p_1, in order to ensure that $p_1(t) + p_2(t) = 1$,

$$\dot{p}_1 = \gamma_\uparrow p_2. \tag{4.8}$$

By adding the decay and the excitation terms, we obtain the two rate equations for p_1 and p_2,

$$\dot{p}_1 = -\gamma_\downarrow p_1 + \gamma_\uparrow p_2, \tag{4.9}$$
$$\dot{p}_2 = \gamma_\downarrow p_1 - \gamma_\uparrow p_2. \tag{4.10}$$

These equations prescribe that in the stationary state, such that $\dot{p}_1 = 0$, $\dot{p}_2 = 0$, one has

$$\frac{(p_1)_s}{(p_2)_s} = \frac{\gamma_\uparrow}{\gamma_\downarrow}. \tag{4.11}$$

In a passive medium we have that $\gamma_\uparrow/\gamma_\downarrow = e^{-\hbar\omega_a/(k_B T)} \ll 1$, and in the following we will approximate this as zero, neglecting the thermal effect. In an active medium we have by definition $\gamma_\uparrow/\gamma_\downarrow > 1$, which one can formally interpret as a thermal equilibrium state with negative temperature. For this reason the laser is a system far from thermal equilibrium. Let us now derive from Eqs. (4.9) and (4.10) the time evolution equation for the inversion variable r_3. We observe that $r_3 = p_1 - p_2 = 2p_1 - 1$, hence

$$\begin{aligned}
\dot{r}_3 = 2\dot{p}_1 &= -2\gamma_\downarrow p_1 + 2\gamma_\uparrow(1 - p_1) = -2(\gamma_\downarrow + \gamma_\uparrow)p_1 + 2\gamma_\uparrow \\
&= -(\gamma_\downarrow + \gamma_\uparrow)(r_3 + 1) + 2\gamma_\uparrow = -(\gamma_\downarrow + \gamma_\uparrow)r_3 + \gamma_\uparrow - \gamma_\downarrow \\
&= -\gamma_\parallel(r_3 - \sigma),
\end{aligned} \tag{4.12}$$

where we define

$$\gamma_\parallel = \gamma_\downarrow + \gamma_\uparrow, \tag{4.13}$$

$$\sigma = \frac{\gamma_\uparrow - \gamma_\downarrow}{\gamma_\downarrow + \gamma_\uparrow}. \tag{4.14}$$

Therefore, if we consider only the decay and the excitation terms, i.e. the situation in the absence of a propagating radiation field, the population inversion r_3 relaxes exponentially to the stationary value σ:

$$r_3(t) = \sigma + [r_3(0) - \sigma]e^{-\gamma_\parallel t}. \tag{4.15}$$

We remark that in the case of a passive medium σ can be approximated by -1, whereas in an active medium it is a number between 0 and 1. In the latter case the upper level is more populated than the lower one, hence in the interaction with photons the prevailing process is stimulated emission, so the light intensity gets amplified as the radiation propagates along the medium, which is therefore called an *amplifier*. On the other hand, in a passive medium the most populated level is the lower one, which produces absorption, so the light intensity gets attenuated and the medium is called an *absorber*.

4.2 Irreversible decay of the atomic polarization

Insofar as the atomic polarization, i.e. the r variable, is concerned, the stationary state which it approaches is 0. Also in this case the effect arises from two kinds of contribution. On the one hand, the atomic processes of decay and of excitation induce a decay of the polarization, too. On the basis of quantum theory, one can demonstrate that the relaxation rate of the atomic polarization due to such processes is half of the decay rate of the population inversion, i.e. is given by $(\gamma_\uparrow + \gamma_\downarrow)/2 = \gamma_\parallel/2$. In addition, the polarization decays also because of the elastic collisions among atoms, which do not affect the atomic populations, but introduce random Stark shifts in the energy levels and therefore in the atomic transition frequency ω_a [22]. If we again neglect the interaction with the propagating radiation field and assume resonance ($\delta = 0$) for simplicity, the equation for r can be written in the form

$$\dot{r} = -\left[i\,\delta\omega(t) + \frac{\gamma_\parallel}{2}\right]r, \tag{4.16}$$

with $\delta\omega(t)$ being the Stark random shift of the transition frequency. The formal solution of this equation reads

$$r(t) = r(0)\exp\left[-\frac{\gamma_\parallel}{2}t - i\int_0^t dt'\,\delta\omega(t')\right]. \tag{4.17}$$

We now evaluate [22] the average of this expression over the random fluctuations $\delta\omega(t)$. This operation affects only the factor containing $\delta\omega(t)$, hence we have

$$r(t) = r(0)\exp\left(-\frac{\gamma_\parallel}{2}t\right)\left\langle\exp\left[-i\int_0^t dt'\,\delta\omega(t')\right]\right\rangle. \tag{4.18}$$

The function $\delta\omega(t)$ fluctuates randomly between positive and negative values, with average value equal to zero. Furthermore, we assume that the variations of $\delta\omega(t)$ are fast in comparison with the other variations which occur on the temporal scale and, more precisely, we assume that the noise is white so that

$$\langle\delta\omega(t)\delta\omega(t')\rangle = 2\gamma_{\mathrm{el}}\delta(t-t'), \tag{4.19}$$

where γ_{el} is a positive constant. Such an approximation, which assumes a vanishingly small memory, is called the Markov approximation. Finally, we assume that the fluctuations $\delta\omega(t)$ correspond to a Gaussian stochastic process. Under these approximations we obtain[1]

$$\left\langle\exp\left[-i\int_0^t dt'\,\delta\omega(t')\right]\right\rangle = e^{-\gamma_{\mathrm{el}}t}. \tag{4.20}$$

Therefore the average of r reads

$$r(t) = r(0)\exp\left[-\left(\frac{\gamma_\parallel}{2}+\gamma_{\mathrm{el}}\right)t\right], \tag{4.21}$$

which is the solution of the equation

$$\dot{r} = -\gamma_\perp r, \tag{4.22}$$

---

[1] If we develop the exponential we obtain

$$\left\langle\exp\left[-i\int_0^t dt'\,\delta\omega(t')\right]\right\rangle = 1 - i\int_0^t dt'\langle\delta\omega(t')\rangle - \frac{1}{2}\int_0^t dt'\int_0^t dt''\langle\delta\omega(t')\delta\omega(t'')\rangle$$

$$+ \frac{i}{6}\int_0^t dt'\int_0^t dt''\int_0^t dt'''\langle\delta\omega(t')\delta\omega(t'')\delta\omega(t''')\rangle$$

$$+ \frac{1}{24}\int_0^t dt'\int_0^t dt''\int_0^t dt'''\int_0^t dt^{\mathrm{iv}}\langle\delta\omega(t')\delta\omega(t'')\delta\omega(t''')\delta\omega(t^{\mathrm{iv}})\rangle + \cdots.$$

For a Gaussian process the odd-order correlations vanish, whereas the even-order correlations are given by the sum of all the distinct products of factors with the form (4.19), where the fluctuating variables are paired in all possible ways. For example, in the case of the product of four variables we have

$$\langle\delta\omega(t')\delta\omega(t'')\delta\omega(t''')\delta\omega(t^{\mathrm{iv}})\rangle = 4\gamma_{\mathrm{el}}^2[\delta(t'-t'')\delta(t'''-t^{\mathrm{iv}}) + \delta(t'-t''')\delta(t''-t^{\mathrm{iv}})$$
$$+ \delta(t'-t^{\mathrm{iv}})\delta(t''-t''')]$$

so that

$$\left\langle\exp\left[-i\int_0^t dt'\,\delta\omega(t')\right]\right\rangle = 1 - \frac{2\gamma_{\mathrm{el}}t}{2} + \frac{12\gamma_{\mathrm{el}}^2 t^2}{24} + \cdots = 1 - \gamma_{\mathrm{el}}t + \frac{1}{2}\gamma_{\mathrm{el}}^2 t^2 + \cdots = e^{-\gamma_{\mathrm{el}}t}.$$

with

$$\gamma_\perp = \frac{\gamma_\parallel}{2} + \gamma_{el}. \tag{4.23}$$

We observe that by definition $\gamma_\parallel \leq 2\gamma_\perp$. The limit case $\gamma_\parallel = 2\gamma_\perp$ arises when the contribution γ_{el} of elastic collisions is negligible. In particular, this situation arises when the contribution of all collisions, elastic and inelastic, is negligible, and in this case it is called *radiative limit*. The notations $\gamma_\parallel$ and $\gamma_\perp$ originate from Bloch's theory [28] for nuclear magnetic resonance, where parallel and perpendicular refer to the direction of the externally applied magnetic field. Still following Bloch, one also introduces the relaxation times $T_1 = \gamma_\parallel^{-1}$ and $T_2 = \gamma_\perp^{-1}$.

4.3 Damped Rabi oscillations and the approach to a stationary state

At this point we come back to the Bloch equations (2.46) and (2.47) and complete them by adding the irreversible terms which appear in Eqs. (4.22) and (4.12), respectively, so that they take the forms

$$\frac{\partial r}{\partial t} = -i\delta r - \frac{i}{2}\Omega r_3 - \gamma_\perp r, \tag{4.24}$$

$$\frac{\partial r_3}{\partial t} = i(\Omega r^* - \Omega^* r) - \gamma_\parallel(r_3 - \sigma). \tag{4.25}$$

Next, as we did in Section 2.4, we assume that the Rabi frequency Ω is stationary in time and uniform in space, i.e. constant, so that r and r_3 depend on time only and Eqs. (4.24) and (4.25) become

$$\dot{r} = -i\delta r - \frac{i}{2}\Omega r_3 - \gamma_\perp r, \tag{4.26}$$

$$\dot{r}_3 = i(\Omega r^* - \Omega^* r) - \gamma_\parallel(r_3 - \sigma). \tag{4.27}$$

Therefore the scenario of Rabi nutation described in Section 2.4 is valid only for times much smaller than T_1 and T_2. In the long-time limit, the atomic variables r and r_3 approach stationary values, obtained by setting $\dot{r} = 0$ and $\dot{r}_3 = 0$, which are given by

$$r_s = \frac{\sigma}{2i}\sqrt{\frac{\gamma_\parallel}{\gamma_\perp}}\frac{1 - i\Delta}{1 + \Delta^2 + |F|^2}F, \tag{4.28}$$

$$r_{3s} = \sigma\frac{1 + \Delta^2}{1 + \Delta^2 + |F|^2}, \tag{4.29}$$

respectively, where we have set

$$F = \frac{\Omega}{\sqrt{\gamma_\perp\gamma_\parallel}}, \tag{4.30}$$

$$\Delta = \frac{\omega_a - \omega_0}{\gamma_\perp}. \tag{4.31}$$

Note that Δ is the atomic detuning parameter normalized with respect to $\gamma_\perp$ and that Eq. (4.31) is identical to (1.22). Hence the Rabi oscillations are damped.

In this book we neglect the dipole–dipole interaction; see e.g. [48].

4.4 The complete Maxwell–Bloch equations

Finally, let us couple the irreversible atomic equations (4.24) and (4.25) to the field equation (3.34). If we use Eqs. (4.30) and (4.31), set

$$P = \frac{2i}{\sigma} \sqrt{\frac{\gamma_\perp}{\gamma_\parallel}} r, \tag{4.32}$$

$$D = \frac{r_3}{\sigma}, \tag{4.33}$$

and introduce the parameter g which couples the atoms and the field,

$$g = \frac{k_0 N d^2}{2\epsilon_0 \hbar \gamma_\perp n_B^2} \sigma, \tag{4.34}$$

we obtain the final and complete set of Maxwell–Bloch equations

$$\frac{n_B}{c} \frac{\partial F}{\partial t} + \frac{\partial F}{\partial z} = gP, \tag{4.35}$$

$$\frac{\partial P}{\partial t} = \gamma_\perp \left[FD - (1 + i\,\delta)P \right], \tag{4.36}$$

$$\frac{\partial D}{\partial t} = -\gamma_\parallel \left[\frac{1}{2}(FP^* + F^*P) + D - 1 \right]. \tag{4.37}$$

This form of the Maxwell–Bloch equations is especially convenient because all of the three variables F, P and D are dimensionless and the number of parameters appearing in the equations is kept to a minimum. The sign of g is determined by that of σ. Since $\sigma \approx -1$ for a passive medium and $0 < \sigma < 1$ for an active medium, g is negative in the first case and positive in the second.

In the case of a passive system, in Eq. (4.34) we must set

$$g = -\alpha, \qquad \alpha = \frac{k_0 N d^2}{2\epsilon_0 \hbar \gamma_\perp n_B^2}, \tag{4.38}$$

where α is the absorption coefficient per unit length. In the active case ($g > 0$), g is called the *gain coefficient* or *gain parameter*.

5

Propagation in irreversible Maxwell–Bloch equations

In Chapter 3 we examined some outstanding phenomena that arise in the framework of the reversible Maxwell–Bloch equations. After having introduced the irreversible terms in the previous chapter, here we describe the most relevant features of the propagation of light in a two-level medium. We will consider first the linear approximation, which basically refers to pre-laser physics and allows us to take contact with the classical linear response theory in dielectrics. Next, we will examine the full nonlinear configuration, in which the phenomenon of intensity saturation plays a major role. Finally we generalize the field equation to the case in which the background refractive index is frequency-dependent (frequency dispersion) and include the possibility of linear absorption in the background material.

5.1 Linear theory

Let us consider the atomic equations (4.36) and (4.37) at steady state, i.e. for $\partial P/\partial t = \partial D/\partial t = 0$:

$$P = \frac{1 - i\Delta}{1 + \Delta^2 + |F|^2} F, \tag{5.1}$$

$$D = \frac{1 + \Delta^2}{1 + \Delta^2 + |F|^2}. \tag{5.2}$$

We now focus on the limit $|F|^2 \ll 1 + \Delta^2$, in which these expressions reduce to

$$P = \frac{1 - i\Delta}{1 + \Delta^2} F, \quad D = 1. \tag{5.3}$$

This result suggests that we should set $D = 1$ in Eqs. (4.35) and (4.36), which then become a linear set of equations for F and P:

$$\frac{n_B}{c} \frac{\partial F}{\partial t} + \frac{\partial F}{\partial z} = gP, \tag{5.4}$$

$$\frac{\partial P}{\partial t} = \gamma_\perp [F - (1 - i\Delta)P]. \tag{5.5}$$

We will discuss these equations in the spatial domain $z \geq 0$ (see Fig. 5.1), assuming the boundary condition $F(0) = F_0$. Taking Eqs. (2.37), (2.39) and (4.30) into account, we see that this amounts to assuming that at $z = 0$ we inject a monochromatic input field of frequency ω_0. By varying ω_0 we probe the linear response of the system. According to the

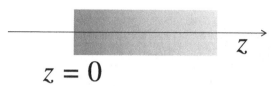

$$z = 0$$

Figure 5.1 The atomic medium occupies the region $z > 0$. We assume that at $z = 0$ we inject a monochromatic field of frequency ω_0.

initial conditions for F and P, the system undergoes a transient evolution which, due to the irreversibility, leads to the stationary solution of Eqs. (5.4) and (5.5), for which

$$P = \tilde{\chi}(\omega_0)F, \quad \tilde{\chi}(\omega_0) = \gamma_\perp \frac{\gamma_\perp + i(\omega_0 - \omega_a)}{\gamma_\perp^2 + (\omega_0 - \omega_a)^2}, \tag{5.6}$$

$$\frac{dF}{dz} = g\tilde{\chi}(\omega_0)F, \tag{5.7}$$

where we have taken Eq. (4.31) into account and $\tilde{\chi}$ is proportional to the linear susceptibility. The solution of Eq. (5.7) is [49]

$$F(z) = F(0)e^{g\tilde{\chi}(\omega_0)z}. \tag{5.8}$$

For the electric field we have therefore

$$E(z, t) \propto F(z)e^{i(k_0z - \omega_0 t)} + \text{c.c.} = F(0)e^{g\chi(\omega_0)z}e^{i(k_0z - \omega_0 t)} + \text{c.c.}, \tag{5.9}$$

so that, by introducing the real and imaginary parts of $\tilde{\chi}(\omega_0)$,

$$\tilde{\chi}(\omega_0) = \tilde{\chi}'(\omega_0) + i\tilde{\chi}''(\omega_0) = \frac{\gamma_\perp^2}{\gamma_\perp^2 + (\omega_0 - \omega_a)^2} + i\frac{\gamma_\perp(\omega_0 - \omega_a)}{\gamma_\perp^2 + (\omega_0 - \omega_a)^2}, \tag{5.10}$$

we obtain

$$E(z, t) \propto F(0)e^{g\tilde{\chi}'(\omega_0)z}e^{i[k_0 + g\tilde{\chi}''(\omega_0)]z - i\omega_0 t} + \text{c.c.} = F(0)e^{-\alpha(\omega_0)z}e^{i[k(\omega_0)z - \omega_0 t]} + \text{c.c.}, \tag{5.11}$$

where

$$\alpha(\omega_0) = -g\tilde{\chi}'(\omega_0) \tag{5.12}$$

is the absorption coefficient, whereas $k(\omega_0)$ is the wave vector in the medium, which is given by the dispersion relation

$$k(\omega_0) = \frac{\omega_0}{c}n_B + g\frac{\gamma_\perp(\omega_0 - \omega_a)}{\gamma_\perp^2 + (\omega_0 - \omega_a)^2}, \tag{5.13}$$

which is plotted in Fig. 5.2 for the passive case $g < 0$. The refractive index $n(\omega_0) = ck(\omega_0)/\omega_0$ is given by

$$n(\omega_0) = n_B + \frac{gc}{\omega_0}\tilde{\chi}''(\omega_0) = n_B + \frac{gc}{\omega_0}\frac{\gamma_\perp(\omega_0 - \omega_a)}{\gamma_\perp^2 + (\omega_0 - \omega_a)^2}. \tag{5.14}$$

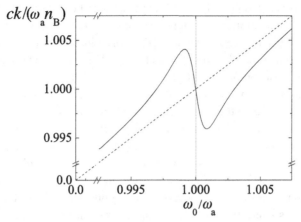

Figure 5.2   The dispersion relation of the wave vector in the medium as a function of the input frequency. The parameters are $\gamma_\perp = 10^{-3}\omega_a$ and $cg/n_B = -10^{-2}\omega_a$.

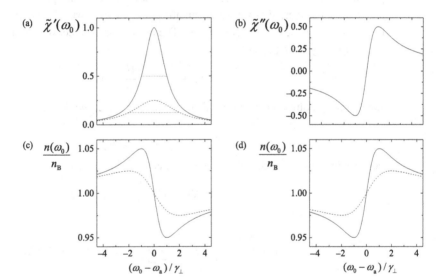

Figure 5.3   The absorption (gain) curve (a) and dispersion curves (b)–(d) for a two-level medium with $\sigma = -1$ (passive case) in (c) and $\sigma > 0$ (active case) in (d). In (c) and (d) we have assumed $|g|c/(n_B\omega_0) \approx |g|c/(n_B\omega_a) = 0.1$. In (a), (c) and (d) the broken lines represent the curves corresponding to the nonlinear case with $|F|^2 = 3$ (see Section 5.2). In (a) the dotted segments indicate the width at half maximum of the absorption curve, i.e. $2\gamma_\perp$ in the linear case, and $2\sqrt{1+|F|^2}\,\gamma_\perp = 4\gamma_\perp$ in the nonlinear case.

Therefore the real part of the susceptibility is linked to absorption in a passive medium. As a matter of fact the radiation intensity along the medium is given by

$$I(z) \propto |F(z)|^2 = |F(0)|^2 e^{-2\alpha(\omega_0)z}, \tag{5.15}$$

which corresponds to Beer's classic exponential-decay law of absorption. The function $\tilde{\chi}'(\omega_0)$ is a Lorentzian centered at the resonance frequency ω_a (Fig. 5.3(a)). In the case of an active medium one does not have absorption but, on the contrary, exponential amplification as the radiation travels along the medium. It must be taken into account, however, that the amplification leads to values of $F(z)$ that violate the condition $|F|^2 \ll 1 + \Delta^2$, therefore in the active case the linear model holds only for a limited range of values of z, beyond which one must consider the stationary solution of the exact equations, discussed in Section 5.3. The quantity $\gamma_\perp$ represents the halfwidth at half maximum (HWHM) of the absorption line or gain line, in the passive case and in the active case, respectively.

On the other hand, the imaginary part $\tilde{\chi}''(\omega_0)$ is linked to the frequency dispersion in the medium; it presents a minimum at $\omega_a - \gamma_\perp$ and a maximum at $\omega_a + \gamma_\perp$, and vanishes at ω_a (Fig. 5.3(b)). As a consequence, in a passive medium one has anomalous dispersion $(dn/d\omega_0 < 0)$ in the region between these two points, whereas in an active medium the anomalous dispersion occurs in the region outside them (Figs. 5.3(c) and (d)).

It is interesting to compare the expression for the linear susceptibility which we have obtained from the two-level model with the well-known result derived from the Lorentz model of forced and damped oscillators. Let us consider an electron bound to its equilibrium position by the elastic force $\mathbf{F}_{el} = -m\omega_a^2\mathbf{x}$, subjected to a dissipative force $\mathbf{F}_d = -2m\gamma\dot{\mathbf{x}}$ and to the force exerted by a monochromatic plane wave polarized along the x axis, described by the electric field $\mathbf{E} = E_0\cos(\omega_0 t)\hat{\mathbf{e}}_x$. The motion of the electron is one-dimensional and is governed by the equation

$$\ddot{x} + 2\gamma\dot{x} + \omega_a^2 x = \frac{eE_0}{m}\cos(\omega_0 t). \tag{5.16}$$

Let us seek a solution of the form $2x(t) = x_0 e^{-i\omega_0 t} + \text{c.c.}$, with x_0 complex. We obtain

$$x_0 = \frac{eE_0/m}{-\omega_0^2 + \omega_a^2 - 2i\gamma\omega_0}. \tag{5.17}$$

For N electrons per unit volume we get a macroscopic polarization

$$P(t) = Nex(t) = \frac{P_0}{2}e^{-i\omega_0 t} + \text{c.c.},$$

where

$$P_0 = \frac{Ne^2}{m}\frac{1}{-\omega_0^2 + \omega_a^2 - 2i\gamma\omega_0}E_0 \equiv \epsilon_0\chi_{cl}(\omega_0)E_0. \tag{5.18}$$

Therefore we can introduce the classical susceptibility

$$\chi_{cl}(\omega_0) = -\frac{Ne^2}{\epsilon_0 m}\frac{1}{\omega_0^2 - \omega_a^2 + 2i\gamma\omega_0}. \tag{5.19}$$

We can obtain a functional dependence on ω_0 identical to that of the semiclassical two-level theory if we introduce an approximation equivalent to the rotating-wave approximation, consisting in considering a situation of quasi-resonance between the field frequency ω_0 and the oscillator frequency ω_a. In this case we can write $\omega_0^2 - \omega_a^2 + 2i\gamma\omega_0 = (\omega_0 - \omega_a)(\omega_0 + \omega_a) + 2i\gamma\omega_0 \approx 2\omega_a(\omega_0 - \omega_a + i\gamma)$, from which it follows that

$$\chi_{cl}(\omega_0) \approx -\frac{Ne^2}{2\epsilon_0 m\omega_a}\frac{1}{\omega_0 - \omega_a + i\gamma}, \tag{5.20}$$

which, if we identify γ with $\gamma_\perp$, differs from Eq. (5.6) only by a multiplicative constant.

In order to complete the comparison, it is useful to calculate the expression for the atomic polarization P_0 as a function of the field E_0 (see Eqs. (2.42) and (2.37)), instead of using the scaled quantities F and P which we have considered in this chapter. On taking into account Eqs. (2.42), (4.32), (5.6), (4.30) and (2.39), we have

$$
\begin{aligned}
P_0 &= 2Ndr = 2Nd\frac{\sigma}{2i}\sqrt{\frac{\gamma_\parallel}{\gamma_\perp}}P = \frac{Nd\sigma}{i}\sqrt{\frac{\gamma_\parallel}{\gamma_\perp}}\frac{\gamma_\perp}{\gamma_\perp + i(\omega_a - \omega_0)}F \\
&= Nod\frac{\sqrt{\gamma_\perp\gamma_\parallel}}{i\gamma_\perp - \omega_a + \omega_0}\frac{\Omega}{\sqrt{\gamma_\perp\gamma_\parallel}} = Nod\frac{1}{\omega_0 - \omega_a + i\gamma_\perp}\frac{dE_0}{\hbar} \\
&= \frac{N\sigma d^2}{\hbar}\frac{1}{\omega_0 - \omega_a + i\gamma_\perp}E_0,
\end{aligned}
\tag{5.21}
$$

in such a way that one can define a semiclassical susceptibility given by

$$
\chi_{sc}(\omega_0) = \frac{N\sigma d^2}{\epsilon_0\hbar}\frac{1}{\omega_0 - \omega_a + i\gamma_\perp}.
\tag{5.22}
$$

There is a perfect correspondence between the semiclassical susceptibility and the classical susceptibility approximated by Eq. (5.20) if we identify $N\sigma$ of the semiclassical case with $-N$ of the classical one (which holds exactly if the two-level atoms are in the lower level) and $d^2/\hbar$ with $e^2/(2m\omega_a)$.

From Eq. (5.10) we have that $\tilde{\chi}''(\omega_0) = \tilde{\chi}'(\omega_0)(\omega_0 - \omega_a)/\gamma_\perp$. This is an example of the general Kramers–Krönig relations which link the real and the imaginary part of the susceptibility in the linear-response theory. Usually the real part of χ is associated with dispersion and the imaginary part with absorption, which is the opposite of what happens in our treatment. This feature arises from the fact that in the definition (4.32) of the quantity P we have introduced an i factor, which exchanges real and imaginary parts.

An important final remark in this section is the following. From Eqs. (5.8) and (5.10) it is clear that in the case of two-level atoms in cavityless configurations considered up to now the longitudinal spatial scale L introduced in Section 3.1 corresponds to g^{-1}. From Eq. (5.21) and Eq. (4.34) we see that the ratio $P_0/(\epsilon_0 E_0)$ for $\omega_0 = \omega_a$ is equal to g/k_0 within a factor of order unity. Now $g/k_0 \propto \lambda/L$, and this confirms that $P_0/(\epsilon_0 E_0)$ is of order $\epsilon^2 = \lambda/L$ as assumed in Section 3.1.

5.2 Saturation and power broadening

Let us discuss the physical meaning of the condition $|F|^2 \ll 1 + \Delta^2$. We observe that from Eqs. (4.30) and (2.39) we obtain

$$
F = \frac{\Omega}{\sqrt{\gamma_\perp\gamma_\parallel}} = \frac{dE_0}{\hbar\sqrt{\gamma_\perp\gamma_\parallel}}.
\tag{5.23}
$$

The radiation intensity, i.e. the energy flux carried by the wave in the propagation direction, is represented by the modulus of the Poynting vector, which in vacuum is given by

$I = \epsilon_0 c |E_0|^2 / 2$. Therefore we can write

$$|F|^2 = \frac{I}{I_{\text{sat}}}, \tag{5.24}$$

with (see Eq. (1.19))

$$I_{\text{sat}} = \frac{\epsilon_0 c \, \hbar^2 \gamma_\perp \gamma_\parallel}{2 \quad d^2} = \frac{c}{12} \frac{\gamma_\perp \gamma_\parallel}{B}, \tag{5.25}$$

hence the radiation intensity is normalized to the saturation intensity on resonance I_{sat}. In the presence of atomic detuning, the saturation intensity is given by $I_{\text{sat},\Delta} = I_{\text{sat},0}(1 + \Delta^2)$. In the linear regime the intensity is much smaller than the saturation intensity, so $|F|^2 \ll 1 + \Delta^2$. Let us now drop this assumption and return to the exact expression (5.1), which leads to the following formula for the susceptibility of the medium:

$$\tilde{\chi}(\omega_0) = \frac{1 - i\Delta}{1 + \Delta^2 + |F|^2}, \tag{5.26}$$

which depends nonlinearly on the normalized intensity $|F|^2$. Note that, for $|F|^2 < 1 + \Delta^2$, $\tilde{\chi}(\omega_0)$ can be expanded in power series of $|F|^2$,

$$\tilde{\chi}(\omega_0) = \sum_{n=0}^{\infty} \tilde{\chi}^{(2n+1)}(\omega_0) |F|^{2n}, \tag{5.27}$$

with

$$\tilde{\chi}^{(2n+1)}(\omega_0) = \frac{1 - i\Delta}{1 + \Delta^2} (-1)^n \frac{1}{\left(1 + \Delta^2\right)^n}, \tag{5.28}$$

and $\tilde{\chi}^{(1)}$ coincides with the susceptibility of the linear theory (see Eqs. (5.3) and (5.6)). This remark is relevant in connection with the discussion of materials with cubic and quadratic nonlinearities in the following two chapters. As a consequence of Eq. (5.26), the real and imaginary parts become

$$\tilde{\chi}'(\omega_0) = \frac{\gamma_\perp^2}{\gamma_\perp^2 (1 + |F|^2) + (\omega_0 - \omega_a)^2}, \tag{5.29}$$

$$\tilde{\chi}''(\omega_0) = \frac{\gamma_\perp(\omega_0 - \omega_a)}{\gamma_\perp^2 (1 + |F|^2) + (\omega_0 - \omega_a)^2}. \tag{5.30}$$

The real part still corresponds to a Lorentzian centered at $\omega_0 = \omega_a$, but now the maximum is $(1 + |F|^2)^{-1}$ and the halfwidth is $\gamma_\perp \sqrt{1 + |F|^2}$. Hence, the larger the intensity, the lower and broader the curve. This phenomenon is called *power broadening*. The same considerations hold for the imaginary part, in which the distance between the maximum and the minimum increases while their values decrease (in modulus). In Fig. 5.3 the broken curves correspond to the case $|F|^2 = 3$, in which the width is twice that for the linear case.

It is also interesting to consider the dependence on frequency of the occupation probability of the upper level p_1, in the absorber case in which $\sigma = -1$ and therefore $D = -r_3$. On taking into account Eqs. (2.71) and (5.2) we have

$$p_1 = \frac{1}{2}(1 + r_3) = \frac{1}{2}(1 - D) = \frac{1}{2} \frac{|F|^2}{1 + \Delta^2 + |F|^2} = \frac{1}{2} |F|^2 \tilde{\chi}'(\omega_0). \tag{5.31}$$

In the absorber case, in the absence of an electric field, all the atoms are in the lower level, hence p_1 coincides with the transition probability to the upper level. From Eq. (5.31) we see that this probability depends on frequency as the real part of the susceptibility and therefore it is affected by power broadening too, attaining the maximum value

$$p_{1\,\text{max}} = \frac{1}{2}\frac{|F|^2}{1+|F|^2}. \tag{5.32}$$

In the small-intensity limit $p_{1\,\text{max}}$ increases as $|F|^2$, but for high intensity one has the phenomenon of saturation, due to which $p_{1\,\text{max}}$ tends asymptotically to the value $1/2$. In this situation the medium becomes transparent because the number of atoms in the upper level is equal to that in the lower level, so stimulated emission and absorption balance each other exactly. This feature demonstrates that one cannot produce a state of population inversion by pumping optically a two-level atom but, at most, a transparency condition. In Chapter 10 we will see that, in order to realize a situation of population inversion by optical pumping, it is necessary to involve at least three energy levels.

5.3 Nonlinear propagation for a monochromatic input field: The role of saturation and nonlinear phase shift

In the fully nonlinear case, the Maxwell–Bloch equations (4.35)–(4.37) can be solved analytically only in the stationary state. By inserting Eq. (5.1) into the field equation (4.35), at steady state ($\partial F/\partial t = 0$) we obtain the following equation which governs the variation of F along the medium, starting from the boundary $z = 0$:

$$\frac{dF}{dz} = g\frac{(1-i\Delta)F}{1+\Delta^2+|F|^2}. \tag{5.33}$$

This equation can be solved by inserting the expression for $F(z)$ in terms of its modulus and its phase $F(z) = \rho(z)e^{i\varphi(z)}$, dividing Eq. (5.33) by $e^{i\varphi(z)}$ and separating real and imaginary terms, giving

$$\frac{d\rho}{dz} = g\frac{\rho}{1+\Delta^2+\rho^2}, \tag{5.34}$$

$$\frac{d\varphi}{dz} = -g\frac{\Delta}{1+\Delta^2+\rho^2}. \tag{5.35}$$

These two equations must be solved with the initial conditions $\rho(0) = \rho_0$ and $\varphi(0) = \varphi_0$. The first is a closed equation for ρ, which has the solution

$$gz = (1+\Delta^2)\ln\left(\frac{\rho(z)}{\rho_0}\right) + \frac{1}{2}\left[\rho^2(z) - \rho_0^2\right]. \tag{5.36}$$

We note that, if we neglect the second term on the r.h.s. of Eq. (5.36), we obtain

$$\rho(z) = \rho_0 \exp\{gz/(1+\Delta^2)\}, \tag{5.37}$$

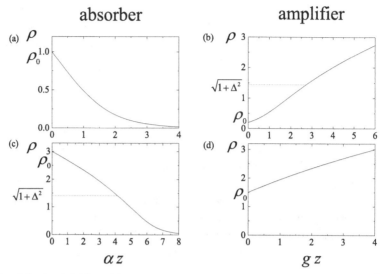

Figure 5.4   The variation of the electric field's modulus ρ along the sample in nonlinear propagation. The value of Δ^2 is 1. Parts (a) and (c) refer to the absorber, parts (b) and (d) to the amplifier. In (a) and (b) the value of ρ_0 is smaller than $\sqrt{1 + \Delta^2}$, whereas in (c) and (d) it is larger.

which, taking Eqs. (5.6) and (4.31) into account, coincides with the modulus of the stationary solution of the linear case Eq. (5.8). On the other hand, the second term on the r.h.s. of Eq. (5.36) provides the contribution due to saturation. The function we want to obtain is $\rho(z)$, whereas Eq. (5.36) corresponds to the inverse of such a function. So the straightforward procedure to obtain the result is to plot the inverse function and, next, rotate the diagram of the function $z(\rho)$ around the bisectrix of the first quadrant. In such a way we obtain the plots shown in Fig. 5.4, which correspond both to the absorber case and to the amplifier case. In the first case α is the absorption coefficient defined by Eq. (4.38). The curve presents an inflection point for $\rho = \sqrt{1 + \Delta^2}$. We see that in the absorber case when $\rho_0 > \sqrt{1 + \Delta^2}$ the system is initially in the saturation regime and then enters an exponentially decaying linear regime (Fig. 5.4(c)), whereas for an amplifier with $\rho_0 < \sqrt{1 + \Delta^2}$ the system starts from the exponentially increasing linear regime and then enters the saturation regime (Fig. 5.4(b)).

Insofar as the phase is concerned, let us return to Eqs. (5.34) and (5.35). By combining them we obtain the equation

$$d\varphi = -\Delta \frac{d\rho}{\rho}, \tag{5.38}$$

the solution of which reads

$$\Delta\varphi(z) = \varphi(z) - \varphi_0 = -\Delta \ln\left(\frac{\rho(z)}{\rho_0}\right), \tag{5.39}$$

where $\rho(z)$ obeys Eq. (5.36). The quantity $\Delta\varphi(z)$ corresponds to the *nonlinear phase shift* which develops in propagation in the detuned configuration. The linear part of $\Delta\varphi(z)$ is obtained by inserting Eq. (5.37) into Eq. (5.39) and coincides with $[k(\omega_0) - \omega_0 n_B/c]z$, where $k(\omega_0)$ is the wave vector given by Eq. (5.13). Note that the nonlinear phase shift is tightly linked to the absorption/amplification $\rho(z)/\rho_0$.

5.4 Background linear dispersion and absorption

In this section we consider the case in which the background medium displays linear dispersion (i.e. there is a frequency dependence of the light velocity, and hence of the background refractive index) or/and linear absorption. The aim is to show how the Maxwell–Bloch equations must be generalized to include them. While the inclusion of linear absorption is trivial, dispersion requires a careful re-examination of the derivation of the equation for the electric field in the paraxial and slowly varying envelope approximations.

5.4.1 Background dispersion

Let us consider a medium in which the light velocity is frequency-dependent. The wave equation is expressed in terms of the Fourier transforms in time of the electric field and of the atomic polarization,

$$\mathbf{E}(\mathbf{x}, \omega) = \int_{-\infty}^{+\infty} dt\, \mathbf{E}(\mathbf{x}, t)e^{i\omega t}, \tag{5.40}$$

$$\mathbf{P}(\mathbf{x}, \omega) = \int_{-\infty}^{+\infty} dt\, \mathbf{P}(\mathbf{x}, t)e^{i\omega t}, \tag{5.41}$$

and reads [50]

$$\nabla^2 \mathbf{E}(\mathbf{x}, \omega) + k^2(\omega)\mathbf{E}(\mathbf{x}, \omega) = -\frac{\omega^2}{\epsilon_0 c^2}\mathbf{P}(\mathbf{x}, \omega), \quad k(\omega) = \frac{\omega}{v(\omega)} = \frac{\omega}{c}n_B(\omega), \tag{5.42}$$

where $v(\omega)$ is the light velocity and $k(\omega)$ is the wavenumber, which depends on frequency via a given dispersion relation. We have neglected the term $\nabla(\nabla \cdot \mathbf{E})$ in accord with the analysis in Section 3.1.

Next, we express the electric field and the atomic polarization in terms of their envelopes using Eqs. (3.28) and (3.29) where now, however, k_0 is given not by $k_0 = \omega_0 n_B/c$ but by $k_0 = k(\omega_0)$. Hence, from Eqs. (5.40) and (5.41), we obtain

$$\mathbf{E}(\mathbf{x}, \omega) = \frac{\hat{\mathbf{e}}}{2}\left[E_0(\mathbf{x}, \omega - \omega_0)e^{ik_0 z} + E_0^*(\mathbf{x}, \omega + \omega_0)e^{-ik_0 z}\right], \tag{5.43}$$

$$\mathbf{P}(\mathbf{x}, \omega) = \frac{\hat{\mathbf{e}}}{2}\left[P_0(\mathbf{x}, \omega - \omega_0)e^{ik_0 z} + P_0^*(\mathbf{x}, \omega + \omega_0)e^{-ik_0 z}\right]. \tag{5.44}$$

By inserting Eqs. (5.43) and (5.44) into the wave equation (5.42), equating the terms which include the factor $e^{ik_0 z}$ and dividing the equation by this exponential, we arrive at

$$\nabla_{\perp}^2 E_0(\mathbf{x}, \omega - \omega_0) - \left(k_0^2 + 2i k_0 \frac{\partial}{\partial z} + \frac{\partial^2}{\partial z^2} \right) E_0(\mathbf{x}, \omega - \omega_0)$$

$$+ \left[k(\omega_0) + k'(\omega_0)(\omega - \omega_0) + \frac{1}{2}k''(\omega_0)(\omega - \omega_0)^2 \right]^2 E_0(\mathbf{x}, \omega - \omega_0)$$

$$= -\frac{1}{\epsilon_0 c^2}[\omega_0 + (\omega - \omega_0)]^2 P_0(\mathbf{x}, \omega - \omega_0), \qquad (5.45)$$

where we have expanded $k(\omega)$ up to second order around ω_0. At this point we scale the spatial variables as in Eq. (3.26). For the frequency variable, we take into account that

$$k'(\omega_0) = \frac{1}{v_g}, \qquad (5.46)$$

where v_g is the group velocity, and we assume that $(\omega - \omega_0)L/v_g$ is at most of order unity. Furthermore, we assume that the dispersion parameter $k''(\omega_0)$ has the order of magnitude

$$k''(\omega_0) \sim \frac{L}{v_g^2}. \qquad (5.47)$$

As in Section 3.1, we keep only the terms of order ϵ^0, where ϵ is given by Eq. (3.20). In terms of the original variables x, y, z and ω we obtain the equation

$$\frac{1}{2i k_0} \nabla_{\perp}^2 E_0(\mathbf{x}, \omega - \omega_0) + \frac{\partial E_0(\mathbf{x}, \omega - \omega_0)}{\partial z} - \frac{i}{v_g}(\omega - \omega_0)E_0(\mathbf{x}, \omega - \omega_0)$$

$$- i\frac{k''(\omega_0)}{2}(\omega - \omega_0)^2 E_0(\mathbf{x}, \omega - \omega_0) = i\frac{\omega_0^2}{2\epsilon_0 k_0 c^2} P_0(\mathbf{x}, \omega - \omega_0). \qquad (5.48)$$

By Fourier transforming back to $E_0(\mathbf{x}, t)$ and $P_0(\mathbf{x}, t)$, we have

$$\frac{1}{2i k_0} \nabla_{\perp}^2 E_0(\mathbf{x}, t) + \frac{\partial E_0(\mathbf{x}, t)}{\partial z} + \frac{1}{v_g} \frac{\partial E_0(\mathbf{x}, t)}{\partial t} + i\frac{k''(\omega_0)}{2} \frac{\partial^2 E_0(\mathbf{x}, t)}{\partial t^2}$$

$$= i\frac{k_0}{2\epsilon_0 n_B^2(\omega_0)} P_0(\mathbf{x}, t), \qquad (5.49)$$

where we have taken into account that, from Eq. (5.42),

$$\frac{\omega_0^2}{k_0 c^2} = \frac{k_0}{n_B^2(\omega_0)}. \qquad (5.50)$$

The term with k'' describes the *group velocity dispersion*.

5.4.2 Background linear absorption

The possible presence of background linear absorption is taken into account by adding a linear background damping term to the equation above:

$$\frac{1}{2ik_0}\nabla_\perp^2 E_0(\mathbf{x}, t) + \frac{\partial E_0(\mathbf{x}, t)}{\partial z} + \frac{1}{v_g}\frac{\partial E_0(\mathbf{x}, t)}{\partial t} + i\frac{k''(\omega_0)}{2}\frac{\partial^2 E_0(\mathbf{x}, t)}{\partial t^2}$$

$$= -\alpha_B E_0(\mathbf{x}, t) + i\frac{k_0}{2\epsilon_0 n_B^2(\omega_0)}P_0(\mathbf{x}, t). \tag{5.51}$$

6

Optical nonlinearities. Materials with quadratic nonlinearities

In the previous chapter we examined nonlinear propagation in a two-level medium, which leads to amplification or absorption of the radiation field. As we will see further on in this volume, when the medium is placed in an optical cavity one realizes a laser in the case of amplification or an optically bistable system in the absorptive configuration. On the other hand, the nonlinear response of the atomic polarization to the electric field leads to further fascinating scenarios [29, 51–60]. For example, a field with central ω_0 can generate harmonic components with central frequencies $2\omega_0$, $3\omega_0$ etc., processes that are called second-harmonic generation, third-harmonic generation etc. Furthermore, the interaction of two fields with central frequencies ω_1 and ω_2, with $\omega_1 > \omega_2$, can generate components with central frequency $\omega_1 + \omega_2$ (sum-frequency generation) and $\omega_1 - \omega_2$ (difference-frequency generation). Such phenomena have not yet appeared in this volume because we have assumed that the electric field corresponds to a narrow frequency band around a central frequency ω_0 only, whereas to describe them it is necessary to assume that the electric field is the superposition of a number of frequency bands. In this chapter we will provide a description of the basic elements which are necessary to describe these phenomena and we will discuss a small number of them for the case of materials with a quadratic nonlinearity, referring to textbooks in nonlinear optics for the description of other phenomena. The following chapter is devoted, instead, to the case of materials with cubic nonlinearities.

The starting point of nonlinear optics is the expansion of the atomic polarization in powers of the electric field, when the electric field is not too intense. In our treatment in Chapter 5 we saw that the expansion of the nonlinear susceptibility contains only even powers of the electric field (see Eq. (5.27)), and therefore the expansion of the polarization includes only odd powers. It turns out, however, that in the case of non-centrosymmetric materials also even powers of the electric field appear in the polarization expansion, and this leads to the possibility of such phenomena as, for example, second-harmonic generation, which appears in materials with quadratic nonlinearities.

In order to keep our treatment at a reasonable level of simplicity and conciseness, we will introduce some important simplifications.

- First of all, nonlinear materials may be anisotropic, which in principle requires a tensorial description. In particular, for materials with quadratic nonlinearities, in order to achieve efficient phenomena of wave mixing the birefringence of the material must be exploited.[1] In the following, we shall assume that the linear polarization directions of the

---

[1] In order to achieve the conservation of momentum (phase matching) in wave-mixing phenomena such as second-harmonic or difference-frequency generation, at least one of the waves must be extraordinarily polarized.

waves participating in these phenomena are properly chosen in order to ensure efficient nonlinear mixing, and we shall treat the electric-field amplitudes along these directions as scalar quantities. Accordingly, the susceptibilities will be described as single numbers representing the appropriate tensor elements. For the sake of simplicity we shall also neglect small effects such as the spatial walk-off that arises because of the material birefringence.

- In such phenomena as, for example, spontaneous parametric down-conversion, the paraxial and the slowly varying envelope approximations are not strictly valid because the emission occurs over an extremely broad band of both temporal and spatial frequencies. However, for the sake of coherence with the rest of the volume we will utilize these approximations, as is done in most of the literature.
- As in all of this book, we will stick to the semiclassical theory, and only when necessary will we include comments on the role of quantum theory in these phenomena, similarly to what we did in Chapter 1 for the laser, with the phenomenological introduction of spontaneous emission.

6.1 Linear and nonlinear polarization

In nonlinear optics it is traditional to split the atomic polarization into a linear and a nonlinear part,

$$P = P_{\mathrm{L}} + P_{\mathrm{NL}}, \tag{6.1}$$

where we have not indicated vectors because, as we said, we will pursue a scalar description. The part P_{L} is the contribution to the polarization which depends linearly on the electric field. On the other hand, for weak enough values of the electric field P_{NL} can be expanded in the following way:

$$P_{\mathrm{NL}}(\mathbf{x}, t) = \epsilon_0 \left(\chi^{(2)} E^2(\mathbf{x}, t) + \chi^{(3)} E^3(\mathbf{x}, t) + \cdots \right), \tag{6.2}$$

where the nonlinear susceptibilities $\chi^{(2)}$, $\chi^{(3)}$, etc. are real one-dimensional numbers. In this book we do not include the quantum-mechanical derivation and discussion of the microscopic expressions of the susceptibilities $\chi^{(i)}$, which can be found in [58,59]. By treating the parameters $\chi^{(i)}$, $i \geq 2$, as constant we neglect the nonlinear frequency dispersion, whereas we take dispersion into account in the linear susceptibility $\chi^{(1)} \equiv \chi$. This step is introduced, again, for the sake of simplicity.

Before continuing, we would like to add two appropriate remarks in order to avoid confusion with the treatment of the previous chapters.

- In this chapter, we identify P_{L} with the linear background polarization introduced in Section 3.1, which is already incorporated in the wave equation. For the sake of simplicity, in the following we indicate the refractive index by $n(\omega)$ instead of $n_{\mathrm{B}}(\omega)$.
- The variables F, P and D which appear in Eqs. (4.35)–(4.37) and in Chapter 5 are normalized variables, whereas the variables E and P we use in this chapter and in

the following chapter are the original variables. In particular, the polarization P which appears in the equations of Chapter 5 is obtained by a normalization that involves the imaginary unit (see Eq. (4.32)), which produces an exchange between the real and imaginary parts of the two-level-atom susceptibility.

In the treatment of the processes that we describe in the remainder of this chapter, it is mandatory to take into account the frequency dependence of the refractive index, because one must consider simultaneously optical frequencies that differ substantially from one another. Therefore our starting point is an equation that is basically identical to Eq. (5.42),

$$\nabla^2 E(\mathbf{x}, \omega) + k^2(\omega) E(\mathbf{x}, \omega) = -\frac{\omega^2}{\epsilon_0 c^2} P_{\text{NL}}(\mathbf{x}, \omega), \quad k(\omega) = \frac{\omega n(\omega)}{c}, \tag{6.3}$$

where $E(\mathbf{x}, \omega)$ and $P_{\text{NL}}(\mathbf{x}, \omega)$ are the Fourier transforms of $E(\mathbf{x}, t)$ and $P_{\text{NL}}(\mathbf{x}, t)$, respectively

$$E(\mathbf{x}, \omega) = \int_{-\infty}^{+\infty} dt \, E(\mathbf{x}, t) e^{i\omega t}, \quad P_{\text{NL}}(\mathbf{x}, \omega) = \int_{-\infty}^{+\infty} dt \, P_{\text{NL}}(\mathbf{x}, t) e^{i\omega t}, \tag{6.4}$$

and the linear part of the atomic polarization P_{L} is incorporated in $k(\omega)$ via the relation $k(\omega)^2 = \omega^2[1 + \chi(\omega)]/c^2$, where $\chi(\omega)$ is the refractive part of the linear susceptibility. Let us now express the electric field as the superposition of a number of frequency bands with optical frequencies ω_p

$$E(\mathbf{x}, t) = \sum_p \frac{1}{2} \left[E_p(\mathbf{x}, t) e^{-i(\omega_p t - k_p z)} + \text{c.c.} \right], \tag{6.5}$$

where E_p are slowly varying envelopes and $k_p = k(\omega_p)$. Hence we obtain

$$E(\mathbf{x}, \omega) = \sum_p \frac{1}{2} \left[E_p(\mathbf{x}, \omega - \omega_p) e^{ik_p z} + E_p^*(\mathbf{x}, \omega + \omega_p) e^{-ik_p z} \right]. \tag{6.6}$$

Next, let us analyze separately the left-hand and right-hand sides of Eq. (6.3). By inserting Eq. (6.6) into the l.h.s. of Eq. (6.3) and expanding $k(\omega)$ we obtain

$$\frac{1}{2} \sum_p \left\{ \nabla_\perp^2 E_p(\mathbf{x}, \omega - \omega_p) - k_p^2 E_p(\mathbf{x}, \omega - \omega_p) + 2ik_p \frac{\partial E_p(\mathbf{x}, \omega - \omega_p)}{\partial z} \right.$$

$$+ \frac{\partial^2 E_p(\mathbf{x}, \omega - \omega_p)}{\partial z^2} + \left[k_p + k'(\omega_p)(\omega - \omega_p) \right.$$

$$\left. \left. + \frac{1}{2} k''(\omega_p)(\omega - \omega_p)^2 + \cdots \right]^2 E_p(\mathbf{x}, \omega - \omega_p) \right\} e^{ik_p z} + \overline{\text{c.c.}}, \tag{6.7}$$

where $\overline{\text{c.c.}}$ means the same expressions, but with $E_p(\mathbf{x}, \omega - \omega_p)$ replaced by $E_p^*(\mathbf{x}, \omega + \omega_p)$ and i replaced by $(-i)$. By keeping only the relevant terms as in Section 5.4 and performing the back Fourier transformation we obtain

$$
\frac{1}{2} \sum_p \left\{ \nabla_\perp^2 E_p(\mathbf{x}, t) + 2i k_p \frac{\partial E_p(\mathbf{x}, t)}{\partial z} + 2i k_p k'(\omega_p) \frac{\partial E_p(\mathbf{x}, t)}{\partial t} \right.
$$
$$
\left. - k_p k''(\omega_p) \frac{\partial^2 E_p(\mathbf{x}, t)}{\partial t^2} \right\} e^{-i(\omega_p t - k_p z)} + \text{c.c.}, \tag{6.8}
$$

where $k'(\omega_p) = 1/v_{\mathrm{gp}}$, with v_{gp} being the group velocity of the envelope with frequency ω_p. If one neglects the contribution of the frequency variation of $n(\omega)$ to the group velocity in the band, one can write $n(\omega_p)/c$ instead of $1/v_{\mathrm{gp}}$.

Let us now turn our attention to the r.h.s. of Eq. (6.3) and carry out the back Fourier transformation

$$
\frac{1}{\epsilon_0 c^2} \frac{\partial^2 P_{\mathrm{NL}}(\mathbf{x}, t)}{\partial t^2}. \tag{6.9}
$$

As we will see in the following sections, $P_{\mathrm{NL}}(\mathbf{x}, t)$ turns out to be the sum of contributions with different frequencies ω_r. One approximates Eq. (6.9) by this sum, in which each contribution is multiplied by $(-\omega_r^2/(c^2 \epsilon_0))$, in accord with the analysis of Section 3.1. Of this sum we consider only the terms with the frequencies ω_p which appear in Eq. (6.5),[2] consistently with the fact that only these frequency bands have been included in the l.h.s. of Eq. (6.3), and equate this sum to Eq. (6.8). Finally we equate separately the terms with the same exponential factor $\exp[-i(\omega_p t - k_p z)]$ and obtain one equation for each frequency band. The set of such equations constitutes our model. In the following sections this procedure will be applied to the fundamental cases of materials with quadratic nonlinearities.

6.2 Media with a quadratic nonlinearity

Let us consider the case of materials displaying a quadratic nonlinearity, so that

$$
P_{\mathrm{NL}}(\mathbf{x}, t) = \epsilon_0 \chi^{(2)} E^2(\mathbf{x}, t), \tag{6.10}
$$

and let us first assume that E is composed by the superposition of two frequency bands,

$$
E(\mathbf{x}, t) = \frac{1}{2} \left[E_1(\mathbf{x}, t) e^{-i(\omega_1 t - k_1 z)} + E_2(\mathbf{x}, t) e^{-i(\omega_2 t - k_2 z)} \right.
$$
$$
\left. + E_1^*(\mathbf{x}, t) e^{i(\omega_1 t - k_1 z)} + E_2^*(\mathbf{x}, t) e^{i(\omega_2 t - k_2 z)} \right], \tag{6.11}
$$

with $\omega_2 = 2\omega_1$. By inserting expression (6.11) into Eq. (6.10), we obtain various kinds of terms.

- The terms $[E_1^2(\mathbf{x}, t)/4] e^{-2i\omega_1 t + 2i k_1 z}$ and $[E_2^2(\mathbf{x}, t)/4] e^{-2i\omega_2 t + 2i k_2 z}$ describe the generation of the second-harmonic frequencies $2\omega_1$ and $2\omega_2$, respectively.

---

[2] For consistency, we assume that other frequency bands, which appear in the expression for P_{NL} but not in the expansion (6.6), are irrelevant because the phase-matching condition (see the following sections) is badly violated for such processes.

- The term $[E_1(\mathbf{x},t)E_2(\mathbf{x},t)/2]e^{-i(\omega_1+\omega_2)t+i(k_1+k_2)z}$ describes the generation of the sum frequency $\omega_1+\omega_2$.
- The terms $E_1(\mathbf{x},t)E_1^*(\mathbf{x},t)/2$ and $E_2(\mathbf{x},t)E_2^*(\mathbf{x},t)/2$ describe the generation of a d.c. field (zero frequency). This kind of process is called *optical rectification*.
- The term $[E_2(\mathbf{x},t)E_1^*(\mathbf{x},t)/2]e^{-i(\omega_2-\omega_1)t+i(k_2-k_1)z}$, taking into account that $\omega_2-\omega_1=\omega_1$ since $\omega_2=2\omega_1$, describes a process of amplification or deamplification of the field component of frequency ω_1.

Now if, as described in the final part of the previous section, we equate the terms of frequency ω_1 in Eq. (6.8) with the term of frequency ω_1 in Eq. (6.10) multiplied by $(-\omega_1^2/(c^2\epsilon_0))$, we obtain the equation [61]

$$\frac{1}{2ik_1}\nabla_\perp^2 E_1(\mathbf{x},t)+\frac{\partial E_1(\mathbf{x},t)}{\partial z}+\frac{1}{v_{g1}}\frac{\partial E_1(\mathbf{x},t)}{\partial t}+i\frac{k''(\omega_1)}{2}\frac{\partial^2 E_1(\mathbf{x},t)}{\partial t^2}$$
$$=i\frac{\omega_1}{2cn(\omega_1)}\chi^{(2)}E_2(\mathbf{x},t)E_1^*(\mathbf{x},t)e^{i\Delta kz}, \tag{6.12}$$

where we define the *phase-mismatch parameter*

$$\Delta k=k_2-2k_1, \tag{6.13}$$

Similarly, if we equate the terms of frequency ω_2 we obtain the equation

$$\frac{1}{2ik_2}\nabla_\perp^2 E_2(\mathbf{x},t)+\frac{\partial E_2(\mathbf{x},t)}{\partial z}+\frac{1}{v_{g2}}\frac{\partial E_2(\mathbf{x},t)}{\partial t}+i\frac{k''(\omega_2)}{2}\frac{\partial^2 E_2(\mathbf{x},t)}{\partial t^2}$$
$$=i\frac{\omega_2}{4cn(\omega_2)}\chi^{(2)}E_1^2(\mathbf{x},t)e^{-i\Delta kz}. \tag{6.14}$$

As already said in the introduction of this chapter, E_1 and E_2 may be field amplitudes along different polarization directions, properly chosen to maximize the efficiency of the nonlinear process. If we consider the parameter (6.13) and impose perfect phase matching, $\Delta k=0$, by setting $k_2=k(\omega_1)=n(\omega_2)\omega_2/c$ and $k_1=k(\omega_2)=n(\omega_1)\omega_1/c$ with $\omega_2=2\omega_1$, one immediately obtains that the phase-matching condition $k_2=2k_1$ is satisfied when $n(2\omega_1)=n(\omega_1)$. This condition cannot be satisfied in a homogeneous non-birefringent medium, because the refractive index far from absorption lines is a monotonic function of the frequency. For this reason it is necessary to use a birefringent medium, choose one of the waves ordinary and the other extraordinary, and choose the propagation direction of the waves with respect to the optical axis of the crystal in such a way that the two refractive indices are equal. For this reason, strictly speaking we should write $n_1(\omega_1)$ in Eq. (6.12) and $n_2(\omega_2)$ in Eq. (6.14). However, after this clarification we leave Eqs. (6.12) and (6.14) as they are, for the sake of simplicity.

Next, let us now turn to the case of three frequency bands, i.e.

$$E(\mathbf{x},t)=\frac{1}{2}\Big[E_1(\mathbf{x},t)e^{-i(\omega_1 t-k_1 z)}+E_2(\mathbf{x},t)e^{-i(\omega_2 t-k_2 z)}$$
$$+E_3(\mathbf{x},t)e^{-i(\omega_3 t-k_3 z)}+E_1^*(\mathbf{x},t)e^{i(\omega_1 t-k_1 z)}$$
$$+E_2^*(\mathbf{x},t)e^{i(\omega_2 t-k_2 z)}+E_3^*(\mathbf{x},t)e^{i(\omega_3 t-k_3 z)}\Big], \tag{6.15}$$

assuming that $\omega_1 + \omega_2 = \omega_3$. By inserting Eq. (6.15) into Eq. (6.10) we obtain, again, a number of terms. Limiting ourselves to the terms with frequencies ω_1, ω_2 and ω_3 we have the following.

- The term $[E_1(\mathbf{x}, t)E_2(\mathbf{x}, t)/2]\,e^{-i(\omega_1+\omega_2)t+i(k_1+k_2)z}$ describes the generation of frequency ω_3 by sum generation from frequencies ω_1 and ω_2.
- The term $\left[E_3(\mathbf{x}, t)E_1^*(\mathbf{x}, t)/2\right]e^{-i(\omega_3-\omega_1)t+i(k_3-k_1)z}$, taking into account that $\omega_3 - \omega_1 = \omega_2$, describes the generation of frequency ω_2 by difference-frequency generation or the amplification of frequency ω_2.
- The term $\left[E_3(\mathbf{x}, t)E_2^*(\mathbf{x}, t)/2\right]e^{-i(\omega_3-\omega_2)t+i(k_3-k_2)z}$, taking into account that $\omega_3 - \omega_2 = \omega_1$, describes the generation of frequency ω_1 by difference-frequency generation or the amplification of frequency ω_1.

By proceeding as before, we obtain the following set of three coupled equations:

$$
\frac{1}{2ik_1}\nabla_\perp^2 E_1(\mathbf{x}, t) + \frac{\partial E_1(\mathbf{x}, t)}{\partial z} + \frac{1}{v_{g1}}\frac{\partial E_1(\mathbf{x}, t)}{\partial t} + i\frac{k''(\omega_1)}{2}\frac{\partial^2 E_1(\mathbf{x}, t)}{\partial t^2}
$$
$$
= i\frac{\omega_1}{2cn(\omega_1)}\chi^{(2)}E_3(\mathbf{x}, t)E_2^*(\mathbf{x}, t)e^{i\Delta kz}, \tag{6.16}
$$

$$
\frac{1}{2ik_2}\nabla_\perp^2 E_2(\mathbf{x}, t) + \frac{\partial E_2(\mathbf{x}, t)}{\partial z} + \frac{1}{v_{g2}}\frac{\partial E_2(\mathbf{x}, t)}{\partial t} + i\frac{k''(\omega_2)}{2}\frac{\partial^2 E_2(\mathbf{x}, t)}{\partial t^2}
$$
$$
= i\frac{\omega_2}{2cn(\omega_2)}\chi^{(2)}E_3(\mathbf{x}, t)E_1^*(\mathbf{x}, t)e^{i\Delta kz}, \tag{6.17}
$$

$$
\frac{1}{2ik_3}\nabla_\perp^2 E_3(\mathbf{x}, t) + \frac{\partial E_3(\mathbf{x}, t)}{\partial z} + \frac{1}{v_{g3}}\frac{\partial E_3(\mathbf{x}, t)}{\partial t} + i\frac{k''(\omega_3)}{2}\frac{\partial^2 E_3(\mathbf{x}, t)}{\partial t^2}
$$
$$
= i\frac{\omega_3}{2cn(\omega_3)}\chi^{(2)}E_1(\mathbf{x}, t)E_2(\mathbf{x}, t)e^{-i\Delta kz}, \tag{6.18}
$$

where the phase-mismatch parameter is

$$
\Delta k = k_3 - k_1 - k_2. \tag{6.19}
$$

We note that, if we take $\omega_2 = \omega_1$ (and therefore $E_2 = E_1$), Eqs. (6.16) and (6.17) become identical, of course, and identical to Eq. (6.12) if we write E_2 instead of E_3. In addition, if in Eq. (6.18) we write E_1 instead of E_2 and E_2 instead of E_3 (and, correspondingly, ω_2 instead of ω_3), Eq. (6.18) becomes identical to Eq. (6.14) apart from the absence of a factor of 2 on the r.h.s. of Eq. (6.14). For these reasons Eqs. (6.12) and (6.14) are called the *frequency-degenerate cases* of Eqs. (6.16)–(6.18).

Before solving these models in a few selected cases, we would like to insert some important remarks concerning the quantum-theoretical description of nonlinear optical processes. In the quantum picture the electric field is quantized and the processes are described in terms of creation and annihilation of photons. For example, the processes which appear in Eqs. (6.12) and (6.14) are described as arising from two fundamental and opposite quantum processes, one in which two photons of frequency ω_1 are annihilated and one photon of frequency ω_2 is created (second-harmonic generation) and the converse in which one photon of frequency ω_2 is annihilated and two photons of frequency ω_1 are created (amplification of frequency ω_1). Both processes conserve the total energy of the system, and conserve also the total photon momentum when the phase-matching condition $\Delta k = 0$ is satisfied exactly. Similarly, in the case of Eqs. (6.16)–(6.18) the two fundamental processes are the annihilation of two photons of frequencies ω_1 and ω_2 and the creation of one photon of frequency ω_3 and, conversely, the annihilation of one photon of frequency ω_3 and the creation of two photons with frequencies ω_1 and ω_2. The creation and annihilation of photons is described by the application of creation and annihilation operator fields. In the semiclassical approximation in which one replaces the mean value of products of operators with the product of mean values, the envelopes $E_i(\mathbf{x}, t)$ are proportional to the mean values of the annihilation operator fields. It is interesting to remark that, if one rescales the envelopes $E_i(\mathbf{x}, t)$ as

$$E_p(\mathbf{x}, t) = \sqrt{\frac{2\hbar\omega_p}{\epsilon_0 cn(\omega_p)}} A_p(\mathbf{x}, t),\tag{6.20}$$

the rescaled envelopes just coincide, in the semiclassical approximation, with the mean values of the annihilation operator fields. Precisely, from Eq. (6.20) we have that

$$|A_p(\mathbf{x}, t)|^2 = \left(\frac{1}{2}\epsilon_0 n^2(\omega_p)|E_p(\mathbf{x}, t)|^2\right)\left(\frac{1}{\hbar\omega_p}\right)\left(\frac{c}{n(\omega_p)}\right).\tag{6.21}$$

The term in the first bracket is the average energy density of the electromagnetic field, as we can see from Eq. (1.13) and the following line, if we take into account that in the present case the energy density has an additional factor $\epsilon_r = n^2$, where ϵ_r is the relative permittivity of the medium. Therefore, the product of the first two brackets corresponds to the photon number density for envelope p, and $|A_p(\mathbf{x}, t|^2$ corresponds to the photon number flux (in the semiclassical approximation).

In terms of the rescaled fields (6.20), Eqs. (6.12) and (6.14) take the forms

$$\frac{1}{2ik_1}\nabla_\perp^2 A_1(\mathbf{x}, t) + \frac{\partial A_1(\mathbf{x}, t)}{\partial z} + \frac{1}{v_{g1}}\frac{\partial A_1(\mathbf{x}, t)}{\partial t} + i\frac{k''(\omega_1)}{2}\frac{\partial^2 A_1(\mathbf{x}, t)}{\partial t^2}$$
$$= i\tilde{g}A_2(\mathbf{x}, t)A_1^*(\mathbf{x}, t)e^{i\Delta kz},\tag{6.22}$$

$$\frac{1}{2ik_2}\nabla_\perp^2 A_2(\mathbf{x}, t) + \frac{\partial A_2(\mathbf{x}, t)}{\partial z} + \frac{1}{v_{g2}}\frac{\partial A_2(\mathbf{x}, t)}{\partial t} + i\frac{k''(\omega_2)}{2}\frac{\partial^2 A_2(\mathbf{x}, t)}{\partial t^2}$$
$$= i\frac{\tilde{g}}{2}A_1^2(\mathbf{x}, t)e^{-i\Delta kz},\tag{6.23}$$

with

$$\tilde{g} = 2\pi\sqrt{\frac{2\hbar\omega_1^2\omega_2}{\epsilon_0 c^3 n^2(\omega_1)n(\omega_2)}}\chi^{(2)}.\tag{6.24}$$

Similarly Eqs. (6.16)–(6.18) assume the form

$$\frac{1}{2ik_1}\nabla_\perp^2 A_1(\mathbf{x},t) + \frac{\partial A_1(\mathbf{x},t)}{\partial z} + \frac{1}{v_{g1}}\frac{\partial A_1(\mathbf{x},t)}{\partial t} + i\frac{k''(\omega_1)}{2}\frac{\partial^2 A_1(\mathbf{x},t)}{\partial t^2}$$
$$= i\tilde{g}' A_3(\mathbf{x},t)A_2^*(\mathbf{x},t)e^{i\Delta kz}, \tag{6.25}$$

$$\frac{1}{2ik_2}\nabla_\perp^2 A_2(\mathbf{x},t) + \frac{\partial A_2(\mathbf{x},t)}{\partial z} + \frac{1}{v_{g2}}\frac{\partial A_2(\mathbf{x},t)}{\partial t} + i\frac{k''(\omega_2)}{2}\frac{\partial^2 A_2(\mathbf{x},t)}{\partial t^2}$$
$$= i\tilde{g}' A_3(\mathbf{x},t)A_1^*(\mathbf{x},t)e^{i\Delta kz}, \tag{6.26}$$

$$\frac{1}{2ik_3}\nabla_\perp^2 A_3(\mathbf{x},t) + \frac{\partial A_3(\mathbf{x},t)}{\partial z} + \frac{1}{v_{g3}}\frac{\partial A_3(\mathbf{x},t)}{\partial t} + i\frac{k''(\omega_3)}{2}\frac{\partial^2 A_3(\mathbf{x},t)}{\partial t^2}$$
$$= i\tilde{g}' A_1(\mathbf{x},t)A_2(\mathbf{x},t)e^{-i\Delta kz}, \tag{6.27}$$

with

$$\tilde{g}' = 2\pi\sqrt{\frac{2\hbar\omega_1\omega_2\omega_3}{\epsilon_0 c^3 n(\omega_1)n(\omega_2)n(\omega_3)}}\chi^{(2)}. \tag{6.28}$$

The remarkable feature is that the coefficient which appears in the model is the same ($\tilde{g}$ or $\tilde{g}'$) in all the equations of the model (differently from the equations of the original model for the field envelopes) and, in addition, it turns out to be proportional to the coupling constant which appears in the quantum-mechanical interaction Hamiltonian.

6.3 The stationary state in the plane-wave approximation

Let us now analyze Eqs. (6.22) and (6.23) and Eqs. (6.25)–(6.27) in the stationary state, in which the time derivatives vanish, in the plane-wave approximation in which the fields are independent also of the transverse variables x and y. Hence Eqs. (6.22) and (6.23) and Eqs. (6.25)–(6.27) reduce to

$$\frac{dA_1(z)}{dz} = i\tilde{g} A_2(z)A_1^*(z)e^{i\Delta kz}, \tag{6.29}$$

$$\frac{dA_2(z)}{dz} = i\frac{\tilde{g}}{2} A_1^2(z)e^{-i\Delta kz}, \tag{6.30}$$

and

$$\frac{dA_1(z)}{dz} = i\tilde{g}' A_3(z)A_2^*(z)e^{i\Delta kz}, \tag{6.31}$$

$$\frac{dA_2(z)}{dz} = i\tilde{g}' A_3(z)A_1^*(z)e^{i\Delta kz}, \tag{6.32}$$

$$\frac{dA_3(z)}{dz} = i\frac{\tilde{g}'}{2} A_1(z)A_2(z)e^{-i\Delta kz}, \tag{6.33}$$

respectively. Since the quantities $A_i^*(z)A_i(z)$ are the respective photon fluxes, by taking into account the fundamental quantum mechanisms of photon creation and annihilation which lie at the basis of each set of equations we can immediately guess the existence of constants of motion with respect to the z variable.

In the case of Eqs. (6.29) and (6.30) a constant of motion is evidently

$$A_1^*(z)A_1(z) + 2A_2^*(z)A_2(z), \tag{6.34}$$

and one immediately verifies using Eqs. (6.29) and (6.30) that the derivative with respect to z of the combination (6.34) is identically equal to zero. On the other hand, in the case of Eqs. (6.31)–(6.33) three independent constants of motion are

$$\begin{aligned}
A_1^*(z)A_1(z) - A_2^*(z)A_2(z), \\
A_1^*(z)A_1(z) + A_3^*(z)A_3(z), \\
A_2^*(z)A_2(z) + A_3^*(z)A_3(z),
\end{aligned} \tag{6.35}$$

and this result is equivalent to what in the literature of nonlinear optics is called the Manley–Rowe relations [59].

6.3.1 Degenerate configuration, phase-sensitive amplification/deamplification of the fundamental field

Let us focus first on the degenerate case of Eqs. (6.29) and (6.30). The process which is described by such equations depends on the values of the two fields at the boundary $z = 0$. If the fundamental-frequency field A_1 is different from zero at $z = 0$, whereas the second-harmonic field vanishes at $z = 0$, the process which is described is second-harmonic generation. In order to calculate the two fields $A_1(z)$ and $A_2(z)$ for $z > 0$, the constant of motion (6.34) turns out to be useful. Because the process of second-harmonic generation is described in detail in textbooks on nonlinear optics [29, 59], we will not discuss it in this volume and concentrate, instead, on the case in which both $A_1(0)$ and $A_2(0)$ are different from zero, but in a strongly asymmetrical way. Namely, we assume that the fundamental field $A_1(0)$ is weak whereas the second-harmonic field $A_2(0)$ is intense, so intense that its value is unaffected by the interaction with the other field throughout the range of values of z which is considered. Hence, on dropping the r.h.s. in Eq. (6.30), we see that A_2 becomes independent of z and therefore can be treated as a constant in Eq. (6.29) which therefore becomes linear in A_1. In the literature this is called the *undepleted-pump approximation*. The physical process which is described in this case is the amplification or deamplification of the fundamental field A_1.

We calculate the solution of Eq. (6.29) (in the undepleted-pump approximation) focusing first on the case of perfect phase matching $\Delta k = 0$. By expressing A_2 in the form $|A_2| e^{i \arg(A_2)}$ and defining

$$r = \tilde{g}|A_2|, \quad e^{i\zeta} = i e^{i \arg(A_2)}, \tag{6.36}$$

Eq. (6.29) can be written as

$$\frac{dA_1(z)}{dz} = r e^{i\zeta} A_1^*(z). \tag{6.37}$$

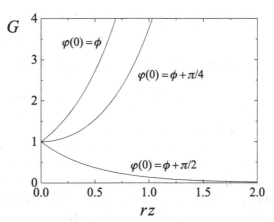

Figure 6.1 Phase-sensitive amplification/deamplification (see Eq. (6.46)). The intensity amplification/deamplification factor G is plotted as a function of rz for three values of the phase $\varphi(0)$ of the input field $A_1(0)$.

In order to solve this equation in the most elegant way and also to illustrate in the clearest possible way the special features of the physical process, it is convenient to express the field A_1 in terms of its quadrature components. Therefore, let us introduce generic orthogonal quadrature components

$$X_\phi = \frac{A_1 e^{-i\phi} + A_1^* e^{i\phi}}{2}, \quad Y_\phi = \frac{A_1 e^{-i\phi} - A_1^* e^{i\phi}}{2i}, \tag{6.38}$$

so that conversely

$$A_1 = e^{i\phi}(X_\phi + iY_\phi), \quad A_1^* = e^{-i\phi}(X_\phi - iY_\phi), \tag{6.39}$$

where ϕ is an arbitrary phase. Note that one has

$$\begin{aligned}\mathcal{E}_1 &= \frac{1}{2}\left(A_1 e^{-i(\omega_1 t - k_1 z)} + A_1^* e^{i(\omega_1 t - k_1 z)}\right) \\ &= X_\phi \cos(k_1 z - \omega_1 t - \phi) + Y_\phi \sin(k_1 z - \omega_1 t - \phi),\end{aligned} \tag{6.40}$$

which explains the name quadrature components. If we now choose

$$\phi = \frac{\zeta}{2}, \tag{6.41}$$

we immediately get from Eq. (6.37)

$$\frac{dX_\phi(z)}{dz} = rX_\phi(z) \quad \Rightarrow \quad X_\phi(z) = e^{rz}X_\phi(0), \tag{6.42}$$

$$\frac{dY_\phi(z)}{dz} = -rY_\phi(z) \quad \Rightarrow \quad Y_\phi(z) = e^{-rz}Y_\phi(0). \tag{6.43}$$

Therefore the quadrature X_ϕ is amplified, whereas the orthogonal quadrature Y_ϕ gets deamplified. This phenomenon is called *phase-sensitive amplification/deamplification* (Fig. 6.1). The full picture for such a phenomenon is obtained from the expression for $A_1(z)$ obtained by using Eq. (6.39),

$$A_1(z) = \cosh(rz)A_1(0) + e^{2i\phi}\sinh(rz)A_1^*(0). \tag{6.44}$$

This result is best described by introducing the modulus and phase of A_1,

$$A_1(z) = |A_1(z)|e^{i\varphi(z)}, \tag{6.45}$$

and evaluating the ratio G of the intensity at position z to the intensity at position $z = 0$,

$$G(z, \phi - \varphi(0)) \equiv \frac{|A_1(z)|^2}{|A_1(0)|^2} = e^{2rz} \cos^2[\phi - \varphi(0)] + e^{-2rz} \sin^2[\phi - \varphi(0)]. \tag{6.46}$$

Even apart from the phase-sensitivity, it is remarkable that in this context we have the possibility of amplification, i.e. gain, without transferring atoms from a lower state to a higher-energy state (apart from a short residence in a virtual level) [59]. This kind of gain is called *parametric gain*, and these processes are called parametric processes.

Let us now turn to the case of imperfect phase matching assuming that Δk is different from zero, so that the equation for A_1 reads

$$\frac{dA_1(z)}{dz} = re^{i\zeta} A_1^*(z)e^{i\Delta kz}. \tag{6.47}$$

By setting

$$\tilde{A}_1(z) = A_1(z)e^{-i(\Delta k/2)z} \tag{6.48}$$

we obtain

$$\frac{d\tilde{A}_1(z)}{dz} = -i\frac{\Delta k}{2}\tilde{A}_1(z) + re^{i\zeta}\tilde{A}_1^*(z). \tag{6.49}$$

Next, we introduce the quadrature components of $\tilde{A}_1$,

$$\tilde{X}_\phi = \frac{\tilde{A}_1 e^{-i\phi} + \tilde{A}_1^* e^{i\phi}}{2}, \quad \tilde{Y}_\phi = \frac{\tilde{A}_1 e^{-i\phi} - \tilde{A}_1^* e^{i\phi}}{2i}, \tag{6.50}$$

in terms of which Eq. (6.49) can be recast in the form

$$\frac{d\tilde{X}_\phi(z)}{dz} = r\tilde{X}_\phi + \frac{\Delta k}{2}\tilde{Y}_\phi, \tag{6.51}$$

$$\frac{d\tilde{Y}_\phi(z)}{dz} = -r\tilde{X}_\phi - \frac{\Delta k}{2}\tilde{Y}_\phi. \tag{6.52}$$

The solutions of these two equations are linear combinations of the two exponentials $e^{\lambda z}$ and $e^{-\lambda z}$, where

$$\lambda = \left(r^2 - \frac{\Delta k^2}{4}\right)^{1/2}. \tag{6.53}$$

Since $\tilde{A}_1$ is a linear combination of the two quadrature components, also $\tilde{A}_1$ is a linear combination of these two exponentials, and we can write

$$\tilde{A}_1(z) = \alpha e^{\lambda z} + \beta e^{-\lambda z}, \tag{6.54}$$

where the constants α and β are obtained from the boundary values $A_1(0)$ and

$$\left.\frac{d\tilde{A}_1(z)}{dz}\right|_{z=0} = -i\frac{\Delta k}{2}\tilde{A}_1(0) + re^{i\zeta}\tilde{A}_1^*(0), \tag{6.55}$$

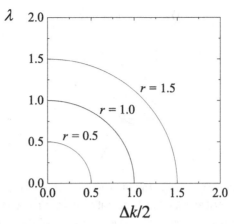

Figure 6.2 A plot of the amplification factor λ (see Eq. (6.53)) as a function of the phase-matching parameter Δk for three values of the parametric gain factor r.

as follows from Eq. (6.49). In such a way, by using also Eq. (6.48) we get

$$A_1(z) = \left\{ \left[\cosh(\lambda z) - i\frac{\Delta k}{2\lambda} \sinh(\lambda z) \right] A_1(0) + \frac{r}{\lambda} e^{i\phi} \sinh(\lambda z) A_1^*(0) \right\} e^{i(\Delta k/2)z}. \quad (6.56)$$

We see therefore that, when the phase matching is imperfect, the amplification/deamplification is governed by the exponentials $e^{\pm \lambda z}$ instead of $e^{\pm rz}$. The amplification/deamplification factor λ is smaller than r according to Eq. (6.53). For $\Delta k > 2r$, λ becomes imaginary and any amplification/deamplification disappears. In Fig. 6.2 we plot the factor λ as a function of Δk for various values of r, in the amplification/deamplification range.

To finish the discussion of the degenerate configuration some important remarks concerning the quantum theory are in order. Let us assume that in the input $z = 0$ there is an intense field A_2, but the field A_1 vanishes. In this situation, according to the semiclassical theory, A_1 remains zero for all values of z. However according to the quantum theory an input A_1 in the vacuum state, interacting with the field A_2, can generate a field $A_1(z)$ with a non-vanishing number of photons, because a fraction of the photons of the field A_2 can generate photon pairs of frequency one half the frequency of the parent photons. This process is called *spontaneous parametric down-conversion* or *subharmonic generation* and, just like the process of spontaneous emission, requires the quantum theory for its description. In contrast with spontaneous emission, in spontaneous parametric down-conversion the emission of photons occurs in pairs.

It is interesting to observe that, in its simplest picture in which one neglects the variation of the fields in the transverse plane and the group velocity dispersion, the process of spontaneous parametric down-conversion can be described as the phase-sensitive amplification/deamplification of the vacuum fluctuations. Let us consider for simplicity the case of perfect phase matching, $\Delta k = 0$. As a matter of fact, in the vacuum state the fluctuations are identical in all quadrature components [49, 62–66]. However, in the interaction with the field A_2 the fluctuations in the quadrature component X_ϕ get amplified whereas

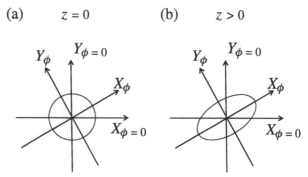

(a) The vacuum input state, in which the variance of the vacuum noise is identical in all quadrature components, i.e. for all values of the phase ϕ. (b) The quadratic interaction increases the fluctuations in the quadrature X_ϕ by a factor of e^r and decreases the fluctuations in the orthogonal quadrature Y_ϕ by a factor of e^{-r} (see Eqs. (6.42) and (6.43)), leading to a squeezed vacuum state.

the fluctuations in the quadrature component Y_ϕ are deamplified by the same amount. This leads to a *squeezed vacuum state* [49, 62–66] (see Fig. 6.3).

6.3.2 Non-degenerate configuration, phase-insensitive amplification

Let us now turn our attention to the non-degenerate case of Eqs. (6.31)–(6.33). Again, let us assume that in the input $z = 0$ the highest-frequency field A_3 is intense, whereas the fields A_1 and A_2 are weak. In this case the pump field A_3 is undepleted and can be treated as a constant in Eqs. (6.31) and (6.32). On setting

$$\eta = \tilde{g}'|A_3|, \quad e^{i\zeta} = i e^{i \arg(A_3)}, \tag{6.57}$$

these equations take the form

$$\frac{d A_1(z)}{dz} = \eta e^{i\zeta} A_2^*(z) e^{i \Delta kz}, \tag{6.58}$$

$$\frac{d A_2(z)}{dz} = \eta e^{i\zeta} A_1^*(z) e^{i \Delta kz}. \tag{6.59}$$

To solve these equations it is convenient to introduce the combinations

$$B_1 = A_1 + A_2, \quad B_2 = A_1 - A_2, \tag{6.60}$$

with the converse transformation

$$A_1 = \frac{B_1 + B_2}{2}, \quad A_2 = \frac{B_1 - B_2}{2}, \tag{6.61}$$

in terms of which Eqs. (6.58) and (6.59) assume the form

$$\frac{d B_1}{dz} = \eta e^{i\zeta} B_1^* e^{i \Delta kz}, \tag{6.62}$$

$$\frac{d B_2}{dz} = -\eta e^{i\zeta} B_2^* e^{i \Delta kz} = \eta e^{i(\zeta + \pi)} B_2^* e^{i \Delta kz}. \tag{6.63}$$

The key point is that each of the two equations (6.62) and (6.63) is self-contained and has the same form of Eq. (6.47) for the degenerate case, the second with $\zeta + \pi$ instead of ζ. Therefore we can immediately write the solutions of Eqs. (6.62) and (6.63) following Eq. (6.56), and by using Eq. (6.61) we arrive at the result

$$A_1(z) = \left\{ \left[\cosh(\lambda z) - i \frac{\Delta k}{2\lambda} \sinh(\lambda z) \right] A_1(0) + \frac{\eta}{\lambda} e^{i\zeta} \sinh(\lambda z) A_2^*(0) \right\} e^{i(\Delta k/2)z}, \quad (6.64)$$

$$A_2(z) = \left\{ \left[\cosh(\lambda z) - i \frac{\Delta k}{2\lambda} \sinh(\lambda z) \right] A_2(0) + \frac{\eta}{\lambda} e^{i\zeta} \sinh(\lambda z) A_1^*(0) \right\} e^{i(\Delta k/2)z}, \quad (6.65)$$

with

$$\lambda = \left(\eta^2 - \frac{\Delta k^2}{4} \right)^{1/2}. \quad (6.66)$$

In the case of perfect phase matching, $\Delta k = 0$, these expressions reduce to

$$A_1(z) = \cosh(\eta z) A_1(0) + e^{i\zeta} \sinh(\eta z) A_2^*(0), \quad (6.67)$$

$$A_2(z) = \cosh(\eta z) A_2(0) + e^{i\zeta} \sinh(\eta z) A_1^*(0). \quad (6.68)$$

Therefore, if in the input $A_1(0)$ is different from zero whereas $A_2(0) = 0$, the amplification is *phase-insensitive*,

$$|A_1(z)|^2 = \cosh^2(\eta z) |A_1(0)|^2, \quad (6.69)$$

while for the difference-frequency generation we obtain

$$|A_2(z)|^2 = \sinh^2(\eta z) |A_1(0)|^2, \quad (6.70)$$

such that for $\eta z \gg 1$ the fields A_1 and A_2 acquire the same intensity.

On the other hand, if the two fields are equal in the input, i.e. $A_1(0) = A_2(0)$, they remain equal for all values of z,

$$A_1(z) = A_2(z) = \cosh(\eta z) A_1(0) + e^{i\zeta} \sinh(\eta z) A_1^*(0), \quad (6.71)$$

and one has phase-sensitive amplification/deamplification because the result is identical to that of the degenerate case Eq. (6.44) since $2\phi = \zeta$ (see Eq. (6.41)).

When both fields A_1 and A_2 vanish in the input, the quantum theory predicts spontaneous parametric down-conversion. Both in the degenerate and in the non-degenerate configurations, a complete treatment of this phenomenon requires consideration of the full models including also the time derivatives (with group velocity dispersion) and the transverse Laplacian [67].

7 Optical nonlinearities. Materials with cubic nonlinearities

In the previous chapter, we illustrated the formalism employed to treat weak optical non-linearities and applied this theory to the case of quadratic nonlinearities. In this chapter we will continue this discussion with the case of cubic nonlinearities. In the literature, this case is often described as four-wave mixing, whereas the case of quadratic nonlinearities is called three-wave mixing.

The main body of this chapter deals with the Kerr medium nonlinearity, discussing first the phenomenon of self-phase modulation (or an intensity-dependent refractive index) in the stationary plane-wave configuration. Next, the vision is enlarged with the inclusion of temporal and spatial effects, namely with the description of *temporal Kerr solitons* and *spatial Kerr solitons*.

The following section is devoted to a very concise discussion of the cubic nonlinearity for a superposition of three frequency bands, especially in connection with the results on amplification discussed in the previous chapter. In the final section we consider the bidirectional configuration and illustrate the phenomenon of *optical phase conjugation*.

7.1 The Kerr medium nonlinearity. Self-phase modulation

In the case of materials with a cubic nonlinearity, the atomic polarization has the form

$$P_{NL}(\mathbf{x}, t) = \epsilon_0 \chi^{(3)} E^3(\mathbf{x}, t), \tag{7.1}$$

where $\chi^{(3)}$ is a real constant. Let us consider first an electric field that consists in one frequency band, i.e.

$$E(\mathbf{x}, t) = \frac{1}{2} \left[E_0(\mathbf{x}, t) e^{-i(\omega_0 t - k_0 z)} + E_0^*(\mathbf{x}, t) e^{i(\omega_0 t - k_0 z)} \right], \tag{7.2}$$

If we insert Eq. (7.2) into (7.1), we obtain

$$P_{NL}(\mathbf{x}, t) = \frac{1}{8} \epsilon_0 \chi^{(3)} \left[E_0^3(\mathbf{x}, t) e^{-3i(\omega_0 t - k_0 z)} + 3|E_0(\mathbf{x}, t)|^2 E_0(\mathbf{x}, t) e^{-i(\omega_0 t - k_0 z)} \right.$$
$$\left. + \text{c.c.} \right]. \tag{7.3}$$

The first term describes the process of third-harmonic generation, while the second term is the only one relevant for the case of one frequency band. In this case the dynamics is governed by a single equation for the envelope E_0. By using the procedure outlined by

Eq. (6.8) and the following part of Section 6.1, we obtain the following equation:[1]

$$\frac{1}{2ik_0}\nabla_\perp^2 E_0(\mathbf{x}, t) + \frac{\partial E_0(\mathbf{x}, t)}{\partial z} + \frac{1}{v_g}\frac{\partial E_0(\mathbf{x}, t)}{\partial t} + i\frac{k''}{2}\frac{\partial^2 E_0(\mathbf{x}, t)}{\partial t^2}$$
$$= i\frac{3}{32}\frac{\omega_0}{cn_0}\chi^{(3)}|E_0(\mathbf{x}, t)|^2 E_0(\mathbf{x}, t), \tag{7.4}$$

where $n_0 = n(\omega_0)$ and $k'' = k''(\omega_0)$. It turns out to be convenient to reason in terms of a nonlinear susceptibility. We observe, first of all, that the cubic nonlinearity which appears in Eq. (7.4) coincides (apart from a constant factor) with the refractive part of the cubic term of the two-level-atom equation (5.33), if we introduce the expansion indicated in Eqs. (5.26)–(5.28). If, similarly to Eq. (7.2), we set

$$P(\mathbf{x}, t) = \frac{1}{2}\left[P_0(\mathbf{x}, t)e^{-i(\omega_0 t - k_0 z)} + \text{c.c.}\right], \tag{7.5}$$

by taking Eq. (7.3) into account, we can write

$$P_0(\mathbf{x}, t) = \epsilon_0\left[\chi + \frac{3}{4}\chi^{(3)}|E_0(\mathbf{x}, t)|^2\right]E_0(\mathbf{x}, t), \tag{7.6}$$

where χ is the refractive part of the linear susceptibility. By inserting the nonlinear susceptibility which appears in Eq. (7.6) into the expression for the refractive index, we have an intensity-dependent refractive index (ϵ_r is the relative permittivity of the medium)

$$n = \sqrt{\epsilon_r} = \sqrt{1 + \chi + \frac{3}{4}\chi^{(3)}|E_0|^2} \approx \sqrt{1 + \chi}\left(1 + \frac{3}{8}\frac{\chi^{(3)}}{1 + \chi}|E_0|^2\right) = n_0 + \bar{n}_2|E_0|^2, \tag{7.7}$$

with

$$\bar{n}_2 = \frac{3}{8}\frac{\chi^{(3)}}{n_0}. \tag{7.8}$$

The contribution to the refractive index described by Eq. (7.8) is usually called the *optical Kerr effect*, in analogy with the electro-optic Kerr effect in which the refractive index changes proportionally to the intensity of a static electric field. In the following, we will call those materials which display the cubic nonlinearity of Eq. (7.4) *Kerr media*. It is straightforward to calculate the stationary solution of Eq. (7.4) in the plane-wave approximation, in which the equation reduces to

$$\frac{\partial E_0(z)}{\partial z} = i\frac{3}{32}\frac{\omega_0}{cn_0}\chi^{(3)}|E_0(z)|^2 E_0(z). \tag{7.9}$$

By setting

$$E_0(z) = \rho(z)e^{i\phi(z)} \tag{7.10}$$

we obtain from Eq. (7.9) the two equations

$$\frac{d\rho(z)}{dz} = 0, \tag{7.11}$$

$$\frac{d\phi(z)}{dz} = \frac{3}{32}\frac{\omega_0}{cn_0}\chi^{(3)}\rho^2, \tag{7.12}$$

---

[1] Note that no phase-matching condition is involved in Eq. (7.4).

from which we see that the modulus (and therefore the intensity) of the field remains constant whereas the phase varies linearly with z,

$$\rho(z) = \rho(0) = \rho_0 \tag{7.13}$$

and

$$\phi(z) = \phi(0) + \frac{3}{32} \frac{\omega_0}{cn_0} \chi^{(3)} \rho_0^2 z. \tag{7.14}$$

Hence this phenomenon is called *self-phase modulation*.

We note that Eq. (7.4) includes neither amplification nor absorption, and represents a *purely conservative* configuration.

7.2 Temporal Kerr solitons

In this section we illustrate an intriguing time-dependent solution called the temporal Kerr soliton. A central role in this phenomenon is played by group velocity dispersion, which is described by the term with second-order time derivative in Eq. (7.4). Assuming the plane-wave approximation again, this equation reduces to

$$\frac{\partial E_0(z,t)}{\partial z} + \frac{1}{v_g} \frac{\partial E_0(z,t)}{\partial t} + i \frac{k''}{2} \frac{\partial^2 E_0(z,t)}{\partial t^2} = i \frac{3}{32} \frac{\omega_0}{cn_0} \chi^{(3)} |E_0(z,t)|^2 E_0(z,t). \tag{7.15}$$

The coefficient k'' of the group-velocity-dispersion term is given by

$$k'' = \left(\frac{d^2 k}{d\omega^2} \right)_{\omega=\omega_0} = \frac{d}{d\omega} \left(\frac{1}{v_g(\omega)} \right)_{\omega=\omega_0} = -\left(\frac{1}{v_g^2} \frac{dv_g}{d\omega} \right)_{\omega=\omega_0}. \tag{7.16}$$

Therefore for $k'' < 0$ the frequency components of a pulse propagate faster the higher the frequency, whereas the opposite happens for $k'' > 0$. In both cases this circumstance leads to a broadening of the pulse in the z direction during the propagation. One can show that also the cubic nonlinear term can lead to similar effects. The interesting point is that, by selecting opposite signs for k'' and $\chi^{(3)}$, one can have that group velocity dispersion makes the higher frequencies propagate faster whereas the cubic nonlinearity makes the higher-frequency components propagate slower, or vice versa. One can obtain that the two effects compensate for each other exactly, so that the pulse propagates without changing its shape at all. To see this, let us introduce the new variables

$$\tau = t - \frac{z}{v_g}, \qquad \eta = z, \tag{7.17}$$

where τ is the retarded time. Hence we have

$$\frac{\partial E_0(z, t)}{\partial t} = \frac{\partial E_0(\eta, \tau)}{\partial \eta}\frac{d\eta}{dt} + \frac{\partial E_0(\eta, \tau)}{\partial \tau}\frac{d\tau}{dt} = \frac{\partial E_0(\eta, \tau)}{\partial \tau},$$

$$\frac{\partial^2 E_0(z, t)}{\partial t^2} = \frac{\partial^2 E_0(\eta, \tau)}{\partial \tau^2}, \tag{7.18}$$

$$\frac{\partial E_0(z, t)}{\partial z} = \frac{\partial E_0(\eta, \tau)}{\partial \eta}\frac{d\eta}{dz} + \frac{\partial E_0(\eta, \tau)}{\partial \tau}\frac{d\tau}{dz} = \frac{\partial E_0(\eta, \tau)}{\partial \eta} - \frac{1}{v_g}\frac{\partial E_0(\eta, \tau)}{\partial \tau},$$

so Eq. (7.15) becomes

$$\frac{\partial E_0(\eta, \tau)}{\partial \eta} + i\frac{k''}{2}\frac{\partial^2 E_0(\eta, \tau)}{\partial \tau^2} = i\tilde{\zeta}|E_0(\eta, \tau)|^2 E_0(\eta, \tau), \tag{7.19}$$

where we have set

$$\tilde{\zeta} = \frac{3}{32}\frac{\omega_0}{cn_0}\chi^{(3)}. \tag{7.20}$$

Equation (7.19) has the form of a nonlinear Schrödinger equation, in which time and space have exchanged roles, because the first-order derivative is in the spatial variable whereas the second-order derivative (which replaces the Laplacian) is in the (retarded) time variable. If we now set

$$\bar{\eta} = \frac{\eta}{L}, \qquad \bar{\tau} = \frac{\tau}{\tau_0}, \tag{7.21}$$

Eq. (7.19) becomes

$$\frac{\partial E_0}{\partial \bar{\eta}} + i\frac{k''L}{2\tau_0^2}\frac{\partial^2 E_0}{\partial \bar{\tau}^2} = i\tilde{\zeta}L|E_0|^2 E_0. \tag{7.22}$$

Let us now assume that $\tilde{\zeta}$ and k'' have opposite signs, and let us define

$$L = \frac{2\tau_0^2}{k''}, \qquad \bar{E}_0 = E_0|\tilde{\zeta}L|^{1/2}, \tag{7.23}$$

so that, taking into account that $\tilde{\zeta}L < 0$, Eq. (7.22) reduces to

$$\frac{\partial \bar{E}_0}{\partial \bar{\eta}} + i\frac{\partial^2 \bar{E}_0}{\partial \bar{\tau}^2} = -i|\bar{E}_0|^2\bar{E}_0. \tag{7.24}$$

One can verify that the function

$$\bar{E}_0(\bar{\eta}, \bar{\tau}) = \sqrt{2}\,\text{sech}(\tau)e^{-i\bar{\eta}} \tag{7.25}$$

satisfies Eq. (7.24). The corresponding expression for E_0 is[2]

$$E_0(z, t) = \sqrt{\left|\frac{k''}{\tilde{\zeta}}\right|}\frac{1}{\tau_0}\,\text{sech}\left[\frac{1}{\tau_0}\left(t - \frac{z}{v_g}\right)\right]e^{-i(k''/(2\tau_0^2))z}, \tag{7.26}$$

---

[2] Similarly to the end of Section 5.1, we observe that the ratio $\epsilon^2 = \lambda/L$ is proportional to $(1/n_0^2)|\chi^{(3)}||E_0|^2$, as one obtains using, in sequence, Eqs. (7.23), (7.26) and (7.20). Within a factor of order unity, the last expression is equal to $|P_{NL}/(\epsilon_0 E_0)|$ (see Eqs. (7.1) and (7.2)), as assumed in Section 3.1.

where we have used Eq. (7.23), and this solution is called a *temporal Kerr soliton*. The condition $\tilde{\zeta} k'' < 0$ is necessary in order to achieve that the group velocity dispersion and cubic nonlinearity balance each other. Solution (7.26) and other solutions corresponding to higher-order temporal Kerr solitons were obtained by Zacharov and Shabat [68] using inverse-scattering methods. Temporal Kerr solitons have been observed experimentally by Mollenauer and his group [69].

Note that the pulse duration τ_0 remains as a free parameter. As in the case of self-induced transparency solitons described in Section 3.3, Kerr temporal solitons have a larger peak height the shorter they are. On the other hand, for Kerr temporal solitons the velocity coincides with the group velocity and does not depend on the pulse width.

In the fiber-optics literature the case $k'' > 0$ is named *normal dispersion*, and the case $k'' < 0$ is called *anomalous dispersion*. By using the definition $k(\omega) = \omega n(\omega)/c$, it is easy to show that this definition coincides with the standard definition $dn/d\omega > 0$ of normal dispersion when the curve $n(\omega)$ has an inflection point at the value of ω at which k is evaluated; we assume that this is the case.

7.3 Spatial Kerr solitons

This section is complementary to the previous one, because we stay in the stationary configuration, but allow spatial variation of the field in the transverse variables, so that Eq. (7.4) becomes

$$\frac{\partial E_0(x, y, z)}{\partial z} - \frac{i}{2k_0} \nabla_\perp^2 E_0(x, y, z) = i\tilde{\zeta} |E_0|^2 E_0(x, y, z), \tag{7.27}$$

with $\tilde{\zeta}$ defined by Eq. (7.20). This is a nonlinear Schrödinger equation again. Let us now focus on the case that the field E_0 is independent of the transverse variable y. This configuration can be realized by making light propagate in a planar waveguide. We can immediately note that the resulting equation,

$$\frac{\partial E_0(x, z)}{\partial z} - \frac{i}{2k_0} \frac{\partial^2 E_0(x, z)}{\partial x^2} = i\tilde{\zeta} |E_0|^2 E_0(x, z), \tag{7.28}$$

coincides with Eq. (7.19) if we replace η by z, τ by x and k'' by $-k_0^{-1}$. Note that k'' is replaced by a negative quantity, so in this case the solitonic solution exists only for $\tilde{\zeta} > 0$. Hence, by using Eq. (7.26), we can write the solitonic solution of Eq. (7.28) in the form

$$E_0(x, z) = \frac{1}{\sqrt{k_0 \tilde{\zeta}} \, x_0} \operatorname{sech}\left(\frac{x}{x_0}\right) e^{iz/(2k_0 x_0^2)}. \tag{7.29}$$

The solution (7.29) is called a *spatial Kerr soliton*, and exists for $\tilde{\zeta} > 0$, i.e. for $\chi^{(3)} > 0$. Note that in this solution the field intensity is constant along z, i.e. the transverse profile is invariant in propagation, and x_0 is a free parameter. It is interesting to note that the power emitted by the soliton, which is proportional to the integral of $|E_0(x, z)|^2$ over x from $-\infty$ to $+\infty$, is independent of the value of x_0. Experimental observations of this phenomenon have been obtained by Barthelemy and collaborators [70] and by Aitchison and collaborators [71].

Let us now turn to the physical interpretation of the mechanism which generates spatial solitons. As in the case of temporal solitons, this phenomenon originates from an exact balance between the cubic nonlinearity and, in this case, diffraction instead of group velocity dispersion. To be precise, diffraction tends to make the beam spread in the transverse plane. On the other hand, if we look at Eqs. (7.7) and (7.8) we see that when $\chi^{(3)}$ is positive the refractive index increases where the intensity increases, which implies that the medium behaves like a positive lens. This effect is called *self-focussing* in the literature. The spatial soliton arises from the balance between the self-focussing effect induced by the cubic nonlinearity and the spreading action exerted by diffraction.

If we take both transverse dimensions x and y into account considering Eq. (7.27), there is a solution that describes a two-dimensional spatial soliton which, in this case, consists in a cylindrically symmetrical "tube" of radiation. However, this solution cannot be described analytically. Even more important is the fact that this two-dimensional solution turns out to be unstable in the case of a Kerr medium, whereas it becomes stable for a saturable nonlinearity or for photorefractive media, as has been shown experimentally by Bjorkholm and Ashkin [72] and by Segev and collaborators [73], respectively.

Further reading in the field of conservative optical solitons is indicated in [29, 74–79].

7.4 The case of three frequency bands. Cross-phase modulation and four-wave mixing

Let us now assume that the electric field is the superposition of three frequency bands as in Eq. (6.15). If we insert this equation into Eq. (7.1) we obtain a large number of terms describing a variety of combinations of frequencies. We assume also that the frequencies ω_1, ω_2 and ω_3 satisfy the relation

$$\omega_1 + \omega_2 = 2\omega_3. \tag{7.30}$$

In the stationary state and in the plane-wave approximation the equations for the three fields E_i ($i = 1, 2, 3$) read

$$\frac{dE_1}{dz} = i\frac{3}{32}\frac{\omega_1\chi^{(3)}}{n_1 c}\left[\left(|E_1|^2 + 2|E_2|^2 + 2|E_3|^2\right)E_1 + E_3^2 E_2^* e^{i\Delta kz}\right], \tag{7.31}$$

$$\frac{dE_2}{dz} = i\frac{3}{32}\frac{\omega_2\chi^{(3)}}{n_2 c}\left[\left(|E_2|^2 + 2|E_1|^2 + 2|E_3|^2\right)E_2 + E_3^2 E_1^* e^{i\Delta kz}\right], \tag{7.32}$$

$$\frac{dE_3}{dz} = i\frac{3}{32}\frac{\omega_3\chi^{(3)}}{n_3 c}\left[\left(|E_3|^2 + 2|E_1|^2 + 2|E_2|^2\right)E_3 + 2E_1 E_2 E_3^* e^{-i\Delta kz}\right], \tag{7.33}$$

where $n_p = n(\omega_p)$, $p = 1, 2, 3$, and

$$\Delta k = 2k_3 - k_1 - k_2. \tag{7.34}$$

In these equations, the first term on the r.h.s. describes self-phase modulation, whereas the second and third terms describe *cross-phase modulation*, because the phase modulation is induced by the intensity of another frequency component. The last term describes four-wave mixing. In the quantum picture, the four-wave mixing is described as arising from two opposite processes, one in which two photons, one of frequency ω_1 and the other of frequency ω_2, are annihilated and two photons of frequency ω_3 are created, and the converse in which two photons of frequency ω_3 are annihilated and two photons with frequencies ω_1 and ω_2 are created.

Similarly to what we did in the quadratic case (Section 6.3.2), let us focus on the case in which the field E_3 is intense whereas E_1 and E_2 are weak, so that E_3 can be treated as a constant in the undepleted pump approximation. On setting

$$E_3 = |E_3| e^{i\zeta/2}, \quad \eta' = \frac{3\chi^{(3)}}{32c} |E_3|^2, \tag{7.35}$$

Eqs. (7.31) and (7.32) reduce to

$$\frac{dE_1}{dz} = i\eta' \frac{\omega_1}{n_1} \left(2E_1 + e^{i\zeta} E_2^*\right), \tag{7.36}$$

$$\frac{dE_2}{dz} = i\eta' \frac{\omega_2}{n_2} \left(2E_2 + e^{i\zeta} E_1^*\right), \tag{7.37}$$

where we have neglected the terms of third order in E_1 and E_2 and assumed perfect phase matching, $\Delta k = 0$. It is interesting to note that the four-wave mixing terms are identical in form to those of the quadratic case, see Eqs. (6.58) and (6.59) for $\Delta k = 0$. This feature is general whenever the nonlinear equations reduce to linear ones. Equation (7.36) and the complex conjugate of (7.37) form a self-contained set of linear equations from which we find that E_1 is given by a linear combination of the form

$$E_1(z) = \alpha \cos\left[\eta' \left(\frac{3\omega_1\omega_2}{n_1n_2}\right)^{1/2} z\right] + \beta \sin\left[\eta' \left(\frac{3\omega_1\omega_2}{n_1n_2}\right)^{1/2} z\right]. \tag{7.38}$$

Therefore we see that cross-phase modulation destroys the parametric gain. By imposing the initial conditions

$$E_1(0) = E_{10}, \quad E_2(0) = E_{20}, \quad \left.\frac{dE_1}{dz}\right|_{z=0} = i\eta' \frac{\omega_1}{n_1} \left(2E_{10} + e^{i\zeta} E_{20}^*\right), \tag{7.39}$$

where we have used Eq. (7.36), we arrive at the final expression

$$E_1(z) = E_{10} \cos\left[\eta' \left(\frac{3\omega_1\omega_2}{n_1n_2}\right)^{1/2} z\right]$$
$$+ i \left(\frac{\omega_1 n_2}{3\omega_2 n_1}\right)^{1/2} \left(2E_{10} + e^{i\zeta} E_{20}^*\right) \sin\left[\eta' \left(\frac{3\omega_1\omega_2}{n_1n_2}\right)^{1/2} z\right]. \tag{7.40}$$

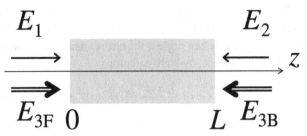

Figure 7.1 In a sample of length L the envelopes E_1 and E_{3F} propagate in the forward direction, while the envelopes E_2 and E_{3B} propagate in the backward direction. The two envelopes E_{3B} and E_{3F} are assumed to be intense, and are called pump beams. The two envelopes E_1 and E_2, instead, are assumed to be weak and are called the probe field and the phase-conjugate field, respectively.

Similarly for E_2 we obtain

$$E_2(z) = E_{20} \cos\left[\eta' \left(\frac{3\omega_1\omega_2}{n_1 n_2} \right)^{1/2} z \right]$$
$$+ i \left(\frac{\omega_2 n_1}{3\omega_1 n_2} \right)^{1/2} \left(2 E_{20} + e^{i\zeta} E_{10}^* \right) \sin\left[\eta' \left(\frac{3\omega_1\omega_2}{n_1 n_2} \right)^{1/2} z \right]. \quad (7.41)$$

The most interesting feature is that, when one of the two fields E_1 and E_2 vanishes in the input, e.g. $E_{20} = 0$, the same field is different from zero in the output, i.e. $E_2(z) \neq 0$. This is the process of frequency generation by four-wave mixing.

7.5 Optical phase conjugation

Quite interesting phenomena appear if we drop the limitation to unidirectional propagation that we have followed up to this point, and address the bidirectional configuration [59]. Let us consider again the case of three frequencies ω_1, ω_2 and ω_3 defined in Eq. (7.30), but with the following modifications (see Fig. 7.1).

- The field of frequency ω_3 is composed by two envelopes E_{3F} and E_{3B} propagating in the forward positive-z direction and in the backward negative-z direction, respectively. The wavenumber is the same for the two envelopes.
- The envelope E_1 propagates in the positive-z direction, whereas the envelope E_2 propagates in the negative-z direction.

In the plane-wave approximation the electric field is composed in the following way:

$$E(z, t) = \frac{1}{2} \left[E_1(z, t) e^{-i(\omega_1 t - k_1 z)} + E_2(z, t) e^{-i(\omega_2 t + k_2 z)} + E_{3F}(z, t) e^{-i(\omega_3 t - k_3 z)} \right.$$
$$+ E_{3B}(z, t) e^{-i(\omega_3 t + k_3 z)} + E_1^*(z, t) e^{i(\omega_1 t - k_1 z)} + E_2^*(z, t) e^{i(\omega_2 t + k_2 z)}$$
$$\left. + E_{3F}^*(z, t) e^{i(\omega_3 t - k_3 z)} + E_{3B}^*(z, t) e^{i(\omega_3 t + k_3 z)} \right]. \quad (7.42)$$

We assume further that the two envelopes E_{3F} and E_{3B}, which we call *pump beams*, are intense, whereas the envelopes E_1 and E_2 are weak. Therefore in the equations for the pump envelopes we can neglect the terms arising from the interaction with the weak fields, and we obtain the steady-state equations

$$\frac{dE_{3F}}{dz} = i\frac{3}{32}\frac{\omega_3\chi^{(3)}}{n_3c}[|E_{3F}(z)|^2 E_{3F}(z) + 2|E_{3B}(z)|^2 E_{3F}(z)], \tag{7.43}$$

$$\frac{dE_{3B}}{dz} = -i\frac{3}{32}\frac{\omega_3\chi^{(3)}}{n_3c}[|E_{3B}(z)|^2 E_{3B}(z) + 2|E_{3F}(z)|^2 E_{3B}(z)], \tag{7.44}$$

in which the first term describes self-phase modulation and the second cross-phase modulation. We have immediately that $|E_{3F}|^2$ and $|E_{3B}|^2$ are independent of z. If we assume that on both sides of the sample we inject fields of equal intensity we have that the two counterpropagating envelopes have the same modulus $|E|^2$, so that by setting

$$\kappa_3 = \frac{9}{32}\frac{\omega_3\chi^{(3)}}{n_3c}|E|^2 \tag{7.45}$$

we obtain

$$E_{3F}(z) = e^{i\kappa_3 z}E_{3F}(0), \qquad E_{3B}(z) = e^{-i\kappa_3 z}E_{3B}(0). \tag{7.46}$$

Therefore we have that also the product $E_{3F}(z)E_{3B}(z)$ is independent of z [29],

$$E_{3F}(z)E_{3B}(z) = E_{3F}(0)E_{3B}(0) \equiv E_F E_B. \tag{7.47}$$

Let us now turn our attention to the weak fields E_1 and E_2. By neglecting the terms which are negligible in the weak-field regime, they obey the steady-state equations

$$\frac{dE_1}{dz} = i\frac{3}{16}\frac{\omega_1\chi^{(3)}}{n_1c}[2|E|^2 E_1(z) + E_F E_B E_2^*(z)e^{i(k_2-k_1)z}], \tag{7.48}$$

$$\frac{dE_2}{dz} = -i\frac{3}{16}\frac{\omega_2\chi^{(3)}}{n_2c}[2|E|^2 E_2(z) + E_F E_B E_1^*(z)e^{-i(k_2-k_1)z}], \tag{7.49}$$

where we have taken Eq. (7.47) into account. In these equations the first term arises from cross-phase modulation and the second from four-wave mixing. We assume that the envelopes 1 and 2 obey the perfect phase-matching condition

$$k_2 = k_1. \tag{7.50}$$

The simplest way to realize this condition is to assume that the envelopes E_1 and E_2 have the same frequency. However, because of the energy conservation, Eq. (7.30), this implies that all of the four envelopes in play have the same frequency (the frequency-degenerate case) $\omega_1 = \omega_2 = \omega_3 \equiv \omega$. In order to be able to distinguish pump envelopes and weak envelopes, one can assume that the direction of the pump beams is slightly tilted with respect to that of the weak beams, as shown in Fig. 7.2. By defining

$$\kappa_1 = \frac{3}{16}\frac{\omega\chi^{(3)}}{nc}|E|^2, \tag{7.51}$$

$$\kappa = \frac{3}{32}\frac{\omega\chi^{(3)}}{nc}E_F E_B, \tag{7.52}$$

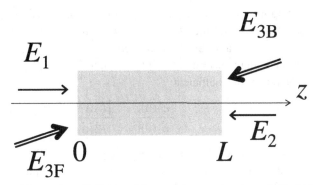

Figure 7.2 When the frequencies of the envelopes E_1, E_2, E_{3F} and E_{3B} are all the same, in order to distinguish between pump beams and weak beams one can assume that the pump beams are slightly tilted with respect to the weak beams.

where $n = n(\omega)$, we can cast Eqs. (7.48) and (7.49) in the form

$$\frac{dE_1}{dz} = i\kappa_1 E_1(z) + i\kappa E_2^*(z), \tag{7.53}$$

$$\frac{dE_2}{dz} = -i\kappa_1 E_2(z) - i\kappa E_1^*(z). \tag{7.54}$$

These equations can be simplified by introducing new variables, i.e. by setting [59]

$$E_1(z) = E_1'(z)e^{i\kappa_1 z}, \tag{7.55}$$

$$E_2(z) = E_2'(z)e^{-i\kappa_1 z}, \tag{7.56}$$

so that Eqs. (7.53) and (7.54) can be written as

$$\frac{dE_1'}{dz} = i\kappa E_2'^*(z), \tag{7.57}$$

$$\frac{dE_2'^*}{dz} = i\kappa E_1'(z), \tag{7.58}$$

from which we see that the general solution is given by

$$E_1'(z) = A \sin|\kappa|z + B \cos|\kappa|z, \tag{7.59}$$

$$E_2'^*(z) = A \frac{|\kappa|}{i\kappa} \cos|\kappa|z - B \frac{|\kappa|}{i\kappa} \sin|\kappa|z. \tag{7.60}$$

By imposing the boundary conditions $E_1'(0)$ and $E_2'^*(L)$, we arrive at the expressions [59]

$$E_1'(z) = i \frac{\kappa}{|\kappa|} \frac{\sin(|\kappa|z)}{\cos(|\kappa|L)} E_2'^*(L) + \frac{\cos[|\kappa|(z-L)]}{\cos(|\kappa|L)} E_1'(0), \tag{7.61}$$

$$E_2'^*(z) = \frac{\cos(|\kappa|z)}{\cos(|\kappa|L)} E_2'^*(L) + i \frac{|\kappa|}{\kappa} \frac{\sin[|\kappa|(z-L)]}{\cos(|\kappa|L)} E_1'(0), \tag{7.62}$$

The most interesting result is that one has a nonzero $E_2(z)$ even when no field E_2 is injected, i.e. for $E_2(L) = 0$. This phenomenon arises from the circumstance that the sum of the two pump beams' envelopes described by Eqs. (7.46) forms a grating of intensity proportional to $\cos(k_3 z + \kappa_3 z + \bar{\phi})$, where $\bar{\phi}$ is the average of the phases of $E_{3F}(0)$ and $E_{3B}(0)$. Such a grating is capable of reflecting the field E_1 backwards. For this reason one calls E_1 the *probe beam* and E_2 the *phase-conjugate beam* which is generated by reflection, and one calls this device a *phase-conjugate mirror*.

By inserting the condition $E'_2(L) = 0$ into Eq. (7.62) one finds the following expression for the reflection coefficient:

$$\frac{|E_2(0)|^2}{|E_1(0)|^2} = \frac{|E'_2(0)|^2}{|E'_1(0)|^2} = \tan^2(|\kappa|L), \tag{7.63}$$

from which we see that in a phase-conjugate mirror the reflection coefficient can be even larger than unity. Similarly, by using Eq. (7.61) one finds the following expression for the transmission coefficient (which is larger than unity):

$$\frac{|E_1(L)|^2}{|E_1(0)|^2} = \frac{|E'_1(L)|^2}{|E'_1(0)|^2} = \frac{1}{\cos^2(|\kappa|L)}. \tag{7.64}$$

An extensive illustration of the topic of phase-conjugate mirrors can be found in [59].

8

Optical resonators. The planar ring cavity. Empty cavity. Linear cavity

With the exception of Chapter 1, up to this point we have considered cavityless optical systems. With this chapter, instead, we start focussing on the case that the atomic medium is located in an optical cavity. We describe in detail the case of a unidirectional ring resonator, the boundary condition which characterizes it and the relations between its output fields and the intracavity field (Sections 8.1–8.3). We discuss the transmission of an empty cavity (Section 8.4) and of a cavity that contains a linear medium (Section 8.5).

8.1 Optical cavities

An *optical cavity* or *optical resonator* is formed by at least two mirrors,[1] appropriately arranged in such a way that the radiation beam follows a closed path. The beam splitters allow the radiation to escape from the cavity (output) and allow one to inject radiation into the cavity (input). In order to confine the radiation transversally with respect to the direction of propagation, the mirrors are usually spherical (see Fig. 1.2). The case of cavities with spherical mirrors will be described in Part III of this volume. On the other hand, an analytical treatment of the problems is possible only within the plane-wave approximation which we will use in the remainder of Part I and in Part II. In accord with this, we assume here that the mirrors are planar.

The two most common types of optical resonator are the *Fabry–Perot cavity* and the *ring cavity*. The planar Fabry–Perot cavity consists in two beam splitters in a parallel configuration as shown in Fig. 8.1. Within the cavity there are two counterpropagating fields, E_F propagating in the forward direction and E_B propagating in the backward direction with respect to the input field E_I. Because we have derived the Maxwell–Bloch equations for the unidirectional configuration, in this and the following chapters we focus on the case of a ring cavity, which allows unidirectionality. The Fabry–Perot case will be considered in Chapter 14.

A ring resonator is formed by two partially reflecting mirrors and at least one completely reflecting mirror, which define a closed path of length $\mathcal{L}$. In this case it is possible to select one propagation direction by locating in the path a non-reciprocal element, such as a Faraday rotator, which suppresses the backward-propagating field. This is not possible in a Fabry–Perot cavity because the forward and backward fields feed each other.

---

[1] We use the expressions mirrors and beam splitters synonymously.

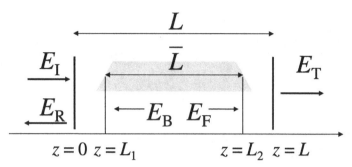

Figure 8.1   A planar Fabry–Perot cavity, E_I, E_T and E_R are the incident, transmitted and reflected fields, respectively. The sample is cut at the Brewster angle in order to select the polarization plane and to avoid reflection at the endfaces.

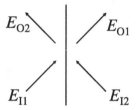

Figure 8.2   A beam splitter. E_{I1} and E_{I2} are the input fields. E_{O1} and E_{O2} are the output fields.

8.2 Beam splitters

A beam splitter has a complex structure that consists in a large number of parallel dielectric layers. We will describe it, however, only via its input–output relations. Let us consider for definiteness the beam splitter of Fig. 8.2, in which we show two input fields and two output fields, and possibly one of the two input fields is zero. If we express the fields in terms of their envelopes,

$$\mathcal{E}_{Ii} = \frac{1}{2}\left(E_{Ii}e^{-i\omega_0 t} + \text{c.c.}\right), \quad \mathcal{E}_{Oi} = \frac{1}{2}\left(E_{Oi}e^{-i\omega_0 t} + \text{c.c.}\right) \quad (i = 1, 2), \tag{8.1}$$

the input–output relations read

$$E_{O1} = tE_{I1} + rE_{I2}, \quad E_{O2} = tE_{I2} + r'E_{I1}, \tag{8.2}$$

where the transmission and reflection coefficients t, r and r' are complex numbers because they involve also a change of phase. The transmission coefficient is the same from the left to the right of the mirror and vice versa, whereas the reflection coefficients may be different in general for the phase [80–84], and for this reason we have two reflection coefficients r and r'. By setting

$$t = \sqrt{T}e^{i\varphi_t}, \quad r = \sqrt{R}e^{i\varphi_r}, \quad r' = \sqrt{R}e^{i\varphi_r'}, \tag{8.3}$$

where T and R are the transmission and reflection coefficients for the intensity, respectively, we assume that $T + R = 1$, i.e. that there is no loss in the beam splitter. Further, it is possible

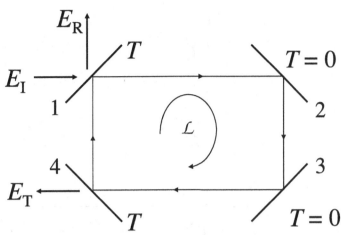

Figure 8.3   A planar ring cavity. Mirrors 1 and 4 have intensity transmissivity coefficient T; mirrors 2 and 3 have reflection coefficient $R = 1$. E_I, E_T and E_R are the incident, transmitted and reflected fields, respectively.

to demonstrate that [80, 81]

$$\varphi_r + \varphi_r' - 2\varphi_t = \pm \pi, \tag{8.4}$$

so the transformation (8.2) is unitary. In order not to overburden the equations with many arbitrary coefficients, let us make for simplicity the following choice, which is adopted in most of the literature and allows one to avoid using complex coefficients:

$$\varphi_t = 0, \quad \varphi_r = 0, \quad \varphi_r' = \pi, \tag{8.5}$$

so that Eqs. (8.2) become

$$E_{O1} = \sqrt{T} E_{I1} + \sqrt{R} E_{I2}, \quad E_{O2} = \sqrt{T} E_{I2} - \sqrt{R} E_{I1}. \tag{8.6}$$

8.3 The planar ring cavity. Boundary condition, input and output fields. Transmission of the cavity

We consider a planar ring cavity with two beam splitters and two mirrors (Fig. 8.3). The input field E_I is injected through mirror 1, whereas the transmitted field E_T comes from mirror 4 and the reflected field comes from mirror 1. Totally reflecting mirrors 2 and 3 are necessary in order to create a closed path of length $\mathcal{L}$.

In cavities, the phase plays a dominant role because the mirrors create a feedback loop that gives rise to multiple interference of light. Let us focus first on an empty cavity. If we consider, for instance, the light at the point of the cavity immediately adjacent to the input mirror, we see that during the first roundtrip it is partially transmitted by mirror 4, but in part is fed back to the same point, where it interferes with the light which enters the cavity, and so on after each roundtrip. The interference is fully constructive when the

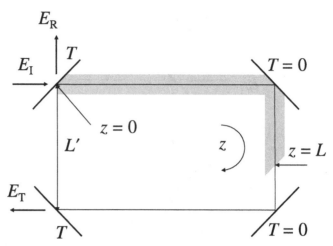

Figure 8.4   A ring cavity containing an atomic sample of length L. The endfaces of the sample are cut at Brewster angles.

phase $2\pi\mathcal{L}/\lambda_0$, where λ_0 is the wavelength of light in vacuum (which is different from the $\lambda = \lambda_0/n_B$ introduced in Chapter 3), accumulated during one roundtrip, is an integer multiple of 2π, i.e.

$$\mathcal{L} = \lambda_0 j, \quad j = 1, 2, \ldots. \tag{8.7}$$

If we take into account that $\omega_0 = 2\pi c/\lambda_0$ we have that if the input light has a frequency

$$\omega_j = \frac{2\pi c}{\mathcal{L}} j \tag{8.8}$$

the interference is fully constructive.

These frequencies are called *cavity frequencies* or *cavity resonances*. We remark that in the case of a Fabry–Perot cavity Eqs. (8.7) and (8.8) are still valid, provided that one replaces $\mathcal{L}$ by the roundtrip length $2L$ in the Fabry–Perot cavity.

Let us now consider the general case of a ring cavity containing an atomic medium (Fig. 8.4). Within the cavity, the coordinate z follows the closed path of the light; $z = 0$ corresponds to the position of the input mirror immediately after it. We introduce envelopes for the input, transmitted and reflected fields

$$\mathcal{E}_I(t) = \frac{1}{2}\left[E_I(t)e^{-i\omega_0 t} + \text{c.c.}\right],$$

$$\mathcal{E}_T(t) = \frac{1}{2}\left[E_T(t)e^{-i\omega_0 t} + \text{c.c.}\right], \tag{8.9}$$

$$\mathcal{E}_R(t) = \frac{1}{2}\left[E_R(t)e^{-i\omega_0 t} + \text{c.c.}\right].$$

Insofar as the intracavity field is concerned, in the part of the cavity occupied by the medium we consider the field envelope E_0 defined by Eq. (2.37) with $k_0 = \omega_0 n_B/c$.

The ring cavity is characterized by a boundary condition that links the field envelope at $z = 0$ with the envelope at position $z = L$, where L is the length of the atomic sample. As a matter of fact, the value of the field at $z = 0$ at time t is given by the sum of two contributions: the part of the incident field which is transmitted by mirror 1 at time t plus the feedback contribution of the field at position $z = L$ at time $t - \Delta t$, reflected by mirror 3 (for the case described by Fig. 8.4) and by mirrors 4 and 1. Δt is the time the light takes to travel from $z = L$ to $z = 0$, i.e.

$$\Delta t = \frac{\mathcal{L} - L}{c}. \tag{8.10}$$

By taking into account that for mirror 3 the reflection coefficient is $R = 1$, the boundary condition reads explicitly

$$E_0(0, t)e^{-i\omega_0 t} = \sqrt{T} E_\mathrm{I}(t)e^{-i\omega_0 t} + R E_0(L, t - \Delta t)e^{-i\omega_0(t-\Delta t)}e^{2\pi i n_\mathrm{B} L/\lambda_0}$$
$$\Rightarrow \quad E_0(0, t) = \sqrt{T} E_\mathrm{I}(t) + R E_0(L, t - \Delta t)e^{2\pi i \Lambda/\lambda_0}, \tag{8.11}$$

where we have taken into account that $\omega_0 = 2\pi c/\lambda_0$ and Λ is the optical length of the cavity, including the presence of the background refractive index n_B (but not the contribution to the refractive index from the two-level atomic system),

$$\Lambda = \mathcal{L} - L + n_\mathrm{B} L. \tag{8.12}$$

In addition, the transmitted field is given by

$$E_\mathrm{T}(t) = \sqrt{T} E_0(L, t - \Delta t')e^{2\pi i(\Lambda - L')/\lambda_0}, \quad \Delta t' = \frac{\mathcal{L} - L - L'}{c}, \tag{8.13}$$

where L' is the distance between mirror 4 and mirror 1 (see Fig. 8.4) and the reflected field corresponds to

$$E_\mathrm{R}(t) = -\sqrt{R} E_\mathrm{I}(t) + \sqrt{TR} E_0(L, t - \Delta t)e^{2\pi i \Lambda/\lambda_0}. \tag{8.14}$$

On passing to the normalized envelopes

$$F(z, t) = \frac{d E_0(z, t)}{\hbar \sqrt{\gamma_\perp \gamma_\parallel}}, \quad y(t) = \frac{d E_\mathrm{I}(t)}{\hbar \sqrt{\gamma_\perp \gamma_\parallel T}},$$
$$x(t) = \frac{d E_\mathrm{T}(t)e^{-2\pi i(\Lambda - L')/\lambda_0}}{\hbar \sqrt{\gamma_\perp \gamma_\parallel T}}, \quad F_\mathrm{R}(t) = \frac{d E_\mathrm{R}(t)}{\hbar \sqrt{\gamma_\perp \gamma_\parallel T}}, \tag{8.15}$$

Eqs. (8.11) and (8.13) take the forms [85, 86]

$$F(0, t) = T y(t) + R F(L, t - \Delta t)e^{2\pi i \Lambda/\lambda_0}, \tag{8.16}$$
$$x(t) = F(L, t - \Delta t'). \tag{8.17}$$

respectively.[2] The input field envelope E_I may be time-varying in general, e.g. when one injects into the cavity a modulated field or a pulse. Let us now focus, however, on the case in which E_I is constant and on the stationary state, in which the envelopes are independent of time, so that we obtain

$$F(0) = T y + R F(L)e^{2\pi i \Lambda/\lambda_0}, \tag{8.18}$$
$$x = F(L). \tag{8.19}$$

---

[2] An experimental parameter related to the transmissivity T is the so-called *finesse* of the cavity, defined as π/T.

Note that from Eqs. (8.14), (8.15) and (8.19) we have in the stationary state for the normalized reflected field

$$F_R = \sqrt{R}\left(xe^{2\pi i\Lambda/\lambda_0} - y\right). \tag{8.20}$$

Next, we derive from these equations a useful expression for the transmission $\mathcal{T}$ of the cavity, which is defined as the ratio between the transmitted intensity and the input intensity and therefore the ratio of $|x|^2$ to $|y|^2$. From Eqs. (8.18) and (8.19) we obtain

$$\mathcal{T} = \frac{|x|^2}{|y|^2} = \frac{T^2}{\left|F(0)/F(L) - Re^{2\pi i\Lambda/\lambda_0}\right|^2}. \tag{8.21}$$

On setting $F(z) = \rho(z)e^{i\varphi(z)}$ as usual, the ratio $F(0)/F(L)$ can be written in the form

$$\frac{F(0)}{F(L)} = \beta e^{-i\Delta\varphi}, \qquad \beta = \frac{\rho(0)}{\rho(L)}, \qquad \Delta\varphi = \varphi(L) - \varphi(0), \tag{8.22}$$

and the transmission becomes

$$\mathcal{T} = \frac{T^2}{\left|\beta - Re^{i(2\pi\Lambda/\lambda_0 + \Delta\varphi)}\right|^2} = \frac{T^2}{\beta^2 + R^2 - 2\beta R\cos\left(2\pi\dfrac{\Lambda}{\lambda_0} + \Delta\varphi\right)}$$

$$= \frac{T^2}{(\beta - R)^2 + 4\beta R\sin^2\left(\pi\dfrac{\Lambda}{\lambda_0} + \dfrac{\Delta\varphi}{2}\right)}. \tag{8.23}$$

The quantities β and $\Delta\varphi$ can be obtained from the solution of the stationary equations for the medium. In the remainder of this chapter we discuss the two cases of an empty cavity and a cavity containing a linear medium.

8.4 The empty cavity

In this case we have that $L = 0$, so $\beta = 1$, $\Delta\varphi = 0$ and $\Lambda = \mathcal{L}$. Therefore Eq. (8.23) reduces to

$$\mathcal{T} = \frac{1}{1 + \dfrac{4R}{T^2}\sin^2\left(\pi\dfrac{\mathcal{L}}{\lambda_0}\right)}, \tag{8.24}$$

where we have taken into account that $1 - R = T$. This is called the *Airy function*, and the graph of $\mathcal{T}$ as a function of ω_0 is shown in Fig. 8.5 for three values of T.

The transmission is maximum and equal to 1 when $\sin^2(\pi\mathcal{L}/\lambda_0) = 0$, i.e. when condition (8.7) is satisfied. When this is not the case the imperfect constructive interference produces a decrease of the transmission, which takes its minimum value $\mathcal{T}_{\min} = (1 + 4R/T^2)^{-1}$ when the cavity length is equal to an odd multiple of $\lambda_0/2$. The minimum approaches zero when R tends to 1 (and therefore T tends to 0).

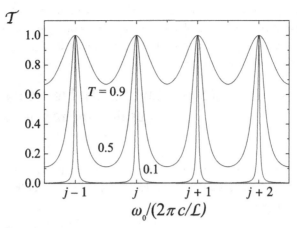

Figure 8.5 The transmission $\mathcal{T}$ of the cavity is shown as a function of the ratio of the input frequency ω_0 to the free spectral range $2\pi c/\mathcal{L}$ for $T = 0.9$, $T = 0.5$ and $T = 0.1$.

The cavity frequencies, cavity resonances or *longitudinal cavity modes* are given by Eq. (8.8). They are equispaced, and the frequency difference between two adjacent cavity frequencies $2\pi c/\mathcal{L}$ is called the *free spectral range*.

It is now interesting to evaluate the halfwidth of the cavity resonances or *cavity linewidth*. This is defined as the frequency $\bar{\kappa}$ at which the transmission of the cavity for $\omega_0 = \omega_j \pm \bar{\kappa}$ is equal to $1/2$. From Eq. (8.24) we obtain

$$\sin^2\left(\frac{\pi \mathcal{L}}{\lambda_0}\right) = \sin^2\left[(\omega_j \pm \bar{\kappa})\frac{\mathcal{L}}{2c}\right] = \sin^2\left(\frac{\mathcal{L}\bar{\kappa}}{2c}\right) = \frac{T^2}{4R}, \tag{8.25}$$

i.e.

$$\bar{\kappa} = \frac{2c}{\mathcal{L}}\arcsin\left(\frac{T}{2\sqrt{R}}\right). \tag{8.26}$$

This expression becomes notably simpler when R approaches unity, i.e. for $T \ll 1$, and becomes equal to

$$\kappa = \frac{cT}{\mathcal{L}}. \tag{8.27}$$

As we will show in Chapter 12, the expression (8.27) coincides with the decay rate of the cavity field in the limit $T \ll 1$. The parameter $2\kappa = 2cT/\mathcal{L}$ represents the escape rate of photons from the cavity, i.e. the decay rate of the light intensity in the cavity.

A final relevant point emerges if we return to Eq. (8.13) at steady state, which gives

$$|E_0|^2 = \frac{|E_T|^2}{T}, \tag{8.28}$$

where we have written E_0 instead of $E_0(L)$ because the intracavity intensity is constant along the cavity. From Eq. (8.28) we see that for $T \ll 1$ the intracavity intensity is much larger than the transmitted intensity. If the input field frequency is equal to one of the cavity resonances, the transmitted intensity equals the incident intensity, and therefore the intracavity intensity is much larger than the input intensity. This result demonstrates another advantage of using optical cavities: One can obtain in the cavity intensities much larger than that of the field which one injects into the cavity, and therefore the atomic sample in the cavity experiences this much more intense field.

8.5 The linear cavity. Frequency pulling and pushing, mode splitting

Let us turn to the case in which the cavity contains a linear medium described by the linear susceptibility specified by Eq. (5.10). The variation of the electric field along the sample is given by Eq. (5.8), so

$$F(L) = F(0)e^{g\tilde{\chi}'(\omega_0)L + ig\tilde{\chi}''(\omega_0)L}, \tag{8.29}$$

which implies that $\beta = e^{-g\tilde{\chi}'(\omega_0)L}$ and $\Delta\varphi = g\tilde{\chi}''(\omega_0)L$. Therefore, using Eq. (8.23), the transmission of the linear cavity is given by

$$\mathcal{T} = \frac{T^2}{\left[e^{-g\tilde{\chi}'(\omega_0)L} - R\right]^2 + 4Re^{-g\tilde{\chi}'(\omega_0)L}\sin^2\left[\pi\frac{\Lambda}{\lambda_0} + \frac{g}{2}\tilde{\chi}''(\omega_0)L\right]}. \tag{8.30}$$

Also in this case the argument of $\sin^2$ amounts to half of the phase accumulated by the wave in a roundtrip of the cavity, which contains a medium with refractive index $n(\omega_0)$ for a segment of length L and is empty in the remaining part of length $\mathcal{L} - L$. As a matter of fact, $2\pi\Lambda/\lambda_0 + g\tilde{\chi}''(\omega_0)L = (2\pi/\lambda_0)[\mathcal{L} - L + Ln(\omega_0)]$ as one sees from Eq. (5.14), and the expression in square brackets is the optical length of the cavity including not only the background refractive index but also the contribution from the two-level atomic system.

If we compare Eqs. (8.23) and (8.24), we see that the presence of the sample replaces the cavity length $\mathcal{L}$ by the optical length Λ. Therefore Eq. (8.7) becomes

$$\Lambda = \lambda j, \quad j = 1, 2, \ldots \tag{8.31}$$

and in the following we will call the frequencies

$$\omega_j = \frac{2\pi c}{\Lambda} j \tag{8.32}$$

cavity frequencies, instead of the empty cavity frequencies (8.8). The free spectral range, or frequency difference between adjacent cavity modes, is now given by $2\pi c/\Lambda$. The presence of the linear medium modifies the position of the transmission peaks of the cavity, i.e. the mode frequencies, with respect to that of the frequencies (8.32). According to Eq. (8.30), the $\sin^2$ term vanishes when

$$2\pi j = \frac{2\pi\Lambda}{\lambda_0} + g\tilde{\chi}''(\omega_0)L = \frac{\omega_0}{c}\Lambda + g\tilde{\chi}''(\omega_0)L. \tag{8.33}$$

By using Eq. (8.32) we can write

$$\omega_j - \omega_0 = gc\frac{L}{\Lambda}\tilde{\chi}''(\omega_0). \tag{8.34}$$

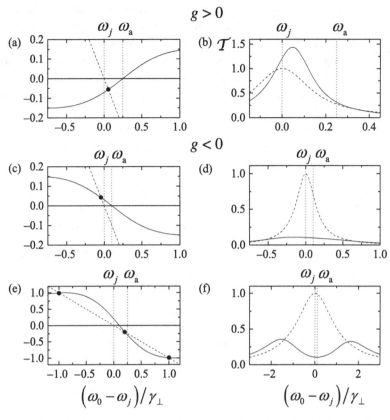

Figure 8.6 Parts (a) and (b) refer to the active case $g > 0$, the others to the passive case $g < 0$. The left curves show the graph of the function $[gcL/(\Lambda \gamma_\perp)] \tilde{\chi}''[(\omega_0 - \omega_j)/\gamma_\perp]$ (see Eq. (5.10)) for $gcL/\Lambda \gamma_\perp = 0.3$ (a), -0.3 (c) and -2 (e) and for $(\omega_a - \omega_j)/\gamma_\perp = 0.25$ in (a) and 0.1 in (c) and (e). ω_j here denotes the cavity frequency (8.32) nearest to the atomic frequency ω_a. The right curves show the transmissivity of the cavity as a function of the input frequency ω_0 (solid line) for $T = 0.05$ and $gL = 0.1$ in (b) and -0.1 in (d) and (f) and for the same values of $gcL/(\Lambda \gamma_\perp)$ and $(\omega_a - \omega_j)/\gamma_\perp$ as in the corresponding left figure. The broken curves show the transmissivity when $\tilde{\chi}''$ is neglected. Parts (a) and (b) illustrate mode pulling, (c) and (d) mode pushing, and (e) and (f) mode splitting.

This highly nontrivial nonlinear equation in the variable ω_0 can be solved graphically by seeking, for each value of j, the intersection point between the straight line $y = \omega_j - \omega_0$ and the curve $(gcL/\Lambda)\tilde{\chi}''(\omega_0)$. This procedure is illustrated in Fig. 8.6 on the left. From Fig. 8.6(a) one sees that the frequency of the linear cavity is pulled closer to the atomic frequency ω_a with respect to the empty-cavity frequencies. This phenomenon is called *frequency pulling*. From Fig. 8.6(c) one sees that, in contrast, in the passive case the frequencies of the linear cavity get pushed away from the atomic transition frequency with respect to the empty-cavity frequencies, which is called *frequency pushing*. Under special conditions, there is a cavity frequency for which one finds three distinct intersections with the curve, i.e. three modes for the linear cavity. An example of this phenomenon, which is called *mode splitting*, is shown in Fig. 8.6(e).

In Fig. 8.6 on the right we see the transmission of the linear cavity. The transmission peaks lie in the positions of the pulled/pushed frequencies, and in the passive case are significantly lower and broader than the empty-cavity peaks. In contrast, for the active configurations the peaks are higher and narrower, in particular they are of amplitude higher than unity because there is light amplification instead of absorption.

Phenomena of frequency pulling and pushing arise also in the case of a nonlinear medium, as we will see in the following chapter.

In view of the following chapters, it is convenient to replace the quantity Λ/λ_0, which appears in the exponential in Eq. (8.18) and is a large number, with a number of order unity. To this end, we first introduce a parameter δ_0 defined as

$$\delta_0 = 2\pi \bar{j} - 2\pi \frac{\Lambda}{\lambda_0}, \tag{8.35}$$

where $\bar{j}$ is a positive integer so that Eq. (8.16) can be written in the form

$$F(0, t) = T y(t) + R F(L, t - \Delta t) e^{-i\delta_0}. \tag{8.36}$$

Note that δ_0 can be recast as follows:

$$\delta_0 = \frac{2\pi \bar{j} c/\Lambda - 2\pi c/\lambda_0}{c/\Lambda} = \frac{\omega_c - \omega_0}{c/\Lambda}, \qquad \omega_0 = \frac{2\pi c}{\lambda_0}, \tag{8.37}$$

where ω_c is the cavity frequency $\omega_c = 2\pi \bar{j} c/\Lambda = \omega_{\bar{j}}$ (see Eq. (8.32)). It is convenient to fix $\bar{j}$ in such a way that

$$\left| 2\pi \frac{\Lambda}{\lambda_0} - 2\pi \bar{j} \right| \leq \pi, \tag{8.38}$$

so that ω_c is the cavity frequency closest to ω_0 and δ_0 is indeed of order unity. The parameter δ_0 measures the difference between the input field frequency and the closest cavity frequency scaled to the free spectral range (divided by 2π).

Equations (8.23) and (8.20) can now be written in the forms

$$\mathcal{T} = \frac{T^2}{(\beta - R)^2 + 4\beta R \sin^2 \left(\dfrac{\Delta\varphi - \delta_0}{2} \right)}, \tag{8.39}$$

$$F_R = \sqrt{R} \left(x e^{-i\delta_0} - y \right). \tag{8.40}$$

The final remark is that we have assumed for definiteness that the atomic sample occupies the segment of the cavity between $z = 0$ and $z = L$. However, we must keep in mind that a translation of the sample along the ring cavity does not affect the transmitted intensity and the reflected intensity when the input intensity is constant in time. The equations of this chapter remain unchanged, provided that one takes the origin $z = 0$ to correspond to the entrance face of the atomic sample.

9

A nonlinear active ring cavity: the ring laser, stationary states

In the nonlinear framework, we treat first the case of the laser because it is the most important nonlinear optical system. It suffices to mention that, whenever in this book we consider an electric field, we always assume that it is coherent, as is the case for the field emitted by a laser sufficiently above threshold. It also turns out that the theoretical treatment of the laser is simpler than that of the passive case, as we will see from the comparison of Chapters 9 and 11.

Before starting the discussion, we underline that the laser case presents some basic differences with respect to the treatment in the previous section.

- The laser is a *source* of radiation and therefore emits light without need of injecting light in the cavity. Therefore, in Eq. (8.16) we must set $y = 0$, giving

$$F(0, t) = RF(L, t - \Delta t)e^{2\pi i \Lambda / \lambda_0}. \tag{9.1}$$

- Because there is no input field, the frequency ω_0 which appears in the carrier factor of Eq. (2.37) needs a definition. The natural choice is to define ω_0 as the emission frequency of the laser in the stationary state. This frequency is not known a priori and therefore must be treated as an unknown that will be determined by the calculations, together with the intensity of the emitted field.
- The laser equations (4.35)–(4.37) together with the boundary condition (9.1) admit the trivial stationary solution

$$F(z) = 0, \qquad P(z) = 0, \qquad D(z) = 1, \tag{9.2}$$

which corresponds to a laser that does not emit, i.e. below threshold.

In the following we calculate the nontrivial stationary solutions, in which the emitted intensity is non-vanishing. Next, we introduce a multiple limit, which plays a relevant role in this book because it allows one to simplify the equations, namely the *low-transmission limit*. Finally, we illustrate the analogy between the behavior of the laser and second-order phase transitions.

9.1 Calculation of the nontrivial stationary solutions

By expressing the stationary field $F(z)$ in terms of its modulus and phase as usual, i.e. $F(z) = \rho(z)\exp(i\phi(z))$, we can split the boundary condition (9.1) into two parts [87, 88],

$$\rho_0 = R\rho(L), \tag{9.3}$$

$$\varphi(L) - \varphi_0 = 2\pi \left(\frac{\Lambda}{\lambda_0} - j \right) = \frac{\omega_0 - \omega_j}{c/\Lambda}, \tag{9.4}$$

where j is an arbitrary positive integer. We have used Eq. (8.32) and we have written ρ_0 and φ_0 instead of $\rho(0)$ and $\varphi(0)$. Equation (9.3) proves that the internal field is amplified along the medium, because $\rho(L)/\rho_0 = 1/R > 1$. By combining these two boundary conditions with Eqs. (5.36) and (5.39) with $z = L$, which describe the behavior of ρ and φ at the two extremes of the sample, we obtain [87, 88][1]

$$(1 + \Delta^2)|\ln R| + \frac{\rho^2(L)}{2}(1 - R^2) = gL, \tag{9.5}$$

$$\frac{\omega_0 - \omega_j}{c/\Lambda} = -\Delta|\ln R|. \tag{9.6}$$

Let us analyze first Eq. (9.6), which, taking into account the definition of Δ, reads

$$\frac{\omega_0 - \omega_j}{c/\Lambda} = -\frac{\omega_a - \omega_0}{\gamma_\perp}|\ln R|, \tag{9.7}$$

and determines the frequency ω_0 of the emitted field. The appearance of the cavity frequencies ω_j in Eq. (9.7) implies that the output frequency depends on the mode index j and therefore we will write ω_{0j} instead of ω_0. By solving Eq. (9.7) with respect to ω_{0j}, we obtain

$$\omega_{0j} = \frac{\kappa'\omega_a + \gamma_\perp\omega_j}{\kappa' + \gamma_\perp}, \tag{9.8}$$

with

$$\kappa' = \frac{c}{\Lambda}|\ln R|. \tag{9.9}$$

Therefore there is an infinity of stationary solutions that differ for the emission frequency ω_{0j}. In the jth solution the laser frequency corresponds to the cavity frequency ω_j pulled towards the atomic transition frequency ω_a. This phenomenon is called frequency pulling as in the linear case. The pulled frequency corresponds precisely to the weighted mean of the frequencies ω_a and ω_j, with weights proportional to κ' and $\gamma_\perp$, respectively. Therefore the laser frequency is intermediate between the two. Note that κ' for $n_B = 1$ is close to the cavity linewidth $\bar{\kappa}$ (Eq. (8.26)) for the empty-cavity case and coincides with it in the limit of vanishing transmissivity T because $|\ln R| = |\ln(1 - T)| \approx T$. For $\kappa' < \gamma_\perp$, ω_{0j} is closer to the cavity frequency ω_j, whereas for $\kappa' > \gamma_\perp$ it is closer to the atomic transition frequency ω_a.

---

[1] We note from Eq. (4.34) that the constant g depends on ω_0 because $k_0 = \omega_0 n_B/c$. Therefore, strictly speaking, we should write g_j corresponding to the frequency ω_{0j} defined by Eq. (9.8). However, this dependence is negligible and, in the expression (4.34) of g, in the laser case, we replace ω_0 by the atomic frequency ω_a.

Because ω_0 depends on j, the same is true for the parameter Δ, which we therefore denote Δ_j. By using Eq. (9.8) we obtain

$$\Delta_j = \frac{\omega_a - \omega_{0j}}{\gamma_\perp} = \frac{\omega_a - \omega_j}{\kappa' + \gamma_\perp}. \tag{9.10}$$

Next, let us turn our attention to Eq. (9.5), which we can solve with respect to the output intensity, obtaining

$$|x_j|^2 = \rho_j^2(L) = \frac{2}{1 - R^2} \left[gL - (1 + \Delta_j^2)|\ln R| \right] = \frac{2|\ln R|}{1 - R^2} \left(A - 1 - \Delta_j^2 \right), \tag{9.11}$$

where we have used Eq. (8.19) and we have set

$$A = \frac{gL}{|\ln R|}. \tag{9.12}$$

In the following we call A the *pump parameter*. Now let us fix our attention on a particular cavity frequency that we denote ω_c, which corresponds to

$$\omega_c = 2\pi \frac{c}{\Lambda} \bar{j}, \tag{9.13}$$

where we leave the choice of $\bar{j}$ somewhat arbitrary, with the condition that ω_c is reasonably close to the atomic transition frequency ω_a. We note that the boundary condition (9.1) can be written in the form of Eq. (8.36) also in the laser case, with δ_0 defined by Eq. (8.37) and $y(t) = 0$. We can write Δ_j in the form

$$\Delta_j = \Delta - (j - \bar{j}) \frac{2\pi c/\Lambda}{\kappa' + \gamma_\perp}, \quad \Delta \equiv \Delta_{\bar{j}} = \frac{\omega_a - \omega_c}{\kappa' + \gamma_\perp}, \tag{9.14}$$

so that Eq. (9.11) becomes

$$|x_j|^2 = \frac{2|\ln R|}{1 - R^2} \left\{ A - 1 - \left[\Delta - (j - \bar{j}) \frac{2\pi c/\Lambda}{\kappa' + \gamma_\perp} \right]^2 \right\}. \tag{9.15}$$

We remark that, even if the stationary solutions are infinite in principle, only a small number of them can be accessed in a reasonable range of variation of the parameters in play. Since $|x_j|^2$ is positive by definition, each stationary solution exists for $A > 1 + \Delta_j^2$, as one sees from Eq. (9.11).

In Fig. 9.1 we plot the intensity of the stationary solutions as a function of the pump parameter A, assuming that the mode ω_c is exactly resonant with the atomic transition frequency, so that $\Delta = 0$. As explained at the end of Chapter 1, the trivial stationary solution (9.2), in which the output intensity vanishes, is stable for $A < 1$ and becomes unstable for $A > 1$, so that, when A is increased beyond the laser threshold $A = 1$, the system follows the nontrivial stationary solution (9.15) for $j = \bar{j}$ and $\Delta = 0$. When the atomic detuning parameter Δ is different from zero and ω_c is the cavity frequency closest to ω_a, the laser threshold occurs for $A_{\text{thr}} = 1 + \Delta^2$ instead of 1.

For $\Delta = 0$ the other stationary solutions for $j \neq \bar{j}$ cannot be reached by increasing A. One can reach them, instead, by varying the parameter Δ, as shown by Fig. 9.2, which exhibits the stationary solutions as functions of Δ for A fixed. As we can see from Eq. (9.15), for each value of j the stationary curve is an inverted parabola with maximum at

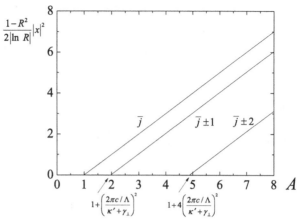

Figure 9.1 The stationary output intensity as a function of the pump parameter A for $\Delta = 0$, $\gamma_\perp = 2\pi c/\Lambda$ and $T = 0.1$ ($R = 0.9$). For the stationary solution with the lowest threshold, the laser frequency is equal to the atomic frequency. For the other stationary solutions, the laser frequency is equal to the pulled frequency of the cavity modes adjacent to the atomic frequency.

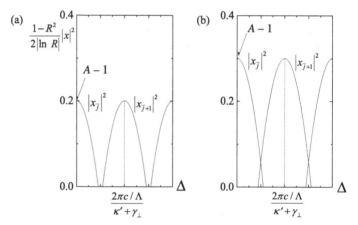

Figure 9.2 The stationary output intensity as a function of the atomic detuning parameter $\Delta = \Delta_j = (\omega_a - \omega_j)/(\kappa' + \gamma_\perp)$, where ω_j is the cavity frequency closest to the atomic transition frequency ω_a. Here (a) refers to a case with $A < 1 + [\pi c/\Lambda(\kappa' + \gamma_\perp)]^2$ ($A = 1.2$) and (b) refers to a case with $A > 1 + [\pi c/\Lambda(\kappa' + \gamma_\perp)]^2$ ($A = 1.3$). The values of the other parameters are as in Fig. 9.1.

$\Delta = (j - \bar{j})2\pi c/\Lambda(\kappa' + \gamma_\perp)$. The ordinate of the maximum is $(2|\ln R|/(1 - R^2))(A - 1)$, while each parabola vanishes for

$$\Delta = \pm\sqrt{A - 1} + (j - \bar{j})\frac{2\pi c/\Lambda}{\kappa' + \gamma_\perp}. \tag{9.16}$$

In particular, when the condition

$$\sqrt{A - 1} > -\sqrt{A - 1} + \frac{2\pi c/\Lambda}{\kappa' + \gamma_\perp} \tag{9.17}$$

is satisfied, adjacent parabolas cross each other as shown in Fig. 9.2(b). In this case there is the possibility of *bistable behavior*, i.e. the coexistence of two different stationary solutions for the same values of the parameters. The stability of the stationary solutions, and in particular the bistability, will be discussed in Chapter 22.

A general remark is that in the stationary solutions the intensity and the oscillation frequency are fixed, but the phase remains arbitrary. Equation (9.4) fixes the difference between the phase of the field at $z = L$ and that at $z = 0$, but the two values remain arbitrary. Therefore the phase of the field emitted by the laser is arbitrary and depends only on the field fluctuations when the laser is switched on. In the laser no privileged value of the phase exists. In the quantum theory one considers the fluctuations of quantum origin. Whatever initial value of the phase is chosen, a phase fluctuation changes this value and there is no mechanism that restores the previous value, so the phase performs a random walk in time, which can be described as a diffusion process. This behavior of the phase determines the coherence time of the light emitted by the laser and therefore the linewidth of the laser, which corresponds to the inverse of the coherence time [6–8].

9.2 The low-transmission limit

The low-transmission limit is a multiple limit that introduces an essential simplification into the equations. In the case of the stationary solutions of the laser, the simplification is not spectacular because the exact solutions themselves are remarkably simple. The multiple limit involves the parameters gL and T, and reads

$$gL \to 0, \quad T \to 0, \quad \text{with} \quad A = \frac{gL}{|\ln(1-T)|} \approx \frac{gL}{T} \quad \text{constant.} \qquad (9.18)$$

In this limit the variation of the stationary fields with z vanishes. As a matter of fact, the boundary conditions (9.3) and (9.4) with Eq. (9.6) guarantee that when T tends to zero (i.e. R tends to unity) the values of $\rho(0)$ and $\rho(L)$ become identical, and the same is true for $\varphi(0)$ and $\varphi(L)$. On the other hand, Eqs. (5.34) and (5.35) guarantee that the functions $\rho(z)$ and $\varphi(z)$ are monotonic in z, therefore in the limit (9.18) the field F becomes uniform in z. In the stationary state the atomic polarization and population difference $P(z)$ and $D(z)$ are given by Eqs. (5.1) and (5.2), and therefore they are also uniform. In the limit (9.18) the stationary solutions (9.15) become

$$|x_j|^2 = A - 1 - \left[\Delta - (j - \bar{j})\frac{2\pi c/\Lambda}{\kappa + \gamma_\perp}\right]^2, \qquad (9.19)$$

where κ is given by

$$\kappa = \frac{cT}{\Lambda}, \qquad (9.20)$$

instead of Eq. (9.9). The same replacement of κ' by κ must be implemented in Eq. (9.8), which becomes

$$\omega_{0j} = \frac{\kappa \omega_a + \gamma_\perp \omega_j}{\kappa + \gamma_\perp}. \qquad (9.21)$$

In the laser case, the limit $gL \to 0$ must not be introduced to make the field uniform, but to keep the pump parameter constant and finite when T tends to 0. In practical terms, the limit (9.18) amounts to assuming that

$$gL \ll 1, \quad T \ll 1, \quad \text{with} \quad A \approx \frac{gL}{T} \quad \text{arbitrary.} \tag{9.22}$$

The physical meaning of the double limit (9.18) is that the gain per pass gL is small, but since T is also small the light travels several times through the medium before escaping from the cavity, and therefore the effect of the interaction of the radiation with the medium is sizable. The condition $gL \ll 1$ can be realized by using either a thin atomic sample or a sample of low density. Another relevant limit is the *single-mode limit* in which, in addition to the limits (9.18), one applies the limit

$$\frac{\gamma_\perp}{c/\Lambda} \to 0, \quad \text{with} \quad \frac{\gamma_\perp}{cT/\Lambda} = \frac{\gamma_\perp}{k} \quad \text{constant.} \tag{9.23}$$

In this limit we can consider only the stationary solution with $j = \bar{j}$,

$$|x_{\bar{j}}|^2 = A - 1 - \Delta^2, \quad \Delta = \frac{\omega_a - \omega_c}{\kappa + \gamma_\perp}, \tag{9.24}$$

because all the other stationary solutions exist only for unreasonably large values of A. In other words, a cavity mode far from the atomic-gain line cannot support a lasing state. In this case ω_c is strictly the cavity frequency closest to the atomic transition frequency ω_a. The second part of Eq. (9.23) does not necessarily prescribe that the atomic linewidth and the cavity linewidth are of the same order of magnitude, for example in Chapter 19 we will consider the cases $\kappa \ll \gamma_\perp$ and $\kappa \gg \gamma_\perp$. Note that in the case of Fig. 9.2, in which several modes come into play simultaneously, we are clearly outside the single-mode limit, because $2\pi c/\Lambda$ is smaller than or on the order of $\kappa + \gamma_\perp$ and, for $T \ll 1$, $\kappa \ll 2\pi c/\Lambda$.

In the passive case (and also in the case of a laser with an injected signal, see Section 13.1), as we will see in Chapter 11, the low-transmission limit includes a third limit, namely

$$\delta_0 = \frac{\omega_c - \omega_0}{c/\Lambda} \to 0, \quad \text{with} \quad \theta = \frac{\delta_0}{T} \quad \text{constant.} \tag{9.25}$$

In the laser case, (9.25) is automatically satisfied for $T \to 0$ thanks to Eq. (9.6).

In the literature (see e.g. [85, 86]) the multiple limit (9.18) plus (9.25) is called the uniform field-limit or mean-field limit. In this book we prefer to call it the *low-transmission limit* because it guarantees the uniformity of the field only in the stationary states or only when it is applied together with the single-mode limit (9.23). We conclude this section with two remarks.

The first remark concerns the stationary values for the normalized atomic polarization P and for the normalized population inversion D. By taking into account that $|x|^2 = |F|^2$ and using Eqs. (9.24), (5.1) and (5.2) we obtain for the stationary state

$$P = \frac{1 - i\Delta}{A} F, \tag{9.26}$$

$$D = \frac{1 + \Delta^2}{A}, \tag{9.27}$$

where the phase of P is linked to the phase of F and remains arbitrary just like the phase of F.

The second remark is that the connection between the analysis of this chapter and that of Chapter 1 is simple but not completely straightforward even if in both cases we consider the laser, because in Chapter 1 we refer to a Fabry–Perot cavity whereas in this chapter we refer to a ring cavity. A detailed comparison will be presented in Section 19.3.1.

9.3 The analogy with second-order phase transitions

Following the theory formulated by Degiorgio and Scully [89] and by Graham and Haken [90], one can establish an interesting correspondence between the transition of the laser from below threshold to above threshold and phase transitions of second order in ferromagnets or in superconductors. Let us consider first a paramagnetic material. By using statistical mechanics one can demonstrate that the magnetization M in the presence of an external magnetic field H depends on H and on the absolute temperature T according to the law

$$M = N\mu_B \tanh\left(\frac{\mu_B H}{k_B T}\right), \tag{9.28}$$

where N denotes the atomic density, μ_B is the Bohr magneton and k_B is Boltzmann's constant. According to Eq. (9.28) the magnetization disappears when the magnetic field vanishes. In order to describe ferromagnetic materials in which, instead, the magnetization persists even in the absence of an applied magnetic field, Weiss introduced the hypothesis that in these materials there exist small domains in which the interaction among the spins of the single atoms leads to an alignment of their magnetic dipoles, generating an internal molecular field H_m, the intensity of which can surpass by many orders of magnitude the intensity of the external magnetic field. One assumes that this field is proportional to the magnetization

$$H_m = \lambda M, \tag{9.29}$$

with a proportionality constant λ that depends on the material. As a consequence the total internal field is $H + \lambda M$ and Eq. (9.28) becomes an equation for M,

$$M = N\mu_B \tanh\left[\frac{\mu_B(H + \lambda M)}{k_B T}\right], \tag{9.30}$$

the solutions of which can be obtained by intersecting the curve $M(\alpha)$,

$$\frac{M}{N\mu_B} = \tanh\alpha, \tag{9.31}$$

and the straight line

$$\frac{M}{N\mu_B} = \frac{1}{\lambda N\mu_B}\left(\frac{k_B T}{\mu_B}\alpha - H\right), \tag{9.32}$$

which intercepts the ordinate axis at the point $-H/(\lambda N \mu_B)$ and has slope proportional to T. In the absence of an external magnetic field the straight line passes through the origin. As a consequence one has spontaneous magnetization if the slope of the straight line is smaller than the slope of the curve (9.31) at the origin, which is 1,

$$\frac{k_B T}{\lambda N \mu_B^2} < 1, \tag{9.33}$$

i.e. if the temperature is lower than the critical Curie temperature,

$$T_C = \frac{\lambda N \mu_B^2}{k_B}. \tag{9.34}$$

Therefore for $T < T_C$ the material behaves as a ferromagnet; and for $T > T_C$ it behaves as a paramagnet. At $T = T_C$ there is a phase transition, which can be thought of as a transition from an ordered state (aligned spins) to a disordered state. One can obtain a simple expression for the residual magnetization for $H = 0$ as a function of temperature by inverting Eq. (9.30) with $H = 0$ and by developing the inverse tanh up to third order in M, assuming that $M \ll N \mu_B$,

$$\frac{\lambda \mu_B}{k_B T} M = \tanh^{-1}\left(\frac{M}{N \mu_B}\right) \approx \frac{M}{N \mu_B} + \frac{1}{3}\left(\frac{M}{N \mu_B}\right)^3, \tag{9.35}$$

from which it follows that

$$0 = \left(\frac{k_B T}{N \mu_B^2} - \lambda\right) M + \frac{k_B T}{3 N^3 \mu_B^4} M^3 = \frac{k_B T}{N \mu_B^2} M \left[\frac{T - T_C}{T} + \frac{M^2}{3 N^2 \mu_B^2}\right]. \tag{9.36}$$

For $T > T_C$ the expression in square brackets is always positive and therefore the only possible solution is $M = 0$, i.e. the material behaves as a paramagnet. For $T < T_C$ there are two symmetrical solutions different from zero, i.e. the material exhibits a residual magnetization in the absence of an external magnetic field, hence we have a ferromagnet. We can write

$$M = \begin{cases} \pm N \mu_B [3(T_C - T)/T]^{1/2} \approx \pm 3 N \mu_B (1 - T/T_C)^{1/2} & T < T_C, \\ 0 & T > T_C. \end{cases} \tag{9.37}$$

where we replace T by T_C in the denominator because the cubic approximation (9.35) holds when M is small, which is true for $T \approx T_C$.

On the other hand, we know from Eq. (9.24) that the stationary solution for the laser in the single-mode approximation is

$$x = \begin{cases} 0 & A < 1 + \Delta^2, \\ \pm [A - (1 + \Delta^2)]^{1/2} & A > 1 + \Delta^2. \end{cases} \tag{9.38}$$

where for the sake of analogy we have taken the output field x real, selecting the two values 0 and π for the arbitrary phase of x. The parameter which in the case of the laser plays a role analogous to that of the temperature in the ferromagnet is σ, the population inversion per atom prescribed by the laser pump in the absence of an electric field (see Eq. (4.14)).

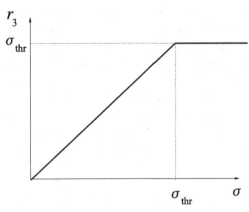

Figure 9.3 Behavior of the population inversion per atom r_3 as a function of the parameter σ.

This parameter and the pump parameter A are proportional; by using the fact that $A = gL/(1 - R)$ with g defined by Eq. (4.34) we obtain

$$\sigma = \frac{\hbar \gamma_\perp n_B^2}{2\pi k_0 N d^2} \frac{1 - R}{L} A. \tag{9.39}$$

Note that in this subsection we indicate the transmissivity of the beam splitter by $(1 - R)$ in order to avoid confusion with the temperature. If we now in Eq. (9.39) replace A by its threshold value $A_\text{thr} = 1 + \Delta^2$, we obtain the expression for σ_thr. We have immediately that

$$\frac{A}{1 + \Delta^2} = \frac{\sigma}{\sigma_\text{thr}}. \tag{9.40}$$

Hence we can cast Eq. (9.38) in the form

$$x = \begin{cases} 0 & \sigma < \sigma_\text{thr}, \\ \pm [1 + \Delta^2]^{1/2} (\sigma/\sigma_\text{thr} - 1)^{1/2} & \sigma > \sigma_\text{thr}. \end{cases} \tag{9.41}$$

We see therefore that in the case of the laser the field plays the role of the order parameter. Equation (9.41) exhibits a clear analogy with Eq. (9.37), which means that there is an analogy between the behavior of the laser around threshold and the behavior of the ferromagnet around the critical temperature. There is a change of sign in the argument of the square root because the laser case corresponds to a situation of negative temperature. In the case of the ferromagnet there is a transition from disorder to order when the temperature is decreased below the critical temperature; in the laser case there is a transition from incoherence to coherence when the laser threshold is passed.

We remind the reader that in a second-order phase transition the order parameter is continuous around the critical temperature, but displays a discontinuity in the derivative. In a first-order phase transition, instead, the order parameter itself displays a discontinuity.

The fact that one can establish this analogy must not come as a surprise, because in both cases the system is described by a self-consistent field theory. In the laser each atom under the action of the macroscopic electric field produced by the other atoms develops an oscillating dipole. The macroscopic polarization given by the sum of the dipoles modifies the macroscopic electric field, which is calculated in a self-consistent way. This is closely similar to the case of a ferromagnetic material in which each spin is subjected to the magnetic field produced by the other spins and aligns to it, thus reinforcing the total magnetic field.

Coming back to the laser case, from Eq. (4.33) we have that $r_3 = \sigma D$. Below threshold $D = 1$ (see Eq. (9.2)), so $r_3 = \sigma$. Above threshold D is given by Eq. (9.27) and, by using Eq. (9.40), we obtain that $r_3 = \sigma_{thr}$, so the graph of r_3 as a function of σ is as given by Fig. 9.3, and shows a discontinuity in the derivative at threshold.

Hence, r_3 grows up to threshold, and beyond threshold it remains clamped to its threshold value. Beyond threshold the energy provided by the pump does not further increase the inversion, but is converted into laser emission.

10

The adiabatic elimination principle

The adiabatic elimination principle [91, 92] is universal in the area of nonlinear dissipative dynamical systems, and its importance arises from the fact that it allows one to reduce the number of the equations which govern the dynamics of the system. We introduce it at this stage in the book in order to be able to describe the optical pumping mechanisms which can be exerted to attain population inversion. However, this principle will be applied repeatedly in the following chapters for other purposes.

In the first subsection we describe the principle in general, while the following subsections concern the issue of optical pumping. We show first that it is not possible to obtain population inversion between the two levels of the lasing transition if the pump involves only these two levels. Next we illustrate the three-level pumping scheme and the four-level pumping scheme. We demonstrate that, by adiabatically eliminating the probabilities of the additional energy levels, one arrives at an equation identical to Eq. (4.12), which, for appropriate ranges of values of the parameters involved in the pumping, allows one to obtain population inversion.

10.1 General formulation of the principle

In a dissipative dynamical system the dynamical variables undergo relaxation processes associated with suitable rate constants such as, for example, the rate $\gamma_\perp$ for the normalized atomic polarization P and the rate $\gamma_\parallel$ for the normalized population difference D. The inverses of these rates provide the time scales which characterize the evolution of each variable. Therefore we can distinguish a set of *fast variables*, which evolve over short time scales and are therefore characterized by large relaxation rates, and a set of *slow variables*, which display small relaxation rates. A remarkable simplification in the analysis of dynamical systems is obtained by using the technique of *adiabatic elimination* of fast variables, which Haken considered as a basis for the discipline which he formulated and called Synergetics [91, 92]. The method is based on the identification of two distinct stages in the time evolution of the system. The first develops over the fast time scales which characterize the fast variables; the second develops over the slow time scales. During the first stage only the fast variables evolve significantly, while the slow variables remain practically unchanged. Independently of their initial values, the fast variables tend to assume, at the end of the first stage, values that are uniquely determined by the values of the slow variables. As a result of the first stage one has a functional dependence of the fast variables on the slow variables, which can be obtained mathematically by setting the derivatives of the slow variables equal to zero, a sort of "stationary state" of the fast variables conditioned by the values of the slow variables. This procedure leads to an algebraic system of equations that can be solved with respect to the fast variables and provides their functional dependence on the slow variables. Haken calls this process "slaving" of the fast variables by the slow variables. Next, one introduces into the dynamical equations for the slow variables the functional expressions of the fast variables, and in this way one obtains a closed set of dynamical equations for the slow variables. This set, which has a smaller dimension than the original set of equations, governs the evolution of the system during the second stage, which occurs over the long time scales. In this way one has brought about the elimination of the fast variables, which follow adiabatically, i.e. without retardation, the evolution of the slow variables.

In order to illustrate how the adiabatic elimination principle works in practice, let us consider the simplest case of a dynamical system described by two dynamical variables x_1 and x_2 that obey the equations

$$\dot{x}_1 = \gamma_1 f_1(x_1, x_2), \tag{10.1}$$

$$\dot{x}_2 = \gamma_2 f_2(x_1, x_2), \tag{10.2}$$

where f_1 and f_2 are in general nonlinear functions. Let us assume that x_2 is the fast variable, i.e. that $\gamma_2 \gg \gamma_1$. By neglecting what happens in the first stage of the evolution let us go directly to consider the second, which is characterized by the temporal scale $1/\gamma_1$. By introducing the dimensionless time $t_1 = \gamma_1 t$ we recast the equations in the forms

$$\frac{dx_1}{dt_1} = f_1(x_1, x_2), \tag{10.3}$$

$$\varepsilon \frac{dx_2}{dt_1} = f_2(x_1, x_2), \tag{10.4}$$

with $\varepsilon = \gamma_1/\gamma_2$. Since $\varepsilon \ll 1$ while all other quantities are of order unity, we can neglect the l.h.s. term in Eq. (10.4). By solving the algebraic equation $f_2(x_1, x_2) = 0$ with respect to the fast variable x_2 we get $x_2 = x_2(x_1)$ and then a closed dynamical equation for the slow variable x_1,

$$\frac{dx_1}{dt_1} = f_1[x_1, x_2(x_1)]. \tag{10.5}$$

An alternative way of proceeding, which illustrates also what happens in the first stage of the evolution, including the loss of memory of the initial conditions for the fast variables, is the following. In this context, it is useful to rewrite the dynamical equations in such a way as to make evident the relaxation terms of the two variables,

$$\dot{x}_1 = -\gamma_1 x_1 + g_1(x_1, x_2), \tag{10.6}$$

$$\dot{x}_2 = -\gamma_2 x_2 + g_2(x_1, x_2), \tag{10.7}$$

where we have set $g_i(x_1, x_2) = \gamma_i f_i(x_1, x_2) + \gamma_i x_i$, $i = 1, 2$. We can cast the solution of Eq. (10.7) in the form of an integral equation as follows:

$$x_2(t) = x_2(0)e^{-\gamma_2 t} + \int_0^t e^{-\gamma_2(t-t')} g_2[x_1(t'), x_2(t')] dt'. \tag{10.8}$$

We see that for times on the order of $1/\gamma_2$ there is memory of the initial condition $x_2(0)$, but for longer times this is lost because the first term on the r.h.s. vanishes. In this limit, by passing to the integration variable $\tau = t - t'$ we can write

$$x_2(t) = \int_0^t e^{-\gamma_2 \tau} g_2[x_1(t - \tau), x_2(t - \tau)] d\tau. \tag{10.9}$$

The integrand still contains memory of the past evolution of the system because it depends on the values of the dynamical variables for times $t - \tau$ with $0 \le \tau \le t$. However, in the integral the function g_2 is multiplied by an exponential term that tends to zero for values of the delay τ much larger than $1/\gamma_2$. Therefore the meaningful contribution to the integral arises for $0 < \tau \lesssim 1/\gamma_2$. On the other hand, over this time scale the slow variable x_1 is practically constant and the same can be said about the fast variable x_2 because it is slaved by x_1. Hence we can conclude that g_2 varies slowly in the time interval $t - \tau$ in which the exponential is significantly different from 0, and therefore we can take g_2 out of the integral, evaluating it at time t,

$$x_2(t) = g_2[x_1(t), x_2(t)] \int_0^t e^{-\gamma_2 \tau} d\tau. \tag{10.10}$$

For $t \gg 1/\gamma_2$ we can safely extend the integral to infinity, obtaining the equation

$$x_2(t) = \frac{g_2[x_1(t), x_2(t)]}{\gamma_2}, \tag{10.11}$$

which coincides with the equation $f_2(x_1, x_2) = 0$ because of the definition $g_2(x_1, x_2) = \gamma_2 f_2(x_1, x_2) + \gamma_2 x_2$.

We observe finally that this technique does not work in all cases, because it may happen that the equation $f_2(x_1, x_2) = 0$ does not provide a functional relation between x_2 and x_1. In these cases one must utilize more sophisticated techniques to realize the adiabatic elimination, such as, for example, the central-manifold technique [93].

10.2 Adiabatic elimination of the atomic polarization in the Bloch equations. Limits of the optical pumping between two levels

A common procedure to realize population inversion consists in optically pumping the medium with an external electric field. Let us consider first an external field resonant with the laser transition.

Let us return to Eqs. (4.36) and (4.37) in the laser case $\sigma > 0$ assuming perfect resonance between the frequency of the field and the atomic transition frequency, so that $\Delta = 0$, and assuming that the normalized field envelope is constant in time and real. Under these conditions, if P is initially real it remains real throughout the time evolution, and the dynamical equations reduce to

$$\dot{P} = \gamma_\perp (FD - P), \tag{10.12}$$

$$\dot{D} = -\gamma_\parallel (FP + D - 1), \tag{10.13}$$

Next, let us assume that $\gamma_\perp \gg \gamma_\parallel$, so that P becomes the fast variable and D the slow variable. Hence by setting $\dot{P} = 0$ we obtain the functional expression for P,

$$P = FD. \tag{10.14}$$

By inserting Eq. (10.14) into Eq. (10.13) we adiabatically eliminate the variable P, obtaining the following closed equation for the slow variable D:

$$\dot{D} = -\gamma_\parallel [D(1 + F^2) - 1] = -\gamma_\parallel' \left(D - \frac{1}{1 + F^2} \right), \tag{10.15}$$

with

$$\gamma_\parallel' = \gamma_\parallel (1 + F^2). \tag{10.16}$$

Let us focus on the strong saturation limit in which $F^2 \gg 1$ so that Eq. (10.15) becomes

$$\dot{D} = -\gamma_\parallel' D, \qquad \gamma_\parallel' = \gamma_\parallel F^2. \tag{10.17}$$

Note that the damped Rabi oscillations of Section 4.3 exist only when $\gamma_\perp$ and $\gamma_\parallel$ are of the same order of magnitude. In the adiabatic elimination limit, instead, Rabi oscillations, which arise from the coherence of atom–field interaction, disappear, the time evolution of D is monotonic and, in the strong-saturation limit, D decays exponentially to zero. By taking Eq. (4.33) into account we see that the population inversion per atom r_3 also tends to zero and therefore, since $r_3 = p_1 - p_2$, the two levels are equally populated. This demonstrates that it is not possible to achieve population inversion if the pumping involves only the two levels 1 and 2 of the laser transition.

For the following section it is useful to recast Eq. (10.17) in the form of coupled rate equations for the probabilities p_1 and p_2. Since r_3 is proportional to D, it obeys the same equation, (10.17), as D. By taking into account that $r_3 = p_1 - p_2 = 2p_1 - 1 = 1 - 2p_2$ we obtain the rate equations

$$\dot{p}_1 = wp_2 - wp_1, \qquad \dot{p}_2 = -wp_2 + wp_1, \tag{10.18}$$

with

$$w = \frac{\gamma_\parallel'}{2} = \frac{\gamma_\parallel}{2} F^2. \tag{10.19}$$

By comparing this with the rate equations (4.9) and (4.10), we see that the adiabatic elimination of the atomic polarization, combined with the strong-saturation limit $F^2 \gg 1$, leads to rate equations with identical upward and downward transition rates between the two levels.

10.3 The three-level optical-pumping scheme

In order to achieve population inversion it is necessary that the pump field is resonant with two levels, at least one of which is different from the two levels of the laser transition [6].

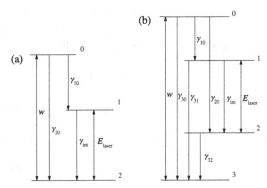

Figure 10.1   Pumping schemes involving (a) three energy levels and (b) four energy levels.

The energy-level scheme for a three-level laser is illustrated in Fig. 10.1(a). Let us indicate the three levels by 0, 1 and 2, and let us assume that $E_0 > E_1 > E_2$. The laser transition is between levels 1 and 2, while the pumping involves levels 0 and 2. Level 0 decays spontaneously to level 1 and to level 2 with rates γ_{10} and γ_{20}, respectively. Level 1 decays spontaneously to level 2, and we call the rate of this decay γ_{int}, to indicate that it is a decay that occurs internally between the two levels of the laser transition. This scheme describes well some lasers such as the ruby laser (the first laser that was realized, in 1960 by Maiman [94]) and the erbium-doped fiber laser. The optical pumping produces a transition probability w from level 2 to level 0 and an equal transition probability from level 0 to level 2, similarly to what is described by Eqs. (10.18).

In order to attain population inversion between levels 1 and 2 of the laser transition it is necessary that the condition

$$\gamma_{10} \gg \gamma_{20}, \; \gamma_{int}, \; w \qquad (10.20)$$

is satisfied, because in this case the atoms which are transferred by optical pumping from level 2 to level 0 decay rapidly to level 1, where they stay for a much longer time because γ_{int} is much smaller than γ_{10}. In this way level 1 can attain a population larger than level 2, which is depopulated by the optical pumping.

Let us now describe mathematically the dynamics of the populations in terms of the following three rate equations for the probabilities p_0, p_1 and p_2 of the three levels:

$$\dot{p}_0 = -(\gamma_{10} + \gamma_{20} + w)p_0 + wp_2, \qquad (10.21)$$

$$\dot{p}_1 = -\gamma_{int}p_1 + \gamma_{10}p_0, \qquad (10.22)$$

$$\dot{p}_2 = -wp_2 + \gamma_{int}p_1 + (w + \gamma_{20})p_0. \qquad (10.23)$$

Obviously $\dot{p}_0 + \dot{p}_1 + \dot{p}_2 = 0$ because the sum of the three probabilities is constant and equal to 1. The decay constants of the three variables are $\gamma_{10} + \gamma_{20} + w$, γ_{int} and w, respectively. From condition (10.20) we see that the variable p_0 is much faster than the others. Hence we eliminate it adiabatically by setting $\dot{p}_0 = 0$, which gives

$$p_0 = \frac{w}{\gamma_{10} + \gamma_{20} + w} p_2 \simeq \frac{w}{\gamma_{10}} p_2. \qquad (10.24)$$

Therefore as a consequence of condition (10.20) the occupation probability of level 0 is very small; in practice the pumping transfers atoms from level 2 to level 1. Next we insert Eq. (10.24) into Eqs. (10.22) and (10.23), obtaining

$$\dot{p}_1 = wp_2 - \gamma_{\text{int}}p_1, \tag{10.25}$$

$$\dot{p}_2 = -w\left(1 - \frac{w + \gamma_{20}}{\gamma_{10}}\right)p_2 + \gamma_{\text{int}}p_1. \tag{10.26}$$

By taking Eq. (10.20) into account again, we see that Eq. (10.26) reduces to

$$\dot{p}_2 = -wp_2 + \gamma_{\text{int}}p_1, \tag{10.27}$$

and it is immediately evident that Eqs. (10.27) and (10.25) are identical to Eqs. (4.9) and (4.10) if we take $\gamma_\uparrow = w$ and $\gamma_\downarrow = \gamma_{\text{int}}$.

Thus, we recover Eq. (4.12) for the inversion $r_3 = p_1 - p_2$, with $\gamma_\| = w + \gamma_{\text{int}}$ and

$$\sigma = \frac{w - \gamma_{\text{int}}}{w + \gamma_{\text{int}}}. \tag{10.28}$$

Therefore one has population inversion, provided that $w > \gamma_{\text{int}}$.

10.4 The four-level optical-pumping scheme

Let us now turn to a four-level pumping scheme [6], in which the laser transition occurs between levels 1 and 2 as before, but the pumping involves also levels 0 and 3, with $E_0 > E_1 > E_2 > E_3$. A scheme of this type, described in Fig. 10.1(b), applies for example to the case of a neodymium laser. In this case, in addition to the pumping rate w and the decay rates γ_{int} from level 1 to level 2, and γ_{10} and γ_{20} from level 0 to levels 1 and 2, respectively, we must consider also the rates γ_{32}, γ_{31} and γ_{30}, which describe the decay to the lowest level, level 3, from, respectively, levels 0, 1 and 2, respectively.

The rate equations for the four populations are

$$\dot{p}_0 = -(w + \gamma_{30} + \gamma_{20} + \gamma_{10})p_0 + wp_3, \tag{10.29}$$

$$\dot{p}_1 = -(\gamma_{31} + \gamma_{\text{int}})p_1 + \gamma_{10}p_0, \tag{10.30}$$

$$\dot{p}_2 = -\gamma_{32}p_2 + \gamma_{\text{int}}p_1 + \gamma_{20}p_0, \tag{10.31}$$

$$\dot{p}_3 = -wp_3 + \gamma_{32}p_2 + \gamma_{31}p_1 + (w + \gamma_{30})p_0. \tag{10.32}$$

Let us assume that the rates γ_{10} and γ_{32} are much larger than all others. Physically this means that level 0 (the upper level of the pump transition) quickly becomes empty, decaying to level 1 (the upper level of the laser transition), and that level 2 (the lower level of the laser transition) quickly becomes empty, decaying to level 3 (the lower level of the pump transition). These conditions ensure the population inversion between the two levels of the laser transition, because they grant an accumulation of population in the upper level, while the lower level is continuously emptied.

In this limit the probabilities p_0 and p_2 can be adiabatically eliminated, giving

$$p_0 = \frac{w}{w + \gamma_{30} + \gamma_{20} + \gamma_{10}} p_3 \approx \frac{w p_3}{\gamma_{10}}, \tag{10.33}$$

$$p_2 = \frac{\gamma_{\text{int}} p_1 + \gamma_{20} p_0}{\gamma_{32}}. \tag{10.34}$$

These equations show that p_0 and p_2 are very small. Hence we can think that the population is concentrated in levels 1 and 3, and write

$$1 = p_1 + p_3. \tag{10.35}$$

Furthermore, we can identify $r_3 = p_1 - p_2$ with p_1. By taking into account Eqs. (10.30), (10.33) and (10.35) we can therefore write

$$\dot{r}_3 = \dot{p}_1 = -(\gamma_{31} + \gamma_{\text{int}})r_3 + w p_3 = -(\gamma_{31} + \gamma_{\text{int}})r_3 + w(1 - r_3)$$

$$= w - (w + \gamma_{31} + \gamma_{\text{int}})r_3 = -\gamma_\parallel(r_3 - \sigma), \tag{10.36}$$

with

$$\gamma_\parallel = w + \gamma_{31} + \gamma_{\text{int}}, \tag{10.37}$$

$$\sigma = \frac{w}{w + \gamma_{31} + \gamma_{\text{int}}}, \tag{10.38}$$

in agreement with Eq. (4.12). In this case σ is always positive, i.e. one always has population inversion irrespective of the value of the pump rate w. We can note that both in the case of the three-level laser and in the case of the four-level laser the stationary value of the population difference σ tends to 1 for large values of w.

11 A nonlinear passive ring cavity: optical bistability

In Chapter 9 we discussed the case of a nonlinear cavity containing an active medium. In this framework, the outstanding phenomenon which arises is laser emission, and we calculated the stationary solutions for the ring laser. In this chapter, instead, we turn our attention to the passive nonlinear cavity and in this framework the outstanding phenomenon is *optical bistability*, i.e. the existence of two distinct stable stationary states that coexist for the same fixed values of the parameters in play. This phenomenon can arise also in the case of an active system contained in a cavity, as we have already noted in Section 9.1 and as we will discuss further in connection with the laser with injected signal, but it became popular from the studies of nonlinear passive optical cavities in the 1970s and 1980s.

In contrast to the laser case, in the passive configuration we use an input coherent field, which is injected into the cavity. We discussed this case in Chapter 8 and derived a general equation for the transmission of the cavity (Eq. (8.23)), which we used for an empty cavity and for a linear cavity in Sections 8.4 and 8.5, respectively. In this chapter we will discuss the full nonlinear configuration.

It is customary to identify two extreme cases of optical bistability called *absorptive optical bistability* and *dispersive* or *refractive optical bistability*. In both cases an essential ingredient to obtain the bistable behavior is the feedback action exerted by the mirrors of the cavity. In the purely absorptive configuration, in which there is exact resonance among the frequencies of the input field, of the atoms and of the cavity, the other essential ingredient is the saturable nonlinearity of the absorption. In the dispersive case, instead, the main feature is the nonlinearity of the refractive index.

These two cases will be illustrated in Sections 11.1 and 11.2, respectively, both "exactly" and in the low-transmission limit. In Section 11.3 we will analyze the general mixed absorptive/dispersive case for two-level atoms.

In this chapter we will illustrate also the analogy between optical bistability and first-order phase transitions in Section 11.1.1 and some applied aspects of optical bistability in Section 11.4.

11.1 Absorptive optical bistability

In general, the name absorptive optical bistability is used to designate the case in which the input field is exactly on resonance with the atomic transition frequency i.e. $\omega_0 = \omega_a$, so that the atomic detuning parameter Δ is zero. An especially simple treatment is possible under complete resonance conditions, i.e. when the input frequency coincides also with

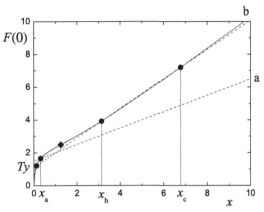

Figure 11.1 The normalized field $F(0)$ at $z = 0$ as a function of the field $F(L) \equiv x$ at $z = L$ (the transfer function of the atomic medium at steady state) for $\alpha L = 3$ (solid curve). For $R = 0$ one has $F(0) = y$. The quantities x and y are proportional to the transmitted and incident fields, respectively. For a generic R, the function $x = x(y)$ is obtained by intersecting the curve with the straight line $F(0) = Ty + Rx$ (broken curve). The intersections are represented by circles, whereas the diamond is the inflection point of the transfer function. (a) $Ty = 1.5$, $R = 0.5$; (b) $Ty = 1.1$, $R = 0.9$.

the empty cavity frequency ω_c, so that also the parameter δ_0 defined by Eq. (8.37) is equal to zero. Under such conditions the stationary equation (5.33) and the boundary condition (8.36) when y and F are time-independent read

$$\frac{dF}{dz} = -\alpha \frac{F}{1 + F^2}, \tag{11.1}$$

$$F(0) = Ty + RF(L), \tag{11.2}$$

where the absorption parameter α is defined by Eq. (4.38) and we have assumed that the input field y is real (and positive, for the sake of definiteness), so that no imaginary unit appears in Eqs. (11.1) and (11.2), hence also $F(z)$ is real and we have omitted the modulus square in the denominator of Eq. (11.1).

Equation (11.1) gives, similarly to Eq. (5.36),

$$\ln \left(\frac{F(0)}{x} \right) + \frac{1}{2} \left[F(0)^2 - x^2 \right] = \alpha L, \tag{11.3}$$

where we have taken Eq. (8.17) into account. By combining this with Eq. (11.2) and remembering that $R = 1 - T$ we obtain [85]

$$\ln \left[1 + T \left(\frac{y}{x} - 1 \right) \right] + \frac{x^2}{2} \left\{ \left[1 + T \left(\frac{y}{x} - 1 \right) \right]^2 - 1 \right\} = \alpha L. \tag{11.4}$$

This equation gives an exact relation between the transmitted field x and the incident field y. It depends on the two parameters αL and T. The structure of Eq. (11.4) can be intuitively understood by considering a graphical representation of Eqs. (11.2) and (11.3) as shown by Fig. 11.1. The steady-state values of x are the intersections of the straight line (11.2) and the curve (11.3). The first is the boundary condition of the cavity. The second is the

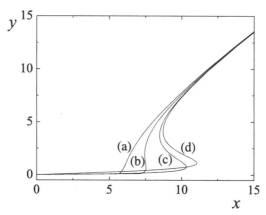

Figure 11.2   A plot of transmitted light versus incident light at steady state for $C = \alpha L/(2T)$ fixed equal to 10, and different values of αL and T. For $\alpha L \to 0$ one approaches the behavior predicted by the low-transmission limit (see Eq. (11.9)). (a) $\alpha L = 20, T = 1$; (b) $\alpha L = 10, T = 0.5$; (c) $\alpha L = 2, T = 0.1$; (d) low-transmission limit, $C = 10$.

transfer function of the medium, which expresses the field at $z = 0$ as a function of the field at $z = L$ (and vice versa). It has neither maxima nor minima, but has an inflection point. The angular coefficient R_c of the tangent at the inflection point is such that $0 < R_c < 1$. R_c depends only on αL. For $R < R_c$ there is only one intersection point for all values of y. For $R > R_c$ there is a range of values of y in correspondence with which one finds three intersection points $x_a < x_b < x_c$. Points x_b are unstable; hence, there is a bistable situation. If we plot the steady-state solutions x as a function of the incident field y, we obtain an S-shaped curve (Fig. 11.2), which, as we will see in the remainder of this section, leads to a hysteresis cycle.

From this analysis we see that bistability arises from the combined action of the nonlinear transfer of the medium (Eq. (11.3)) and from the feedback of the mirrors (Eq. (11.2)) [86]. This feedback action is essential, because, as one sees from Figs. 11.1 and 11.2, there is no bistability for $R = 0$ (i.e. $T = 1$).

Equations (11.2) with $F(L) = x$ and (11.3) can be solved parametrically using $\beta = F(0)/x$ as a parameter. From Eq. (11.3) one obtains the function $x(\beta)$. From Eq. (11.2) divided by x (which is equal to $F(L)$) one obtains the expression of y as a function of β and $x(\beta)$. The functions $x(\beta)$ and $y(\beta)$, taken together, provide the parametric expression of the function $x(y)$.

In order to simplify the equations, let us introduce the low-transmission limit which, similarly to Eq. (9.18), is defined by the double limit [85]

$$\alpha L \to 0, \quad T \to 0, \quad \text{with} \quad 2C = \frac{\alpha L}{T} \quad \text{constant}, \tag{11.5}$$

while also x and y are kept constant in the limit. C is called the *bistability parameter*. Note that in this case the uniformity of the field along the atomic medium is guaranteed by the condition $\alpha L \ll 1$ rather than by the condition $T \ll 1$ (as happens, instead, in the laser case). As a matter of fact, from Eq. (11.1) we have

$$|F(z) - F(0)| = \alpha \int_0^z dz' \, \frac{F(z')}{1 + F^2(z')} < \alpha z \leq \alpha L, \tag{11.6}$$

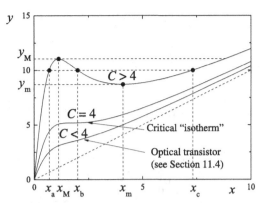

Figure 11.3 A plot of the state equation (11.9) for purely absorptive bistability for different values of the bistability parameter C.

so that for $\alpha L \ll 1$ we conclude that $|F(z) - F(0)| \ll 1$. If we set

$$T\left(\frac{y}{x} - 1\right) = \epsilon \tag{11.7}$$

we have that $\epsilon \to 0$ in the limit (11.5) and, keeping only the terms of order ϵ in Eq. (11.4), we obtain

$$\epsilon(1 + x^2) = \alpha L \to \epsilon = \frac{\alpha L}{1 + x^2}, \tag{11.8}$$

which, using Eq. (11.7), leads to the steady-state relation [95–97]

$$y = x + \frac{2Cx}{1 + x^2}. \tag{11.9}$$

This equation expresses the incident field as a function of the transmitted field. In order to obtain the inverse function $x = x(y)$ in which we are interested, the x and y axes must be swapped.

If we calculate dy/dx and look for the maxima and minima of the function $y(x)$, the condition $dy/dx = 0$ leads to a quadratic equation for x^2, which has two real solutions for $C > 4$, which is the *bistability condition*. The plot of the function $y(x)$ is shown in Fig. 11.3. For $C = 4$ the maximum and the minimum coalesce into an inflection point with a horizontal tangent.

Figure 11.4 shows the curve of the transmitted field x vs. the incident field y for $C = 20$. When $C > 4$ there is a range $y_m < y < y_M$ for which the transmitted field displays three possible stationary values. As we will demonstrate in Part II (Section 19.2.2) of this book, the stationary states along the segment of the steady-state curve with negative-slope are unstable, as one can easily guess from the circumstance that, following this negative-slope segment, an increase of the incident field is accompanied by a decrease of the transmitted field.

Imagine that we start from the value $y = 0$ of the input field. If we increase y, x increases, following the lower transmission branch of the steady-state curve. When y becomes larger than y_m there are two stable states, but for reasons of continuity the system remains in

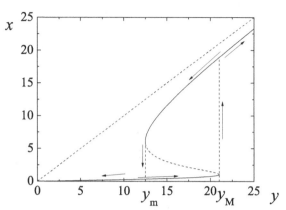

Figure 11.4   The hysteresis cycle of the transmitted field x as a function of the input field y according to Eq. (11.9) for $C = 20$. The broken segment of the curve is unstable.

the lower transmission branch up to the value $y = y_M$, where the system is forced to discontinuously jump to the upper branch following the upward arrow. If we increase y further, the system follows the upper transmission branch of the steady-state curve. At this stage, if we start decreasing the incident field, the transmitted field decreases, following the upper branch down to the point $y = y_m$, where the system discontinuously jumps to the lower branch following the downward arrow. In this way, when the input field is varied up and down the system covers a *hysteresis cycle*.

In purely absorptive optical bistability the feature which distinguishes the upper branch from the lower branch is that the internal field intensity is unsaturated over most of the lower branch and saturated over most of the upper branch. The mechanism which gives rise to switching and hysteresis in absorptive optical bistability can be described as follows. As we see from Eq. (11.1), absorptive optical bistability is characterized by an intensity-dependent effective absorption coefficient

$$\alpha_{\text{eff}} = \frac{\alpha}{1 + F^2} \, . \tag{11.10}$$

In the lower transmission branch, when the input field is increased, the intracavity field F increases. As shown by Eq. (11.10), this makes the absorption decrease, which in turn produces an increase of the intracavity field, which in turn makes the absorption decrease and so on. Hence there is a positive feedback loop that leads the intensity to saturate, producing the switching to the upper branch. On the other hand, when the system is in the upper transmission branch and the input field is decreased, the field internal to the cavity is already strong enough to keep the absorber saturated, and therefore the transmitted light switches "off" at an incident field lower than that necessary to switch "on", thereby producing hysteresis.

11.1.1 The analogy with first-order phase transitions

Optical bistability is a phenomenon that occurs in a system driven far from thermal equilibrium by the external input field. However, it bears an immediate analogy with first-order phase transitions in equilibrium systems [96, 97], as one has in a ferromagnet or in the liquid–vapor phase transition. As a matter of fact, the curves $y(x)$ obtained by varying C in Fig. 11.3 are analogous to the van der Waals curves for the liquid–vapor phase transition, with y, x and C playing the role of pressure, volume and temperature, respectively. In particular, the curve for $C = 4$ is analogous to the critical isotherm. The analogy of optical bistability with first-order phase transitions was further developed in the framework of the quantum-statistical treatment of optical bistability, with the introduction of a generalized Maxwell rule [86, 98, 99] for the absorptive case.

11.2 Dispersive optical bistability

The bistable behavior is said to be *dispersive* or *refractive* when it arises from the dispersive part of the nonlinear susceptibility.

The simplest configuration is that of a cavity containing a Kerr medium as nonlinear material. The steady-state behavior in the medium is governed by Eq. (7.9).

In the case of a two-level medium, the dispersive case arises when the modulus of the atomic detuning parameter Δ is very large, because, as one sees from Eq. (5.26), the dispersive imaginary part of the susceptibility is larger than the absorptive real part by a factor of Δ. In this limit, one can approximate

$$\tilde{\chi}(\omega_0) = -\frac{i}{\Delta} + \frac{i}{\Delta} \frac{|F|^2}{\Delta^2}, \tag{11.11}$$

where we have neglected the real part of the susceptibility, we have neglected 1 with respect to Δ^2 and we have expanded the imaginary part of the susceptibility up to first order in $|F|^2/\Delta^2$. In this way one obtains a model that is simply related to Eq. (7.9).

The two-level model in the dispersive limit will be precisely analyzed in Section 13.3. Here we focus on the Kerr-medium model Eq. (7.9). As in the two-level case, it is convenient to introduce normalized variables for the input and output fields, similarly to Eqs. (8.15):

$$\tilde{F}(z) = E_0(z), \quad \tilde{y} = \frac{E_I}{\sqrt{T}}, \quad \tilde{x} = \frac{E_T}{\sqrt{T}} e^{-2\pi i (\Lambda - L')/\lambda_0}. \tag{11.12}$$

Note that $\tilde{F}(z)$, $\tilde{y}$ and $\tilde{x}$ are not dimensionless in this case, but Eqs. (8.18) and (8.19) and (8.21)–(8.23), with F, x and y replaced by $\tilde{F}$, $\tilde{x}$ and $\tilde{y}$, respectively, are still valid. By using Eqs. (7.13), (7.14), (8.22) and (8.39), we obtain

$$\mathcal{T} = \frac{1}{1 + (4R/T^2)\sin^2[(\tilde{\zeta} L |\tilde{x}|^2 - \delta_0)/2]}, \tag{11.13}$$

where we have used the definition (7.20) and taken into account that $|\tilde{x}|^2 = \rho_0^2$, as one obtains using Eqs. (8.17), (11.12), (7.10) and (7.13).

Following [100], we observe that the plot of $\mathcal{T}$ versus $|\tilde{x}|^2$ has the same shape as the usual graph of the transmissivity of an empty cavity as a function of the normalized input frequency, i.e. the Airy function in Fig. 8.5. Since $\mathcal{T} = |\tilde{x}|^2/|\tilde{y}|^2$, for a fixed value of $|\tilde{y}|^2$

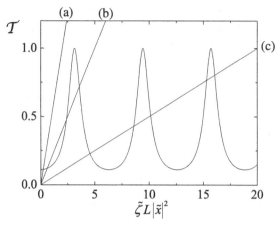

Figure 11.5 The steady-state transmission $\mathcal{T}$ of the cavity, filled with a Kerr medium, is plotted as a function of the normalized transmitted intensity $\tilde{\zeta} L|\tilde{x}|^2$. Lines (a), (b) and (c) show the straight line $\mathcal{T} = |\tilde{x}|^2/|\tilde{y}|^2$ for different values of the normalized incident power $|\tilde{y}|^2$. The intersections with the curve correspond to the (stable or unstable) stationary states of the system for a fixed value of $|\tilde{y}|^2$. $R = 0.5$, $\delta_0 = \pi$ and $\tilde{\zeta} L|\tilde{y}|^2 = 2$ (a), $\tilde{\zeta} L|\tilde{y}|^2 = 5$ (b) and $\tilde{\zeta} L|\tilde{y}|^2 = 20$ (c).

the stationary values of $|\tilde{x}|^2$ can be found by intersecting the line given by (11.13) with the straight line $\mathcal{T} = |\tilde{x}|^2/|\tilde{y}|^2$ (see Fig. 11.5). The value of $|\tilde{y}|^2$ controls the angular coefficient of the straight line. In the case of line (a) in Fig. 11.5 one has only one intersection. Upon increasing $|\tilde{y}|^2$, the intersections become three in number (line (b)), with the middle one being unstable. Thus, we obtain bistability and the plot of $|\tilde{x}|^2$ versus $|\tilde{y}|^2$ shows a hysteresis cycle. For even larger values of $|\tilde{y}|^2$ one obtains multiple solutions (line (c)), leading to multistability and multiple hysteresis cycles. When T approaches unity, the curve (11.13) flattens and bistability disappears as usual.

As in previous cases, let us now consider the low-transmission limit. In order to deal with dimensionless quantities, let us consider the number of photons in the volume V of the Kerr-medium sample. Since the electromagnetic energy density is equal to $\epsilon_0 n_0^2 |E_0|^2/2$ (see Section 1.2) this photon number is given by $\epsilon_0 n_0^2 |E_0|^2 V/(2\hbar\omega_0)$. Hence, if we define the quantities

$$x = \tilde{x} \left(\frac{\epsilon_0 n_0^2 V}{2\hbar\omega_0} \right)^{1/2}, \qquad y = \tilde{y} \left(\frac{\epsilon_0 n_0^2 V}{2\hbar\omega_0} \right)^{1/2}, \qquad (11.14)$$

we have from Eq. (11.12) and (8.19) that the photon number is given by $|x|^2$ and that $\mathcal{T} = |x|^2/|y|^2$. We also define

$$\zeta = \tilde{\zeta} \frac{2\hbar\omega_0}{\epsilon_0 n_0^2 V} \qquad (11.15)$$

so that

$$\tilde{\zeta} |\tilde{x}|^2 = \zeta |x|^2. \qquad (11.16)$$

The low-transmission limit is defined by the conditions

$$\zeta L \to 0, \quad T \to 0, \qquad \text{with} \quad B \equiv \frac{\zeta L}{T} \quad \text{constant,} \tag{11.17}$$

where the first limit ensures the uniformity of the field phase along the sample (see Eqs. (7.14), (7.20), (11.15) and (11.16)). To this we add the limit

$$\delta_0 = \frac{\omega_c - \omega_0}{c/\Lambda} \to 0, \quad \text{with} \quad \theta = \frac{\delta_0}{T} = \frac{\omega_c - \omega_0}{\kappa} \quad \text{constant}, \tag{11.18}$$

where θ is the *cavity-detuning* (or *cavity-mistuning*) parameter. The physical meaning of this limit is that the cavity detuning ($\omega_c - \omega_0$) must be much smaller than the free spectral range $2\pi c/\Lambda$, but of the same order of magnitude as the cavity linewidth $\kappa = cT/\Lambda$. This limit implies that in the stationary state the system operates only with the cavity mode ω_c closest to resonance with the input field.

In the limits defined by Eqs. (11.17) and (11.18), by taking Eq. (11.16) into account we see that Eq. (11.13) reduces to

$$\frac{|x|^2}{|y|^2} = \frac{1}{1 + (\theta - B|x|^2)^2} . \tag{11.19}$$

Hence, by introducing the definitions

$$X = \bar{\eta} B |x|^2, \quad Y = \bar{\eta} B |y|^2, \quad \bar{\theta} = \bar{\eta}\theta \implies \theta = \bar{\eta}\bar{\theta}, \tag{11.20}$$

$$\bar{\eta} = \begin{cases} 1 & \text{for} \quad B > 0, \\ -1 & \text{for} \quad B < 0, \end{cases} \tag{11.21}$$

we obtain the final equation

$$Y = X\left[1 + (\bar{\theta} - X)^2\right], \tag{11.22}$$

where X and Y are positive by definition.

The cubic equation (11.22) depends on just one parameter ($\bar{\theta}$), just as the steady-state equation (11.9) for absorptive optical bistability depends solely on the parameter C. The model (11.22) was first introduced by Gibbs, McCall and Venkatesan [101] in connection with their first experimental observation of optical bistability in 1976.

It is simple to verify that the steady-state curve (11.22) displays a maximum and a minimum when $\bar{\theta} > \sqrt{3}$ (the bistability condition). Figure 11.6 shows the hysteresis cycle of transmitted vs. incident normalized intensities for $\bar{\theta} = 5$. Note that the cycle has a qualitatively different shape from that of absorptive optical bistability shown in Fig. 11.4; in particular the curve is tangent to the straight line $X = Y$ for the value $X = \bar{\theta}$.

The mechanism which produces switching and hysteresis in the dispersive case of Eq. (11.22) was elucidated in [101] and is quite different from that of absorptive optical bistability described in the previous section. In the lower branch of the cycle the transmission is low because the cavity frequency ω_c is detuned from the input frequency ω_0. If the parameter B has the appropriate sign, then upon increasing the incident field the nonlinear refractive index changes the effective optical length of the cavity, driving the cavity frequency towards resonance. This in turn increases the incident field, which further drives the effective cavity frequency $\omega_c' = \omega_c - \kappa X$ towards the incident frequency ω_0 and so on, creating again a

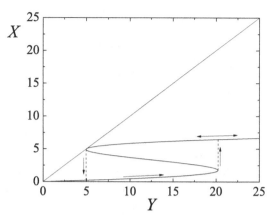

The hysteresis cycle of transmitted vs. incident normalized intensity (Eq. (11.22)) for $\bar{\theta} = 5$.

positive feedback loop, which causes the switching to the higher-transmission branch. On the other hand, when the system is in the higher branch and the input intensity is decreased, the internal field is already strong enough to maintain resonance, which produces hysteresis.

We note finally that the case $\bar{\eta} = +1$ corresponds to $\chi^{(3)} > 0$, as one obtains using Eqs. (11.21), (11.17), (11.15) and (7.20), and therefore to the self-focussing configuration (see Section 7.3).

11.3 Optical bistability in two-level systems: the general case

In the previous sections we have analyzed purely absorptive optical bistability for two-level atoms and purely dispersive bistability for Kerr media. The case of purely dispersive bistability for two-level atoms will be considered in Section 13.3.

In this section we return to the case of two-level atoms, but for general values of the parameters [86, 99] (see also [102]). Let us first consider Eq. (5.36) for $z = L$, writing $-\alpha$ instead of g:

$$\alpha L = (1 + \Delta^2)\ln \beta + \frac{1}{2}\rho^2(L)(\beta^2 - 1), \tag{11.23}$$

where we have used the definition (8.22) of the parameter β. By defining in this case (see Eq. (8.19))

$$X = \rho^2(L) = |x|^2, \qquad Y = |y|^2, \tag{11.24}$$

Eq. (11.23) can be rewritten in the form

$$X(\beta) = \frac{2}{\beta^2 - 1}\left[\alpha L - (1 + \Delta^2)\ln \beta\right]. \tag{11.25}$$

On the other hand, by taking into account that $\mathcal{T} = X/Y$ and Eq. (5.39), Eq. (8.39) can be rewritten in the form

$$Y(\beta) = \frac{X(\beta)}{\mathcal{T}^2}\left[(\beta - R)^2 + 4R\beta \sin^2\left(\frac{\Delta \ln \beta - \delta_0}{2}\right)\right]. \tag{11.26}$$

Equations (11.25) and (11.26) together provide a parametric expression of the steady-state curve $X(Y)$ in the plane (Y, X), because by letting the parameter β vary from 1 to ∞ one gets, for each value of β, the corresponding values of Y and X.

Let us now simplify the result by using the low-transmission limit, which in practical terms amounts to

$$\alpha L \ll 1, \quad \mathcal{T} \ll 1, \quad |\delta_0| \ll 1, \quad \text{with}$$
$$2C = \frac{\alpha L}{\mathcal{T}} \quad \text{arbitrary}, \quad \theta = \frac{\delta_0}{\mathcal{T}} = \frac{\omega_c - \omega_0}{\kappa} \quad \text{arbitrary}. \tag{11.27}$$

Let us rewrite Eqs. (5.34) and (5.35) in the form

$$\frac{d\rho}{dz} = -\alpha\rho\tilde{\chi}'(\rho^2), \tag{11.28}$$

$$\frac{d\varphi}{dz} = -\alpha\tilde{\chi}''(\rho^2), \tag{11.29}$$

with

$$\tilde{\chi}'(\rho^2) = \frac{1}{1 + \Delta^2 + \rho^2}, \quad \tilde{\chi}''(\rho^2) = \frac{-\Delta}{1 + \Delta^2 + \rho^2}. \tag{11.30}$$

From Eq. (11.28) we get

$$\rho(L) - \rho(0) = -\alpha\int_0^L dz\, \tilde{\chi}'(\rho^2(z))\rho(z) \approx -\alpha L\tilde{\chi}'(\rho^2(L))\rho(L), \tag{11.31}$$

which gives

$$\beta \approx 1 + \alpha L\tilde{\chi}'(X). \tag{11.32}$$

Similarly, Eq. (11.29) leads to

$$\Delta\varphi \approx -\alpha L\tilde{\chi}''(X). \tag{11.33}$$

If we insert Eqs. (11.32) and (11.33) into Eq. (11.26), in the square bracketes of Eq. (11.26) keep only the contributions on the same order as $\mathcal{T}^2$ and take the limits (11.27) we arrive at the steady-state equation

$$Y = X\left[\left(1 + \frac{2C}{1 + \Delta^2 + X}\right)^2 + \left(\theta - \frac{2C\Delta}{1 + \Delta^2 + X}\right)^2\right]. \tag{11.34}$$

As we will show in the next chapter, in terms of the normalized fields x and y the steady-state equation reads

$$y = x\left[\left(1 + \frac{2C}{1 + \Delta^2 + |x|^2}\right) + i\left(\theta - \frac{2C\Delta}{1 + \Delta^2 + |x|^2}\right)\right]. \tag{11.35}$$

In the completely resonant case $\Delta = 0$ and $\theta = 0$, Eq. (11.34) reduces to

$$Y = X\left(1 + \frac{2C}{1+X}\right)^2, \tag{11.36}$$

which, on taking into account Eq. (11.24) and that x and y are real in this case, coincides with the square of Eq. (11.9). In the general case, the steady-state curve $X(Y)$ in the limit of large Y is asymptotic to the straight line $X = Y/(1 + \theta^2)$ instead of to $X = Y$ as in the completely resonant case. In this limit the medium is saturated (i.e. transparent), and part of the incident energy is not transmitted but reflected (see Eq. (8.40)) because the input field is not resonant with the cavity. By using Eq. (8.40) with $R = 1$ and $\delta_0 = 0$, one can immediately check that $|F_R|^2 = Y\theta^2/(1 + \theta^2)$, since, in the limit of Y and X large, from Eq. (11.35) one has that $x = y/(1 + i\theta)$, so that the sum of transmitted and reflected intensities $X + |F_R|^2$ is equal to the input intensity Y.

The identification of the bistability conditions for Eq. (11.34) in the space of the three parameters C, Δ and θ is complex and the main conclusions [103] are as follows.

(i) Bistability is impossible for $C < 4$.
(ii) For a fixed value of $C > 4$, the largest hysteresis cycle is obtained for $\Delta = \theta = 0$ and bistability exists only in a finite domain of the (Δ, θ) plane around the origin.
(iii) If we keep $C > 4$ fixed and increase Δ and θ simultaneously from zero, with the ratio Δ/θ fixed, the hysteresis cycle of the transmitted intensity X as a function of the input intensity Y shifts to the left and decreases in size, until it disappears.

An especially simple case is $\theta = -\Delta$, in which the input frequency lies between the atomic transition frequency ω_a and the cavity frequency ω_c, in such a position that the two detunings are equal in modulus. In this case Eq. (11.34) reduces to

$$Y = X(1 + \Delta^2)\left(1 + \frac{2C}{1 + \Delta^2 + X}\right)^2, \tag{11.37}$$

which can be rewritten in the form

$$\overline{Y} = \overline{X}\left(1 + \frac{2\overline{C}}{1 + \overline{X}^2}\right)^2, \tag{11.38}$$

where we defined

$$\overline{C} = \frac{C}{1 + \Delta^2}, \qquad \overline{Y} = \frac{Y}{(1 + \Delta^2)^2}, \qquad \overline{X} = \frac{X}{1 + \Delta^2}. \tag{11.39}$$

Equation (11.38) is formally identical to Eq. (11.36) for the absorptive case, hence there is bistability for $\overline{C} > 4$, i.e. for $C > 4(1 + \Delta^2)$.

It is important that the scenario described for Eq. (11.34) holds only in the case of homogeneous broadening implicitly assumed up to now. In the case of inhomogeneous broadening (see Chapter 15), when the inhomogeneous broadening dominates there are values of C such that the system is not bistable for $\Delta = \theta = 0$, but becomes bistable when Δ and θ are large enough [104, 105].

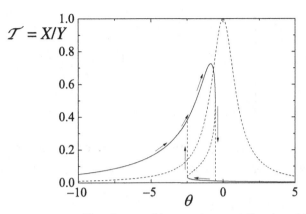

Figure 11.7 The hysteresis cycle of the transmissivity $\mathcal{T}$ as a function of the cavity detuning θ. The values of the other parameters are $C = 30$, $\Delta = -5$ and $Y = 441$. The dashed curve shows the empty-cavity result $X/Y = 1/(1 + \theta^2)$.

We remark that we have considered only hysteresis cycles obtained by varying the incident field intensity Y and keeping the parameters C, Δ and θ fixed but one can also consider cycles obtained by keeping Y fixed and varying C, Δ or θ, or some of these parameters simultaneously [106]. For instance, one can sweep back and forth the cavity length, i.e. the cavity detuning θ, or the input field frequency, i.e. the two parameters Δ and θ simultaneously.

Figure 11.7 shows the hysteresis cycle of the transmitted intensity X as a function of θ, for Y, C and Δ fixed. Note that when Y is kept constant X and $\mathcal{T}$ are simply proportional. The broken line shows the transmissivity peak for the empty cavity. The nonlinearity bends the stationary curve, and the segment with negative slope is unstable as usual. The hysteresis cycle is covered clockwise in this case. Hysteresis cycles of this kind were first observed experimentally in [107, 108].

For further reading on optical bistability, we refer the reader especially to [86, 109, 110]. The case of optical bistability in a Fabry–Perot cavity will be briefly discussed in Section 14.3. The literature on this topic is based on [111], and encompasses at least [97, 112–119]. General discussions can be found in [86, 109]. Absorptive bistability was experimentally observed in [120].

11.4 Functionalities of optically bistable systems

Let us analyze again the behavior of the transmitted intensity I_T as a function of the input intensity I_I, considering as an example the case of purely absorptive optical bistability for $\Delta = \theta = 0$, in which the behavior is governed by the single parameter C (Fig. 11.8). In the case of an empty cavity ($C = 0$) the steady-state curve is the straight line $I_T = I_I$. For C different from zero the response is nonlinear, of course. We find first that the curve develops a portion in which the so-called differential gain dI_T/dI_I is larger than unity. In this condition, the system works as an *optical transistor*. If we modulate the incident

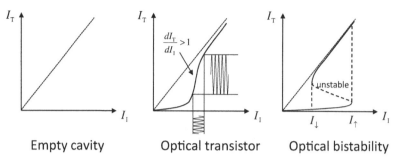

Figure 11.8 Transmitted intensity versus incident intensity for an empty cavity, optical transistor operation and bistable operation.

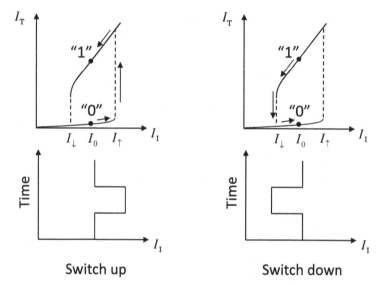

Figure 11.9 Optical memory. I_0 is the holding intensity. In switch-up operation the system is initially in the lower branch at $I_I = I_0$. By means of the pulse shown in the figure, the system switches to the upper branch. Eventually the system is in the upper branch at $I_I = I_0$. In switch-down operation the system is initially in the upper branch at $I_I = I_0$. By means of the pulse shown in the figure, the system switches to the lower branch. Eventually the system is in the lower branch at $I_I = I_0$.

intensity slowly enough to allow the system to follow the steady-state curve (adiabatic variation of the incident intensity), this modulation is transferred to the transmitted beam and the modulation depth turns out to be larger in the transmitted than in the input intensity. This is precisely the transistor action, in which the incident intensity plays the role of the base-emitter voltage.

If we further increase the parameter C, the steady-state curve develops a portion with negative slope and we have bistability. In the bistable regime the system works as an *optical memory*, in which information can be stored. In the range $I_\downarrow < I_I < I_\uparrow$ the lower and the upper states are the 0 and the 1 of a memory element, and by suitably varying the input intensity one can switch from one state to the other as illustrated in Fig. 11.9. These devices can also work as pulse shapers that tailor the incident light pulses in many different ways, by amplifying parts of them and eliminating others, suppressing the noisy parts (clippers, discriminators, limiters and so on). All-optical logic gates using such devices have also been realized. Detailed descriptions can be found in the book by Gibbs [109].

Because of the large bandwidth of optical systems, with the perspective of realizing fast miniaturized devices, in the 1980s there was an organized effort aimed at the realization of an optical computer based on the physical principles of optical bistability. The prototypes that were realized were not competitive mainly because of the trade-off between the size of the nonlinearity in available materials and the speed of operation of the devices. The advent of new materials with the required properties might bring new interest in this approach. Anyway, that research contributed to the birth of the discipline called *photonics*.

The devices we have described in this section are usually called all-optical (or intrinsic) bistable devices. On the other hand, there are also the so-called hybrid electro-optic systems, of which many variants have been devised. A typical device of this kind is obtained by replacing the medium by an electro-optic crystal, which is monitored by the output beam and produces changes in the refractive index proportional to the output power, thus producing a synthetic Kerr nonlinearity.

12 Modal equations for the ring cavity. The single-mode model

In the previous chapters we considered the Maxwell–Bloch equations with the boundary condition appropriate for a unidirectional ring cavity and calculated the stationary solutions for laser and optical bistability, respectively. Next, we showed that the exact stationary solutions simplify appreciably in the low-transmission limit.

In this chapter we want, first of all, to apply the low-transmission approximation on the dynamical equations directly, beyond the stationary case. We start from the Maxwell–Bloch equations (4.35)–(4.37) and the boundary condition (8.36). The latter is complex and includes the retardation Δt and some parameters such as the transmissivity and reflectivity coefficients T and R, the input field y and the cavity detuning δ_0.

We introduce a transformation of the coordinates z and t and a transformation of the variables F and P [121–123]. As a result, the boundary condition reduces to a simple periodic boundary condition, and the electric-field equation incorporates the parameters which previously appeared in the boundary condition. The virtue of the periodic boundary condition is that it allows one to expand the electric field in terms of modal amplitudes and to recast the field equation in the form of a set of time-evolution equations for the modal amplitudes. All of this is shown in Section 12.1.

In Section 12.2 we apply the low-transmission approximation and the Maxwell–Bloch equations assume a form that is a generalization of a set of equations commonly used in the literature to describe the ring laser [124]. Together with the Bloch equations for the atomic variables, the modal equations provide a model that lends itself ideally to a numerical resolution of the equations.

In Section 12.3 we show that, if we assume that only the resonant cavity mode is active, this model reduces to a simple single-mode model consisting in three coupled ordinary differential equations, which will be analyzed in Part II of this book. Next, in the same section we derive the single-mode model in a more direct way, without passing through the multimodal equations. We start again from the original equations, Eqs. (4.35)–(4.37), and boundary condition (8.36), and implement not only the low-transmission approximation but also the single-mode approximation, by means of which we arrive directly at the single-mode model. Finally in Section 12.4 we perform a concise analysis of the single-mode stationary state, in relation to results in Chapters 9 and 11.

The models derived in this chapter apply to three kinds of system: a passive system (optical bistability), the laser and a laser with an input coherent field that is injected into the cavity as in the case of optical bistability. Indeed, the last system, which is called a *laser with an injected signal*, is the exact active counterpart of optical bistability. To distinguish the standard laser from the laser with an injected signal, in the following we will sometimes call the ordinary laser a *free-running laser*.

12.1 Transformation of coordinates and transformation of variables. Modal equations

As we said, we start from the Maxwell–Bloch equations (4.35)–(4.37) and from the boundary condition (8.36), in which we assume that the input field y is time-independent. These equations correspond to the configuration of a laser with an injected signal. In the case of a standard free-running laser we must set $y = 0$ in the boundary condition. In the passive case of optical bistability, we must instead leave y different from zero, and in the field equation (4.35) we must replace the gain parameter g by $-\alpha$, where α is the absorption coefficient (see Eq. (4.38)).

We now introduce the following transformation of coordinates z and t [121–123]:

$$\eta = z, \qquad \tau = t + \frac{z}{L}\,\Delta t, \tag{12.1}$$

where Δt is defined by Eq. (8.10). Thanks to this transformation the retardation Δt disappears from the boundary condition. In fact, the couple $(z = 0,\, t)$ is transformed into the couple $(\eta = 0,\, \tau = t)$ while the couple $(z = L,\, t - \Delta t)$ is transformed into the couple $(\eta = L,\, \tau = t)$ so that the boundary condition (8.36) becomes isochronous,

$$F(0, \tau) = Ty + Re^{-i\delta_0} F(L, \tau). \tag{12.2}$$

Furthermore, since

$$\frac{\partial}{\partial z} = \frac{\partial \eta}{\partial z}\frac{\partial}{\partial \eta} + \frac{\partial \tau}{\partial z}\frac{\partial}{\partial \tau} = \frac{\partial}{\partial \eta} + \frac{\Delta t}{L}\frac{\partial}{\partial \tau}, \tag{12.3}$$

$$\frac{\partial}{\partial t} = \frac{\partial \eta}{\partial t}\frac{\partial}{\partial \eta} + \frac{\partial \tau}{\partial t}\frac{\partial}{\partial \tau} = \frac{\partial}{\partial \tau}, \tag{12.4}$$

the Maxwell–Bloch equations (4.35)–(4.37) take the forms

$$\frac{\partial F}{\partial z} + \frac{1}{c}\frac{\Lambda}{L}\frac{\partial F}{\partial \tau} = gP, \tag{12.5}$$

$$\frac{\partial P}{\partial \tau} = \gamma_\perp [FD - (1 + i\Delta)P], \tag{12.6}$$

$$\frac{\partial D}{\partial \tau} = -\gamma_\parallel \left[\frac{1}{2}(FP^* + F^*P) + D - 1\right], \tag{12.7}$$

where, for simplicity, we write z instead of η and the optical length of the cavity Λ is defined by Eq. (8.12). Next, we introduce the following transformation of the variables F and P [121–123]:

$$F'(z, \tau) = F(z, \tau)e^{(\ln R - i\delta_0)z/L} + Ty\frac{z}{L}, \tag{12.8}$$

$$P'(z, \tau) = P(z, \tau)e^{(\ln R - i\delta_0)z/L}. \tag{12.9}$$

In such a way the boundary condition (12.2) becomes a periodic boundary condition,

$$F'(L, \tau) = Re^{-i\delta_0} F(L, \tau) + Ty = F(0, \tau) = F'(0, \tau), \tag{12.10}$$

while the equations for F', P' and D read

$$\frac{\partial F'}{\partial z} + \frac{1}{c}\frac{\Lambda}{L}\frac{\partial F'}{\partial \tau} = \frac{1}{L}(\ln R - i\delta_0)\left(F' - Ty\frac{z}{L}\right) + \frac{Ty}{L} + gP', \tag{12.11}$$

$$\frac{\partial P'}{\partial \tau} = \gamma_\perp\left[\left(F' - Ty\frac{z}{L}\right)D - (1+i\Delta)P'\right], \tag{12.12}$$

$$\frac{\partial D}{\partial \tau} = -\gamma_\parallel\left\{\frac{1}{2}\left[\left(F' - Ty\frac{z}{L}\right)P'^* + \left(F'^* - Ty\frac{z}{L}\right)P'\right]e^{-2(\ln R)z/L} + D - 1\right\}. \tag{12.13}$$

We note that Eqs. (12.11)–(12.13) are exact. At this stage, we can profit from the periodic boundary condition (12.10) to expand F' in the following way:

$$F'(z,\tau) = \sum_{n=-\infty}^{+\infty} e^{2\pi inz/L}f_n(\tau), \tag{12.14}$$

where

$$f_n(\tau) = \frac{1}{L}\int_0^L dz\, e^{-2\pi inz/L}F'(z,\tau). \tag{12.15}$$

The Fourier coefficients $f_n(\tau)$ have the physical meaning of modal amplitudes. By using Eqs. (12.11) and (12.15), we get the modal equations

$$\frac{df_n}{d\tau} + i\alpha_n f_n = \frac{c}{\Lambda}\left[(\ln R - i\delta_0)\left(f_n - \frac{1}{2}Ty\delta_{n,0}\right) + Ty\delta_{n,0}\right]$$
$$+ cg\frac{L}{\Lambda}\frac{1}{L}\int_0^L dz\, e^{-2\pi inz/L}P'(z,\tau), \tag{12.16}$$

where

$$\alpha_n = 2\pi n\frac{c}{\Lambda}, \qquad n = 0, \pm 1, \pm 2, \dots. \tag{12.17}$$

We emphasize that, even if we have indicated that $\sum_n$ is extended from $-\infty$ to $+\infty$, in practice the sum must be kept restricted to a limited range of values of n to ensure that $F'(z,t)$ varies with z much more slowly than $\exp(ik_0 z)$.

After having considered the effects of the transformations (12.1), (12.8) and (12.9) on the boundary condition and on the equations, let us discuss the effects on the solutions.

We remark, first of all, that F' given by Eq. (12.8) exhibits a much smaller percentage variation with respect to z than does F, as one may expect from the circumstance that F' obeys periodic boundary conditions [123].

Next, in order to be able to carry out simple analytical calculations, let us assume that the cavity contains a medium that contributes only the background refractive index, so that the field equation (4.35) reduces to

$$\frac{\partial F}{\partial z} + \frac{n_B}{c}\frac{\partial F}{\partial t} = 0. \tag{12.18}$$

Let us analyze first the effect of the transformation of variables (12.8) independently of the transformation of coordinates (12.1). To do this, let us formally set $L = \mathcal{L}$. More precisely, we apply the following assumptions.

(i) The atomic sample fills the whole part of the cavity between the two beam splitters from $z = 0$ to $z = \mathcal{L} - L'$ (see Fig. 8.4), so the boundary condition is

$$F(0, t) = Re^{-i\delta_0} F(\mathcal{L} - L', t - L'/c). \tag{12.19}$$

(ii) Let us suppose that $L' \ll \mathcal{L}$, so that Eq. (12.19) reduces to

$$F(0, t) = Re^{-i\delta_0} F(\mathcal{L}, t), \tag{12.20}$$

and hence the retardation disappears from the original boundary condition and the time τ defined by Eq. (12.1) becomes identical to t.

Let us assume that for negative times we have injected into the cavity a field with frequency ω_0, so that the function F, which is proportional to the envelope E_0 which appears in Eq. (2.37), has reached a stationary and uniform configuration. At time $t = 0$ we switch the input field off; hence the intracavity field decays to zero with a time evolution that is governed by the cavity frequencies instead of by the input frequency ω_0. In order to describe this, let us start from Eqs. (12.18) and (12.20) and look for a solution of the form

$$F(z, t) = e^{-\mu[t-(z/c)n_B]}. \tag{12.21}$$

By inserting Eq. (12.21) into Eq. (12.20) we obtain

$$1 = Re^{-i\delta_0} e^{\mu(\mathcal{L}/c)n_B} \rightarrow 2\pi i n = \ln R - i\delta_0 + \mu\frac{\Lambda}{c}, \tag{12.22}$$

where n is a positive or negative integer, and Λ is the optical length of the cavity when $L = \mathcal{L}$, i.e. $\Lambda = n_B L$. From Eqs. (12.20), (12.22), (8.32) and (8.37) we obtain

$$F(z, t) = e^{[(c/\Lambda)\ln R + i(\omega_0 - \omega_j)][t-(z/c)n_B]}, \tag{12.23}$$

where $j = \bar{j} + n$, with $\bar{j}$ being the value of j which appears in the definition of δ_0 (Eq. (8.37)) and corresponds to the resonance frequency ω_c. Hence, by using Eq. (2.37) and taking into account that E_0 is proportional to F, we arrive at the result

$$\mathcal{E}(z, t) \propto e^{[-i\omega_j + (c/\Lambda)\ln R][t-(z/c)n_B]} + \text{c.c.}, \tag{12.24}$$

so that the decay of the electric field $\mathcal{E}$ is characterized by the rate $c \ln(1/R)/\Lambda$, which for $n_B = 1$ and $T \ll 1$ coincides with the linewidth of the empty cavity defined by Eq. (8.27), and is characterized by a cavity frequency ω_j as expected, and the frequency ω_0 has disappeared.

Now let us repeat the calculation starting from Eq. (12.11) with $g = 0$, $y = 0$, $L = \mathcal{L}$, $\Lambda = n_B L$ and $\tau = t$,

$$\frac{\partial F'}{\partial z} + \frac{n_B}{c}\frac{\partial F'}{\partial t} = \frac{1}{\mathcal{L}}(\ln R - i\delta_0)F', \tag{12.25}$$

and let us look for a solution of the form

$$F'(z, t) = e^{-\tilde{\mu}[t-(n_B/c)z]} e^{(c/\Lambda)(\ln R - i\delta_0)t}. \tag{12.26}$$

By inserting Eq. (12.26) into the boundary condition (12.10) one obtains

$$\tilde{\mu} = 2\pi i \frac{c}{\Lambda} n,\tag{12.27}$$

where, again, n is a positive or negative integer, which implies that

$$F'(z, t) = e^{-2\pi i (c/\Lambda)n[t-(n_B/c)z]}e^{(c/\Lambda)(\ln R + i\delta_0)t}.\tag{12.28}$$

By carrying out the transformation (12.8) with $y = 0$ and $L = \mathcal{L}$ we arrive at a result that coincides with Eq. (12.23). Therefore, by using the transformation (12.8) one multiplies F' by a factor that is functional to produce the disappearance of the frequency ω_0 from the expression for the complete electric field $\mathcal{E}$ (see Eq. (12.24)).

We observe also that if we express F' in terms of modal amplitudes we have

$$F'(z, t) = e^{2\pi i n z/\mathcal{L}} f_n(t),\tag{12.29}$$

where we have used Eq. (12.14) with $L = \mathcal{L}$ and $\tau = t$, and we have reduced the sum over n to one term to facilitate comparison with the case considered above. If we take into account Eq. (12.16) with $\tau = t$, $y = 0$ and $g = 0$, and use Eq. (12.17) and the fact that $\Lambda = \mathcal{L}n_B$, we obtain a result identical to Eq. (12.28). Therefore the mode with $n = 0$ corresponds to the resonant mode with frequency ω_c, and α_n is the frequency offset of the generic mode $j = \bar{j} + n$ from the resonant mode, i.e.

$$\omega_j = \omega_c + \alpha_n.\tag{12.30}$$

Next, let us consider the effect of the transformation of coordinates (12.1), assuming now that $L < \mathcal{L}$. Let us consider Eq. (12.16) with $y = 0$ and $g = 0$, and let us neglect also the remaining term on the right-hand side because it arises from the transformation of variables Eq. (12.8) and now we want to identify the effect of the coordinate transformation (12.1) only. Therefore we obtain

$$f_n(\tau) = e^{-i\alpha_n \tau} f_n(0) \quad \Rightarrow \quad F'(z, \tau) = \sum_{n=-\infty}^{+\infty} e^{2\pi i n z/L} e^{-i\alpha_n \tau} f_n(0),\tag{12.31}$$

and, by using Eqs. (12.1) and (12.17) and the definition of the optical length of the cavity $\Lambda = \mathcal{L} + (n_B - 1)L$, we arrive at the result

$$F'(z, t) = \sum_{n=-\infty}^{+\infty} e^{-i\alpha_n[t-(n_B/c)z]} f_n(0).\tag{12.32}$$

Therefore the transformation (12.1) converts the factor $e^{i2\pi n z/L}$ into $e^{i2\pi n z n_B/\Lambda}$, producing the correct propagation with velocity c/n_B. Obviously the result holds only for $0 < z < L$ and must not be applied to the empty part of the cavity.

We observe finally that there is some arbitrariness in the transformation (12.8) plus (12.9). As a matter of fact, any transformation of the form

$$F'(z, \tau) = F(z, \tau)e^{g_1(z)}F(z, \tau) + g_2(z),\tag{12.33}$$

$$P'(z, \tau) = P(z, \tau)e^{g_1(z)}\tag{12.34}$$

leads to a periodic boundary condition for $F'(z, \tau)$ (Eq. (12.10)), provided that $g_1(0) = g_2(0) = 0$, $g_1(L) = \ln R - i\delta_0$ and $g_2(L) = Ty$. In selecting the transformation (12.8) plus (12.9) we have assumed that g_1 and g_2 are linear in z. This is the simplest choice to ensure that the equations of the model have constant coefficients in the low-transmission limit (see Eqs. (12.35)–(12.37) below).

12.2 Introduction of the low-transmission approximation

At this point, after applying the transformations (12.1), (12.8) and (12.9), let us introduce the low-transmission approximation defined by Eq. (11.27) with α replaced by g and $2C$ replaced by A (see Eq. (9.18)). In this limit F' and P' defined by Eqs. (12.8) and (12.9) coincide with F and P, respectively, and the dynamical equations (12.11)–(12.13) reduce to

$$c\frac{L}{\Lambda}\frac{\partial F}{\partial z} + \frac{\partial F}{\partial \tau} = -\kappa[(1+i\theta)F - y - AP], \tag{12.35}$$

$$\frac{\partial P}{\partial \tau} = \gamma_\perp[FD - (1+i\Delta)P], \tag{12.36}$$

$$\frac{\partial D}{\partial \tau} = -\gamma_\parallel\left[\frac{1}{2}(FP^* + F^*P) + D - 1\right], \tag{12.37}$$

with

$$\kappa = \frac{cT}{\Lambda} \tag{12.38}$$

being the decay rate of the field which, for $n_B = 1$, coincides with the cavity linewidth defined by Eq. (8.27). We underline that in our approach the damping term $-\kappa F$ is not added phenomenologically, but arises naturally from the boundary conditions of the cavity, via the transformations (12.8) and (12.9) and the systematic utilization of the low-transmission limit. The damping term has the same form as the linear background term in Eq. (5.51), and in the presence of background linear absorption the total damping rate is $\kappa + c\alpha_B/n_B$.

On the other hand, the modal equations (12.16) reduce to

$$\frac{df_n}{d\tau} + i\alpha_n f_n = -\kappa\left[(1+i\theta)f_n - y\delta_{n,0} - A\frac{1}{L}\int_0^L dz\, e^{-2\pi inz/L}P(z, \tau)\right], \tag{12.39}$$

and must be coupled with Eqs. (12.36) and (12.37) taking into account that

$$F(z, \tau) = \sum_{n=-\infty}^{+\infty} e^{2\pi inz/L}f_n(\tau), \qquad f_n(\tau) = \frac{1}{L}\int_0^L dz\, e^{-2\pi inz/L}F(z, \tau). \tag{12.40}$$

The modal equations (12.39) with Eqs. (12.36) and (12.37) are optimal for the numerical resolution.

An important remark is now that in a cavity the length scale is increased by a factor of T^{-1} with respect to the length scale in a cavityless system. In order to make a simple and fair comparison with the scaling used in Eq. (3.26), let us consider Eq. (12.35) with

$A = \theta = y = 0$ and $\Lambda = n_B L$ (i.e. $\mathcal{L} = L$). Under such conditions, if we define

$$\bar{z} = z\frac{T}{L}, \qquad \bar{t} = t\frac{cT}{\Lambda} = t\frac{cT}{n_B L}, \tag{12.41}$$

Eq. (12.35) reduces to

$$\frac{\partial F}{\partial \bar{z}} + \frac{\partial F}{\partial \bar{t}} = -F, \tag{12.42}$$

where the r.h.s. describes the escape of photons from the cavity. Since Eq. (12.42) does not contain parameters, Eq. (12.41) does indeed define the correct scaling of space and time coordinates in the cavity case. In the cavityless case, instead, the scaling is $\bar{z} = z/L$ and $\bar{t} = tc/(n_B L)$ (see Eq. (3.26)), so, with respect to the cavityless case, the length scale is L/T instead of L.

12.3 The single-mode model

When the free spectral range α_1 is much larger than the atomic linewidth $\gamma_\perp$, one can reasonably assume that only the resonant mode f_0 is active. Therefore $F = f_0$ is independent of z, and, in addition, the time τ can be identified with t because their difference is smaller than the roundtrip time $\mathcal{L}/c$ and the variables F, P and D vary in time with temporal scales equal to the inverse of the rates κ, $\gamma_\perp$ and $\gamma_\parallel$, which are much larger than the roundtrip time. For the same reason, we can neglect the retardation $\Delta t'$ in Eq. (8.17) and write $x(t)$ instead of $F(t)$. Also the variables P and D become independent of the longitudinal coordinate z. Hence the dynamical equations (12.39) for $n = 0$, (12.36) and (12.37) reduce to

$$\dot{x} = -\kappa[(1 + i\theta)x - y - AP], \tag{12.43}$$

$$\dot{P} = \gamma_\perp[xD - (1 + i\Delta)P], \tag{12.44}$$

$$\dot{D} = -\gamma_\parallel\left[\frac{1}{2}(xP^* + x^*P) + D - 1\right], \tag{12.45}$$

where the overdot $\dot{}$ indicates d/dt as usual. These equations constitute the *single-mode model* which will be analyzed in the stationary state in the following section.

It is interesting to show that it is possible to derive the single-mode model directly from the original Maxwell–Bloch equations plus the boundary condition by using the low-transmission limit together with the single-mode limit.

Let us start from the Maxwell–Bloch equations (4.35)–(4.37) and let us operate on both sides of these equations with the integral $(1/L)\int_0^L dz$. If we introduce the spatial averages of the variables F, P and D,

$$\overline{F}(t) = \frac{1}{L}\int_0^L dz\, F(z, t), \quad \overline{P}(t) = \frac{1}{L}\int_0^L dz\, P(z, t), \quad \overline{D}(t) = \frac{1}{L}\int_0^L dz\, D(z, t), \tag{12.46}$$

we obtain the following equations

$$\frac{n_\mathrm{B}}{c}\frac{d\overline{F}}{dt} + \frac{1}{L}[F(L,t) - F(0,t)] = g\overline{P}, \tag{12.47}$$

$$\frac{d\overline{P}}{dt} = \gamma_\perp \left[\overline{FD} - (1 + i\Delta)\overline{P}\right], \tag{12.48}$$

$$\frac{d\overline{D}}{dt} = -\gamma_\parallel \left[\frac{1}{2}\left(\overline{FP^*} + \overline{F^*P}\right) + \overline{D} - 1\right]. \tag{12.49}$$

Let us now return to the low-transmission limit defined by Eqs. (9.18) and (11.18) and jointly consider the single-mode limit defined by Eq. (9.23).

Under such conditions the variables F, P and D become uniform in the interval $0 < z < L$, so in Eqs. (12.48) and (12.49) we can replace the spatial averages of products by the corresponding products of spatial averages. In addition, the spatial average of F can be replaced by $F(L,t)$ and hence by $x(t)$. Therefore, if we indicate the spatial averages of the atomic variables simply by P and D, respectively, Eqs. (12.48) and (12.49) become identical to Eqs. (12.44) and (12.45), respectively.

Next, let us consider Eq. (12.47) and the boundary condition (8.36) expressing t in terms of the normalized time,

$$t' = \gamma_\perp t, \tag{12.50}$$

so that we obtain

$$F(L,t') - F(0,t') + \frac{\gamma_\perp n_\mathrm{B} L}{c}\frac{d\overline{F}}{dt'} = gL\overline{P} \tag{12.51}$$

and

$$F(0,t') = Ty + Re^{-i\delta_0}F(L, t' - \gamma_\perp \Delta t), \tag{12.52}$$

respectively. Since $\Delta t < \Lambda/c$ we have that $\gamma_\perp \Delta t$ is small because of Eq. (9.23), and by using Eq. (12.52) we can write

$$F(0,t') \simeq Ty + (1 - T)(1 - i\delta_0)\left[F(L,t') - \gamma_\perp \Delta t \frac{dF(L,t')}{dt'}\right]$$

$$\Rightarrow F(L,t') - F(0,t') \simeq -Ty + (T + i\delta_0)F(L,t') + \gamma_\perp \Delta t \frac{dF(L,t')}{dt'}. \tag{12.53}$$

If we insert Eq. (12.53) into Eq. (12.51) and identify $F(L,t')$ and $\overline{F}(t')$ with $x(t')$, and $dF(L,t')/dt'$ and $d\overline{F}(t')/dt'$ with dx/dt', we obtain

$$\frac{\gamma_\perp}{\kappa}\frac{dx}{dt'} = -(1 + i\theta)x + y + AP, \tag{12.54}$$

where we have used the definitions of Λ (Eq. (8.12)), κ (Eq. (8.27)), θ (Eq. (11.27)) and A (Eq. (9.18)). Finally, by returning to the original time t we obtain Eq. (12.43). In conclusion, the full single-mode model given by Eqs. (12.43)–(12.45) is recovered.

12.4 Stationary solutions of the single-mode model

If we consider the equations of the single-mode model in the stationary state by setting the time derivatives equal to zero, we arrive at the equation

$$y = x \left[\left(1 - \frac{A}{1 + \Delta^2 + |x|^2} \right) + i \left(\theta + \frac{A\Delta}{1 + \Delta^2 + |x|^2} \right) \right]. \tag{12.55}$$

Let us now discuss separately the cases of the free-running laser and of optical bistability. The case of the laser with an injected signal will be studied in the following chapter.

For the free-running laser we set the input field y equal to zero in Eq. (12.55). The stationary solution $x = 0$ corresponds to the laser below threshold. If we divide Eq. (12.55) by x and separate real and imaginary parts, we obtain

$$1 = \frac{A}{1 + \Delta^2 + |x|^2}, \tag{12.56}$$

$$\theta = -\Delta. \tag{12.57}$$

From Eq. (12.56) we recover Eq. (9.24), which was obtained in the low-transmission and single-mode limits. From Eq. (12.57), by inserting the definitions (1.22) and (11.27) of Δ and θ, we obtain the mode-pulling formula

$$\omega_0 = \frac{\kappa \omega_a + \gamma_\perp \omega_c}{\kappa + \gamma_\perp}, \tag{12.58}$$

which coincides with Eq. (9.21) when in the latter one replaces ω_{0j} by ω_0. From Eq. (12.57) we see that in the case of the free-running laser θ and Δ are not independent parameters, and the equations of the single-mode model can be rewritten as

$$\dot{x} = -\kappa[(1 - i\Delta)x - AP], \tag{12.59}$$

$$\dot{P} = \gamma_\perp[xD - (1 + i\Delta)P], \tag{12.60}$$

$$\dot{D} = -\gamma_\parallel \left[\frac{1}{2}(xP^* + x^*P) + D - 1 \right]. \tag{12.61}$$

In the case of optical bistability, if in Eq. (12.55) we replace the pump parameter A by $-2C$, where C is the bistability parameter, we recover Eq. (11.35). The stationary solutions for optical bistability have been discussed in Chapter 11.

13 Single- and two-mode models

In the first section we discuss the stationary solutions for a laser with an injected signal, the active counterpart of the optical bistability analyzed in Chapter 11.

Next, in Section 13.2 we consider another very interesting system, namely a laser with a saturable absorber, a free-running laser in which the optical cavity contains a passive medium in addition to the active medium. The single-mode model for this system can be constructed by combining the single-mode models for the free-running laser and for optical bistability. The discussion is then focussed on its stationary states.

In Section 13.3 we return to the single-mode model for optical bistability and, by introducing an appropriate "cubic limit" on the parameters in play, we derive a cubic model for dispersive optical bistability with two-level atoms. The stationary equation for such a model formally coincides with that derived and discussed for a Kerr medium in Section 11.2.

In the last section we start from the equations for a quadratic nonlinear medium in the degenerate configuration derived in Chapter 6 and combine them with the appropriate boundary conditions for the two field envelopes in play, assuming that the quadratic medium is contained in a ring cavity as happens in an optical parametric oscillator or in second-harmonic generation in a cavity. Next, by introducing the low-transmission limit and the single-mode limit we derive a two-mode model that describes the dynamics of a degenerate optical parametric oscillator or of second-harmonic generation in a cavity. Finally, for the case of a parametric oscillator we illustrate extensively the stationary solutions of the equations.

13.1 A laser with an injected signal

From a technical point of view, this system is interesting because of the possibility of stabilizing the frequency and the phase of separate lasers by taking advantage of the known affinity of nonlinear oscillators to lock onto a periodic forcing element.

In 1972 Spencer and Lamb [125] predicted the possibility of bistable behavior in a laser with an injected signal before the surge of interest in optical bistability in the passive configuration. However, while in the passive case optical bistability is the dominant phenomenon, in the case of a laser with an injected signal a full hysteresis cycle, unaffected by the presence of oscillatory instabilities, is possible only when the corresponding free-running laser, which is obtained by setting the value y of the field injected into the cavity equal to zero, is operating below threshold.

From Eq. (12.55) we obtain the following stationary equation for $Y = y^2$ and $X = |x|^2$:

$$Y = X \left[\left(1 - \frac{A}{1 + \Delta^2 + X} \right)^2 + \left(\theta + \frac{A\Delta}{1 + \Delta^2 + X} \right)^2 \right].$$
(13.1)

It is important to keep in mind that the parameter ω_0 in this section has a different meaning from that in Chapter 9, and consequently the same is true for the parameter Δ. Here ω_0 is the frequency of the input field, whereas in Chapter 9 it corresponds to the emission frequency of the free-running laser in the stationary state. In this section we denote the emission frequency of the free-running laser by ω_L. By using Eq. (12.58) and expressing the frequencies ω_a and ω_c in terms of the parameters Δ and θ (see Eqs. (1.22) and (11.18)) we obtain

$$\omega_L = \omega_0 + \frac{\kappa \gamma_\perp}{\kappa + \gamma_\perp} (\Delta + \theta),$$
(13.2)

such that the frequency ω_L coincides with the input frequency ω_0 only for $\Delta + \theta = 0$. When $\omega_a = \omega_c$ the laser frequency ω_L coincides obviously with ω_a and ω_c, as can be verified from Eq. (13.2) by setting $\Delta = \kappa \theta / \gamma_\perp$, which is equivalent to $\omega_a = \omega_c$.

Correspondingly, also the threshold value of the pump parameter A is given by (see Eq. (9.24) together with Eqs. (1.22) and (11.18))

$$A_{\text{thr}} = 1 + \left(\frac{\omega_a - \omega_c}{\kappa + \gamma_\perp} \right)^2 = 1 + \left(\frac{\gamma_\perp \Delta - \theta \kappa}{\gamma_\perp + \kappa} \right)^2.$$
(13.3)

In the discussion of the stationary curve (13.1) it is convenient to keep in mind the stationary solutions for the free-running laser ($Y = 0$), which are

$$X = 0 \quad \text{(trivial stationary solution)},$$

$$X = A - 1 - \left(\frac{\gamma_\perp \Delta - \theta \kappa}{\gamma_\perp + \kappa} \right)^2 \quad \text{(nontrivial stationary solution)}.$$
(13.4)

For $Y = 0$ the trivial stationary solution is also a point of the stationary curve of the laser with injected signal and is stable when the free-running laser is operating below threshold, whereas it is unstable when the free-running laser is operating above threshold. On the other hand, as we shall see, the nontrivial stationary solution for $Y = 0$ is also a point of the stationary curve of the laser with injected signal only when ω_L coincides with ω_0, i.e. for $\Delta = -\theta$.

Let us consider first the resonant case $\omega_L = \omega_c = \omega_a$, i.e. $\Delta = \theta = 0$, in which the stationary equation reduces to

$$Y = X \left(1 - \frac{A}{1 + X} \right)^2.$$
(13.5)

The stationary curve is shown in Fig. 13.1 for the case in which the free-running laser is operating above threshold, i.e. $A > 1$. The curve is S-shaped, but the entire lower branch is unstable. The negative-slope segment is unstable as always. Hence there is no bistability and no hysteresis cycle [126]. For $Y = 0$ the system approaches the nontrivial stationary solution for the free-running laser, which is stable, and, when Y is increased, the output intensity grows monotonically following the stable upper branch of the stationary curve.

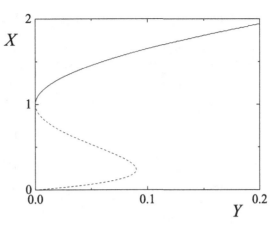

Figure 13.1 A laser with an injected signal in the resonant case $\Delta = \theta = 0$ when the pump parameter A is larger than the threshold for the free-running laser $A_{\text{thr}} = 1$ (here $A = 2$). Not only the negative-slope segment, but also the entire lower branch, is unstable.

Next, let us consider the case $\Delta = -\theta$ in which the free-running laser frequency still coincides with the input frequency. The situation is closely similar to that considered in Eq. (11.39) in the passive case. On setting

$$\overline{A} = \frac{A}{1 + \Delta^2}, \qquad \overline{Y} = \frac{Y}{(1 + \Delta^2)^2}, \qquad \overline{X} = \frac{X}{1 + \Delta^2}, \qquad (13.6)$$

Eq. (13.1), when expressed in terms of $\overline{Y}$, $\overline{X}$ and $\overline{A}$, becomes identical to Eq. (13.5), and therefore also the scenario for this case is the same.

Let us now turn to the case in which the frequency of the free-running laser ω_L differs from the input frequency ω_0. For values of the parameters such that the free-running laser is operating above threshold, the branch of the stationary curve which starts from the origin of the Y–X plane displays an unstable segment, and this is true both when the stationary curve is single-valued and when it is S-shaped. On the other hand, there is a competition between the frequency of the free-running laser and the input frequency, which leads to the onset of spontaneous oscillations in the output intensity, with a frequency that, for small values of the input intensity, is equal to the beat note between the two competing frequencies. On increasing the input intensity, at a certain point this becomes strong enough to force the laser to oscillate at the input frequency, so the output intensity becomes stationary. The value of the input intensity for which this happens is called the *injection-locking threshold*, because the laser frequency gets locked to the input frequency. The stationary curve does not touch the X axis, in contrast with what happens in the case $\Delta = -\theta$ (see Fig. 13.1).

The shape of the stationary curve is determined by the values of the parameters A, Δ and θ only, whereas the stability of the stationary states is governed also by the ratios $\kappa/\gamma_\perp$ and $\gamma_\parallel/\gamma_\perp$. In Fig. 13.2 we consider values of A, Δ and θ such that the stationary curve is S-shaped, and gradually increase the ratio $\tilde{\kappa} = \kappa/\gamma_\perp$ while keeping $\gamma_\parallel/\gamma_\perp$ fixed. For $\tilde{\kappa} = 0.01$ (Fig. 13.2(a)) the entire stationary curve (with the exception of the part with

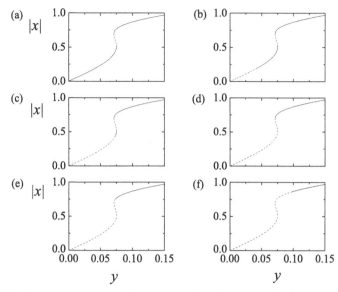

Figure 13.2 A laser with an injected signal for $A = 6$, $\Delta = 2.24$, $\theta = -2$, $\gamma_\parallel/\gamma_\perp = 0.1$ and (a) $\tilde{\kappa} = \kappa/\gamma_\perp = 0.01$, (b) $\tilde{\kappa} = 0.1$, (c) $\tilde{\kappa} = 0.53$, (d) $\tilde{\kappa} = 1$, (e) $\tilde{\kappa} = 1.4$ and (f) $\tilde{\kappa} = 2$. The broken part of the steady-state curve is unstable.

negative slope) is stable, so the picture is the same as for optical bistability of dispersive type in the passive configuration. In this case the free-running laser is operating (slightly) below threshold. If we increase $\tilde{\kappa}$ the free-running laser operates more and more above threshold, and a portion of the steady-state curve with positive slope becomes unstable. In that region, whose size increases with $\tilde{\kappa}$, spontaneous oscillations appear in the output intensity.

In Fig. 13.2(b) the unstable segment is limited to a part of the lower branch to the left of the hysteresis cycle. In Fig. 13.2(c) the instability partially invades the lower part of the hysteresis cycle, such that there is a range of values of Y for which the system displays bistability between an oscillatory state and a stationary state in the upper branch. In Fig. 13.2(d) the entire lower branch is unstable, whereas the entire upper branch is stable, such that the injection-locking threshold coincides with the left turning point of the stationary curve. In Fig. 13.2(e) the injection-locking threshold is beyond the left turning point but inside the hysteresis cycle, and in Fig. 13.2(f) the instability invades the whole hysteresis cycle.

The stability of the stationary states, as well as the spontaneous oscillations which arise in the laser with an injected signal, will be analyzed in Section 25.1 in Part II.

We have seen that, when the free-running laser is operating below threshold, the entire stationary curve may be stable. It is appropriate to remark, however, that this is true in the plane-wave approximation. If we allow for a variation of the field in the transverse plane by including the transverse Laplacian, a segment of the stationary curve may become unstable and one has the onset of a spatial structure in the transverse plane. This will be described in Part III of this volume.

13.2 A laser with a saturable absorber

A difference from the configuration of optical bistability described in Chapter 11 is that in the case of a laser with a saturable absorber the coherent light is not emitted by an external laser and injected into the optical cavity which contains the absorber, but is generated in the same cavity, which includes also an active element with population inversion. So it is not surprising that this system too can display a bistable response, as we will show in this section.

This possibility was predicted by Salomaa and Stenholm in 1973 [127], again before the surge of interest in optical bistability in passive systems.

In order to treat this system theoretically in a reasonably simple way, we choose the approach of generalizing the single-mode model of Eqs. (12.59)–(12.61) with $y = 0$ for the laser to include also a passive element. The generalization is, however, not straightforward because the variables which appear in such equations are normalized with respect to atomic parameters, which differ for the amplifier and for the absorber. Therefore we must, first of all, reformulate Eqs. (12.59)–(12.61) in terms of the original variables E_0, r and r_3. By using the fact that $x = F$ and Eqs. (4.30), (2.39), (4.32), (4.33), (4.34) and (12.38) we obtain the equations

$$\dot{E}_0 = -\kappa(1 + i\theta)E_0 + ic\frac{L}{\mathcal{L}}\frac{k_0}{\epsilon_0}Ndr, \tag{13.7}$$

$$\dot{r} = -\frac{i}{2}\frac{dE_0}{\hbar}r_3 - \gamma_\perp(1 + i\Delta)r, \tag{13.8}$$

$$\dot{r}_3 = i\left(\frac{dE_0}{\hbar}r^* - \text{c.c.}\right) - \gamma_\parallel(r_3 - \sigma), \tag{13.9}$$

where, for simplicity, we have assumed that $n_B = 1$ so that $\Lambda = \mathcal{L}$.

In order to introduce the absorber, we stipulate in general that we indicate the quantities (variables, parameters) which refer to the absorber with the same symbols as used for the amplifier but with an overbar. Therefore Eq. (13.7) becomes

$$\dot{E}_0 = -\kappa(1 + i\theta)E_0 + ic\frac{L}{\mathcal{L}}\frac{k_0}{\epsilon_0}Ndr + ic\frac{\bar{L}}{\mathcal{L}}\frac{k_0}{\epsilon_0}\bar{N}\bar{d}\bar{r}, \tag{13.10}$$

where, for example, $\bar{L}$ and $\bar{N}$ are the length and the atomic density in the passive segment. Furthermore, we must add to Eqs. (13.8)–(13.10) two equations for $\bar{r}$ and $\bar{r}_3$, which read

$$\dot{\bar{r}} = -\frac{i}{2}\frac{\bar{d}E_0}{\hbar}\bar{r}_3 - \bar{\gamma}_\perp(1 + i\bar{\Delta})\bar{r}, \tag{13.11}$$

$$\dot{\bar{r}}_3 = i\left(\frac{\bar{d}E_0}{\hbar}\bar{r}^* - \text{c.c.}\right) - \bar{\gamma}_\parallel(\bar{r}_3 + 1), \tag{13.12}$$

where we have taken into account that in an absorber $\bar{\sigma} = -1$.

Next, we return to the normalized variables $x = F, P, D, \bar{P}$ and $\bar{D}$, where the last two variables are defined in the same way as P and D in Eqs. (4.32) and (4.33), respectively, but with barred quantities. By using the same equations as those mentioned above in the passage from Eqs. (12.59)–(12.61) to Eqs. (13.7)–(13.9), we arrive at the final model for the laser with a saturable absorber [128, 129],

$$\dot{x} = -\kappa \left[(1 + i\theta)x - AP + \frac{2C}{\sqrt{s}} \bar{P} \right],$$ (13.13)

$$\dot{P} = \gamma_\perp [xD - (1 + i\Delta)P],$$ (13.14)

$$\dot{D} = -\gamma_\| \left[\frac{1}{2}(xP^* + x^*P) + D - 1 \right],$$ (13.15)

$$\dot{\bar{P}} = \bar{\gamma}_\perp \left[\sqrt{s}x\bar{D} - (1 + i\bar{\Delta})\bar{P} \right],$$ (13.16)

$$\dot{\bar{D}} = -\bar{\gamma}_\| \left[\frac{\sqrt{s}}{2}(x\bar{P}^* + x^*\bar{P}) + \bar{D} - 1 \right],$$ (13.17)

where the parameter s is the ratio of the saturation intensity (see Eq. (5.25)) of the amplifier to that of the absorber

$$s = \frac{\gamma_\perp \gamma_\|}{\bar{\gamma}_\perp \bar{\gamma}_\|} \frac{\bar{d}^2}{d^2},$$ (13.18)

and C is defined by Eq. (11.5) with L replaced by $\bar{L}$.

Let us now focus on the stationary states, limiting ourselves to the resonant case in which the atomic transition frequencies of the amplifier and of the absorber coincide, and coincide with a cavity frequency, so that $\Delta = \bar{\Delta} = \theta = 0$. By setting the derivatives equal to zero we obtain

$$P = \frac{x}{1 + x^2}, \qquad D = \frac{1}{1 + x^2},$$ (13.19)

$$\bar{P} = \frac{\sqrt{s}x}{1 + sx^2}, \qquad \bar{D} = \frac{1}{1 + sx^2},$$ (13.20)

and the stationary-state equation

$$0 = x \left(1 - \frac{A}{1 + X} + \frac{2C}{1 + sX} \right),$$ (13.21)

where $X = x^2$ as usual. One stationary solution, in which the system does not emit, is the trivial stationary solution $x = 0$. For $s = 1$ the nontrivial stationary solution is $X = A - 2C - 1$. Otherwise, the nontrivial stationary solutions are

$$X_\pm = \frac{1}{2s} \left\{ s(A - 1) - 2C - 1 \pm \left[(s(A - 1) - 2C - 1)^2 - 4s(2C + 1 - A) \right]^{1/2} \right\}.$$ (13.22)

For $s > 1$ they are real for $A \le A_-$ and $A \ge A_+$, with

$$A_\pm = \frac{1}{s} \left[(2C)^{1/2} \pm (s - 1)^{1/2} \right]^2.$$ (13.23)

For these two values of A the two solutions coalesce and we have

$$X_\pm(A_-) = -\frac{[2C(s - 1)]^{1/2} + 1}{s}, \qquad X_\pm(A_+) = \frac{[2C(s - 1)]^{1/2} - 1}{s}.$$ (13.24)

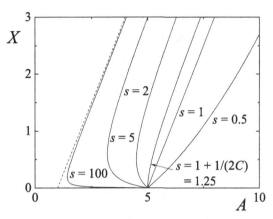

Figure 13.3 A laser with a saturable absorber. Stationary curves for $2C = 4$ and the indicated values of s. The dashed line is the stationary state $X = A - 1$ of the laser without an absorber.

The solutions with $A \leq A_-$ are negative and must be discarded. On the other hand, $X_\pm(A_+)$ is positive when

$$s > 1 + \frac{1}{2C}, \tag{13.25}$$

which is the bistability condition. Figure 13.3 shows the stationary solutions (13.22) as a function of the pump parameter A with C fixed and for various values of s. The intercept of the stationary curve with the A axis occurs for $A = 2C + 1$, which corresponds to the threshold value for the laser with a saturable absorber, which is larger than that in the laser without a saturable absorber.

In this section we discuss the static scenario for a laser with a saturable absorber, while the oscillatory instabilities will be discussed in Section 25.2. A peculiarity of the laser with a saturable absorber is that, in addition to the nontrivial stationary solution, there may also be other stationary intensity solutions with a frequency different from that of the stationary solution. This may happen also in the case of exact resonance between the atomic transition frequencies and a cavity frequency [129]. This circumstance makes the scenario of the laser with a saturable absorber quite complex in general [130]. Here we assume that $\gamma_\perp < \bar{\gamma}_\perp$, which condition excludes the existence of stationary intensity solutions different from the usual resonant stationary solution. In this case the scenario is as follows.

When $s < 1 + 1/(2C)$ the stationary curve is single-valued and one has the same scenario of the laser, with a second-order phase transition at threshold, because (as one can show by performing the linear-stability analysis illustrated in Section 25.2) the trivial stationary solution is stable below threshold and unstable above threshold.

When condition (13.25) is satisfied, there is bistability between the nontrivial stationary solution X_+ given by Eq. (13.22) and the trivial stationary solution $X = 0$, whereas the solution X_- with negative slope is unstable as usual. Thus one has the hysteresis cycle shown in Fig. 13.4, and the scenario corresponds to that of a first-order phase transition. For $s = 1 + 1/(2C)$ one has the critical curve, which displays a vertical tangent at threshold.

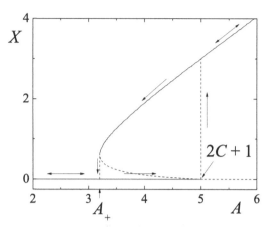

Figure 13.4 A laser with a saturable absorber. The hysteresis cycle of the normalized output intensity as a function of the pump parameter A for $2C = 4$ and $s = 5$.

The bistability arises from the presence of the saturable absorber. When s is small the saturation parameter of the absorber is much larger than that of the amplifier and the behaviour is similar to that of the laser, with the threshold shifted from $A = 1$ to $A = 1 + 2C$. When s is increased, the saturation parameter of the absorber becomes smaller than that of the amplifier and the absorber can be saturated more and more easily. Beyond the value of s specified by condition (13.25), the saturation of the absorber occurs in a discontinuous way, leading to a hysteresis cycle. Finally, when s becomes very large the hysteresis cycle gets compressed towards the A axis and its left turning point approaches the threshold $A = 1$, $X = 0$ of the resonant laser without an absorber (case $s = 100$ in Fig. 13.3). This implies that in this limit, when the value $A = 1$ is surpassed, any small fluctuation is capable of bringing the system to the nontrivial stationary solution, so that the behavior of the system, when the pump parameter is gradually increased, basically coincides with that of the laser without an absorber. This happens because in the limit of parameter s being very large the absorber very easily gets saturated and therefore loses any influence it may have had.

Hysteretic effects in lasers with saturable absorbers were observed experimentally in [131, 132], while full hysteresis cycles have been reported in [133–135].

13.3 The cubic model for dispersive optical bistability

In this section we return to the single-mode model for optical bistability in passive systems of two-level atoms, and focus on the limit of large dispersion, i.e. large atomic detuning Δ. We introduce an appropriate multiple limit [103, 136] that leads to a cubic model for dispersive optical bistability that is essentially identical, in the stationary state, to Eq. (11.22) derived in Section 11.2 for a cavity containing a Kerr medium.

The starting point is given by Eqs. (12.43)–(12.45) with A replaced by the opposite of the bistability parameter $2C$. If we assume that $\kappa \ll \gamma_\perp, \gamma_\parallel$ we can adiabatically eliminate the atomic variables by setting $dP/dt = 0$ and $dD/dt = 0$, so that, by normalizing the time with respect to the inverse of the cavity decay rate κ, we obtain

$$\frac{dx}{dt''} = y - \left[1 + i\theta + \frac{2C(1 - i\Delta)}{1 + \Delta^2 + |x|^2}\right] x, \tag{13.26}$$

where $t'' = \kappa t$. For $\Delta^2 \gg 1$ we can neglect the 1 in the denominator and expand the factor $1 + |x|^2/\Delta^2$, obtaining

$$\frac{dx}{dt''} = y - \left[1 + i\theta + \frac{2C}{\Delta^2}(1 - i\Delta) \sum_{n=0}^{\infty} \left(-\frac{|x|^2}{\Delta^2}\right)^n\right] x. \tag{13.27}$$

If we now define

$$C' = \frac{C}{\Delta\theta}, \quad x' = x\frac{|\theta|^{1/2}}{\Delta}, \quad y' = y\frac{|\theta|^{1/2}}{\Delta}, \tag{13.28}$$

Eq. (13.27) becomes

$$\frac{dx'}{dt''} = y' - \left\{1 + i\theta + 2C'\left(\frac{\theta}{\Delta} - i\theta\right)\left[1 - \frac{|x'|^2}{|\theta|} + \sum_{n=2}^{\infty}\left(-\frac{|x'|^2}{|\theta|}\right)^n\right]\right\} x'. \tag{13.29}$$

Let us now introduce a multiple limit that we call the *cubic limit* [103, 136]. This is defined by

$$\Delta \to \infty, \ \theta \to \infty, \ C \to \infty,$$
$$\text{with} \quad \frac{\theta}{\Delta} \to 0, \ 2C' = \frac{2C}{\Delta\theta} \to 1, \ \theta(1 - 2C') \to \bar{\theta} \quad \text{arbitrary}, \tag{13.30}$$

while we keep x' and y' finite. Note that, since C is positive, the limit $2C/\Delta\theta \to 1$ implies that Δ and θ have the same sign. In the multiple limit (13.30), Eq. (13.29) reduces to the cubic equation

$$\frac{dx'}{dt''} = y' - \left[1 - i\bar{\eta}\left(|x'|^2 - \bar{\theta}\right)\right] x', \tag{13.31}$$

where we have set

$$\bar{\eta} = -\frac{\theta}{|\theta|}, \quad \bar{\bar{\theta}} = \bar{\eta}\bar{\theta}. \tag{13.32}$$

Finally, if we define

$$Y = |y'|^2, \quad X = |x'|^2, \tag{13.33}$$

the stationary equation takes the form

$$Y = \left[1 + \left(\bar{\bar{\theta}} - X\right)^2\right] X, \tag{13.34}$$

which formally coincides with Eq. (11.22) obtained in the case of a Kerr medium instead of two-level atoms. The solutions of this stationary equation have been discussed in Section 11.2. Here we note only that on using Eq. (13.28) we have that $|x|^2/\Delta^2 = |x'|^2/|\theta|$ tends to zero in the cubic limit (13.30) because θ tends to infinity while x' remains finite. Therefore in the dispersive limit (13.30) the entire hysteresis cycle corresponds to a situation of an unsaturated system, whereas in the case of absorptive bistability (see Section 11.1) the lower branch is unsaturated and the upper branch is saturated.

We note finally that the case $\bar{\eta} = +1$ corresponds to $\theta < 0$, hence $\Delta < 0$ since Δ and θ have the same sign. In the two-level case the parameter $\chi^{(3)}$ which appears in Eq. (7.9) is replaced by $g\chi^{(3)''}$ (see Eqs. (5.33), (5.26) and (5.27)). From Eq. (5.28) we see that $g\chi^{(3)''}$ has the opposite sign of Δ since in the passive case $g < 0$. Therefore $\bar{\eta} = +1$ corresponds to $g\chi^{(3)''} > 0$, i.e. to the case $\chi^{(3)} > 0$ of the Kerr medium, which is the self-focusing configuration (see Section 7.3)).

13.4 A model for the degenerate optical parametric oscillator (and harmonic generation in a cavity) and its stationary solutions

In this section we turn to the case of quadratic materials discussed in Chapter 6 and consider the case in which an optical ring cavity contains a sample of such a material. We derive, first of all, a model that is capable of describing the dynamics of a degenerate optical parametric oscillator (or subharmonic generation in a cavity) and of second-harmonic generation in a cavity. For the latter case we refer the reader to other books [29, 59], whereas we focus on the case of the degenerate parametric oscillator, for which we describe, in this chapter, the plane-wave stationary states.

In Section 6.3.1 we considered a cavityless sample into which we inject a weak coherent field of frequency ω_1 and a strong coherent pump field of frequency $\omega_2 = 2\omega_1$, and we saw that the weak field is amplified in a phase-sensitive way (optical parametric amplifier). In the degenerate optical parametric oscillator there is the interaction between two cavity modes, one of frequency ω_2 and the other with the subharmonic frequency $\omega_1 = \omega_2/2$, and one injects into the cavity a coherent field of frequency ω_2 (the pump field) and no field of frequency ω_1 (the signal field). Beyond a certain threshold for the intensity of the input field, in the output of the cavity there is not only a field of frequency ω_2, but also a signal field of frequency ω_1. Thus, with respect to the signal field the degenerate optical parametric oscillator behaves in a way similar to that of a laser. Exactly as the laser emission is triggered by spontaneous emission, in the degenerate optical parametric oscillator the emission of the signal field is triggered by spontaneous parametric down-conversion (see Section 6.3.1). An important difference is that in the parametric-oscillator case the signal photons are generated in pairs. This difference becomes of paramount importance in the quantum domain because the two signal photons are in a state of quantum entanglement, and quantum entanglement is a basic resource in the field of quantum information and quantum communication.

An important remark is that the two-mode model expressed by the following two equations (13.51) and (13.52) is derived, in the literature, from a quantum-mechanical master equation in which the electromagnetic field is quantized (the same is true for the model for the laser with a saturable absorber discussed in Section 13.2). In accord with the style of this book we will derive it, instead, in the framework of a semiclassical theory.

In order to derive the two-mode model, let us start from Eqs. (6.22) and (6.23) assuming that there is perfect phase matching so that $\Delta k = 0$ (see Eq. (6.13)). The condition $k(\omega_1) = k(\omega_2)$, with $k(\omega) = \omega n(\omega)/c$, and $\omega_2 = 2\omega_1$, implies that $n(\omega_2) = n(\omega_1)$ (see the discussion after Eq. (6.14)), and we call their common value simply n.

As we saw in Section 6.2, the two fields A_1 and A_2 in Eqs. (6.22) and (6.23) are such that $|A_1|^2$ and $|A_2|^2$ correspond to the photon number fluxes of the two fields, respectively. It is now convenient to pass to variables $\tilde{A}_1$ and $\tilde{A}_2$ such that their squared moduli correspond, in the low-transmission limit, to the photon numbers of the two fields in the sample of volume V contained in the cavity, respectively, so that the variables $\tilde{A}_1$ and $\tilde{A}_2$ are dimensionless. The two variables $\tilde{A}_p$ ($p = 1, 2$) are linked to A_p by the relation

$$A_p = \left(\frac{c}{nV}\right)^{1/2} \tilde{A}_p \quad (p = 1, 2). \tag{13.35}$$

If we reformulate Eqs. (6.22) and (6.23) in terms of the variables $\tilde{A}_p$ we obtain

$$\frac{n}{c}\frac{\partial \tilde{A}_1}{\partial t} + \frac{\partial \tilde{A}_1}{\partial z} = i\bar{g}\tilde{A}_2\tilde{A}_1^*, \tag{13.36}$$

$$\frac{n}{c}\frac{\partial \tilde{A}_2}{\partial t} + \frac{\partial \tilde{A}_2}{\partial z} = i\frac{\bar{g}}{2}\tilde{A}_1^2, \tag{13.37}$$

where we have neglected the terms which describe diffraction and group velocity dispersion, we have approximated $1/v_{g1}$ and $1/v_{g2}$ by n/c and we have defined

$$\bar{g} = 2\pi \left[\frac{2\hbar\omega_1^2\omega_2}{\epsilon_0 V}\right]^{1/2} \frac{1}{cn^2}\chi^{(2)}. \tag{13.38}$$

On the other hand, the boundary conditions for the two fields are

$$\tilde{A}_p(0, t) = T_p y_p + R_p \tilde{A}_p(L, t - \Delta t)e^{-i\delta_{0p}} \quad (p = 1, 2), \tag{13.39}$$

where, in complete analogy with Eqs. (8.36) and (8.37), we have the following.

- T_p and $R_p = 1 - T_p$ ($p = 1, 2$) are the transmissivity and reflectivity coefficients of the beam splitters of the cavity for frequencies ω_1 and ω_2, respectively.
- y_p are the input fields of frequency ω_1 and ω_2, respectively, injected into the cavity. For the moment we consider the general case of two input fields.
- δ_{0p} are the cavity detuning parameters for the two fields, defined as

$$\delta_{0p} = \frac{\omega_{cp} - \omega_p}{c/\Lambda} \quad (p = 1, 2), \tag{13.40}$$

and ω_{cp} is the cavity frequency closest to ω_p ($p = 1, 2$). Λ is defined as in Eq. (8.12), with n_B replaced by n.

Next, let us impose the low-transmission limit and the single-mode limit in the same way as we did in Section 12.3 to derive the single-mode model for the two-level case.[1]

---

[1] In the present case the single-mode limit leads to a single cavity mode for each of the two frequency bands centered at frequencies ω_1 and ω_2.

First, let us introduce spatial averages of the fields

$$\bar{A}_p(t) = \frac{1}{L} \int_0^L dz \, \tilde{A}_p(z, t), \tag{13.41}$$

and let us take the spatial average over Eqs. (13.36) and (13.37), giving

$$\frac{n}{c} \frac{d\bar{A}_1}{dt} + \frac{1}{L} \left[\bar{A}_1(L, t) - \bar{A}_1(0, t) \right] = i\bar{g}\bar{A}_2\bar{A}_1^*, \tag{13.42}$$

$$\frac{n}{c} \frac{d\bar{A}_2}{dt} + \frac{1}{L} \left[\bar{A}_2(L, t) - \bar{A}_2(0, t) \right] = i\frac{\bar{g}}{2}\bar{A}_1^2, \tag{13.43}$$

where we have factorized spatial averages of products into products of spatial averages because in the low-transmission plus single-mode limits the fields become uniform along the sample.

In order to apply the single-mode limit, in Section 12.3 we normalized the time with respect to the inverse of the atomic rate $\gamma_\perp$ (Eq. (12.50)). Since in the single-mode limit the rates $\gamma_\perp$ and κ have comparable orders of magnitude, an equivalent step is to normalize time with respect to the inverse of the cavity linewidth κ. In the present case we introduce the damping rates for the two cavity modes,

$$\kappa_p = \frac{cT_p}{\Lambda} \quad (p = 1, 2), \tag{13.44}$$

and normalize t with respect to the inverse of κ_1

$$t_1 = \kappa_1 t. \tag{13.45}$$

Hence Eq. (13.39) becomes

$$\tilde{A}_p(0, t_1) \simeq T_p y_p + (1 - T_p)(1 - i\delta_{0p}) \left[\tilde{A}_p(L, t_1) - \kappa_1 \, \Delta t \, \frac{\partial \tilde{A}_p(L, t_1)}{\partial t_1} \right]. \tag{13.46}$$

The low-transmission plus single-mode limit is defined, in this case, by

$$T_p \to 0, \qquad \bar{g}L \to 0, \qquad \delta_{0p} \to 0,$$

$$\text{with} \qquad \frac{T_1}{T_2}, \qquad B = \frac{\bar{g}L}{T_1}, \qquad \theta_p = \frac{\delta_{0p}}{T_p} \qquad \text{constant.} \tag{13.47}$$

In this multiple limit one can express the differences $[\tilde{A}_p(L, t) - \tilde{A}_p(0, t)]$ as a sum of terms of first order in the small quantities T_p, δ_{0p} and $\kappa_1 \Delta t < T_1$, insert these differences into Eqs. (13.42) and (13.43) multiplied by κ_p^{-1}, respectively, and express the time t in terms of the time t_1. Since in the limit (13.47) the fields become uniform, we can identify $\tilde{A}_p(L, t)$ with $\bar{A}_p(t)$ and $d\tilde{A}_p(L, t)/dt$ with $d\bar{A}_p/dt$. By taking into account that $\Lambda = nL + \mathcal{L} - L$, taking into account the definitions of B and θ_p in Eq. (13.47) and returning to the time t we arrive at these two coupled equations for sub/second-harmonic generation in a cavity [137, 138]:

$$\frac{d\bar{A}_1}{dt} = \kappa_1 \left[y_1 - (1 + i\theta_1)\bar{A}_1 + iB\bar{A}_2\bar{A}_1^* \right], \tag{13.48}$$

$$\frac{d\bar{A}_2}{dt} = \kappa_2 \left[y_2 - (1 + i\theta_2)\bar{A}_2 \right] + i\kappa_1 \frac{B}{2} \bar{A}_1^2. \tag{13.49}$$

Finally, the transformation of variables

$$x_1 = [\kappa_1/(2\kappa_2)]^{1/2} B \bar{A}_1, \quad x_2 = i B \bar{A}_2, \quad y_1^{(0)} = [\kappa_1/(2\kappa_2)]^{1/2} B y_1, \quad y_2^{(0)} = i B y_2,$$
(13.50)

brings us to the final forms of the equations of the two-mode model for sub/second-harmonic generation [139]:

$$\dot{x}_1 = \kappa_1 \left[y_1^{(0)} - (1 + i\theta_1)x_1 + x_2 x_1^* \right],$$
(13.51)

$$\dot{x}_2 = \kappa_2 \left[y_2^{(0)} - (1 + i\theta_2)x_2 - x_1^2 \right].$$
(13.52)

It is remarkable that this model displays only four free parameters, i.e. $y_1^{(0)}, y_2^{(0)}, \theta_1$ and θ_2. In the case $y_2^{(0)} = 0$, in which one injects into the cavity the fundamental frequency, the model (13.51) and (13.52) describes second-harmonic generation in a ring cavity. When, instead, $y_1^{(0)} = 0$, one injects the second-harmonic frequency and the model describes subharmonic down-conversion in a ring cavity, i.e. a degenerate optical parametric oscillator. In the following we focus on this second case and describe the behavior of the stationary solutions.

By setting $\dot{x}_1 = 0$ and $\dot{x}_2 = 0$ and defining

$$X_1 = \frac{|x_1|^2}{1 + \theta_2^2}, \quad X_2 = \frac{|x_2|^2}{1 + \theta_2^2}, \quad Y = \frac{y_2^{(0)2}}{1 + \theta_2^2},$$
(13.53)

where we have taken $y_2^{(0)}$ real, one finds two different kinds of stationary solutions:

(i) the trivial stationary solution

$$X_1 = 0, \quad X_2 = \frac{Y}{1 + \theta_2^2},$$
(13.54)

in which no signal field is emitted; and

(ii) the nontrivial solutions, which obey the stationary equation

$$Y = 1 + \theta_1^2 + 2(1 - \theta_1\theta_2)X_1 + (1 + \theta_2^2)X_1^2,$$
(13.55)

while X_2 is given by

$$X_2 = \frac{1 + \theta_1^2}{1 + \theta_2^2}.$$
(13.56)

Note that in the trivial stationary solution (13.54) the phase of x_1 is arbitrary while the phase of x_2 is linked to that of $y_2^{(0)}$ by the relation $x_2 = y_2^{(0)}/(1 + i\theta_2)$. From Eq. (13.55) we see that the threshold of the degenerate parametric oscillator for the emission of the signal field, corresponding to $X_1 = 0$, is $Y = 1 + \theta_1^2$. One can prove that below threshold the trivial stationary solution is stable and above threshold it is unstable, exactly as in the laser (see Section 25.3).

If we solve the quadratic equation (13.55) with respect to X_1 and plot X_1 as a function of Y, for $\theta_1\theta_2 < 1$ we find the behavior shown in Fig. 13.5(a) and for $\theta_1\theta_2 > 1$ the scenario of Fig. 13.5(b). In general, the picture is closely similar to that of the laser with a saturable absorber, with an analogy to second-order phase transitions for $\theta_1\theta_2 < 1$ and with

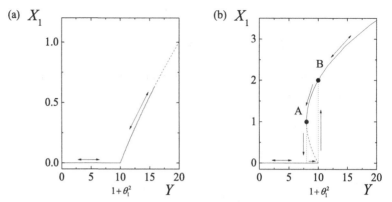

Figure 13.5 The normalized intensity X_1 of the subharmonic is plotted as a function of the normalized input intensity Y for $\theta_1 = 3$ and (a) $\theta_2 = -1$ and (b) $\theta_2 = 1$. The broken part in (a) is unstable for $\kappa_1 = \kappa_2$; the broken (negative-slope) portion in (b) is always unstable.

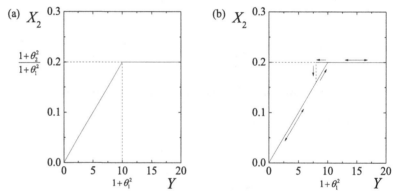

Figure 13.6 The behavior of the intensity X_2 of the pump field as a function of Y, with the same values for θ_1 and θ_2 as in Fig. 13.5 for parts (a) and (b).

bistability and an analogy to first-order phase transitions for $\theta_1\theta_2 > 1$. Therefore the bistability condition is

$$\theta_1\theta_2 > 1, \tag{13.57}$$

and the bistability is between a nontrivial stationary solution and the trivial stationary solution in the interval

$$1 + \theta_1^2 - \frac{(\theta_1\theta_2 - 1)^2}{1 + \theta_2^2} < Y < 1 + \theta_1^2, \tag{13.58}$$

which is indicated by the points A and B in Fig. 13.5(b). As usual, one can prove that the part of the stationary curve with negative slope is unstable. Note that the bistability condition (13.57) requires that both fields are detuned.

Figures 13.6(a) and (b) show the behavior of the normalized intensity X_2 of the second-harmonic field as a function of the normalized input intensity Y for the same values of the parameters as in Figs. 13.5(a) and (b), respectively. The shape of Fig. 13.6(a) is reminiscent of the behavior of the inversion per atom r_3 as a function of the pump inversion σ (see Fig. 9.3), with a clamping of X_2 to its threshold value (13.56) (X_2 remains constant for $Y > 1 + \theta_1^2$). In the bistable case the hysteresis cycle for X_2 has a triangular shape.

Bistability in the optical parametric oscillator has been demonstrated experimentally in [140].

In Section 25.3 we will show that in the case of Fig. 13.5(a) the broken part of the curve is unstable against the onset of spontaneous oscillations.

14

In Chapters 8–13 we always assumed that the nonlinear medium is contained in a uni-directional ring cavity. The case of a Fabry–Perot cavity (Fig. 8.1) is substantially more complex because the interference of the two counterpropagating fields gives rise to a standing-wave pattern that modulates the population difference and the atomic polarization to form gratings with a spatial frequency equal to the inverse of the half-wavelength and its harmonics. Therefore the atomic variables exhibit, in addition to the slow spatial variation which characterizes the envelopes, also a fast spatial variation on the wavelength scale, and the coexistence of these two spatial scales poses a highly nontrivial problem that must be treated with care.

In Chapters 9 and 11 we solved exactly the Maxwell–Bloch equations with the ring-cavity boundary conditions for both the active and the passive configuration. Exact solutions of this kind have been obtained also for the equations and boundary conditions of a Fabry–Perot cavity [117–119]. However, for the sake of simplicity we will pass directly to models that are valid in the low-transmission approximation. Again for the sake of simplicity, in this chapter we assume that $n_B = 1$, so that the light velocity in the empty part of the cavity is the same as that in the atomic sample.

By following the same line of reasoning as that which we pursued in Chapter 12, in Section 14.1 we introduce a spatial transformation of the electric-field envelopes, which leads to boundary conditions without parameters, while the parameters present in the boundary conditions of the original variables appear in the envelope equations. As in the ring-cavity case, this approach allows one to introduce the cavity modes in a natural way, and to derive a set of modal equations coupled to the equations for the atomic variables. Sets of equations of this kind are common in the laser literature, but are not derived from the true boundary conditions for the cavity. The modes for a closed cavity are forced into the model from the very beginning and the damping term of the field equations is added phenomenologically, instead of arising naturally from the boundary conditions. These modal equations have been studied under rather substantial approximations such as, for example, the single-mode approximation and the cubic approximation. The single-mode model is discussed in Section 14.2, in which we derive an analytic form of the stationary equation and discuss the stationary solutions, comparing them with those for the ring-cavity case.

In Section 14.3 we pursue another approach in which one distinguishes the two coexisting spatial scales in play right from the very beginning. This approach allows one to formulate separate equations for the forward and backward propagating fields, which is most advanta-geous, for example, for the linear-stability analysis of the single-mode stationary solution. This approach is also most appropriate for discussing the case in which the atomic diffusion washes out the grating and to demonstrate that in this situation the single-mode model for the Fabry–Perot cavity reduces to a model equivalent to the single-mode model for a ring cavity.

In Section 14.4 we return to the case of a ring cavity, in order to show that in the case $n_B = 1$ one can avoid introducing the transformation of coordinates used in Chapter 12, and in this way we simplify the resulting modal equations.

The analysis of Chapter 12 and of Sections 14.1–14.3 assumes that the atomic sample is much longer than a wavelength. In Section 14.5 we focus, instead, on the special case of an ultrathin atomic sample of length much shorter than a wavelength. This case can arise, for example, when the atomic medium is formed by a multiple quantum well. We follow the approach introduced by Ikeda in order to transform the set of Maxwell–Bloch equations into a set of difference-differential equations, which is much more convenient for the numerical resolution than the original set of partial differential equations. The remarkable feature is that, in this special case, the set of difference-differential equations can be derived both for a ring and a Fabry–Perot cavity, and that the equations in the Fabry–Perot case are only slightly more complex than those in the ring-cavity case. In this chapter we calculate in a straightforward way the stationary solutions for this ultrathin medium case, while the discussion of the dynamical behaviors will be presented in Part II in Section 22.7.

14.1 Modal equations for the Fabry–Perot cavity

In this chapter we assume that there is no background refractive index, i.e. $n_B = 1$. In the case of bidirectional propagation, the definition of the envelopes for the electric field and for the atomic polarization is different from that used for the unidirectional ring cavity. Returning to Eq. (2.37), for the electric field we write

$$\mathcal{E}(z, t) = \frac{1}{2}[e^{-i\omega_0 t}(E_F(z, t)e^{ik_0 z} + E_B(z, t)e^{-ik_0 z}) + \text{c.c.}], \qquad (14.1)$$

where E_F and E_B are the envelopes of the forward- and backward-propagating fields, respectively (see Fig. 8.1), which vary slowly both in time and in space. The Rabi frequency Ω_R (see Eq. (2.38)) is written as

$$\Omega_R(z, t) = e^{-i\omega_0 t}\left(\Omega_F e^{ik_0 z} + \Omega_B e^{-ik_0 z}\right) + \text{c.c.},$$

$$\Omega_F(z, t) = \frac{dE_F}{\hbar}, \qquad \Omega_B(z, t) = \frac{dE_B}{\hbar} \qquad (14.2)$$

and $k_0 = \omega_0/c$.

On the other hand, for the atomic polarization we return to Eqs. (2.40) and (2.41), writing

$$\rho_{12} = \tilde{r}(z, t)e^{-i\omega_0 t}, \qquad \mathcal{P}(z, t) = \frac{\tilde{P}_0(z, t)}{2}e^{-i\omega_0 t} + \text{c.c.}, \qquad (14.3)$$

with

$$\tilde{P}_0(z, t) = 2Nd\tilde{r}(z, t), \qquad (14.4)$$

where now the envelopes $\tilde{r}(z, t)$ and $\tilde{P}_0(z, t)$ vary slowly in time but not in space. If we now insert Eqs. (14.1) and (14.3) into the Bloch equations (2.35) and (2.36), take Eq. (2.43) into account and apply the rotating-wave approximation, we arrive at the equations

$$\frac{\partial \tilde{r}}{\partial t} = -i\delta \tilde{r} - \frac{i}{2}\left(\Omega_\mathrm{F} e^{ik_0 z} + \Omega_\mathrm{B} e^{-ik_0 z}\right) r_3 - \gamma_\perp \tilde{r}, \tag{14.5}$$

$$\frac{\partial r_3}{\partial t} = i\left[\left(\Omega_\mathrm{F} e^{ik_0 z} + \Omega_\mathrm{B} e^{-ik_0 z}\right)\tilde{r}^* + \mathrm{c.c.}\right] - \gamma_\parallel (r_3 - \sigma \chi(z)), \tag{14.6}$$

where δ is defined by Eq. (2.48) and we have added the irreversible atomic terms which were introduced in Chapter 4 (see Eqs. (4.24) and (4.25)). $\chi(z)$ is the characteristic function of the atomic sample, which is defined as (see Fig. 8.1)

$$\chi(z) = 1 \quad \text{for} \quad L_1 < z < L_2, \qquad \chi(z) = 0 \quad \text{for} \quad 0 \le z \le L_1, \ L_2 \le z \le L, \tag{14.7}$$

and $\bar{L} = L_2 - L_1$ is the length of the atomic sample. On the other hand, let us consider the Maxwell wave equation in the plane-wave approximation

$$\frac{\partial^2 \mathcal{E}}{\partial z^2} - \frac{1}{c^2}\frac{\partial^2 \mathcal{E}}{\partial t^2} = \frac{1}{\epsilon_0 c^2}\frac{\partial^2 \mathcal{P}}{\partial t^2}. \tag{14.8}$$

If we insert Eqs. (14.1)–(14.4) into this equation and apply the slowly varying envelope approximation, we arrive at the equation

$$e^{ik_0 z}\left(\frac{\partial \Omega_\mathrm{F}}{\partial z} + \frac{1}{c}\frac{\partial \Omega_\mathrm{F}}{\partial t}\right) + e^{-ik_0 z}\left(-\frac{\partial \Omega_\mathrm{B}}{\partial z} + \frac{1}{c}\frac{\partial \Omega_\mathrm{B}}{\partial t}\right) = i\frac{k_0}{\epsilon_0}N\frac{d^2}{\hbar}\tilde{r}. \tag{14.9}$$

As in Chapter 8 for the ring cavity, it is convenient to introduce normalized dimensionless variables, defined as

$$F_\mathrm{F} = \frac{\Omega_\mathrm{F}}{\sqrt{\gamma_\perp \gamma_\parallel}}, \qquad F_\mathrm{B} = \frac{\Omega_\mathrm{B}}{\sqrt{\gamma_\perp \gamma_\parallel}}, \qquad \tilde{P} = \frac{2i}{\sigma}\sqrt{\frac{\gamma_\perp}{\gamma_\parallel}}\tilde{r}, \qquad D = \frac{r_3}{\sigma}, \tag{14.10}$$

in terms of which the dynamical equations become

$$e^{ik_0 z}\left(\frac{\partial F_\mathrm{F}}{\partial z} + \frac{1}{c}\frac{\partial F_\mathrm{F}}{\partial t}\right) + e^{-ik_0 z}\left(-\frac{\partial F_\mathrm{B}}{\partial z} + \frac{1}{c}\frac{\partial F_\mathrm{B}}{\partial t}\right) = g\tilde{P}, \tag{14.11}$$

$$\frac{\partial \tilde{P}}{\partial t} = \gamma_\perp \left[\left(F_\mathrm{F} e^{ik_0 z} + F_\mathrm{B} e^{-ik_0 z}\right)D - (1 + i\Delta)\tilde{P}\right], \tag{14.12}$$

$$\frac{\partial D}{\partial t} = -\gamma_\parallel \left[\frac{1}{2}\left(\left(F_\mathrm{F} e^{ik_0 z} + F_\mathrm{B} e^{-ik_0 z}\right)\tilde{P}^* + \mathrm{c.c.}\right) + D - \chi(z)\right], \tag{14.13}$$

where the atomic detuning parameter Δ is given by Eq. (4.31) and the parameter g is defined by Eq. (4.34). Equations (14.11)–(14.13) constitute the Maxwell–Bloch equations for the Fabry–Perot cavity [141]. The presence of $\chi(z)$ in Eq. (14.13) ensures that D and $\tilde{P}$ vanish for values of z where $\chi(z)$ vanishes.

In Section 14.3 we will show how one can formulate separate equations for the forward- and backward-propagating envelopes.

If we assume that there is an incident field E_I injected into the cavity (in the forward direction) of frequency ω_0, the boundary conditions at the two mirrors located at $z = 0$ and

$z = L$ are (see Fig. 8.1)

$$E_F(0, t) = \sqrt{R}E_B(0, t) + \sqrt{T}E_I, \tag{14.14}$$

$$E_B(L, t)e^{-ik_0 L} = \sqrt{R}E_F(L, t)e^{ik_0 L}$$

$$\longrightarrow E_B(L, t) = \sqrt{R}e^{-i\delta_0}E_F(L, t), \tag{14.15}$$

where the parameter δ_0 is defined as

$$\delta_0 = 2\pi\bar{j} - 2k_0 L = \frac{\omega_c - \omega_0}{c/(2L)}, \tag{14.16}$$

with $\bar{j}$ being a positive integer and ω_c the cavity frequency closest to the input frequency ω_0. We remind the reader that for a Fabry–Perot cavity the resonance frequencies are given by

$$\omega_j = \frac{\pi c}{L}j, \quad j = 1, 2, \ldots. \tag{14.17}$$

The transmitted and reflected fields are given by

$$E_T(t) = \sqrt{T}E_F(L, t)e^{i2\pi L/\lambda_0}, \tag{14.18}$$

$$E_R(t) = -\sqrt{R}E_I + \sqrt{T}E_B(0, t), \tag{14.19}$$

with $\lambda_0 = 2\pi/k_0$. Let us now introduce normalized variables also for the input and for the transmitted (or emitted, in the case of a free-running laser) field

$$y = \frac{dE_I}{\hbar\sqrt{\gamma_\perp\gamma_\parallel T}}, \quad x = \frac{dE_T}{\hbar\sqrt{\gamma_\perp\gamma_\parallel T}}e^{-i2\pi L/\lambda_0}. \tag{14.20}$$

The transmitted field is linked to the envelope F_F by the relation

$$x(t) = F_F(L, t). \tag{14.21}$$

In terms of normalized variables, the boundary conditions (14.14) and (14.15) read

$$F_F(0, t) = \sqrt{R}F_B(0, t) + Ty, \tag{14.22}$$

$$F_B(L, t) = \sqrt{R}e^{-i\delta_0}F_F(L, t). \tag{14.23}$$

As in the ring cavity, we introduce a transformation on the field envelopes [141],

$$F'_F(z, t) = e^{[(z-L)/(2L)](\ln R - i\delta_0)}\left[F_F(z, t) + e^{i\delta_0/2}\frac{z - L}{2L}\frac{Ty}{\sqrt{R}}\right], \tag{14.24}$$

$$F'_B(z, t) = e^{-[z/(2L)](\ln R - i\delta_0)}\left[e^{i\delta_0/2}F_B(z, t) - e^{i\delta_0/2}\frac{z - L}{2L}\frac{Ty}{\sqrt{R}}\right], \tag{14.25}$$

which puts the boundary conditions into a form that does not contain any parameter,

$$F'_F(0, t) = F'_B(0, t), \qquad F'_B(L, t) = F'_F(L, t). \tag{14.26}$$

If we apply the transformations (14.24) and (14.25) to Eq. (14.11), we obtain

$$
e^{ik_{\bar{j}}z} \left\{ e^{-[(z-L)/(2L)]\ln R} e^{-i\delta_0/2} \left[\frac{\partial F_F'}{\partial z} + \frac{1}{c} \frac{\partial F_F'}{\partial t} - \frac{e^{i\delta_0/2}}{2L} \frac{Ty}{\sqrt{R}} \right. \right.
$$
$$
\left. \left. - \frac{\ln R - i\delta_0}{2L} \left(F_F' - e^{i\delta_0/2} \frac{z-L}{2L} \frac{Ty}{\sqrt{R}} \right) \right] \right\}
$$
$$
+ e^{-ik_{\bar{j}}z} \left\{ e^{[z/(2L)]\ln R} e^{-i\delta_0/2} \left[-\frac{\partial F_B'}{\partial z} + \frac{1}{c} \frac{\partial F_B'}{\partial t} + \frac{e^{i\delta_0/2}}{2L} \frac{Ty}{\sqrt{R}} \right. \right.
$$
$$
\left. \left. - \frac{\ln R - i\delta_0}{2L} \left(e^{-i\delta_0/2} F_B' + \frac{z-L}{2L} \frac{Ty}{\sqrt{R}} \right) \right] \right\} = g\tilde{P},
$$

(14.27)

where $k_{\bar{j}} = \omega_{\bar{j}}/c$ and we have taken into account that

$$
k_0 + \frac{\delta_0}{2L} = \frac{\omega_0}{c} + \frac{\omega_{\bar{j}} - \omega_0}{c/(2L)} \frac{1}{2L} = \frac{\omega_{\bar{j}}}{c} = k_{\bar{j}}.
$$

(14.28)

Next, let us take the boundary conditions (14.26) into account. If we set

$$
F_F'(z,t) = f_F e^{ik_F(z-ct)}, \qquad F_B'(z,t) = f_B e^{-ik_B(z+ct)},
$$

(14.29)

then from the boundary conditions we obtain that

$$
f_F = f_B, \qquad k_F = k_B = k_n, \qquad k_n = \frac{\pi}{L}n, \; n = 0, \pm 1, \dots.
$$

(14.30)

Therefore we can introduce into Eq. (14.27) the expansions

$$
F_F'(z,\tau) = \sum_n f_n(t) e^{i(\pi n/L)z}, \qquad F_B'(z,\tau) = \sum_n f_n(t) e^{-i(\pi n/L)z},
$$

(14.31)

where the sum is extended to a limited range of (positive and negative) values of n in order to ensure that the envelopes vary slowly in space. If we now apply the low-transmission approximation (T tends to zero, $g\bar{L}$ tends to zero with $g\bar{L}/T$ constant) we obtain

$$
\sum_n \left\{ e^{ik_{\bar{j}+n}z} \left[i\alpha_n f_n + \frac{df_n}{dt} + \kappa(1+i\theta)f_n - \kappa y \delta_{n,0} \right] \right.
$$
$$
\left. + e^{-ik_{\bar{j}+n}z} \left[i\alpha_n f_n + \frac{df_n}{dt} + \kappa(1+i\theta)f_n - \kappa y \delta_{n,0} \right] \right\} = gc\tilde{P}
$$
$$
\implies 2\sum_n \cos(k_{\bar{j}+n}z) \left[i\alpha_n f_n + \frac{df_n}{dt} + \kappa(1+i\theta)f_n - \kappa y \delta_{n,0} \right] = gc\tilde{P}, \quad (14.32)
$$

where we define

$$
\alpha_n = \frac{\pi c}{L}n, \qquad n = 0, \pm 1, \dots,
$$

(14.33)

$$
\kappa = \frac{cT}{2L}, \qquad \theta = \frac{\delta_0}{T}.
$$

(14.34)

We note that, as in the ring-cavity case, α_n is the offset between the frequency of the mode of index $(\bar{j}+n)$ and the resonant mode of index $\bar{j}$.

Now we multiply Eq. (14.32) by $\cos(k_{\bar{j}+m}z)$ and integrate over z from 0 to L. Owing to the orthogonality of the functions $\cos(k_{\bar{j}+n}z)$ and $\cos(k_{\bar{j}+m}z)$ for $n \neq m$, one obtains the equation

$$i\alpha_n f_n + \frac{df_n}{dt} + \kappa[(1+i\theta)f_n - y\delta_{n,0}] = \kappa A \frac{1}{\bar{L}} \int_{L_1}^{L_2} dz \, \cos(k_{\bar{j}+n}z)\tilde{P}(z,t), \qquad (14.35)$$

where we have written n instead of m, we have restricted the integration interval $[0, L]$ to $[L_1, L_2]$ because we have taken into account that $\tilde{P}$ vanishes for the values of z where $\chi(z)$ vanishes, i.e. for $z < L_1$ and $z > L_2$, and the pump parameter A is defined as

$$A = \frac{g2\bar{L}}{T}, \qquad (14.36)$$

with a factor of 2 difference from the definition in the ring-cavity case.

Next, we apply the transformation (14.24) and (14.25) to the atomic equations (14.12) and (14.13), and introduce the expansions (14.31) and the low-transmission approximation, arriving at the equations

$$\frac{\partial \tilde{P}}{\partial t} = \gamma_\perp \left[2 \sum_n \cos(k_{\bar{j}+n}z)f_n D - (1+i\Delta)\tilde{P} \right], \qquad (14.37)$$

$$\frac{\partial D}{\partial t} = -\gamma_\parallel \left[\sum_n \cos(k_{\bar{j}+n}z)f_n \tilde{P}^* + \text{c.c.} + D - 1 \right], \qquad (14.38)$$

where in Eq. (14.38) we have written 1 instead of $\chi(z)$ because in Eqs. (14.35), (14.37) and (14.38) one considers only the values of z in the interval $L_1 \leq z \leq L_2$ where $\chi(z) = 1$.

Equations (14.35), (14.37) and (14.38) constitute a closed set of equations for the modal amplitudes f_n and the atomic variables $\tilde{P}$ and D. Equations of this kind are well known in the literature [7, 8]. They are usually derived by introducing from the very beginning the expansion of the electric field over the modes of the closed cavity and adding the damping term of the electric field phenomenologically. In our derivation, instead, we start from the real boundary conditions of the cavity, and the expansion over cavity modes as well as the damping term arise naturally in the low-transmission limit.

Because of the very fast variation of the cosine functions in play ($\bar{j}$ is very large), Eqs. (14.35), (14.37) and (14.38), in the form in which they are formulated, lend themselves easily to numerical resolution only after being subjected to important approximations. By using the cubic approximation, Lamb derived a celebrated two-mode model [8, 142]. By adiabatically eliminating the atomic polarization and by further approximations, Tang, Statz and DeMars derived and analyzed a model that describes the coupled dynamics of the intensities of a number of competing modes [143]. The single-mode model will be analyzed, and analytically solved in the stationary state, in the next section.

We finally observe that the cosine configuration of the cavity modes arises from our choice of the reflection phases in Section 8.1. If one chooses, instead, $\phi_r = \pi$ and $\phi_r' = 0$, then, as one can easily verify, the r.h.s. in the boundary condition (14.15) changes sign, and this leads to a sine configuration of the cavity modes.

14.2 The single-mode model for the Fabry–Perot cavity. Spatial hole-burning

If we assume that the free spectral range $\pi c/L$ is much larger than the atomic linewidth $\gamma_\perp$ we can limit ourselves to the resonant mode $n = 0$, so that Eqs. (14.35), (14.37) and (14.38) reduce to the single-mode model

$$\frac{dx}{dt} = -\kappa \left[(1 + i\theta)x - y - A\frac{1}{L} \int_{L_1}^{L_2} dz \cos^2\phi \, p(z, t) \right], \tag{14.39}$$

$$\frac{\partial p}{\partial t} = \gamma_\perp [2xD - (1 + i\Delta)p], \tag{14.40}$$

$$\frac{\partial D}{\partial t} = -\gamma_\parallel [\cos^2\phi (xp^* + \text{c.c.}) + D - 1], \tag{14.41}$$

where

- we have replaced f_0 by x because of Eq. (14.31), which in the single-mode approximation gives $f_0(t) = F'_F(L, t)$, because in the low-transmission limit we can replace F'_F by F_F and because of Eq. (14.21); and
- we have set

$$\phi = k_{\bar{j}}z, \qquad \tilde{P}(z, t) = \cos(k_{\bar{j}}z)p(z, t). \tag{14.42}$$

Now let us consider the integral in Eq. (14.39) and change the integration variable from z to ϕ, taking into account that p depends on z only via $\cos^2\phi$. The integral becomes

$$\frac{1}{k_{\bar{j}}\bar{L}} \int_{k_jL_1}^{k_jL_2} d\phi \cos^2\phi \, p(\phi, t) = \frac{1}{\pi(m_2 - m_1) + \delta_2 - \delta_1} \int_{\pi m_1+\delta_1}^{\pi m_2+\delta_2} d\phi \cos^2\phi \, p(\phi, t)$$

$$\simeq \frac{1}{\pi} \int_0^\pi d\phi \cos^2\phi \, p(\phi, t), \tag{14.43}$$

where

- we have set $k_{\bar{j}}L_i = \pi m_i + \delta_i$, $i = 1, 2$, where m_i are integers chosen in such a way that $|\delta_i| < \pi$ (hence we have also that $k_{\bar{j}}\bar{L} = k_{\bar{j}}(L_2 - L_1) = \pi(m_2 - m_1) + \delta_2 - \delta_1$);
- we have assumed that the sample length $\bar{L}$ is much larger than the wavelength, so that $m_2 - m_1$ is a large number and δ_1 and δ_2 can be neglected; and
- we have taken into account that the integrand is a function of $\cos^2\phi$, which is periodic, with period π.

Therefore in the single-mode model we can cast Eq. (14.39) into the form

$$\frac{dx}{dt} = -\kappa \left[(1 + i\theta)x - y - \frac{A}{\pi} \int_0^\pi d\phi \cos^2\phi \, p(\phi, t) \right], \tag{14.44}$$

and the model can be easily solved numerically. A noteworthy feature is that the single-mode model does not depend on the position of the atomic sample in the cavity. This is a consequence of the assumption that the length of the sample is much larger than the wavelength.

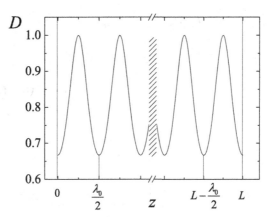

Figure 14.1 Spatial hole-burning. The figure shows the spatial profile of the population inversion for $\Delta = 1$ and $4|x|^2 = 1$.

The stationary equation can even be written in an explicit analytic form. In the stationary state, from Eqs. (14.40) and (14.41) we obtain the following expressions for the atomic variables:

$$D = \frac{1 + \Delta^2}{1 + \Delta^2 + 4|x|^2 \cos^2\phi}, \tag{14.45}$$

$$P = 2\frac{1 - i\Delta}{1 + \Delta^2 + 4|x|^2 \cos^2\phi}x, \tag{14.46}$$

from which, taking into account that $\phi = k_{\bar{j}}z$ and expanding in powers of $\cos^2\phi$, we see that they display a grating with spatial frequency equal to $\lambda_0/2$ and its harmonics. The spatial modulation of the grating for the normalized population inversion D (14.45) is shown in Fig. 14.1. In correspondence to the nodes of the standing-wave pattern, where $\cos\phi = 0$, the inversion attains the value $D = 1$ imposed by the pump in the absence of the laser field. On the other hand, this field burns a hole in the population profile at the locations of the antinodes, where $\cos^2\phi = 1$. This feature is called *spatial hole-burning*.

By inserting Eq. (14.46) into the field equation (14.44) we obtain the stationary equation

$$y = x\left[1 + i\theta - 2A\frac{1 - i\Delta}{\pi}\int_0^\pi d\phi \frac{\cos^2\phi}{1 + \Delta^2 + 4|x|^2 \cos^2\phi}\right]. \tag{14.47}$$

The integral can be reformulated as follows:

$$\frac{1}{\pi}\int_0^\pi d\phi \frac{\cos^2\phi}{1 + \Delta^2 + 4|x|^2 \cos^2\phi} = \frac{1}{4|x|^2}\left[1 - \frac{1}{\pi}\int_0^\pi d\phi \frac{1}{1 + \beta \cos^2\phi}\right]. \tag{14.48}$$

with $\beta = 4|x|^2/(1 + \Delta^2)$. The value of the remaining integral is $\pi/\sqrt{1 + \beta}$, as can be checked with the transformation $z = \sqrt{1 + \beta}\cot\phi$. Hence, the stationary equation reads

$$y = x\left\{1 + i\theta - \frac{A(1 - i\Delta)}{2|x|^2}\left[1 - \frac{1}{\sqrt{1 + 4|x|^2/(1 + \Delta^2)}}\right]\right\}. \tag{14.49}$$

Let us analyze Eq. (14.49) in the case of a free-running laser first. On setting $y = 0$ we see that, as usual, there is the trivial stationary solution $x = 0$. In order to find the nontrivial

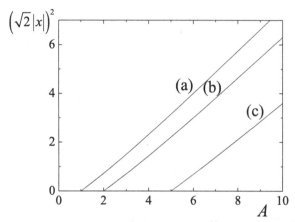

Figure 14.2 A Fabry–Perot cavity. The stationary value of the output intensity as a function of the pump parameter for $\Delta = 0$ (a), $\Delta = 1$ (b) and $\Delta = 2$ (c).

stationary solution one divides Eq. (14.49) by x and separates real and imaginary parts. One finds the equation

$$\frac{A}{2|x|^2}\left[1 - \frac{1}{\sqrt{1 + 4|x|^2/(1 + \Delta^2)}}\right] = 1, \tag{14.50}$$

and, from the imaginary parts, one obtains $\theta = -\Delta$, which leads to the mode-pulling formula Eq. (12.58). Therefore, we can write the dynamical equation (14.44) as

$$\frac{dx}{dt} = -\kappa\left[(1 - i\Delta)x - \frac{A}{\pi}\int_0^\pi d\phi \cos^2\phi \, p(\phi, \tau)\right]. \tag{14.51}$$

By applying the limit $|x|^2 \to 0$ in Eq. (14.50) we obtain the threshold value of the pump parameter A, which is given by

$$A_{\text{thr}} = 1 + \Delta^2 \tag{14.52}$$

and coincides with the threshold for a ring cavity. From Eq. (14.50) we can obtain an analytic expression for the output intensity as a function of the pump parameter,

$$(\sqrt{2}|x|)^2 = A - \frac{1 + \Delta^2}{4}\left(1 + \sqrt{1 + \frac{8A}{1 + \Delta^2}}\right). \tag{14.53}$$

The graph of $|x|^2$ as a function of A is no longer a straight line. It is shown in Fig. 14.2 for three values of Δ.

In the case of optical bistability we replace A by the bistability parameter $-2C$, and, for the sake of simplicity, we consider the completely resonant absorptive case $\theta = \Delta = 0$. The stationary equation reduces to [117, 144]

$$y = x\left[1 + \frac{C}{x^2}\left(1 - \frac{1}{\sqrt{1 + 4x^2}}\right)\right], \tag{14.54}$$

where we have set x real. Note that the asymptotic behavior for small x (i.e. $y = x(1 + 2C)$) and the asymptotic behavior for large x (i.e. $y = x$) are the same as in the ring cavity (see Eq. (11.9)). The bistability threshold can be evaluated numerically and corresponds to $C = 4.968\,78$ [109], which is somewhat larger than in the ring-cavity case.

It turns out in general that in the Fabry–Perot case it is convenient to use the variable $|x|\sqrt{2}$ (see e.g. Eq. (14.53) and Fig. 14.2) instead of $|x|$. The physical interpretation of this fact emerges if we consider the averaged (in space and time) energy density $\bar{w}$ of the electromagnetic field. Using the ring-cavity expression of the electric field Eq. (2.37), one has the result $\bar{w} = \epsilon_0 |E_0|^2/2$. On the other hand, using the Fabry–Perot expression Eq. (14.1) and taking into account that in the low-transmission limit $E_B = E_F$ (see Eq. (14.15)), we have the result $\bar{w} = \epsilon_0 |E_F|^2$. Therefore the Fabry–Perot cavity expression of $\bar{w}$ as a function of $\sqrt{2}|E_F|$ coincides with the ring-cavity expression of $\bar{w}$ as a function of $|E|$. Since in the ring-cavity case $x \propto E$ and in the Fabry–Perot cavity case $x \propto E_F$ (see Eq. (14.21)) with the same proportionality coefficient, the use of $|x|\sqrt{2}$ in the Fabry–Perot case is in line with this correspondence.

14.3 A more convenient set of modal equations

The first goal of this section is to derive separate equations for the two envelopes F_F and F_B. The crucial point is to introduce a clear-cut separation between the wavelength spatial scale and the scale characterized by the cavity length. To materialize this idea, for a cavity of length much larger than a wavelength, we introduce the variable [141, 144]

$$\varphi = k_0 z, \tag{14.55}$$

and indicate explicitly that the atomic variables depend both on the "slow" spatial variable z and on the "fast" spatial variable φ, whereas the two envelopes depend only on the slow scale. Equations (14.11)–(14.13) can be rewritten in the form

$$e^{i\varphi}\left(\frac{\partial F_F(z,t)}{\partial z} + \frac{1}{c}\frac{\partial F_F(z,t)}{\partial t}\right) + e^{-i\varphi}\left(-\frac{\partial F_B(z,t)}{\partial z} + \frac{1}{c}\frac{\partial F_B(z,t)}{\partial t}\right)$$
$$= g\tilde{P}(z,\varphi,t), \tag{14.56}$$

$$\frac{\partial \tilde{P}(z,\varphi,t)}{\partial t} = \gamma_\perp \big[\left(F_F(z,t)e^{i\varphi} + F_B(z,t)e^{-i\varphi}\right) D(z,\varphi,t)$$
$$- (1 + i\Delta)\,\tilde{P}(z,\varphi,t)\big], \tag{14.57}$$

$$\frac{\partial D(z,\varphi,t)}{\partial t} = -\gamma_\parallel \bigg[\frac{1}{2}\left(F_F(z,t)e^{i\varphi} + F_B(z,t)e^{-i\varphi}\right) \tilde{P}^*(z,\varphi,t) + \text{c.c.}$$
$$+ D(z,\varphi,t) - \chi(z)\bigg]. \tag{14.58}$$

Next, we multiply Eq. (14.56) by $e^{-i\varphi}/(2\pi)$ and integrate over φ from $-\pi$ to $+\pi$, obtaining the separate equation for F_F [141, 144],

$$\frac{\partial F_F(z,t)}{\partial z} + \frac{1}{c}\frac{\partial F_F(z,t)}{\partial t} = \frac{g}{2\pi}\int_{-\pi}^{\pi} d\varphi\, e^{-i\varphi}\,\tilde{P}(z,\varphi,t). \tag{14.59}$$

Proceeding in a similar manner, we obtain the following equation for F_B:

$$-\frac{\partial F_B(z,t)}{\partial z} + \frac{1}{c}\frac{\partial F_B(z,t)}{\partial t} = \frac{g}{2\pi}\int_{-\pi}^{\pi}d\varphi\, e^{i\varphi}\tilde{P}(z,\varphi,t). \qquad (14.60)$$

At this point, we implement the transformation (14.24) and (14.25) and the low-transmission approximation and arrive at the set of equations [141]

$$c\frac{\partial F_F'(z,t)}{\partial z} + \frac{\partial F_F'(z,t)}{\partial t} = -\kappa\Bigg[(1+i\theta)F_F'(z,t) - y$$
$$-\frac{A}{2\pi}\frac{L}{\bar{L}}\int_{-\pi}^{\pi}d\varphi\, e^{-i\varphi}\tilde{P}(z,\varphi,t)\Bigg], \qquad (14.61)$$

$$-c\frac{\partial F_B'(z,t)}{\partial z} + \frac{\partial F_B'(z,t)}{\partial t} = -\kappa\Bigg[(1+i\theta)F_B'(z,t) - y$$
$$-\frac{A}{2\pi}\frac{L}{\bar{L}}\int_{-\pi}^{\pi}d\varphi\, e^{i\varphi}\tilde{P}(z,\varphi,t)\Bigg], \qquad (14.62)$$

$$\gamma_\perp^{-1}\frac{\partial\tilde{P}(z,\varphi,t)}{\partial t} = \left(F_F'(z,t)e^{i\varphi} + F_B'(z,t)e^{-i\varphi}\right)D(z,\varphi,t)$$
$$-(1+i\Delta)\tilde{P}(z,\varphi,t), \qquad (14.63)$$

$$\gamma_\parallel^{-1}\frac{\partial D(z,\varphi,t)}{\partial t} = -\frac{1}{2}\left[\left(F_F'(z,t)e^{i\varphi} + F_B'(z,t)e^{-i\varphi}\right)\tilde{P}^*(z,\varphi,t) + \text{c.c.}\right]$$
$$-[D(z,\varphi,t) - \chi(z)], \qquad (14.64)$$

where we have taken into account the definitions (14.34) and (14.36).

Now,

- we introduce into Eqs. (14.61)–(14.64) the expansions (14.31) and take Eq. (14.33) into account;
- we sum the two resulting equations; and
- we multiply the sum equation by $(1/L)\cos(\pi mz/L)$ and integrate over z from 0 to L, using the identity

$$2\cos(\pi mz/L)\cos(\pi nz/L) = \cos[\pi(m+n)z/L] + \cos[\pi(m-n)z/L].$$

In this way we obtain the equation

$$i\alpha_n\frac{1}{2}(f_n - f_{-n}) + \frac{1}{2}\frac{d}{dt}(f_n + f_{-n})$$
$$= -\kappa\Bigg[(1+i\theta)\frac{1}{2}(f_n + f_{-n}) - y\delta_{n,0}$$
$$-\frac{A}{2\pi}\frac{1}{\bar{L}}\int_{L_1}^{L_2}dz\int_{-\pi}^{\pi}d\varphi\,\cos\varphi\cos\left(\frac{\pi n}{L}z\right)\tilde{P}(z,\varphi,t)\Bigg], \qquad (14.65)$$

where we have taken into account the fact that $\tilde{P}(z,\varphi,t)$ vanishes for the values of z for which $\chi(z)$ vanishes, and we have written n instead of m. If, instead, we evaluate the difference between the two resulting equations, we multiply it by $\sin(\pi mz/L)$ with $m \neq 0$, use the identity $2 \sin(\pi mz/L)\sin(\pi nz/L) = \cos[\pi(m-n)z/L] - \cos[\pi(m+n)z/L]$ and integrate over z from 0 to L we obtain the companion equation

$$i\alpha_n \frac{1}{2}(f_n + f_{-n}) + \frac{1}{2}\frac{d}{dt}(f_n - f_{-n})$$
$$= -\kappa\left[(1+i\theta)\frac{1}{2}(f_n - f_{-n})\right.$$
$$\left. + \frac{A}{2\pi}\frac{1}{L}\int_{L_1}^{L_2}dz\int_{-\pi}^{\pi}d\varphi\,\sin\varphi\sin\left(\frac{\pi n}{L}z\right)\tilde{P}(z,\varphi,t)\right]. \quad (14.66)$$

Finally, by summing Eqs. (14.65) and (14.66) we arrive at the equation

$$\frac{df_n(t)}{dt} = -i\alpha_n f_n(t) - \kappa\left[(1+i\theta)f_n(t) - y\delta_{n,0}\right.$$
$$\left. - \frac{A}{2\pi}\frac{1}{L}\int_{L_1}^{L_2}dz\int_{-\pi}^{\pi}d\varphi\,\cos\left(\frac{\pi n}{L}z+\varphi\right)\tilde{P}(z,\varphi,t)\right]. \quad (14.67)$$

On the other hand, the atomic equations become

$$\gamma_\perp^{-1}\frac{\partial\tilde{P}(z,\varphi,t)}{\partial t} = 2\sum_{n'}\cos\left(\frac{\pi n'}{L}z+\varphi\right)f_{n'}(t)D(z,\varphi,t)$$
$$- (1+i\Delta)\tilde{P}(z,\varphi,t), \quad (14.68)$$
$$\gamma_\parallel^{-1}\frac{\partial D(z,\varphi,t)}{\partial t} = -\left[\sum_{n'}\cos\left(\frac{\pi n'}{L}z+\varphi\right)f_{n'}(t)\tilde{P}^*(z,\varphi,t) + \text{c.c.}\right]$$
$$- [D(z,\varphi,t)-1]. \quad (14.69)$$

The set of modal equations (14.67)–(14.69) is much more convenient for the numerical resolution than the set (14.35), (14.37) and (14.38). We note also that in Eqs. (14.61)–(14.64), which are valid in the low-transmission approximation, one can replace F_F' and F_B' by F_F and F_B, respectively.

In this approach it is straightforward to write the single-mode equations. In the single-mode approximation we keep only the modal amplitude f_0 so that the field envelopes do not depend on z and also the atomic variables are z-independent. Therefore we obtain the single-mode equations [141]

$$\frac{dx(t)}{dt} = -\kappa\left[(1+i\theta)x(t) - y - \frac{A}{\pi}\int_0^{\pi}d\varphi\,\cos^2\varphi\,p(\varphi,t)\right], \quad (14.70)$$
$$\gamma_\perp^{-1}\frac{\partial p(\varphi,t)}{\partial t} = -(1+i\Delta)p(\varphi,t) + 2x(t)D(\varphi,t), \quad (14.71)$$
$$\gamma_\parallel^{-1}\frac{\partial D(\varphi,t)}{\partial t} = -[D(\varphi,t)-1] - \cos^2\varphi[x(t)p^*(\varphi,t)+\text{c.c.}], \quad (14.72)$$

where we have set

$$\tilde{P}(\varphi,t) = \cos\varphi\,p(\varphi,t). \quad (14.73)$$

Eqs. (14.70)–(14.72) are identical to Eqs. (14.44), (14.40) and (14.41), which were discussed in Section 14.2.

In the single-mode model, the standing-wave effects arise from the factor $\cos^2\varphi$. It is interesting to analyze the case in which the atoms undergo a diffusion process in the longitudinal direction, which washes out the grating created by the $\cos^2\varphi$ factor, so that the atomic variables become independent of φ. Hence Eqs. (14.70) and (14.71) become

$$\frac{dx}{dt} = -\kappa\left[(1+i\theta)x - y - \frac{A}{2}p\right], \tag{14.74}$$

$$\gamma_\perp^{-1}\frac{dp}{dt} = -(1+i\Delta)p + 2xD. \tag{14.75}$$

On the other hand, by dividing Eq. (14.72) by π, integrating over φ from 0 to π and taking into account that the atomic variables are φ-independent because of the atomic diffusion, we obtain the equation

$$\gamma_\parallel^{-1}\frac{dD}{dt} = -(D-1) - \frac{1}{2}(xp^* + x^*p). \tag{14.76}$$

Finally, by reformulating these equations in terms of the variables

$$x_0 = \sqrt{2}x, \qquad y_0 = \sqrt{2}y, \qquad p_0 = p/\sqrt{2} \tag{14.77}$$

we arrive at the set of equations

$$\frac{dx_0}{dt} = -\kappa[(1+i\theta)x_0 - y_0 - Ap_0], \tag{14.78}$$

$$\gamma_\perp^{-1}\frac{dp_0}{dt} = -(1+i\Delta)p_0 + x_0 D, \tag{14.79}$$

$$\gamma_\parallel^{-1}\frac{dD}{dt} = -(D-1) - \frac{1}{2}(x_0 p_0^* + x_0^* p_0), \tag{14.80}$$

which coincides with the single-mode model for the ring cavity given by Eqs. (12.43)–(12.45). From Eq. (14.77) we note again (see the end of Section 14.2) that in the Fabry–Perot case it is convenient to use the variable $x\sqrt{2}$.

In Fig. 14.3 we compare the stationary curve for purely resonant absorptive bistability ($\Delta = \theta = 0$) in a Fabry–Perot cavity with standing-wave effects, described by Eq. (14.54), with the stationary curve when the standing-wave effects are washed out by diffusion, so that the stationary curve coincides with the ring-cavity equation (11.9) with x and y replaced by x_0 and y_0 defined in Eq. (14.77), so that, when expressed in terms of x and y, the stationary equation becomes $y = x\left(1 + 2C/(1+2x^2)\right)$. In both cases the bistability parameter is defined as $2C = 2\alpha\bar{L}/T$, with the same value $C = 10$. The removal of standing effects leads to a reduction by about 1/3 of the width of the bistability domain.

We underline that the equations derived in this and in the previous section are valid also in the case $n_B > 1$, provided that the cavity is completely filled by the atomic sample. In this case the light velocity is c/n_B instead of c in all formulas.

We note that also the laser model formulated in Chapter 1 neglects the standing-wave effects.

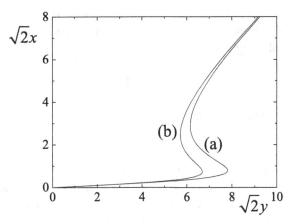

Figure 14.3 A Fabry–Perot cavity. (a) The steady-state S curve for absorptive bistability with $C = 10$. (b) The steady-state curve when atomic diffusion washes out the population grating.

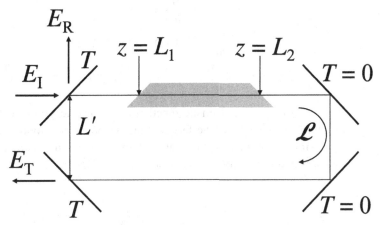

Figure 14.4 This figure is identical to Fig. 8.4 except for the position of the atomic sample in the cavity and the fact that $L' \ll \mathcal{L}$.

14.4 Again the ring cavity: simplified forms of the models

In this section we return to the ring cavity to show that, in the case in which the background refractive index is equal to unity, which we are considering in this chapter, one can derive a simplified version of the dynamical equations. As a matter of fact for $n_B = 1$ the light velocity is the same in all parts of the cavity and one can consider, instead of the boundary condition (8.36), the boundary condition

$$F(0, t) = Ty + Re^{-i\delta_0} F\left(\mathcal{L} - L', t - \frac{L'}{c}\right), \tag{14.81}$$

where $\mathcal{L}$ is the roundtrip cavity length as usual and L' is defined in Fig. 8.4. In addition, if we assume (Fig. 14.4) that

$$L' \ll \mathcal{L}, \tag{14.82}$$

we can not only neglect L' with respect to $\mathcal{L}$ in the spatial argument of F on the r.h.s., but also neglect L'/c with respect to t in the temporal argument. As a matter of fact, if we normalize time with respect to $\mathcal{L}/c$, defining $\bar{t} = ct/\mathcal{L}$, we have

$$\frac{c}{\mathcal{L}}\left(t - \frac{L'}{c}\right) = \bar{t} - \frac{L'}{\mathcal{L}} \approx \bar{t}. \tag{14.83}$$

Therefore the boundary condition can be written in the form

$$F(0, t) = Ty + Re^{-i\delta_0} F(\mathcal{L}, t), \tag{14.84}$$

and no retardation appears any longer.

In addition, while in Chapter 8 and the following chapters we considered an atomic sample that occupies the segment from $z = 0$ to $z = L$, here we consider a sample located at any position $L_1 < z < L_2$ in the cavity (see Fig. 14.4). The Maxwell–Bloch equations can be written in the form (compare with Eqs. (4.35)–(4.37))

$$\frac{1}{c}\frac{\partial F}{\partial t} + \frac{\partial F}{\partial z} = gP, \tag{14.85}$$

$$\frac{\partial P}{\partial t} = \gamma_\perp [FD - (1 + i\Delta)P], \tag{14.86}$$

$$\frac{\partial D}{\partial t} = -\gamma_\parallel \left[\frac{1}{2}(FP^* + F^*P) + D - \chi(z)\right], \tag{14.87}$$

where $\chi(z)$ is the characteristic function of the atomic sample defined by Eq. (14.7). If we return to Chapter 12, the fact that the boundary condition (14.84) does not include any retardation in time implies that the transformation of coordinates (12.1) is no longer necessary and that it suffices to perform the transformations

$$F'(z, t) = F(z, t)e^{(\ln R - i\delta_0)z/\mathcal{L}} + Ty\frac{z}{\mathcal{L}}, \tag{14.88}$$

$$P'(z, t) = P(z, t)e^{(\ln R - i\delta_0)z/\mathcal{L}}, \tag{14.89}$$

which coincide with (12.8) and (12.9) with τ replaced by t and L replaced by $\mathcal{L}$. The boundary condition for F' is the periodic boundary condition $F'(0, t) = F'(\mathcal{L}, t)$. After performing the transformation and taking the low-transmission limit, Eq. (14.85) becomes

$$c\frac{\partial F}{\partial z} + \frac{\partial F}{\partial t} = -\kappa[(1 + i\theta)F - y] + gcP, \tag{14.90}$$

whereas Eqs. (14.86) and (14.87) retain their form. Because of the periodic boundary condition, we can now introduce the expansion

$$F(z, t) = \sum_n f_n(t)e^{i(2\pi/\mathcal{L})nz}, \tag{14.91}$$

which, when inserted into Eq. (14.90), leads, after the same steps as illustrated in Section 12.1, to the set of modal equations

$$i\alpha_n f_n + \frac{df_n}{dt} = -\kappa\left[(1 + i\theta)f_n - y\delta_{n,0} - \frac{A}{\mathcal{L}}\int_{L_1}^{L_2} dz\, e^{-i(2\pi/\mathcal{L})nz} P(z, t)\right], \tag{14.92}$$

where $\bar{L} = L_2 - L_1$ and we have used the definitions $\kappa = cT/\mathcal{L}$, $\alpha_n = 2\pi cn/\mathcal{L}$ and $A = g\bar{L}/T$ and taken into account that P vanishes for the values of z for which $\chi(z)$ vanishes. These equations must be coupled to Eqs. (14.86) and (14.87) (in which one can set $\chi(z) = 1$ because the integral in Eq. (14.92) is limited to the interval where $\chi(z) = 1$), taking into account the expansion Eq. (14.91).

If we now introduce the change of variable $z' = z - L_1$ and define the new field amplitudes $f'_n = f_n e^{i(2\pi/\mathcal{L})nL_1}$, Eqs. (14.91) and (14.92) remain formally identical, with the only difference being that now the integral domain in Eq. (14.92) is $(0, \bar{L})$ instead of (L_1, L_2). This confirms that the position of the medium inside the cavity is not relevant.

As in the Fabry–Perot case, we observe that the modal equations are convenient for numerical integration and remain valid in the case $n_B > 1$, if the atomic sample fills the cavity and the light velocity c is replaced by c/n_B.

14.5 The case of an atomic sample of length much shorter than the wavelength: difference-differential equations

In the late 1970s Ikeda [145–147] reformulated the Maxwell–Bloch equations for the ring cavity plus the boundary condition in the form of a set of difference-differential equations, which is very convenient from a numerical standpoint and allowed him to predict the possibility of chaotic behavior in optical bistability. He derived also a set of difference-differential equations for the Fabry–Perot cavity, but the mathematics was substantially more complex than in the ring-cavity case [147].

In the following we introduce the basic assumption that the length of the atomic sample is much less than the wavelength. This case arises in engineered devices such as, for example, quantum-well structures. We will see that this step brings about a dramatic simplification, and allows one to derive a set of difference-differential equations (precisely, a discrete temporal map for the electric field and differential equations – in time only – for the atomic variables), which in the case of a Fabry–Perot cavity is only slightly more complex than in the case of a ring cavity [148, 149].

We consider a Fabry–Perot cavity of length $L \gg \lambda_0$ containing an atomic sample of thickness $\bar{L} \ll \lambda_0$ centered at $z = \bar{z}$. We start from the Maxwell–Bloch equations (14.11)–(14.13) assuming (as implied by the condition $\bar{L} \ll \lambda_0$) that the atomic variables depend only on the values of the forward and backward fields at $z = \bar{z}$ and that the source term in the wave equation is δ-like in z (see also [150]):

$$e^{ik_0 z}\left(\frac{\partial F_F(z,t)}{\partial z} + \frac{1}{c}\frac{\partial F_F(z,t)}{\partial t}\right) + e^{-ik_0 z}\left(-\frac{\partial F_B(z,t)}{\partial z} + \frac{1}{c}\frac{\partial F_B(z,t)}{\partial t}\right)$$
$$= g\tilde{P}(t)\bar{L}\delta(z - \bar{z}), \tag{14.93}$$

$$\frac{\partial \tilde{P}}{\partial t} = \gamma_\perp\left[\mathcal{F}(t)D(t) - \tilde{P}(t)\right], \tag{14.94}$$

$$\frac{\partial D}{\partial t} = -\gamma_\parallel\left\{\frac{1}{2}\left[\mathcal{F}(t)\tilde{P}^*(t) + \text{c.c.}\right] + D(t) - 1\right\}, \tag{14.95}$$

where we have set $\tilde{P}(z, t) = \hat{P}(t)\bar{L}\delta(z - \bar{z})$,

$$\mathcal{F}(t) = e^{i\phi} F_F(\bar{z}, t) + e^{-i\phi} F_B(\bar{z}, t), \qquad \phi = k_0\bar{z} \pmod{\pi}, \qquad (14.96)$$

and, for the sake of simplicity, we have assumed that $\omega_0 = \omega_a$, i.e. $\Delta = 0$. If we multiply Eq. (14.93) by e^{-ik_0z}, introduce the variables $t_1 = t - z/c$ and $z_1 = z$, and write

$$F_F(z, t) = \tilde{F}_F(z_1, t_1) = F_F(z_1, t_1 + z_1/c),$$
$$F_B(z, t) = \tilde{F}_B(z_1, t_1) = F_B(z_1, t_1 + z_1/c), \qquad (14.97)$$

Eq. (14.93) becomes

$$\frac{\partial \tilde{F}_F(z_1, t_1)}{\partial z_1} + e^{-2ik_0z_1} \left(-\frac{\partial \tilde{F}_B(z_1, t_1)}{\partial z_1} + \frac{2}{c}\frac{\partial \tilde{F}_B(z_1, t_1)}{\partial t_1} \right)$$

$$= g\bar{L}e^{-ik_0z_1} g\tilde{P}\left(t_1 + \frac{z_1}{c} \right)\delta(z_1 - \bar{z}). \qquad (14.98)$$

We integrate this equation in z_1 first from $z_1 = 0$ to $z_1 = \bar{z}$ and then from $z_1 = \bar{z}$ to $z_1 = L$. In both cases the contribution of the terms with $\tilde{F}_B$ vanishes, provided that $\bar{z}$, $L - \bar{z} \gg \lambda_0$, because the variation of those terms is much slower than the variation of the exponential term $\exp(-2ik_0z_1)$. In this way we obtain

$$F_F\left(\bar{z}, t_1 + \frac{L}{c} \right) - F_F\left(\bar{z}, t_1 + \frac{\bar{z}}{c} \right) = F_F\left(\bar{z}, t_1 + \frac{\bar{z}}{c} \right) - F_F(0, t_1)$$

$$= \frac{g\bar{L}}{2}e^{-i\phi}\tilde{P}\left(\bar{z}, t_1 + \frac{\bar{z}}{c} \right). \qquad (14.99)$$

By proceeding in a similar way with the variables $t_2 = t + z/c$ and $z_2 = z$, we have

$$F_B\left(\bar{z}, t_2 - \frac{\bar{z}}{c} \right) - F_B\left(\bar{z}, t_2 - \frac{L}{c} \right) = F_B(0, t_2) - F_B\left(\bar{z}, t_2 - \frac{\bar{z}}{c} \right)$$

$$= \frac{g\bar{L}}{2}e^{i\phi}\tilde{P}\left(\bar{z}, t_2 - \frac{\bar{z}}{c} \right). \qquad (14.100)$$

Now let us return to the boundary conditions (14.22) and (14.23) with $\delta_0 = 0$ (i.e. $k_0L = \pi\bar{j}$ with $\bar{j}$ integer), i.e. we assume that the input field is in resonance also with the cavity. By setting $t_1 + \bar{z}/c \to t$ in Eq. (14.99) and $t_2 - \bar{z}/c \to t$ in Eq. (14.100) and expressing F_F and F_B at $z = 0$ and $z = L$ in terms of F_F and F_B at $z = \bar{z}$, we can transform the two boundary conditions into two coupled maps for $F_F(\bar{z}, t)$ and $F_B(\bar{z}, t)$:

$$F_F(\bar{z}, t) = \sqrt{R}\left[F_B\left(\bar{z}, t - 2\frac{\bar{z}}{c} \right) + \frac{g\bar{L}}{2}e^{i\phi}\tilde{P}\left(t - 2\frac{\bar{z}}{c} \right) \right]$$

$$+ Ty + \frac{g\bar{L}}{2}e^{-i\phi}\tilde{P}(t), \qquad (14.101)$$

$$F_B(\bar{z}, t) = \sqrt{R}\left[F_F\left(\bar{z}, t + 2\frac{\bar{z} - L}{c} \right) + \frac{g\bar{L}}{2}e^{-i\phi}\tilde{P}\left(t - 2\frac{\bar{z} - L}{c} \right) \right]$$

$$+ \frac{g\bar{L}}{2}e^{i\phi}\tilde{P}(t). \qquad (14.102)$$

Once they have been inserted into each other, these maps yield two separate maps for F_F and F_B, respectively, which can be combined to obtain a single map for the total field $\mathcal{F}$ given by Eq. (14.96). In the low-transmission limit $T \ll 1$, $g\bar{L} \ll 1$ with $A = 2g\bar{L}/T$ arbitrary, this map reads

$$\mathcal{F}(t) = R\mathcal{F}(t - t_R) + T\left\{2\cos\phi\, y + \frac{A}{2}\left[\tilde{P}(t) + \tilde{P}(t - t_R) + e^{2i\phi}\tilde{P}(t - \bar{t})\right.\right.$$
$$\left.\left. + e^{-2i\phi}\tilde{P}(t - t_R + \bar{t})\right]\right\}, \qquad (14.103)$$

where we define

$$t_R = \frac{2L}{c}, \qquad \bar{t} = \frac{2\bar{z}}{c}. \qquad (14.104)$$

Here t_R is the cavity roundtrip time and $\bar{t}$ is the time taken by light to travel from the atomic sample to the beam splitter at $z = 0$ and back. The discrete temporal map (14.103) and Eqs. (14.94) and (14.95) constitute our set of difference-differential equations.

In the case of a ring cavity of roundtrip length $\mathcal{L}$ one can obtain a similar map starting from the Maxwell–Bloch equations (see Eqs. (14.85)–(14.87))

$$\frac{1}{c}\frac{\partial F(z, t)}{\partial t} + \frac{\partial F(z, t)}{\partial z} = g\bar{L}\,P(t)\delta(z - \bar{z}), \qquad (14.105)$$

$$\frac{dP(t)}{dt} = \gamma_\perp[F(\bar{z}, t)D(t) - P(t)], \qquad (14.106)$$

$$\frac{dD(t)}{dt} = -\gamma_\parallel\left\{\frac{1}{2}[F(\bar{z}, t)P(t)^* + F(\bar{z}, t)^* P(t)] + D(t) - 1\right\}, \qquad (14.107)$$

and from the boundary condition (14.84) with $\delta_0 = 0$.

By following steps similar to those described for the Fabry–Perot cavity, one arrives at the map

$$F(t) = RF(t - t_R) + T\left\{y + \frac{A}{2}[P(t) + P(t - t_R)]\right\}, \qquad (14.108)$$

with $t_R = \mathcal{L}/c$ and $A = g\bar{L}/T$. In this case the equation does not depend on the position $\bar{z}$ of the atomic sample, and therefore the parameters ϕ and $\bar{t}$ do not appear. Equations (14.106)–(14.108) constitute the set of difference-differential equations for the ring cavity. We note that, differently from the Ikeda models, these equations do not require the adiabatic elimination of the atomic polarization. In addition, the set of difference-differential equations is remarkably simple for both kinds of cavity.

Let us now focus on the special case in which only the resonant mode is active. Since in this case the field evolves over the spatial scale $\kappa^{-1} \gg t_R$, we can approximate

$$\mathcal{F}(t) - R\mathcal{F}(t - t_R) \simeq T\mathcal{F}(t) + t_R\frac{d\mathcal{F}}{dt}, \qquad (14.109)$$

and, assuming that t_R is also much smaller than $\gamma_\perp^{-1}$ (the single-mode approximation), we can neglect the retardation in $\tilde{P}$ in Eq. (14.103), so this difference equation reduces to the differential equation

$$\frac{d\mathcal{F}}{dt} = \kappa[2\cos\phi\, y - \mathcal{F} + A(1 + \cos(2\phi))\tilde{P}], \tag{14.110}$$

where we have taken into account that $T/t_R = \kappa$ (see Eqs. (14.34) and (14.104)). On the other hand, in the ring-cavity case the map (14.108) becomes

$$\frac{dF}{dt} = \kappa(y - F + AP). \tag{14.111}$$

so that we are back to the usual ring-cavity model (12.43)–(12.45) for $\Delta = \theta = 0$. It is important to note that Eqs. (14.110), (14.94) and (14.95) for the Fabry–Perot case are identical to Eqs. (14.111), (14.106) and (14.107) for the ring-cavity case, apart from the following details:

- the field F is replaced by the combination $\mathcal{F}$ defined by Eq. (14.96);
- there is a factor of 2 difference in the definitions of t_R and A; and
- y is multiplied by $2\cos\phi$ and A by $1 + \cos(2\phi) = 2\cos^2\phi$.

In the stationary state we can assume that the variables $\mathcal{F}$, P and D in Eqs. (14.103), (14.94) and (14.95) are time-independent and obtain

$$2\cos\phi\, y = \mathcal{F}\left(1 - 2A\frac{\cos^2\phi}{1 + \mathcal{F}^2}\right). \tag{14.112}$$

In particular, for the free-running laser the nontrivial stationary solution is given by

$$\mathcal{F}^2 = 2A\cos^2\phi - 1, \tag{14.113}$$

so that the threshold value is $A_{\text{thr}} = 1/(2\cos^2\phi)$. Note that the stationary equations (14.112) and (14.113) hold also for the differential approximation (14.110).

In this way, we have established a key connection between the equations for the Fabry–Perot cavity with an ultrathin medium and the ring-cavity model. On the other hand, by using different notation we can establish a connection between the equation for the Fabry–Perot cavity with an ultrathin medium and the single-mode Fabry–Perot model. To this end we observe that, from Eqs. (14.96) and (14.21), taking into account that in the low-transmission limit F_F and F_B are uniform and equal, we have

$$\mathcal{F} = 2\cos\phi\, x, \tag{14.114}$$

so that Eq. (14.110) can be recast in the form

$$\frac{dx}{dt} = \kappa\left[y - x + A\cos\phi\tilde{P}\right], \tag{14.115}$$

and Eqs. (14.94) and (14.95), with Eq. (14.96), become

$$\frac{d\tilde{P}(t)}{dt} = \gamma_\perp\left[2\cos\phi x(t)D(t) - \tilde{P}(t)\right], \tag{14.116}$$

$$\frac{dD(t)}{dt} = -\gamma_\parallel\left\{\cos\phi\left[x(t)\tilde{P}(t)^* + x(t)^*\tilde{P}(t)\right] + D(t) - 1\right\}. \tag{14.117}$$

Note that Eqs. (14.115)–(14.117) are closely similar to Eqs. (14.70)–(14.72) with $\Delta = \theta = 0$ if we set $\tilde{P}(t) = \cos\phi\, p(t)$. The only difference is that in Eqs. (14.70)–(14.72) there is an integration over the variable φ. In the stationary state we obtain from Eq. (14.115)–(14.117)

$$y = x\left(1 - 2A\frac{\cos^2\phi}{1 + 4\cos^2\phi\, x^2}\right). \tag{14.118}$$

The remarkable feature is that, in this case of an ultrathin medium, the stationary equation depends on the position $\bar{z}$ of the atomic sample in the cavity via the factor $\cos^2\phi$. This is in contrast with the case in which the atomic sample is much longer than the wavelength, as we saw in the previous sections.

In the case of optical bistability, the pump parameter A must be replaced by $-2C$ as usual, where C is the bistability parameter.

In the free-running-laser case, on setting $y = 0$ we have, in addition to the trivial stationary solution $x = 0$, the nontrivial solution

$$(\sqrt{2}x)^2 = A - \frac{1}{2\cos^2\phi}. \tag{14.119}$$

When the atomic sample is located at an antinode of the standing-wave pattern ($\phi = 0$), the threshold value for the pump parameter is $A = 1/2$, and increases the nearer to a node the sample is located as expected.

In the case of a ring cavity, the position $\bar{z}$ of the atomic sample in the cavity is immaterial and from Eq. (14.108) the stationary equation is $y = x[1 - A/(1 + x^2)]$, which agrees with Eq. (13.1) with $\theta = 0$ and $\Delta = 0$, taking into account that $X = x^2$ and $Y = y^2$.

15

Inhomogeneous broadening

Up to this point we have always assumed that all the atoms which interact with the radiation field have the same atomic transition frequency, with an atomic linewidth given by $\gamma_\perp$. This configuration is called homogeneous broadening. However, in the great majority of lasers the linewidth is *inhomogeneously broadened*, because the atomic frequency follows a distribution. For example, in solid-state lasers such as the ruby laser and fiber lasers the inhomogeneity is due to the imperfections in the crystalline structure of the material. In gas lasers, where the atoms or molecules are free to move, the inhomogeneous broadening is due to the Doppler effect, for which a molecule moving with velocity $\mathbf{v}$ in a field of frequency ω and wave vector $\mathbf{k}$ experiences an effective frequency $\omega' = \omega - \mathbf{k} \cdot \mathbf{v}$ [8]. Since what matters is the difference between the atomic transition frequency and the field frequency, we can keep the field frequency fixed and imagine that the field interacts with different groups of atoms that move with different velocities and therefore have different transition frequencies. The realistic velocity distribution is a Gaussian, for which the corresponding distribution for the modulus of the velocity is Maxwellian. However, in the following we will mainly analyze the case of a Lorentzian distribution for the atomic transition frequency, because this allows an analytical treatment.

Again for the sake of simplicity, we will limit our analysis to the case of a ring cavity. In Section 15.1 we will generalize to the inhomogeneous-broadening configuration the modal equations for the case of homogeneous broadening, which were formulated in Chapter 12. In Section 15.2 we will specialize to the single-mode model and discuss its stationary solutions for the laser case.

15.1 Multimode dynamical equations

The atomic transition frequency of the generic group of atoms is given by $\omega_a + \delta_D$, where ω_a is the central atomic frequency and δ_D takes positive and negative values. We call ω_c the frequency of the cavity mode nearest to ω_a and denote by δ_{ac} the difference $\omega_a - \omega_c$. It is convenient to normalize with respect to $\gamma_\perp$, introducing the quantities

$$\bar{\Delta} = \frac{\delta_D}{\gamma_\perp}, \quad \tilde{\delta}_{ac} = \frac{\omega_a - \omega_c}{\gamma_\perp}, \quad \tilde{\kappa} = \frac{\kappa}{\gamma_\perp}, \tag{15.1}$$

while $\Delta = (\omega_a - \omega_0)/\gamma_\perp$ as usual. In the following, we label the atomic group by $\bar{\Delta}$ and therefore we indicate the atomic variables by $P(z, t, \bar{\Delta})$ and $D(z, t, \bar{\Delta})$. We designate by $W(\bar{\Delta})$ the frequency distribution of the atomic groups. In the following we assume that W

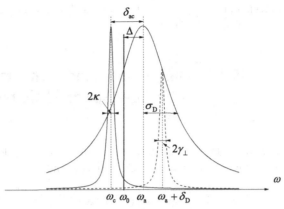

Figure 15.1 A schematic layout showing the spectral information. The narrow line (halfwidth κ) centered at the cavity eigenfrequency ω_c is the single-mode cavity resonance, the broad line (halfwidth σ_D) centered at ω_a represents the atomic profile and δ_{ac} is the atom–cavity detuning parameter. The dashed line within the atomic profile identifies a homogeneous packet of atoms (halfwidth $\gamma_\perp$). The thick vertical line represents the frequency ω_0. Δ is the offset from the atomic line center.

is Lorentzian,

$$W = \frac{\tilde{\sigma}_D}{\pi} \frac{1}{\bar{\Delta}^2 + \tilde{\sigma}_D^2} \equiv \mathcal{L}_{\tilde{\sigma}_D}(\bar{\Delta}), \quad \tilde{\sigma}_D = \frac{\sigma_D}{\gamma_\perp}, \tag{15.2}$$

or Gaussian,

$$W(\bar{\Delta}) = \frac{1}{\sqrt{2\pi}\tilde{\sigma}_D} \exp\left(-\frac{\bar{\Delta}^2}{2\tilde{\sigma}_D^2}\right), \tag{15.3}$$

where σ_D is called the inhomogeneous linewidth and the distribution is normalized to unity.

The qualitative shape of the distribution and other elements referring to the cavity mode are illustrated in Fig. 15.1. It is now straightforward to generalize to the inhomogeneously broadened configuration the modal equations for the homogeneously broadened case, namely Eqs. (12.39), (12.36) and (12.37). The generalized equations read [130]

$$\frac{df_n}{d\tau} + i\alpha_n f_n = -\kappa \left[(1 + i\theta) f_n - y\delta_{n,0} - \frac{A}{L} \int_0^L dz\, e^{-i2\pi nz/L} \int_{-\infty}^{+\infty} d\bar{\Delta}\, P(z, \tau, \bar{\Delta}) \right], \tag{15.4}$$

$$\frac{\partial P(z, \tau, \bar{\Delta})}{\partial \tau} = \gamma_\perp \{ F(z, \tau) D(z, \tau, \bar{\Delta}) - [1 + i(\Delta + \bar{\Delta})]P \}, \tag{15.5}$$

$$\frac{\partial D(z, \tau, \bar{\Delta})}{\partial \tau} = -\gamma_\parallel \left[\frac{F(z, \tau)P^*(z, \tau, \bar{\Delta}) + \text{c.c.}}{2} + D(z, \tau, \bar{\Delta}) - W(\bar{\Delta}) \right], \tag{15.6}$$

where in Eq. (15.5) we have taken into account that the frequency of the atomic group is $\omega_a + \delta_D$, so that the difference between the frequency of the atomic group and the reference frequency ω_0, normalized with respect to $\gamma_\perp$, is equal to $\Delta + \bar{\Delta}$. The time τ is defined by

Eq. (12.1), and the field $F(z, \tau)$ is expressed in terms of the modal amplitudes $f_n(\tau)$ as in Eqs. (12.14) and (12.15).

15.2 The single-mode model. The stationary state for the laser. Spectral hole-burning

Let us now reduce the equations to the single-mode model by considering only the mode $n = 0$. The equations become (see [130, 151] and references quoted therein)

$$\frac{dx}{dt} = -\kappa \left[(1 + i\theta)x - y - A \int_{-\infty}^{+\infty} d\bar{\Delta} \, P(t, \bar{\Delta}) \right], \tag{15.7}$$

$$\frac{dP(t, \bar{\Delta})}{dt} = \gamma_\perp \{ x(t)D(t, \bar{\Delta}) - [1 + i(\Delta + \bar{\Delta})]P(t, \bar{\Delta}) \}, \tag{15.8}$$

$$\frac{dD(t, \bar{\Delta})}{d\tau} = -\gamma_\parallel \left[\frac{x(t)P^*(t, \bar{\Delta}) + \text{c.c.}}{2} + D(t, \bar{\Delta}) - W(\bar{\Delta}) \right], \tag{15.9}$$

and generalize Eqs. (12.43)–(12.45) to the case of inhomogeneous broadening. In writing these equations we have taken into account that in the single-mode limit the atomic variables become independent of z, the time τ can be replaced by the normal time t and we have indicated f_0 by x as in the case of Eqs. (12.43)–(12.45).

Let us now focus on the case of the free-running laser by setting $y = 0$. In the stationary state we set the derivatives equal to zero and obtain

$$P(\bar{\Delta}) = \frac{1 - i(\Delta + \bar{\Delta})}{1 + (\Delta + \bar{\Delta})^2 + |x|^2} W(\bar{\Delta})x, \tag{15.10}$$

$$D(\bar{\Delta}) = \frac{1 + (\Delta + \bar{\Delta})^2}{1 + (\Delta + \bar{\Delta})^2 + |x|^2} W(\bar{\Delta}) \tag{15.11}$$

and the stationary state equation for the field

$$x[1 + i\theta - AI(\Delta, \tilde{\sigma}_D, \xi)] = 0, \tag{15.12}$$

with

$$I(\Delta, \tilde{\sigma}_D, \xi) = \int_{-\infty}^{+\infty} d\bar{\Delta} \, W(\bar{\Delta}) \frac{1 - i(\Delta + \bar{\Delta})}{(\Delta + \bar{\Delta})^2 + \xi^2} \tag{15.13}$$

and

$$\xi = \sqrt{1 + |x|^2}. \tag{15.14}$$

As usual, there is the trivial stationary solution $x = 0$, whereas the nontrivial solution satisfies the two equations $1 = A \, \text{Re}[I(\Delta, \tilde{\sigma}_D, \xi)]$ and $\theta = A \, \text{Im}[I(\Delta, \tilde{\sigma}_D, \xi)]$, which are obtained by dividing Eq. (15.12) by x and equating to zero the real and imaginary parts.

In the integral of Eq. (15.13) we can change the sign of $\bar{\Delta}$ and take into account that $W(-\bar{\Delta}) = W(\bar{\Delta})$ for our two choices (15.2) and (15.3) of the frequency distribution. Hence the integral can be regarded as the convolution of $W(\Delta)$ and $(\pi/\xi)(1 - i\Delta)\mathcal{L}_\xi(\Delta)$, where

$\mathcal{L}_{\xi}(\Delta)$ is defined by Eq. (15.2) with $\tilde{\sigma}_{\mathrm{D}}$ replaced by ξ. By applying the convolution theorem we get

$$I(\Delta, \tilde{\sigma}_{\mathrm{D}}, \xi) = \frac{\pi}{\xi} \mathcal{F}^{-1}\{\mathcal{F}[W(\Delta)]\mathcal{F}[(1 - i\Delta)\mathcal{L}_{\xi}(\Delta)]\}, \tag{15.15}$$

where $\mathcal{F}$ and $\mathcal{F}^{-1}$ are the Fourier and anti-Fourier transform, respectively. If we denote by t the conjugate variable of Δ, we have

$$\mathcal{F}[\mathcal{L}_{\xi}(\Delta)](t) = \int_{-\infty}^{+\infty} d\Delta\, e^{i\Delta t} \mathcal{L}_{\xi}(\Delta) = e^{-\xi |t|}, \tag{15.16}$$

$$\mathcal{F}[i\Delta\mathcal{L}_{\xi}(\Delta)](t) = \frac{d}{dt}\mathcal{F}[\mathcal{L}_{\xi}(\Delta)](t) = -e^{-\xi |t|}\xi \frac{|t|}{t}, \tag{15.17}$$

and Eq. (15.15) can be written as

$$I(\Delta, \tilde{\sigma}_{\mathrm{D}}, \xi) = \frac{\pi}{\xi}\mathcal{F}^{-1}\left\{\mathcal{F}[W(\Delta)](t)e^{-\xi|t|}\left(1 + \xi\frac{|t|}{t}\right)\right\}. \tag{15.18}$$

For the Lorentzian profile (15.2) we have

$$\begin{aligned} I(\Delta, \tilde{\sigma}_{\mathrm{D}}, \xi) &= \frac{\pi}{\xi}\mathcal{F}^{-1}\left[e^{-\tilde{\sigma}_{\mathrm{D}}|t|}e^{-\xi|t|}\left(1 + \xi\frac{|t|}{t}\right)\right] \\ &= \frac{\pi}{\xi}\left[\mathcal{L}_{\tilde{\sigma}_{\mathrm{D}}+\xi}(\Delta) - \frac{\xi}{\tilde{\sigma}_{\mathrm{D}}+\xi}i\Delta\mathcal{L}_{\tilde{\sigma}_{\mathrm{D}}+\xi}(\Delta)\right], \end{aligned} \tag{15.19}$$

where we have applied $\mathcal{F}^{-1}$ to Eqs. (15.16) and (15.17) with $\mathcal{L}_{\xi}$ replaced by $\mathcal{L}_{\tilde{\sigma}_{\mathrm{D}}+\xi}$. The equations for the nontrivial solution with Lorentzian inhomogeneous broadening are then [151]

$$1 = \frac{A}{\xi}\frac{\tilde{\sigma}_{\mathrm{D}}+\xi}{(\tilde{\sigma}_{\mathrm{D}}+\xi)^2 + \Delta^2}, \tag{15.20}$$

$$\theta = -\frac{A\Delta}{(\tilde{\sigma}_{\mathrm{D}}+\xi)^2 + \Delta^2}. \tag{15.21}$$

These two equations must be solved simultaneously to find the values of the unknowns ξ and Δ. By inserting (15.20) into (15.21) we obtain the equation

$$\theta = -\Delta\frac{\xi}{\tilde{\sigma}_{\mathrm{D}}+\xi}. \tag{15.22}$$

By using the definitions (1.22) and (11.27) for Δ and θ, we obtain the following expression for the laser frequency:

$$\omega_0 = \frac{\kappa\omega_{\mathrm{a}} + \gamma_{\perp}(1 + \tilde{\sigma}_{\mathrm{D}}/\xi)\omega_{\mathrm{c}}}{\kappa + \gamma_{\perp}(1 + \tilde{\sigma}_{\mathrm{D}}/\xi)}, \tag{15.23}$$

which generalizes to the inhomogeneous configuration the mode-pulling formula (12.58) of the homogeneous case, which is retrieved by setting $\tilde{\sigma}_{\mathrm{D}} = 0$. The most remarkable feature is that, if we exclude the resonant case $\delta_{\mathrm{ac}} = 0$ in which Δ is always equal to zero, in the case of inhomogeneous broadening the laser emission frequency depends on the output intensity (see Eq. (15.14)) and is no longer a constant as in the homogeneous configuration. The same is true for the mode-pulling effect, which is affected both by ξ and by $\tilde{\sigma}_{\mathrm{D}}$.

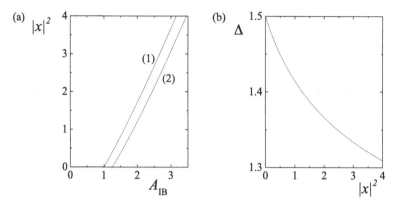

Figure 15.2 (a) The steady-state output intensity for a Lorentzian-broadened laser as a function of the pump parameter for $\tilde{\sigma}_D = 2$, $\tilde{\kappa} = 1$ and (1) $\tilde{\delta}_{ac} = 0$ and (2) $\tilde{\delta}_{ac} = 2$. (b) The behavior of the atomic detuning as a function of the normalized output intensity for the same parameters as in curve (2) of part (a).

In order to obtain the output intensity we observe, first of all, that from the definitions (1.22) and (11.27) of Δ and θ we easily obtain that

$$\theta = \frac{\Delta - \tilde{\delta}_{ac}}{\tilde{\kappa}}, \tag{15.24}$$

where we have used Eq. (15.1), and by inserting (15.24) into (15.22) we obtain the following expression for Δ as a function of ξ:

$$\Delta = \frac{\xi + \tilde{\sigma}_D}{(1 + \tilde{\kappa})\xi + \tilde{\sigma}_D}\tilde{\delta}_{ac}, \tag{15.25}$$

which generalizes Eq. (9.10) (with $\omega_j = \omega_c$ and κ' replaced by κ) to the inhomogeneous case. By solving Eq. (15.20) with respect to A and inserting Eq. (15.25) we obtain finally

$$A_{IB} = \frac{\xi(\xi + \tilde{\sigma}_D)}{1 + \tilde{\sigma}_D}\left\{1 + \left[\frac{\tilde{\delta}_{ac}}{(1 + \tilde{\kappa})\xi + \tilde{\sigma}_D}\right]^2\right\}, \tag{15.26}$$

where we define

$$A_{IB} = \frac{A}{1 + \tilde{\sigma}_D}. \tag{15.27}$$

The threshold value for the pump parameter A_{IB} is obtained by setting $\xi = 1$ (which corresponds to $x = 0$, see Eq. (15.14)) in Eq. (15.26),

$$A_{IB,thr} = 1 + \left(\frac{\tilde{\delta}_{ac}}{1 + \tilde{\kappa} + \tilde{\sigma}_D}\right)^2 \tag{15.28}$$

from which we see that A_{IB} is the correct pump parameter for the inhomogeneous broadening case because on resonance ($\tilde{\delta}_{ac} = 0$) its threshold value is 1. By using Eqs. (15.26) and (15.14) one can obtain the graph of A_{IB} as a function of $|x|^2$. We are interested in the inverse function, which gives the curve of the normalized output intensity as a function of the pump parameter. Note from Eq. (15.26) that this curve depends also on the parameters κ, $\gamma_\perp$ and $\tilde{\sigma}_D$. This curve is shown in Fig. 15.2(a) for two values of $\tilde{\delta}_{ac}$, while Fig. 15.2(b)

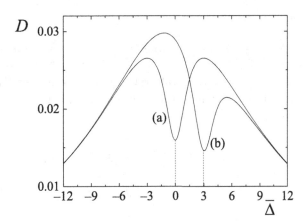

Figure 15.3 Spectral hole-burning. The stationary profile of the atomic population inversion is plotted as a function of $\bar{\Delta}$ for $\tilde{\sigma}_D = 10$, $|x|^2 = 1$ and (a) $\Delta = 0$ and (b) $\Delta = -3$.

shows the variation of Δ as a function of the normalized output intensity $|x|^2$ as described by Eq. (15.25).

In the inhomogeneous limit $\tilde{\sigma}_D \to \infty$, from Eqs. (15.26) and (15.14) we obtain

$$|x|^2 = A_{IB}^2 - 1, \tag{15.29}$$

independently of any detuning, while θ converges to zero (see Eq. (15.22)), i.e. the laser frequency coincides with the cavity frequency. Indeed, in the limit $\tilde{\sigma}_D \to \infty$ any finite detuning becomes negligible with respect to σ_D.

On the other hand, in the homogeneous case $\tilde{\sigma}_D = 0$ one has the usual relation $|x|^2 = A - 1 - \Delta^2$ (since $\Delta = (\omega_a - \omega_c)/(\kappa + \gamma_\perp) = \tilde{\delta}_{ac}/(1 + \tilde{\kappa})$, where we used Eq. (9.24)).

We observe that when the parameter $\tilde{\delta}_{ac}$ is very large (on the order of 100) the curve of the output intensity as a function of A may display bistability [151, 152], but in these cases the threshold value for A is also very large.

Of noteworthy interest is the behavior of the stationary population inversion D as a function of the parameter $\bar{\Delta}$, which is simply related to the frequency of the atomic group $\omega_a + \bar{\Delta}\gamma_\perp$. As described by Eq. (15.11), the expression for D depends on two factors, of which one is the distribution $W(\bar{\Delta})$ of the inhomogeneously broadened atomic line and the other is very close to 1 everywhere except in a neighborhood (of width on the order of $\gamma_\perp$) of the value $\bar{\Delta} = -\Delta$, where this factor burns a hole in the atomic population difference. This phenomenon, which is called *spectral hole-burning* [8, 11], is illustrated in Fig. 15.3.

In the case of Gaussian inhomogeneous broadening (15.3), the integral of Eq. (15.18) reads

$$I(\Delta, \tilde{\sigma}_D, \xi) = \sqrt{\frac{\pi}{2}} \frac{1}{\xi} \mathcal{F}^{-1}\left[e^{-\tilde{\sigma}_D^2 t^2/2} e^{-\xi|t|} \left(1 - \xi \frac{|t|}{t}\right)\right] \tag{15.30}$$

$$= \sqrt{\frac{\pi}{2}} \frac{1}{\xi}[(1 + \xi)\mathcal{I}(\Delta, \tilde{\sigma}_D, \xi) + (1 - \xi)\mathcal{I}^*(\Delta, \tilde{\sigma}_D, \xi)], \tag{15.31}$$

with

$$\mathcal{I}(\Delta, \tilde{\sigma}_D, \xi) = \frac{1}{\sqrt{2\pi}} \int_0^{+\infty} dt\, e^{-\tilde{\sigma}_D^2 t^2/2 - \xi t - i\Delta t} = \frac{e^{-z^2}}{2\tilde{\sigma}_D} \operatorname{erfc}(-iz), \quad z = \frac{\Delta + i\xi}{\sqrt{2}\tilde{\sigma}_D} \quad (15.32)$$

and $\operatorname{erfc}(z) = 1 - \operatorname{erf}(z)$, with erf being the error function. Therefore, the equations for the nontrivial solution with Gaussian inhomogeneous broadening are, instead of Eqs. (15.20) and (15.21),

$$1 = \sqrt{\frac{\pi}{2}} \frac{A}{\xi\tilde{\sigma}_D} \operatorname{Re}[e^{-z^2} \operatorname{erfc}(-iz)], \quad (15.33)$$

$$\theta = -\sqrt{\frac{\pi}{2}} \frac{A}{\tilde{\sigma}_D} \operatorname{Im}[e^{-z^2} \operatorname{erfc}(-iz)]. \quad (15.34)$$

16

The semiconductor laser

This chapter plays a special role in the economy of this book. In the treatment of nonlinear optical systems such as the laser, we have adopted an idealized theoretical description such as the two-level picture. For any kind of real laser the physics behind it is complex. In order to let the reader appreciate this point and evaluate to what extent the models used in this book can describe such a complex reality, we have chosen the case of one of the lasers which is most commonly utilized in applications, namely the semiconductor laser [153–163], for which the physics is especially rich and interesting. A detailed description of this type of laser is not simple, because a semiconductor material cannot be modeled as a collection of identical atoms with homogeneous broadening. Even if the interaction with the electric field involves electron–hole pairs, each of which can be assimilated to a two-level atom, the picture is complicated by various elements. Owing to the Pauli exclusion principle, there may exist at most two electrons, and two holes, having the same energy, with the factor of 2 arising from the spin degeneracy. As a consequence, different electron–hole pairs correspond to different transition frequencies, and the gain line is inhomogeneously broadened. In addition, the carriers (electrons and holes) interact with the crystalline lattice, and this circumstance leads to a complex band structure of the energy levels. Finally the carriers interact with one another via the Coulomb force, i.e. there is a many-body interaction.

We will focus on a simple model that takes the inhomogeneous broadening into account, considers the simplest possible structure for the bands and, among the effects of the many-body interaction, includes only, in a phenomenological way, the so-called band-gap renormalization, i.e. the progressive reduction of the transition frequency when the carrier number increases. In this way we will arrive at a model for the semiconductor laser, the equations of which are formally identical to those which govern the dynamics of class-B lasers out of resonance, i.e. lasers in which the atomic polarization can be adiabatically eliminated, which will be described in Chapter 19. Despite its simplicity, such a model is capable of describing many of the phenomena which arise in semiconductor lasers.

16.1 Some elements of semiconductor physics

In gas lasers the electrons of each atom are spatially well separated from those of the other atoms. Hence it makes sense to consider the nonlinear medium as a collection of independent atoms. One first evaluates the polarization r of the single atom and the slowly varying envelope of the macroscopic atomic polarization is simply given by $P_0 = 2Ndr$ (Eq. (2.42)). This approach holds also for some solid-state lasers such as, for example, the

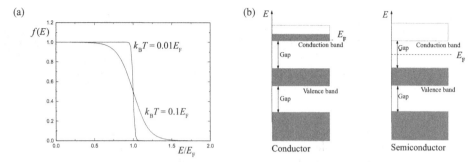

Figure 16.1 (a) The shape of the Fermi–Dirac distribution for two values of temperature. (b) The occupation of the energy bands in a conductor and in a semiconductor.

ruby laser, because the electrons which participate in the lasing process are those of the Cr^{3+} ions, which are localized within a distance of 1 or 2 Å from their nuclei and, for typical values of the doping in Cr, do not interact with electrons of other ions.

The main difference between semiconductor lasers and other lasers is that in a semiconductor all of the participating electrons occupy and share the entire volume of the crystal. Owing to the spatial superposition of their wave packets, two electrons in a crystal cannot be in the same quantum state, because of the Pauli exclusion principle. Hence each electron has its own wave function characterized by specific quantum numbers and an associated energy. If we imagine tracing a horizontal line for any allowed energy level, we discover that these levels are arranged in bands separated by intervals called energy gaps. The uppermost band which is entirely occupied is called the valence band and the band immediately above is called the conduction band.

For a system of N electrons per unit volume at thermal equilibrium with absolute temperature T the probability that the electronic state with energy E is occupied is given by the Fermi–Dirac distribution,

$$f(E) = \frac{1}{e^{(E-E_F)/(k_B T)} + 1}, \tag{16.1}$$

where E_F is the Fermi energy, which can be evaluated by imposing that the sum of the probabilities over all states is equal to the total number of electrons. The behaviour of $f(E)$ is shown in Fig. 16.1(a).

For energies much smaller than the Fermi energy, for which $E_F - E \gg k_B T$, $f(E) \to 1$ and the electronic states are entirely occupied. In the opposite limit $E - E_F \gg k_B T$, $f(E) \propto e^{-E/(k_B T)}$, the Fermi distribution approaches the Boltzmann distribution. At zero temperature, $f(E) = 1$ for $E < E_F$ and $f(E) = 0$ for $E > E_F$, i.e. all the states below the Fermi level are occupied and those above are empty.

If the Fermi energy lies in the conduction band the material is a conductor. However, the most general situation is that the Fermi level lies in the gap between the valence band and the conduction band. This implies that at zero temperature the valence band is entirely occupied and the conduction band is completely empty. Hence the material is an insulator (Fig. 16.1(b)). The value of the energy gap determines the electrical conductivity at finite temperature. In an insulator the gap is large enough, on the order of 3 eV, which makes thermal excitation negligible. If we apply an electric field, there will be no flow of current because all of the levels of the valence band are occupied and the electrons cannot acquire kinetic energy.

If, on the other hand, the gap is small, on the order of 2 eV or smaller, thermal excitation can cause a partial transfer of electrons from the valence band to the conduction band and the crystal can conduct electricity. Such crystals are called semiconductors. The gap energy at a temperature of 300 K is 0.67 eV for Si, 1.11 eV for Ge and 1.42 eV for GaAs. The electrical conductivity of a semiconductor can be controlled not only by temperature but also by doping the material with the inclusion of acceptor atoms, i.e. atoms in which the outermost occupied shell presents missing electrons, or donor atoms, where that shell presents an excess of electrons.

A donor-doped semiconductor is labeled a semiconductor of n type. In this case the donor energy levels are located just below the conduction band, at a distance smaller than $k_B T$, so these levels are effectively ionized and generate conduction electrons, i.e. negatively charged current carriers, hence the label n. On the other hand, a semiconductor of type p is doped with acceptors, the energy levels of which lie just above the valence band. The electrons can easily jump, by thermal excitation, from the valence band to such levels, thus creating a hole in the valence band, i.e. a current carrier with positive charge, from which the label p originates. In strongly doped semiconductors the Fermi energy is shifted to the conduction band for type-n semiconductors and to the valence band for type-p semiconductors, as shown in Fig. 16.2(a). These types of semiconductors are called *degenerate*.

16.2 The p–n junction

The p–n junction is a device in which two semiconductors of opposite type are set in contact. The conduction electrons of the material diffuse across the junction to fill holes and, similarly, the valence holes diffuse across the junction in the opposite direction, annihilating with electrons in the conduction band. The diffusion process generates a situation in which there is a layer of positive charges in the n material and of negative charges in the p material, in the region around the junction. At equilibrium this charge distribution forbids a further diffusion of the charges. The electric potential $V(x)$, as a function of the x coordinate orthogonal to the junction plane, is negative on the p side, increases to zero at the junction and becomes positive on the n side, as shown in Fig. 16.2(b).

The behavior of the bands in the junction region can be approximately described by summing the energy $-eV(x)$ to the bands in the two types of semiconductor. At equilibrium the material is described by a unique Fermi energy, which lies in the valence band of the p material and in the conduction band of the n material (Fig. 16.2(c)). In no region is there the simultaneous presence of electrons in the conduction band and holes in the valence band, i.e. population inversion. Such an inversion, which is necessary for laser emission, however, can be obtained by applying to the junction a potential difference such that the potential is larger on the p side (forward bias). In this way one has a current flow across

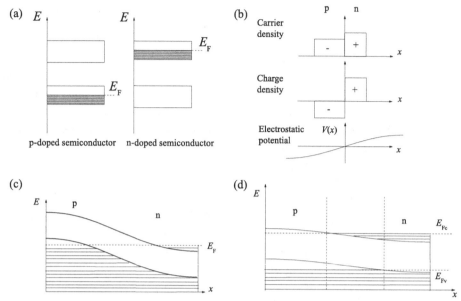

Figure 16.2 (a) The Fermi energy level in degenerate semiconductors of types p and n. (b) The carrier distribution and electric potential as functions of the direction orthogonal to the separation plane between the two materials in a p–n junction. (c) Band occupation in a p–n junction. (d) Band occupation when a direct potential difference is applied to the junction.

the junction, namely electrons and holes are injected into the junction in the n region and in the p region, respectively. When the applied potential is almost equal to the gap between the two bands one has population inversion in the junction region. In such a situation, which is far from thermal equilibrium, the band occupation must be described by two different Fermi energies E_{Fc} and E_{Fv} (Fig. 16.2(d)). One calls them quasi-Fermi energies, because the system is not at thermal equilibrium. The region close to the junction, in which there is population inversion, is the active region of the semiconductor laser.

16.3 The double heterojunction. Optical confinement

The first semiconductor lasers, which were realized at the beginning of the 1960s, were based on a p–n junction constituted by the same material with different doping. In this case one speaks of a homojunction. However, this scheme presents a serious problem, in that the penetration region of electrons in the p layer and of holes in the n layer, i.e. the region in which there is population inversion, has a large and not controllable thickness. This implies that in order to obtain a large enough carrier concentration it is necessary for the applied potential, and therefore the current, to be very high. Owing to the heating associated with the passage of an intense current, these lasers could be operated only in a pulsed regime.

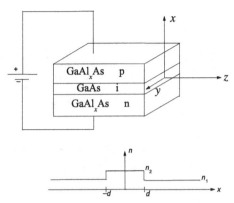

Figure 16.3 The scheme for a semiconductor laser with a double heterojunction.

The realization of continuous-wave (CW) semiconductor lasers was possible thanks to the adoption of a scheme in which a layer of non-doped semiconductor material is located between two layers of different materials with p and n doping. In this way one realizes a *double heterojunction* (Fig. 16.3). This design presents three advantages. First, one can arrange that the gap energy in the central layer is smaller than in the two external layers. This implies spatial confinement of the carriers inside the central layer, the width of which can be controlled in the fabrication process. Second, a material with a smaller energy gap has a larger refractive index; hence the double heterostructure works as a waveguide that confines the radiation in the central layer. Third, the field which invades the external layers propagates without undergoing absorption, since the photons have an energy smaller than the energy gap in such layers, and therefore cannot excite electrons from the valence band to the conduction band.

The heterostructures can be realized only using materials for which the crystals are similar enough. For example, one can use GaAs and AlAs because both of them are cubic face-centered crystals, with very similar lattice periods.

The analysis of the Maxwell equations in the planar waveguide realized by the double heterostructure leads to the identification of transverse propagation modes, which are grouped into two classes, namely TE modes, in which the electric field is orthogonal to the propagation direction, and TM modes, in which the magnetic field is orthogonal to the propagation direction [156, 160]. By considering, for example, TE modes, one can calculate the spatial configuration of these modes as a function of the x coordinate in Fig. 16.3 by solving the Maxwell wave equation for the electric field and by imposing the continuity conditions at the two interfaces which separate the three layers of the heterostructure. The results depend on the values of the refractive index inside and outside the active region, on the thickness of the active region and on the frequency ω_0 of the electric field. On fixing these parameters, one finds a finite number of transverse modes TE_1, TE_2, ..., TE_n, even or odd with respect to the variable x. When ω_0 is smaller than an appropriate value called the *cutoff frequency*, there is only one mode, TE_1, which is even and never assumes a value of zero. In the following we assume that the frequency is below the cutoff value and therefore the laser operates with the TE_1 mode.

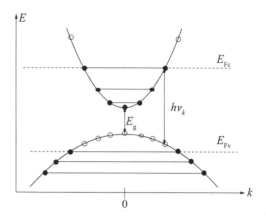

Figure 16.4 Conduction band and valence band in the active region of a semiconductor laser.

16.4 Band structure

The wave function of an electron in a band is a solution of the Schrödinger equation with a periodic potential with a period determined by the crystal structure. It is characterized by the wave vector $\mathbf{k}$ and by the spin s,

$$\psi_{\mathbf{k},s}(\mathbf{r}) = u_{\mathbf{k},s}(\mathbf{r})e^{i\mathbf{k}\cdot\mathbf{r}}, \tag{16.2}$$

and, due to the periodicity of the potential, it must satisfy the condition

$$\psi_{\mathbf{k},s}(\mathbf{r} + \mathbf{R}) = \psi_{\mathbf{k},s}(\mathbf{r}), \tag{16.3}$$

with $\mathbf{R}$ being the lattice constant, so that the wave vector is quantized. On the other hand, the energy of the electron depends, to within a good approximation, on the modulus k of the wave vector only. Such a dependence is very complex in general. However, if we consider a limited enough range of values of k and if the band-gap is of *direct type*, i.e. if the extrema of the conduction band and of the valence band correspond to the same value of k, we can use the *effective-mass approximation*, in which the bands are approximated by parabolas. By fixing the zero of the energy at the maximum of the valence band, we can write

$$E_c(\mathbf{k}) = \frac{\hbar^2 k^2}{2m_c} + E_g, \qquad E_v(\mathbf{k}) = \frac{\hbar^2 k^2}{2m_v}, \tag{16.4}$$

as for a free particle, where, however, the electron mass is replaced by the effective mass of the two bands. Since the valence mass is negative, the two parabolas have opposite concavities and are separated by the energy gap E_g, as shown in Fig. 16.4. In the electron–hole formalism one sets the mass of the electron equal to the effective mass in the conduction band, $m_e = m_c$, and the mass m_h of the hole equal to the modulus of the effective mass in the valence band, and one writes

$$E_c(\mathbf{k}) = E_e(\mathbf{k}) + E_g, \qquad E_v(\mathbf{k}) = -E_h(\mathbf{k}), \tag{16.5}$$

with

$$E_e(\mathbf{k}) = \frac{\hbar^2 k^2}{2m_e}, \qquad E_h(\mathbf{k}) = \frac{\hbar^2 k^2}{2m_h}. \tag{16.6}$$

The energies $E_e(\mathbf{k})$ and $E_h(\mathbf{k})$ are positive and are measured from the edges of the respective bands. Since $m_h > m_e$, the valence band is wider than the conduction band.

We have seen that a p–n junction to which a potential difference is applied must be described by two different Fermi energies for electrons and holes. In the parabolic-band approximation the two Fermi–Dirac distributions are

$$f_{e\mathbf{k}}(E) = \frac{1}{e^{(\hbar^2 k^2/(2m_e) - E_{Fe})/(k_B T)} + 1}, \tag{16.7}$$

$$f_{h\mathbf{k}}(E) = \frac{1}{e^{(\hbar^2 k^2/(2m_h) - E_{Fh})/(k_B T)} + 1} \tag{16.8}$$

and the Fermi energies can be calculated by imposing that the sum of the two distributions over all $\mathbf{k}$ is equal to N. Hence the Fermi energies are functions of N, T and the masses. From this it follows that at a fixed temperature electrons and holes have different Fermi energies even if $N_e = N_h$, because their masses are different.

The passage of a photon through the active region gives rise to a process of stimulated emission, due to electron–hole recombination. The light produced in such a way propagates along the junction and is partially reflected at the surface which separates the material from the air. The reflection coefficient for the field intensity at the interface between a dielectric with refractive index n and the air is $R = (n - 1)^2/(n + 1)^2$. For a typical value of $n = 3.5$ one has $R = 25/81 \simeq 0.31$. Therefore in the case of a semiconductor laser the reflection arises not from the presence of beam splitters, but from the air–medium interface. When the gain is larger than the loss one has laser emission.

It is important to observe that in the recombination processes the value of $\mathbf{k}$ remains constant, because the probability that the process occurs is proportional to an integral over the crystal volume, which contains the product of the final wave-function and the complex conjugate of the initial wavefunction. Since the wave-functions have the form prescribed by Eq. (16.2) this integral is vanishingly small unless the initial and final wave vectors are equal. Hence in the band diagram (Fig. 16.4) the transitions are represented by vertical lines. Therefore one can describe the semiconductor material as a collection of two-level atoms, uncoupled from one another from the optical viewpoint, in which the transition frequency ω_k depends on k according to the law

$$\hbar \omega_k = E_g + \frac{\hbar^2 k^2}{2m_e} + \frac{\hbar^2 k^2}{2m_h}. \tag{16.9}$$

We can phenomenologically include in this formula the effect of the *band-gap renormalization* by setting [160]

$$\hbar \omega_k = E_g + \frac{\hbar^2 k^2}{2m_e} + \frac{\hbar^2 k^2}{2m_h} - \sigma N^{1/3}, \tag{16.10}$$

with $\sigma \simeq 4 \times 10^{-29}$ J m, which prescribes a reduction of the transition frequency when the electron number increases.

16.5 Dynamical equations

On the basis of the considerations of the previous sections we shall derive the dynamical equations for the semiconductor laser starting from the equations for two-level atoms, with the necessary modifications. Let us start from the Bloch equations for bidirectional propagation (14.5) and (14.6) without the irreversible terms, in which we indicate the time derivative with a dot and we write $\tilde{\Omega}(z, t)$ instead of $\Omega_F(z, t)\exp(i k_0 z) + \Omega_B(z, t)\exp(-i k_0 z)$:

$$\dot{\tilde{r}} = -i\delta\tilde{r} - i\frac{r_3}{2}\tilde{\Omega}, \tag{16.11}$$

$$\dot{r}_3 = i\left(\tilde{\Omega}\tilde{r}^* - \tilde{\Omega}^*\tilde{r}\right). \tag{16.12}$$

Since $r_3 = p_1 - p_2 = 2p_1 - 1 = 1 - 2p_2$, with p_1 and p_2 being the occupation probabilities of the upper level and of the lower level, respectively, we have

$$\dot{p}_1 = -\dot{p}_2 = \frac{\dot{r}_3}{2} = \frac{i}{2}\left(\tilde{\Omega}\tilde{r}^* - \tilde{\Omega}^*\tilde{r}\right). \tag{16.13}$$

In a semiconductor laser, due to the band structure of the energy levels, we must assign an index $\mathbf{k}$ to the variables p_1, p_2 and $\tilde{r}$, to the dipole moment d which enters the definition of the Rabi frequency $\tilde{\Omega}$ and to the off-resonance parameter δ. Furthermore, we shall interpret p_1 and p_2 as occupation probabilities of the $\mathbf{k}$ level in the conduction band and in the valence band, respectively. Hence we make the following replacements:

$$p_1 \longrightarrow n_{\mathrm{ek}}, \tag{16.14}$$

$$p_2 \longrightarrow n_{\mathrm{vk}} = 1 - n_{\mathrm{hk}}, \tag{16.15}$$

$$\tilde{r} \longrightarrow \tilde{r}_{\mathbf{k}}, \tag{16.16}$$

$$\tilde{\Omega} = \frac{d\tilde{E}_0}{\hbar} \longrightarrow \tilde{\Omega}_{\mathbf{k}} = \frac{d_{\mathbf{k}}\tilde{E}_0}{\hbar}, \quad \tilde{E}_0 = E_F e^{ik_0 z} + E_B e^{-ik_0 z}, \tag{16.17}$$

$$\delta \longrightarrow \delta_{\mathbf{k}} = \omega_k - \omega_0. \tag{16.18}$$

In Eq. (16.15) we have introduced the hole probability n_{hk}, which by definition is equal to $1 - n_{\mathrm{vk}}$, i.e. the probability of having a hole in the valence band is equal to the probability of not having an electron. The Bloch equations in these variables read

$$\dot{\tilde{r}}_{\mathbf{k}} = -i\delta_{\mathbf{k}}\tilde{r}_{\mathbf{k}} - i\frac{\tilde{\Omega}_{\mathbf{k}}}{2}(n_{\mathrm{ek}} + n_{\mathrm{hk}} - 1), \tag{16.19}$$

$$\dot{n}_{j\mathbf{k}} = \frac{i}{2}\left(\tilde{\Omega}_{\mathbf{k}}\tilde{r}_{\mathbf{k}}^* - \tilde{\Omega}_{\mathbf{k}}^*\tilde{r}_{\mathbf{k}}\right), \qquad j = \mathrm{e, h}. \tag{16.20}$$

The equations for electrons and for holes are identical, but the introduction of the irreversible terms diversifies them, because in the absence of an electric field the two kinds of carriers approach two different quasi-equilibrium distributions f_{ek} and f_{hk}. We denote by γ_{c} the rate which governs the approach of the probabilities to the quasi-equilibrium state. Furthermore, also the pump term $\Lambda_{j\mathbf{k}}$, which is linked to the current injected into the active region, may differ for the two kinds of carriers. In addition to these terms, in the equations for n_{ek} and n_{hk} there is a linear loss term with rate γ_{nr}, which describes the nonradiative recombination rate, i.e. those electron–hole recombination processes which reduce the carrier numbers without producing photons (the energy is transferred to the vibrational modes of the crystal). For the sake of simplicity, we neglect the loss term which describes radiative recombination with emission of photons in directions different from z. Insofar as the dipole $\tilde{r}_k$ is concerned, we assume as usual that, in the absence of a radiation field, it tends to zero with a decay rate that we denote by γ_p. Therefore the Bloch equations read

$$\dot{\tilde{r}}_k = -i\delta_k \tilde{r}_k - i\frac{\tilde{\Omega}_k}{2}(n_{ek} + n_{hk} - 1) - \gamma_p \tilde{r}_k, \tag{16.21}$$

$$\dot{n}_{jk} = \frac{i}{2}\left(\tilde{\Omega}_k \tilde{r}_k^* - \tilde{\Omega}_k^* \tilde{r}_k\right) + \Lambda_{jk} - \gamma_{nr} n_{jk} - \gamma_c(n_{jk} - f_{jk}). \tag{16.22}$$

Next, we take into account that the dipole decay rate is much larger than the evolution rates of the field and of the populations, so that we can adiabatically eliminate the dipole variable $\tilde{r}_k$. By setting the derivative of $\tilde{r}_k$ equal to zero we obtain

$$\tilde{r}_k = -i\frac{n_{ek} + n_{hk} - 1}{\gamma_p + i\delta_k}\frac{\tilde{\Omega}_k}{2}, \tag{16.23}$$

which, once inserted into Eq. (16.22), gives

$$\dot{n}_{jk} = -\frac{1}{2}\frac{\gamma_p}{\gamma_p^2 + \delta_k^2}(n_{ek} + n_{hk} - 1)|\tilde{\Omega}_k|^2 + \Lambda_{jk} - \gamma_{nr} n_{jk} - \gamma_c(n_{jk} - f_{jk}). \tag{16.24}$$

Let us now consider the total number of carriers per unit volume N, defined as

$$N = \frac{1}{V}\sum_k n_{ek} = \frac{1}{V}\sum_k n_{hk} = \frac{1}{V}\sum_k f_{ek} = \frac{1}{V}\sum_k f_{hk}, \tag{16.25}$$

which is the same for electrons and holes because the material which forms the active region is a non-doped semiconductor. The equation for N is obtained from Eq. (16.24) by summing over $\mathbf{k}$. Owing to this operation the relaxation terms which govern the approach to quasi-equilibrium disappear. If we set

$$\frac{1}{V}\sum_k \Lambda_{ek} = \frac{1}{V}\sum_k \Lambda_{hk} = \frac{J}{eV}, \tag{16.26}$$

where J is the injected current and e the modulus of the electron charge, we can write

$$\dot{N} = -\frac{1}{2V}\sum_k \frac{\gamma_p}{\gamma_p^2 + \delta_k^2}(n_{ek} + n_{hk} - 1)|\tilde{\Omega}_k|^2 + \frac{J}{eV} - \gamma_{nr} N. \tag{16.27}$$

This equation is not closed in N, because there are terms with n_{ek} and n_{hk}. However, one introduces the *quasi-equilibrium approximation*, by assuming that also in the presence of a non-vanishing electric field the carriers are distributed as in the quasi-equilibrium state, i.e.

$$n_{ek} \simeq f_{ek}, \qquad n_{hk} \simeq f_{hk}. \tag{16.28}$$

In this way the time evolution equation for N depends only on N, because the distributions f_{ek} and f_{hk} are functions of N, and on $|\tilde{E}_0|$, since $\tilde{\Omega}_k = d_k \tilde{E}_0/\hbar$:

$$\dot{N} = -\frac{1}{2V\hbar^2} \sum_k \frac{\gamma_p}{\gamma_p^2 + \delta_k^2} [f_{ek}(N) + f_{hk}(N) - 1] d_k^2 |\tilde{E}_0|^2 + \frac{J}{eV} - \gamma_{nr} N. \tag{16.29}$$

Now this equation must be coupled with the dynamical equation for the electric field in order to obtain a set of equations that govern the semiconductor laser. Such a field equation will be derived from the Maxwell wave equation for the field envelope by applying the low-transmission and single-mode approximations. The cavity configuration for the semiconductor laser is Fabry–Perot and, as we saw in Chapter 14, the treatment of the Fabry–Perot case is much more complex than that of the ring cavity because of the standing-wave effects arising from the grating in the population-inversion profile. However, we shall assume that this grating is washed out by carrier diffusion in the longitudinal direction z and, as we saw in Section 14.3, in this case the equations of the single-mode model for the Fabry–Perot case reduce to those for the ring cavity, provided that the electric field which appears in the final equations is interpreted as the forward-propagating field multiplied by $\sqrt{2}$ (see Eq. (14.77)). Therefore we will derive the field equation for the semiconductor laser as if the cavity were a ring cavity.

Accordingly, in Eq. (16.29) we replace $\tilde{E}_0$ by the E_0 of the ring cavity and, for the field, we start from Eq. (5.49), where we neglect the diffraction term (transverse Laplacian) and the group-velocity-dispersion term (the second derivative with respect to time), replace the background refractive index $n_B(\omega_0)$ by the constant refractive index n in the active region and replace the group velocity v_g by c/n:

$$\frac{n}{c} \frac{\partial E_0}{\partial t} + \frac{\partial E_0}{\partial z} = i \frac{k_0}{2\epsilon_0 n^2} P_0. \tag{16.30}$$

Next, we want to apply the low-transmission and single-mode approximations. In this connection, we observe that in the case of a semiconductor laser the transmission T (not to be confused with temperature) is not low because, as we discussed in Section 16.4, in this case $R = 1 - T \simeq 0.31$, and therefore the low-transmission approximation does not work well. However, we apply it nonetheless in order to arrive at a reasonably simple model. In addition, the low-transmission approximation applies very well in the case of a vertical-cavity surface-emitting laser, which will be discussed in the final section of this chapter.

In order to apply the low-transmission and single-mode approximations in a straightforward way, we observe that, starting from Eq. (4.35), i.e.

$$\frac{n}{c} \frac{\partial F}{\partial t} + \frac{\partial F}{\partial z} = gP, \tag{16.31}$$

the application of the two approximations (see Section 12.3) leads to the equation (12.59), i.e.

$$\dot{F} = -\kappa[(1 + i\theta)F - AP] = i(\omega_0 - \omega_c)F - \kappa \left(F - \frac{2gL}{T} P \right), \tag{16.32}$$

where we have taken into account that $\Delta = -\theta$ (Eq. (12.57)) and we have used Eq. (11.18). The variable F which appears in Eq. (16.32) is the spatial average over the atomic sample of the field $F(z, t)$ which appears in Eq. (16.31) and depends on time only. L is the sample length, which in the case of a semiconductor laser is one half of the roundtrip cavity length, and we have used the definition of A given by Eq. (14.36), which is appropriate for the Fabry–Perot case (contextually, the definition of κ to be used is given by Eq. (14.34)). Therefore, starting from Eq. (16.30), the application of the low-transmission and single-mode limits leads to the equation

$$\dot{E}_0 = i(\omega_0 - \omega_c)E_0 - \kappa \left(E_0 - i \frac{\omega_0 L}{\epsilon_0 cn T} P_0 \right), \tag{16.33}$$

where we have taken into account that $k_0 = \omega_0 n/c$. As we said, E_0 corresponds to the forward-propagating field E_F (which, in the low-transmission and single-mode approximations, coincides with the backward-propagating field E_B) multiplied by $\sqrt{2}$. Next we set

$$P_0 = -i \frac{\epsilon_0 cn}{\omega_0} G(N, \omega_0)E_0, \tag{16.34}$$

which allows us to make Eqs. (16.33) and (16.29) a closed system of equations for the variables E_0 and N. Similarly to Eq. (5.26), the quantity

$$\tilde{\chi}(N, \omega_0) = -i \frac{cn}{\omega_0} G(N, \omega_0) \tag{16.35}$$

is the dimensionless complex susceptibility of the medium, whereas the function $G(N, \omega_0)$ has the dimensions of inverse length. In order to derive the expression for the function $G(N, \omega_0)$ let us recall that in the case of two-level atoms the relation between the microscopic polarization r and the macroscopic polarization $\mathcal{P}$ was given by

$$\mathcal{P} = \frac{P_0}{2} e^{i[(\omega_0 n_B/c)z - \omega_0 t]} + \text{c.c.} = Nd[re^{i[(\omega_0 n_B/c)z - \omega_0 t]} + \text{c.c.}], \tag{16.36}$$

so that the atomic polarization density was simply the sum of the microscopic polarizations. In the case of a semiconductor laser we must take into account that both r and d depend on **k**. Hence we write

$$\mathcal{P} = \frac{P_0}{2} e^{i[(\omega_0 n/c)z - \omega_0 t]} + \text{c.c.} = \frac{1}{V} \sum_{\mathbf{k}} d_{\mathbf{k}}[\tilde{r}_{\mathbf{k}} e^{i[(\omega_0 n/c)z - \omega_0 t]} + \text{c.c.}], \tag{16.37}$$

from which it follows that

$$P_0 = \frac{2}{V} \sum_{\mathbf{k}} d_{\mathbf{k}} \tilde{r}_{\mathbf{k}}. \tag{16.38}$$

By inserting the expression (16.23) for $\tilde{r}_{\mathbf{k}}$ with $n_{e\mathbf{k}} = f_{e\mathbf{k}}$, $n_{h\mathbf{k}} = f_{e\mathbf{k}}$ and $\tilde{\Omega}_{\mathbf{k}} = d_{\mathbf{k}} E_0/\hbar$ we obtain

$$P_0 = -\frac{i}{V\hbar} \sum_{\mathbf{k}} \frac{f_{e\mathbf{k}} + f_{h\mathbf{k}} - 1}{\gamma_p + i\delta_{\mathbf{k}}} d_{\mathbf{k}}^2 E_0, \tag{16.39}$$

from which it follows that the function $G(N, \omega_0)$ is given by

$$G(N, \omega_0) = \frac{\omega_0}{\epsilon_0 V \hbar c n} \sum_{\mathbf{k}} \frac{\gamma_p - i\delta_\mathbf{k}}{\gamma_p^2 + \delta_\mathbf{k}^2} [f_{e\mathbf{k}}(N) + f_{h\mathbf{k}}(N) - 1] d_\mathbf{k}^2$$

$$\equiv G'(N, \omega_0) + i G''(N, \omega_0). \tag{16.40}$$

In Eq. (16.29) the term which multiplies $|E_0|^2$ is proportional to the real part G', hence we can write the dynamical equations (16.33) and (16.29) in the compact forms

$$\dot{E}_0 = i(\omega_0 - \omega_c)E_0 - \kappa \left[1 - \frac{L}{T} G(N, \omega_0) \right] E_0, \tag{16.41}$$

$$\dot{N} = -\frac{\epsilon_0 c n}{2\hbar \omega_0} G'(N, \omega_0)|E_0|^2 + \frac{J}{eV} - \gamma_{nr} N. \tag{16.42}$$

In the stationary state we have

$$0 = \left[1 + i\frac{\omega_c - \omega_0}{\kappa} - \frac{L}{T} G(N, \omega_0) \right] E_0, \tag{16.43}$$

$$0 = -\frac{\epsilon_0 c n}{2\hbar \omega_0} G'(N, \omega_0)|E_0|^2 + \frac{J}{eV} - \gamma_{nr} N. \tag{16.44}$$

As long as the laser is below threshold and therefore $E_0 = 0$, the first equation is always satisfied, while the second gives the relation $J = eV\gamma_{nr}N$ between the injected current and the carrier density. Above threshold, instead, the first equation can be divided by E_0, and, by separating real and imaginary parts, we arrive at the two equations

$$1 = \frac{L}{T} G'(N_s, \omega_0), \tag{16.45}$$

$$\frac{\omega_c - \omega_0}{\kappa} = \frac{L}{T} G''(N_s, \omega_0). \tag{16.46}$$

These two equations allow one to obtain the stationary value of the carrier density N_s, which is independent of the value J of the injected current, and the laser frequency ω_0. The value of the injected current at threshold is given by $J_s = eV\gamma_{nr}N_s$, as one obtains from Eq. (16.44) by setting $E_0 = 0$. Using Eqs. (16.44) and (16.45) the field intensity above threshold is given by

$$|E_0|^2 = \frac{2\hbar \omega_0 L}{\epsilon_0 c n T e V} (J - J_s). \tag{16.47}$$

Since above threshold the carrier density N is fixed and equal to N_s it makes sense, in order to simplify the problem, to linearize the function $G(N, \omega_0)$ around the value N_s with ω_0 fixed, i.e. to set

$$G(N) = G(N_s) + \left. \frac{\partial G}{\partial N} \right|_{N=N_s} (N - N_s) = G(N_s) + \tilde{a}(1 - i\alpha)(N - N_s). \tag{16.48}$$

The coefficient $\tilde{a}$, which we assume to be positive, coincides with the derivative of G' for $N = N_s$ and has the dimensions of a surface. The dimensionless coefficient α is equal to the opposite to the ratio of the derivative of G'' to the derivative of G' for $N = N_s$. This parameter is typical of the semiconductor laser, and in the literature is called the *linewidth-enhancement factor* [164]. By taking Eqs. (16.45) and (16.46) into account the linearized G function can be re-expressed in the form

$$
\begin{aligned}
G(N) &= G'(N_s) + i\,G''(N_s) + \tilde{a}(1 - i\alpha)(N - N_s) \\
&= \tilde{a}(1 - i\alpha)(N - N_s) + (1 - i\alpha)G'(N_s) + i[\alpha\,G'(N_s) + G''(N_s)] \\
&= (1 - i\alpha)[\tilde{a}(N - N_s) + G'(N_s)] + i\frac{T}{L}\left(\alpha + \frac{\omega_c - \omega_0}{\kappa}\right) \\
&= (1 - i\alpha)\tilde{a}(N - N_0) + i\frac{T}{L}\left(\alpha + \frac{\omega_c - \omega_0}{\kappa}\right),
\end{aligned}
\tag{16.49}
$$

where

$$
N_0 = N_s - \frac{G'(N_s)}{\tilde{a}} = N_s - \frac{T}{\tilde{a}L}
\tag{16.50}
$$

is called the transparency density, i.e. the value of the carrier density for which, in the linear approximation, the real part of the susceptibility vanishes (see Eq. (16.48)) and the material is transparent. Obviously $N_0 < N_s$. The transparency current $J_0 = eV\gamma_{nr}N_0$ is immediately associated with the transparency density. If we now insert Eq. (16.49) into Eq. (16.41) we obtain

$$
\begin{aligned}
\dot{E}_0 &= i(\omega_0 - \omega_c)E_0 - \kappa\left[1 - i\frac{\omega_c - \omega_0}{\kappa} - i\alpha - \frac{\tilde{a}L}{T}(1 - i\alpha)(N - N_0)\right]E_0 \\
&= -\kappa(1 - i\alpha)\left[1 - \frac{\tilde{g}L}{T}\left(\frac{N}{N_0} - 1\right)\right]E_0,
\end{aligned}
\tag{16.51}
$$

where we defined the gain/absorption parameter

$$
\tilde{g} = \tilde{a}N_0.
\tag{16.52}
$$

On the other hand, by inserting into Eq. (16.42) the expression $G'(N) = \tilde{a}(N - N_0)$ which follows from Eq. (16.49), the time-evolution equation for N becomes

$$
\begin{aligned}
\dot{N} &= -\frac{\epsilon_0 cn}{2\hbar\omega_0}\tilde{a}(N - N_0)|E_0|^2 + \frac{J}{eV} - \gamma_{nr}N \\
&= -\gamma_{nr}\left[\frac{\epsilon_0 cn\tilde{a}}{2\hbar\omega_0\gamma_{nr}}(N - N_0)|E_0|^2 - N_0\left(\frac{J}{J_0} - 1\right) + N - N_0\right] \\
&= -\gamma_{nr}\left[(N - N_0)\left(1 + \frac{\epsilon_0 cn\tilde{a}}{2\hbar\omega_0\gamma_{nr}}|E_0|^2\right) - N_0\left(\frac{J}{J_0} - 1\right)\right].
\end{aligned}
\tag{16.53}
$$

Upon introducing the scaled variables

$$
F = \sqrt{\frac{\epsilon_0 cn\tilde{a}}{2\hbar\omega_0\gamma_{nr}}}\,E_0, \qquad \tilde{D} = \frac{\tilde{g}L}{T}\left(\frac{N}{N_0} - 1\right), \qquad \mu = \frac{\tilde{g}L}{T}\left(\frac{J}{J_0} - 1\right),
\tag{16.54}
$$

the dynamical equations (16.51) and (16.53) take the compact forms

$$
\dot{F} = -\kappa(1 - i\alpha)(1 - \tilde{D})F,
\tag{16.55}
$$

$$
\dot{\tilde{D}} = -\gamma_{nr}\left[\tilde{D}\left(1 + |F|^2\right) - \mu\right].
\tag{16.56}
$$

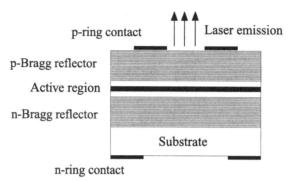

p-ring contact   Laser emission

p-Bragg reflector

Active region

n-Bragg reflector

Substrate

n-ring contact

Figure 16.5 The scheme of a typical vertical-cavity surface-emitting laser consisting of an active layer sandwiched between Bragg mirrors.

As we will see in Section 19.3.1, these equations have the same forms as the dynamical equations which govern class-B lasers, as follows from the single-mode model of Chapter 12 on adiabatically eliminating the atomic polarization. In the two-level case of Section 19.3.1, the parameter α is replaced by the atomic detuning parameter $\Delta = (\omega_a - \omega_0)/\gamma_\perp$. We observe, however, that the α parameter of semiconductor lasers has quite a different physical meaning from the atomic detuning parameter in gas lasers, even if it plays a similar role in the dynamical equations. The value of α depends on the semiconductor material and it typically ranges from 3 to 6. In a semiconductor laser the nonlinear frequency dispersion is always strong and keeps the same sign, whereas in gas lasers Δ can be positive or negative and vanishes when the gain is maximum, i.e. at resonance.

Finally it is important to observe that the model (16.55) and (16.56) holds not only for the type of semiconductor laser that we described in the previous sections but also for any semiconductor laser for which the dynamical equations have the forms (16.41) and (16.42) independently of the precise definition of the function $G(N, \omega_0)$. As a matter of fact, all the steps which follow Eqs. (16.41) and (16.42) do not depend on the form of G.

16.6 Vertical-cavity surface-emitting lasers

A class of semiconductor lasers that has been realized more recently, with construction criteria quite different from those of the semiconductor lasers we described in Section 16.3, is that of *vertical-cavity surface-emitting lasers* (VCSELs). In VCSELs the laser cavity is perpendicular (vertical) to the alternating layers of materials and light emerges from the surface unlike in conventional semiconductor lasers, which are called *edge-emitting lasers* because the cavity is horizontal and light emerges from the edge (Fig. 16.3). The structure of the VCSEL is grown vertically by successively depositing layers of different materials on the substrate. The bottom mirror (Bragg reflector) is first created, then the gain region and finally the top mirror (Bragg reflector, see Fig. 16.5). This geometry allows one to etch easily two-dimensional arrays of VCSELs as well as a single VCSEL from the same sample. The typical diameter of one laser is about 10 μm, although VCSELs of diameter up to 200 μm have been produced, which are particularly suitable for the study of cavity solitons, as we will show in Chapter 30.

The total height of a VCSEL is a few microns, and the cavity volume is almost all occupied by the two mirrors, which are actually distributed Bragg reflectors, formed by alternating $\lambda/4$ layers of semiconductor materials with slightly different refractive indices. Typically each mirror contains about twenty layers, which produce a total reflectivity larger than 99%. Such a high reflectivity is necessary in order to compensate for the relatively small gain experienced by the light in a single passage because the active region is made of one or a few quantum wells [165] that are traversed perpendicularly by light, such that the thickness of the active region is just some tens of nanometers.

With respect to edge emitters, VCSELs have several benefits: miniaturized structure, low threshold and simple modal composition of the emitted light. In fact, the cavity is so short that no more than one longitudinal mode can oscillate, and the transverse modes are similar to those of lasers with spherical mirrors in small-size VCSELs and are those of a planar cavity for broad-area devices.

Curiously, in spite of the greatly different geometrical dimensions, VCSELs and edge emitters have similar photon lifetimes. This is because edge emitters have a very poor reflectivity, unless their facets are cleaved. Since the photon lifetime is given by $\kappa^{-1} = 2L/(c|\ln R|)$, it has the same order of magnitude in the two kinds of lasers because both $|\ln R|$ and the cavity length L are two orders of magnitude smaller in VCSELs.

From the modeling viewpoint, VCSELs are probably the type of lasers where the low-transmission limit is best fulfilled. Moreover, although they are Fabry–Perot lasers, the grating in the active medium can be safely neglected, so the model basically reduces to that of a ring cavity. In the case of edge-emitting semiconductor lasers, the grating was assumed to be washed out by carrier diffusion. In a VCSEL we can observe that the thickness of the active material is much smaller than a wavelength. Therefore we can refer to the two-level model given by Eqs. (14.110), (14.94) and (14.95), which, as noted in Section 14.5, are basically identical to Eqs. (14.111), (14.106) and (14.107) for the ring-cavity case. In this vein, also taking into account the generality of the model (16.55) and (16.56) with respect to the function $G(N, \omega_0)$ noted in the previous section, we can use the same equations even for a free-running VCSEL by simply replacing the field F by the field $\mathcal{F}$ defined in Eq. (14.96), which leads to the model

$$\dot{\mathcal{F}} = -\kappa(1 - i\alpha)(1 - \tilde{D})\mathcal{F}, \tag{16.57}$$

$$\dot{\tilde{D}} = -\gamma_{\mathrm{nr}}[\tilde{D}(1 + |\mathcal{F}|^2) - \mu]. \tag{16.58}$$

In VCSELs, unlike in edge emitters, the laser geometry does not privilege a particular polarization plane for light. Therefore complex dynamical polarization is possible, which requires a description where the electric field is no longer a scalar. The problem is beyond the scope of this book, but we refer the reader to [166, 167].

17

Lasers without inversion and the effects of atomic coherence

Up to now, we have considered multilevel atomic systems only in Chapter 10, in order to describe the mechanisms for optical pumping in the laser in detail. In this chapter we focus on the case of three-level atoms in order to illustrate the striking phenomena that can be generated in multilevel systems by exploiting the *atomic coherence* and *quantum interference*.

First of all, if the atom lies initially in appropriate coherent superpositions of the two lowest energy levels, the atomic population remains trapped in these levels, while the uppermost energy level remains empty. This phenomenon, which arises from the quantum interference between the two transitions from the two lowest state to the uppermost state, is called *coherent population trapping*.

Atomic coherence and quantum interference lie also at the basis of the phenomenon called *electromagnetically induced transparency*. Also in this case there are two possible coherent processes that can bring the atom to the upper state, and their destructive quantum interference can produce the cancellation of absorption in the stationary state. In this situation, the introduction of an additional incoherent pump mechanism, capable of bringing even a small part of the population to the upper state, can generate gain and therefore *amplification without inversion*.

In this case, if the atomic system is located in a resonant cavity and the gain overcomes the losses due to the escape of photons from the cavity, one can obtain *lasing without inversion*, which is not possible in standard laser schemes, where quantum interference is absent.

17.1 Model equations

As general references for this chapter, see [22, 110, 168, 169]. Let us consider a three-level system, called the Λ scheme and shown in Fig. 17.1; $\Omega_\alpha = 2d_{12}E_\alpha/\hbar$ and $\Omega_\beta = 2d_{23}E_\beta/\hbar$ are the Rabi frequencies associated with the fields E_α and E_β quasi-resonant with the transitions $1 \to 2$ and $3 \to 2$, respectively. If we denote by $E_i = \hbar\omega_i$, $i = 1, 2, 3$, the energies of the three levels, the Hamiltonian which governs the atom reads

$$H = \hbar \begin{pmatrix} \omega_1 & -\Omega_\alpha/2 & 0 \\ -\Omega_\alpha/2 & \omega_2 & -\Omega_\beta/2 \\ 0 & -\Omega_\beta/2 & \omega_3 \end{pmatrix}. \tag{17.1}$$

The equations for the elements ρ_{ij} of the density matrix are

$$\dot{\rho}_{ij} = -\frac{i}{\hbar} \sum_k (H_{ik}\rho_{kj} - \rho_{ik}H_{kj}), \tag{17.2}$$

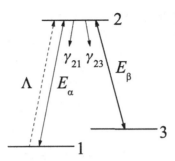

Figure 17.1   The three-level Λ scheme.

and read explicitly

$$\dot{\rho}_{11} = i\frac{\Omega_\alpha}{2}(\rho_{21} - \rho_{21}^*), \tag{17.3}$$

$$\dot{\rho}_{33} = i\frac{\Omega_\beta}{2}(\rho_{23} - \rho_{23}^*), \tag{17.4}$$

$$\dot{\rho}_{21} = -i\omega_{21}\rho_{21} - \frac{i}{2}[\Omega_\alpha(\rho_{22} - \rho_{11}) - \Omega_\beta\rho_{31}], \tag{17.5}$$

$$\dot{\rho}_{23} = -i\omega_{23}\rho_{23} + \frac{i}{2}[\Omega_\beta(\rho_{33} - \rho_{22}) + \Omega_\alpha\rho_{31}^*], \tag{17.6}$$

$$\dot{\rho}_{31} = -i\omega_{31}\rho_{31} - \frac{i}{2}(\Omega_\alpha\rho_{23}^* - \Omega_\beta\rho_{21}), \tag{17.7}$$

where we define $\omega_{ij} = (E_i - E_j)/\hbar$, the transition frequency between level i and level j. We have not written the equation for ρ_{22} because from the relation $1 = \rho_{11} + \rho_{22} + \rho_{33}$ it follows that $\dot{\rho}_{22} = -\dot{\rho}_{11} - \dot{\rho}_{33}$. Next, we introduce the slowly varying envelopes for the two fields,

$$E_\alpha = \frac{E_{\alpha,0}}{2}e^{i(k_\alpha z - \omega_\alpha t)} + \text{c.c.}, \tag{17.8}$$

$$E_\beta = \frac{E_{\beta,0}}{2}e^{i(k_\beta z - \omega_\beta t)} + \text{c.c.}, \tag{17.9}$$

from which it follows that

$$\frac{\Omega_\alpha}{2} = \alpha e^{i(k_\alpha z - \omega_\alpha t)} + \text{c.c.}, \qquad \alpha = \frac{d_{12}E_{\alpha,0}}{2\hbar}, \tag{17.10}$$

$$\frac{\Omega_\beta}{2} = \beta e^{i(k_\beta z - \omega_\beta t)} + \text{c.c.}, \qquad \beta = \frac{d_{23}E_{\beta,0}}{2\hbar}. \tag{17.11}$$

Note that the Rabi frequencies α and β include a factor of 2 in the denominator, which is a difference from the Rabi frequency Ω considered in Chapter 2 (see Eq. (2.38)), and that we assume the background refractive index equal to unity, so that $\omega_\alpha = ck_\alpha$ and $\omega_\beta = ck_\beta$. Furthermore, let us set

$$\rho_{21} = r_{21}e^{i(k_\alpha z - \omega_\alpha t)}, \qquad \rho_{23} = r_{23}e^{i(k_\beta z - \omega_\beta t)}, \qquad \rho_{31} = r_{31}e^{i[(k_\alpha - k_\beta)z - (\omega_\alpha - \omega_\beta)t]}, \tag{17.12}$$

so that in the rotating-wave approximation Eqs. (17.3)–(17.7) become

$$\dot\rho_{11} = i(\alpha^* r_{21} - \alpha r_{21}^*), \tag{17.13}$$

$$\dot\rho_{33} = i(\beta^* r_{23} - \beta r_{23}^*), \tag{17.14}$$

$$\dot r_{21} = -i\delta_\alpha r_{21} + i[\alpha(\rho_{11} - \rho_{22}) + \beta r_{31}], \tag{17.15}$$

$$\dot r_{23} = -i\delta_\beta r_{23} + i[\beta(\rho_{33} - \rho_{22}) + \alpha r_{31}^*], \tag{17.16}$$

$$\dot r_{31} = -i(\delta_\alpha - \delta_\beta) r_{31} + i(\beta^* r_{21} - \alpha r_{23}^*), \tag{17.17}$$

where we introduced the two detuning parameters between the two fields and the respective quasi-resonant transitions

$$\delta_\alpha = \omega_{21} - \omega_\alpha, \qquad \delta_\beta = \omega_{23} - \omega_\beta. \tag{17.18}$$

17.2 Coherent population trapping

For the sake of simplicity, let us consider the case of perfect resonance ($\delta_\alpha = \delta_\beta = 0$) and let us fix equal to zero the phase of one of the two fields, e.g. we set β real. If ζ is the phase of α, we set

$$\alpha = |\alpha|e^{i\zeta}, \quad r_{31} = qe^{i\zeta}, \quad r_{21} = re^{i\zeta}, \quad r_{23} = s, \tag{17.19}$$

so that ζ disappears from Eqs. (17.13)–(17.17), which now read

$$\dot\rho_{11} = i|\alpha|(r - r^*), \tag{17.20}$$

$$\dot\rho_{33} = i\beta(s - s^*), \tag{17.21}$$

$$\dot r = i[|\alpha|(\rho_{11} - \rho_{22}) + \beta q], \tag{17.22}$$

$$\dot s = i[\beta(\rho_{33} - \rho_{22}) + |\alpha|q^*], \tag{17.23}$$

$$\dot q = i(\beta r - |\alpha|s^*). \tag{17.24}$$

Next, we define the two mixed polarization variables

$$R = \beta\frac{r - r^*}{2i} - |\alpha|\frac{s - s^*}{2i} = \beta\,\mathrm{Im}\,r - |\alpha|\,\mathrm{Im}\,s, \tag{17.25}$$

$$S = |\alpha|\frac{r - r^*}{2i} + \beta\frac{s - s^*}{2i} = |\alpha|\,\mathrm{Im}\,r + \beta\,\mathrm{Im}\,s, \tag{17.26}$$

which, according to Eqs. (17.22) and (17.23), obey the equations

$$\dot R = |\alpha|\beta(\rho_{11} - \rho_{33}) + (\beta^2 - |\alpha|^2)\mathrm{Re}\,q, \tag{17.27}$$

$$\dot S = |\alpha|^2(\rho_{11} - \rho_{22}) + \beta^2(\rho_{33} - \rho_{22}) + 2|\alpha|\beta\,\mathrm{Re}\,q. \tag{17.28}$$

If we differentiate again with respect to time and take into account Eqs. (17.20), (17.21) and (17.24) the two equations transform into those of two uncoupled harmonic oscillators,

$$\ddot R + \Omega^2 R = 0, \tag{17.29}$$

$$\ddot S + 4\Omega^2 S = 0, \tag{17.30}$$

with

$$\Omega = \sqrt{|\alpha|^2 + \beta^2}, \tag{17.31}$$

whose general solution is

$$R(t) = A\sin(\Omega t + \varphi), \tag{17.32}$$

$$S(t) = B\sin(2\Omega t + \theta). \tag{17.33}$$

The arbitrary real parameters A and B and phases φ and θ are determined by the initial conditions for R, S and $\dot{R}$, $\dot{S}$. If we set $z = 0$ for simplicity, then from Eq. (17.12) it follows that $r_{ij}(0) = \rho_{ij}(0)$. Assuming that initially the atom lies in a linear superposition of the states $|1\rangle$ and $|3\rangle$, we have $\rho_{22}(0) = 0$ and $\rho_{21}(0) = \rho_{23}(0) = 0$ and, using Eq. (17.19), we obtain $r(0) = s(0) = 0$ and $q(0) = \rho_{31}(0)e^{-i\zeta}$. Therefore, Eqs. (17.25)–(17.28) yield $R(0) = S(0) = 0$, which implies $\varphi = \theta = 0$, and

$$\dot{R}(0) = |\alpha|\beta[\rho_{11}(0) - \rho_{33}(0)] + (\beta^2 - |\alpha|^2)\mathrm{Re}[\rho_{31}(0)e^{-i\zeta}], \tag{17.34}$$

$$\dot{S}(0) = |\alpha|^2\rho_{11}(0) + \beta^2\rho_{33}(0) + 2|\alpha|\beta\mathrm{Re}[\rho_{31}(0)e^{-i\zeta}]. \tag{17.35}$$

Let us now introduce the precise and crucial initial condition of atomic coherence, namely that the state vector of the atom at time $t = 0$ is given by the superposition

$$|\psi\rangle(0) = \frac{1}{\sqrt{2}}(|1\rangle + e^{-i\mu}|3\rangle). \tag{17.36}$$

The corresponding initial condition for the density operator is

$$\rho(0) = \frac{1}{2}(|1\rangle\langle 1| + |3\rangle\langle 3| + e^{i\mu}|1\rangle\langle 3| + e^{-i\mu}|3\rangle\langle 1|), \tag{17.37}$$

which means that $\rho_{11}(0) = \rho_{33}(0) = 1/2$ and $\rho_{31}(0) = e^{-i\mu}/2$. Therefore, in that state

$$\dot{R}(0) = \frac{\beta^2 - |\alpha|^2}{2}\cos(\zeta + \mu), \tag{17.38}$$

$$\dot{S}(0) = \frac{|\alpha|^2 + \beta^2}{2} + |\alpha|\beta\cos(\zeta + \mu). \tag{17.39}$$

If the atoms are prepared in such a way that $\zeta + \mu = \pi$ and the two fields have the same intensity, i.e. if $\beta = |\alpha|$, we have $\dot{R}(0) = \dot{S}(0) = 0$, which implies $A = B = 0$ and then $R(t) = S(t) = 0$ at all times. According to the definition of R and S this means that $\mathrm{Im}\,r = \mathrm{Im}\,s = 0$ at all times, which, inserted into Eqs. (17.20) and (17.21), gives $\dot{\rho}_{11} = \dot{\rho}_{33} = 0$. Therefore $\rho_{11}(t) = \rho_{33}(t) = 1/2$ and $\rho_{22}(t) = 0$ at all times.

The fact that the atom is in the special superposition (17.36) makes it impossible to excite the atom to the uppermost level. For this reason a state like (17.36) is called a *dark state*. This is an example of *quantum interference*, and is similar to what happens in the Young interference experiment. Since the atom can arrive at the uppermost level 2 starting from level 1 or from level 3, the two processes interfere and one can obtain that the total transition probability vanishes. In the Young experiment, the particle can arrive at the screen by passing through one of two slits, and one observes an interference pattern. If one of the two exciting fields is eliminated it becomes possible to excite the atom to level 2, just as in the Young experiment the interference fringes disappear when one of the two slits is closed. One calls this phenomenon *coherent population trapping* [169–180] because the population is trapped in the two lowest levels as a consequence of the coherence (purity) of the atomic state (17.36). The same phenomenon arises also when the initial state of the atom is

$$|\psi\rangle(0) = \frac{\beta|1\rangle - \alpha|3\rangle}{\Omega}. \tag{17.40}$$

so that the initial density operator reads

$$\rho(0) = \frac{1}{\Omega^2} \left(\beta^2 |1\rangle\langle 1| + |\alpha|^2 |3\rangle\langle 3| - \alpha^* \beta |1\rangle\langle 3| - \alpha\beta |3\rangle\langle 1| \right), \tag{17.41}$$

and $\rho_{11}(0) = \beta^2/\Omega^2$, $\rho_{33}(0) = |\alpha|^2/\Omega^2$ and $\rho_{31}(0) = -|\alpha|\beta e^{i\zeta}/\Omega^2$. From Eqs. (17.34) and (17.35) we get, without any further assumptions, $\dot{R}(0) = \dot{S}(0) = 0$. By following the same line of reasoning as before, we conclude that $\rho_{11}(t)$ and $\rho_{33}(t)$ are constant in time and therefore $\rho_{22}(t)$ remains equal to zero at all times.

The initial state we considered before (see Eq. (17.36)) is a special case of Eq. (17.40) with $\beta = |\alpha|$ and $\zeta = \pi - \mu$. The interest of this generalization lies in the fact that it suggests an experimental procedure to realize a trapping state like that described in Eq. (17.36).

In order to illustrate this, it is necessary to assume that there are two decay processes, with rates γ_{21} and γ_{23}, which lead the atom from level 2 to level 1 and level 3, respectively, as shown in Fig. 17.1. Let us assume, however, that the two damping rates are much smaller than β and $|\alpha|$, so that the damping terms in the atomic equations are usually negligible. Now, let us assume that $\alpha = 0$ and $\beta \neq 0$. For any initial condition of the atom, in the long term the atom will approach the state $|1\rangle$ which coincides with Eq. (17.40) with $\alpha = 0$ and $\beta \neq 0$. Because the damping rates are small, this process takes a long time, which is, however, irrelevant. At this stage, one gradually decreases the field β and simultaneously increases the field α, starting from zero and proceeding adiabatically, i.e. the variation of the two fields occurs on a time scale much longer than β^{-1} and $|\alpha|^{-1}$. This produces a gradual transfer of population from level 1 to level 3; by virtue of continuity, the system remains in the superposition (17.40) and, at a certain point, one reaches, for example, the state (17.36). As a matter of fact, with this slow procedure the system remains always in the stationary state of the atomic equations and, because the damping rates are very small, the stationary state differs in a negligible way from the state (17.41), which is a stationary solution of Eqs. (17.20)–(17.24) that do not include the damping terms.

17.3 Electromagnetically induced transparency

In the previous section we saw that it is possible to trap the atomic population in the two lowest energy levels. In this context, one has the inhibition of absorption in the two transitions from level 1 to level 2 and from level 3 to level 2, since the atomic polarizations r_{21} and r_{23} are equal to zero. The problem is that it is necessary to assume that the atoms initially lie in special superpositions of the two lowest energy levels, before they start interacting with the radiation fields. However, if one of the two fields is intense whereas the other is weak and can be treated as a probe field, one can show that the system, independently of the initial condition, under appropriate parametric conditions evolves to a stationary state in which it is transparent to the probe field. This phenomenon, is called *electromagnetically induced transparency* [181–185].

Let us consider again the Λ system shown in Fig. 17.1. We start from Eqs. (17.13)–(17.17) and assume that β is the driving field, resonant with the $3 \leftrightarrow 2$ transition so that $\delta_\beta = 0$. On the other hand we leave δ_α arbitrary because we want to study the response of the medium to the probe beam as a function of its frequency. For the following treatment it is essential to include in the equations the dissipative terms which describe the decay of the various elements of the density matrix. We assume that the transition from level 3 to level 1 is forbidden, while the decays from level 2 to level 1 and from level 2 to level 3 are associated with the rates γ_{21} and γ_{23}, respectively. We assume that the decays are purely radiative, i.e. atomic collisions do not contribute to them, and therefore the decay rate of the generic polarization r_{ij} is given by the sum of all the decay rates which involve levels i and j divided by two. In our case only level 2 is subject to decay processes, so the decay rates for the two polarizations r_{12} and r_{23} are the same while the decay of r_{31} vanishes:

$$\Gamma_{21} = \Gamma_{23} = \Gamma = \frac{\gamma_{21} + \gamma_{23}}{2}, \qquad \Gamma_{31} = 0. \tag{17.42}$$

Therefore the atomic equations read

$$\dot{\rho}_{11} = i(\alpha^* r_{21} - \alpha r_{21}^*) + \gamma_{21}\rho_{22}, \tag{17.43}$$

$$\dot{\rho}_{33} = i\beta(r_{23} - r_{23}^*) + \gamma_{23}\rho_{22}, \tag{17.44}$$

$$\dot{r}_{21} = i[\alpha(\rho_{11} - \rho_{22}) + \beta r_{31}] - (\Gamma + i\delta_\alpha)r_{21}, \tag{17.45}$$

$$\dot{r}_{23} = i[\beta(\rho_{33} - \rho_{22}) + \alpha r_{31}^*] - \Gamma r_{23}, \tag{17.46}$$

$$\dot{r}_{31} = i(\beta r_{21} - \alpha r_{23}^*) - i\delta_\alpha r_{31}. \tag{17.47}$$

We want to analyse the linear response to a weak probe field α when the atoms are driven by a strong field β. In analogy to what we did in the case of two-level atoms in Section 5.1, we write the wave equation for the field α in the slowly varying envelope and plane-wave approximations

$$\frac{\partial \alpha}{\partial z} + \frac{1}{c}\frac{\partial \alpha}{\partial t} = iG\Gamma r_{21}, \tag{17.48}$$

with G being the coupling constant defined by

$$G = \frac{\omega_{21}Nd_{21}^2}{2\epsilon_0\Gamma c\hbar}, \tag{17.49}$$

with respect to the constant which multiplies the atomic polarization r in the two-level case (Eq. (3.34)). G differs by a factor of 2 because in Eq. (17.10) α is not divided by 2; in addition, in Eq. (17.49) the background refractive index n_B is assumed equal to unity. The constant G has the dimensions of inverse length. We are considering the stationary state and the linear limit in α in which r_{12} is proportional to α, so we write Eq. (17.48) in the form

$$\frac{d\alpha}{dz} = iG\left(\frac{\Gamma r_{21}}{\alpha}\right)\alpha = G\chi\alpha, \tag{17.50}$$

where we introduce the dimensionless linear susceptibility, which is independent of α,

$$\chi = i\left(\frac{\Gamma r_{21}}{\alpha}\right), \tag{17.51}$$

in analogy with Eq. (5.6). Its real part is linked to the absorption (or gain) $G\chi'(\omega_\alpha)$, whereas the imaginary part is linked to dispersion because the refractive index is given by $n(\omega) = 1 + (Gc/\omega_\alpha)\chi''(\omega_\alpha)$ (compare this with Eq. (5.14)). In order to calculate the susceptibility χ we must evaluate r_{12} to first order in α. In the stationary state Eqs. (17.45)–(17.47) give

$$r_{21} = i\frac{\alpha(\rho_{11} - \rho_{22}) + \beta r_{31}}{\Gamma + i\delta_\alpha}, \tag{17.52}$$

$$r_{23} = i\frac{\beta(\rho_{33} - \rho_{22}) + \alpha r_{31}^*}{\Gamma}, \tag{17.53}$$

$$0 = \beta r_{21} - \alpha r_{23}^* - \delta_\alpha r_{31}. \tag{17.54}$$

By inserting the first two equations into the third we have

$$\frac{\alpha\beta(\rho_{11} - \rho_{22}) + \beta^2 r_{31}}{\Gamma + i\delta_\alpha} + \frac{\alpha\beta(\rho_{33} - \rho_{22}) + |\alpha|^2 r_{31}}{\Gamma} + i\delta_\alpha r_{31} = 0. \tag{17.55}$$

In the linear approximation we neglect the term with $|\alpha|^2$ and have

$$r_{31} = -\frac{\alpha\beta}{\Gamma}\frac{\Gamma(\rho_{11} + \rho_{33} - 2\rho_{22}) + i\delta_\alpha(\rho_{33} - \rho_{22})}{\beta^2 + i\delta_\alpha(\Gamma + i\delta_\alpha)}. \tag{17.56}$$

Therefore r_{31} and r_{21} are proportional to α, whereas $r_{23} = i\beta(\rho_{33} - \rho_{22})/\Gamma$ in the linear limit. If r_{21} is proportional to α, then from Eq. (17.43) in the stationary state we obtain, to first order in α, $\rho_{22} = 0$. Therefore, from Eq. (17.53), we have that $r_{23} = i\beta\rho_{33}/\Gamma$. Furthermore, since $\rho_{22} = 0$, from Eq. (17.44) in the stationary state it follows that r_{23} must be real, and therefore also that $\rho_{33} = 0$. In conclusion, in the stationary state the whole population lies in level 1. The expression (17.56) for r_{31} reduces to

$$r_{31} = -\frac{\alpha\beta}{\beta^2 + i\delta_\alpha(\Gamma + i\delta_\alpha)}, \tag{17.57}$$

and the polarization r_{21} (17.52) becomes

$$r_{21} = i\frac{\alpha + \beta r_{31}}{\Gamma + i\delta_\alpha} = i\frac{\alpha}{\Gamma + i\delta_\alpha}\left[1 - \frac{\beta^2}{\beta^2 + i\delta_\alpha(\Gamma + i\delta_\alpha)}\right] = -\frac{\delta_\alpha}{\beta^2 + i\delta_\alpha(\Gamma + i\delta_\alpha)}\alpha. \tag{17.58}$$

By inserting this expression into Eq. (17.51) we obtain the linear susceptibility

$$\chi = -i\frac{\Gamma\delta_\alpha}{\beta^2 - \delta_\alpha^2 + i\Gamma\delta_\alpha}, \tag{17.59}$$

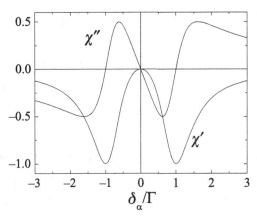

Figure 17.2 The behavior of the real (absorptive) and imaginary (dispersive) parts of the linear susceptibility for $\beta = \Gamma$.

whose real and imaginary parts, linked to absorption and dispersion, respectively, are

$$\chi' = -\frac{\Gamma^2 \delta_\alpha^2}{(\beta^2 - \delta_\alpha^2)^2 + \Gamma^2 \delta_\alpha^2}, \qquad \chi'' = -\frac{\Gamma \delta_\alpha (\beta^2 - \delta_\alpha^2)}{(\beta^2 - \delta_\alpha^2)^2 + \Gamma^2 \delta_\alpha^2}. \tag{17.60}$$

The two functions are plotted in Fig. 17.2. The curve for χ' exhibits the presence of two absorption peaks at $\delta_\alpha = \pm\beta$. The curve for χ'' differs qualitatively from that for the two-level case because it vanishes not only on resonance, but also in correspondence to the two absorption peaks. The key feature is that on resonance ($\delta_\alpha = 0$) one has quantum interference and the absorption is inhibited ($\chi' = 0$).

17.4 Amplification without inversion

References for the topics discussed in this and the next section of this chapter can be found in [186–210]. In electromagnetically induced transparency a probe field resonant with the transition between levels 1 and 2 does not undergo absorption even if the whole atomic population is in level 1. By exploiting this phenomenon it is possible to obtain amplification of the probe field if we can arrange that the uppermost level, level 2, is somewhat populated, even if with a population smaller than that of level 1. In this case one speaks of *amplification without inversion*.

We consider again the Λ scheme of Fig. 17.1, but we add a further essential ingredient, i.e. a pumping mechanism that transfers population from level 1 to level 2, described by the parameter Λ which plays the same role as the parameter $\gamma_\uparrow$ in Chapter 4. The presence of this term modifies the atomic equations for the populations and introduces a loss term into the equations for the atomic polarizations, with the rates

$$\Gamma_{21} = \frac{\gamma_{21} + \gamma_{23} + \Lambda}{2}, \qquad \Gamma_{23} = \frac{\gamma_{21} + \gamma_{23}}{2}, \qquad \Gamma_{31} = \frac{\Lambda}{2}. \tag{17.61}$$

The time-evolution equations for the density-matrix elements read

$$\dot{\rho}_{11} = i(\alpha^* r_{21} - \alpha r_{21}^*) + \gamma_{21}\rho_{22} - \Lambda\rho_{11}, \tag{17.62}$$

$$\dot{\rho}_{33} = i\beta(r_{23} - r_{23}^*) + \gamma_{23}\rho_{22}, \tag{17.63}$$

$$\dot{r}_{21} = i[\alpha(\rho_{11} - \rho_{22}) + \beta r_{31}] - (\Gamma_{21} + i\delta_\alpha)r_{21}, \tag{17.64}$$

$$\dot{r}_{23} = i[\beta(\rho_{33} - \rho_{22}) + \alpha r_{31}^*] - \Gamma_{23}r_{23}, \tag{17.65}$$

$$\dot{r}_{31} = i(\beta r_{21} - \alpha r_{23}^*) - (\Gamma_{31} + i\delta_\alpha)r_{31}. \tag{17.66}$$

Let us proceed as in the previous section. From the polarization equations we obtain

$$r_{21} = i\frac{\alpha(\rho_{11} - \rho_{22}) + \beta r_{31}}{\Gamma_{21} + i\delta_\alpha}, \tag{17.67}$$

$$r_{23} = i\frac{\beta(\rho_{33} - \rho_{22}) + \alpha r_{31}^*}{\Gamma_{23}}, \tag{17.68}$$

$$0 = i(\beta r_{21} - \alpha r_{23}^*) - (\Gamma_{31} + i\delta_\alpha)r_{31}. \tag{17.69}$$

By inserting the first two into the third we obtain

$$\beta\frac{\alpha(\rho_{11} - \rho_{22}) + \beta r_{31}}{\Gamma_{21} + i\delta_\alpha} + \alpha\frac{\beta(\rho_{33} - \rho_{22}) + \alpha^* r_{31}}{\Gamma_{23}} + (\Gamma_{31} + i\delta_\alpha)r_{31} = 0, \tag{17.70}$$

from which, by neglecting the term proportional to $|\alpha|^2$, we get

$$r_{31} = -\frac{\alpha\beta}{\Gamma_{23}}\frac{(\rho_{11} - \rho_{22})\Gamma_{23} + (\rho_{33} - \rho_{22})(\Gamma_{21} + i\delta_\alpha)}{\beta^2 + (\Gamma_{31} + i\delta_\alpha)(\Gamma_{21} + i\delta_\alpha)}. \tag{17.71}$$

As in the previous section we observe that, since r_{31} is proportional to α, in the linear limit r_{23} is given by

$$r_{23} = i\frac{\beta(\rho_{33} - \rho_{22})}{\Gamma_{23}}. \tag{17.72}$$

On the other hand, the polarization r_{21}, which we must evaluate in order to determine the susceptibility, by using Eqs. (17.67) and (17.69) can be expressed as

$$r_{21} = \alpha\frac{i(\rho_{11} - \rho_{22})(\Gamma_{31} + i\delta_\alpha) + \beta r_{23}^*}{\beta^2 + (\Gamma_{31} + i\delta_\alpha)(\Gamma_{21} + i\delta_\alpha)}. \tag{17.73}$$

With respect to the previous section, now all three populations are different from zero. From Eq. (17.62) in the stationary state, by neglecting terms of order $|\alpha|^2$, we obtain

$$\rho_{11} = \frac{\gamma_{21}}{\Lambda}\rho_{22}. \tag{17.74}$$

From Eq. (17.63) in the stationary state, by taking Eq. (17.72) into account we get

$$\rho_{33} = \frac{2\beta^2 + \gamma_{23}\Gamma_{23}}{2\beta^2}\rho_{22}, \tag{17.75}$$

and finally, from the relation $\rho_{11} + \rho_{22} + \rho_{33} = 1$, we obtain

$$\rho_{22} = \frac{2\beta^2\Lambda}{A}, \qquad A = 2\beta^2(2\Lambda + \gamma_{21}) + \gamma_{23}\Gamma_{23}\Lambda. \tag{17.76}$$

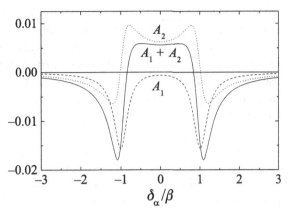

Figure 17.3 The behavior of the real part of the susceptibility ($A_1 + A_2$) and of the two functions A_1 and A_2 for $\Lambda = \gamma_{23} = 0.2\beta$ and $\gamma_{21} = 0.22\beta$.

and therefore

$$\rho_{11} = \frac{2\beta^2\gamma_{21}}{A}, \qquad \rho_{33} = \frac{\Lambda(2\beta^2 + \gamma_{23}\Gamma_{23})}{A}. \tag{17.77}$$

It is important to observe that the population ρ_{22} is smaller than ρ_{11}, i.e. there is no population inversion, if the following condition is satisfied:

$$\Lambda < \gamma_{21}. \tag{17.78}$$

Not surprisingly, this condition is identical to the condition $\gamma_\uparrow < \gamma_\downarrow$ found in Chapter 4 for the absence of population inversion in the two-level case. If we set the pump term Λ equal to zero we are back to the case of the previous section, namely $\rho_{11} = 1$, and therefore $\rho_{22} = \rho_{33} = 0$ (and $r_{23} = 0$). Furthermore, $\Gamma_{31} = 0$, so that Eq. (17.73) reduces to (17.58). If, instead, we keep the pump term, let us analyze the real part of the susceptibility, which is linked to absorption and gain. On setting $r_{23} = iy_{23}$ we can write

$$\chi' = -\text{Im}\left(\frac{\Gamma_{21}r_{21}}{\alpha}\right) = -\Gamma_{21}\,\text{Im}\left[\frac{i(\rho_{11} - \rho_{22})(\Gamma_{31} + i\delta_\alpha) - i\beta y_{23}}{\beta^2 + (\Gamma_{31} + i\delta_\alpha)(\Gamma_{21} + i\delta_\alpha)}\right]$$

$$= \Gamma_{21}(\rho_{22} - \rho_{11})\text{Im}\left[\frac{i\Gamma_{31} - \delta_\alpha}{\beta^2 + (\Gamma_{31} + i\delta_\alpha)(\Gamma_{21} + i\delta_\alpha)}\right]$$

$$+ \Gamma_{21}\beta y_{23}\,\text{Re}\left[\frac{1}{\beta^2 + (\Gamma_{31} + i\delta_\alpha)(\Gamma_{21} + i\delta_\alpha)}\right] = A_1 + A_2, \tag{17.79}$$

where the two terms A_1 and A_2 are given by

$$A_1 = \Gamma_{21}(\rho_{22} - \rho_{11})\frac{\Gamma_{31}\beta^2 + \Gamma_{21}(\Gamma_{31}^2 + \delta_\alpha^2)}{(\beta^2 + \Gamma_{21}\Gamma_{31} - \delta_\alpha^2)^2 + \delta_\alpha^2(\Gamma_{21} + \Gamma_{31})^2}, \tag{17.80}$$

$$A_2 = \Gamma_{21}\beta y_{23}\frac{\beta^2 + \Gamma_{21}\Gamma_{31} - \delta_\alpha^2}{(\beta^2 + \Gamma_{21}\Gamma_{31} - \delta_\alpha^2)^2 + \delta_\alpha^2(\Gamma_{21} + \Gamma_{31})^2}. \tag{17.81}$$

The functions A_1 and A_2 and their sum, which represents the total gain (or absorption), are represented in Fig. 17.3. A_1 has the same sign as the population inversion $\rho_{22} - \rho_{11}$

with which the field α interacts. When condition (17.78) is satisfied it is always negative; therefore it is an absorption term. On the other hand, in the interval of δ_α between the two values $\mp\beta_0$, with $\beta_0 = \sqrt{\beta^2 + \Gamma_{21}\Gamma_{31}}$, A_2 is positive because y_{23} is positive since there is no population inversion between levels 2 and 3. Hence this is a gain term, which is there because the imaginary part of the coherence between levels 2 and 3 exists and is positive. In the central region of the spectrum the sum of the two terms may be positive. In particular, in the resonant case $\delta_\alpha = 0$ the condition $\chi' > 0$ is equivalent to

$$(\rho_{22} - \rho_{11})\Gamma_{31}\Gamma_{23} + \beta^2(\rho_{33} - \rho_{22}) > 0, \tag{17.82}$$

as one can see by using Eqs. (17.80), (17.81) and (17.72) and taking into account that $r_{23} = iy_{23}$. By substituting the expressions (17.76) and (17.77) for the stationary populations into (17.82) we obtain the condition

$$\Lambda y_{23} > 2(\gamma_{21} - \Lambda)\Gamma_{31}, \tag{17.83}$$

and, by taking into account that $\Gamma_{31} = \Lambda/2$, we get the conditions for amplification without inversion, namely

$$\gamma_{21} - \gamma_{23} < \Lambda < \gamma_{21}, \tag{17.84}$$

where the r.h.s. guarantees the absence of inversion (see Eq. (17.78)).

17.5 Lasing without inversion

If a medium composed of three-level atoms with a Λ configuration amplifies the probe field resonant with the transition between levels 1 and 2 and is located in an optical cavity, and if the gain overcomes the losses due to the escape of photons from the cavity, one can obtain *lasing without inversion*.

In the case of two-level atoms, considering a ring cavity of length $\mathcal{L}$ and an atomic medium with length L and background refractive index $n_B = 1$, in Chapter 12 we saw that in the low-transmission limit the field equation

$$\frac{\partial F}{\partial z} + \frac{1}{c}\frac{\partial F}{\partial t} = gP, \tag{17.85}$$

reduces to

$$\dot{F} = -\kappa(F - AP) = -\kappa F + \frac{gcL}{\mathcal{L}}P, \tag{17.86}$$

where we assumed that the radiation field is in resonance with a cavity frequency, in addition to being resonant with the atomic transition frequency, and we used Eqs. (9.20) and (9.22). Also in the three-level case we assume exact resonance of the field α with a cavity frequency and with the transition frequency between levels 1 and 2, i.e. $\delta_\alpha = 0$ and, as before, we assume that $\delta_\beta = 0$. If we start from the equation (compare this with Eq. (17.48))

$$\frac{\partial \alpha}{\partial z} + \frac{1}{c}\frac{\partial \alpha}{\partial t} = iG\Gamma_{21}r_{21}, \tag{17.87}$$

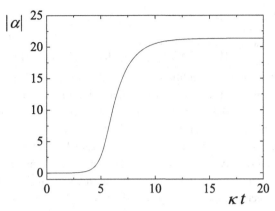

Figure 17.4   Dynamical evolution towards the lasing stationary state for $\Lambda = \beta = 10\kappa, \gamma_{21} = 11\kappa, \gamma_{23} = 2\kappa$ and $G' = 5000\kappa^2$, which implies $\sigma_{\mathrm{LWI}} \simeq 0.0115$ and $A_{\mathrm{LWI}} \simeq 5$.

in the low-transmission limit we obtain the equation

$$\dot{\alpha} = -\kappa\alpha + \frac{iG\Gamma_{21}cL}{\mathcal{L}}r_{21} = -\kappa\alpha + iG'r_{21}, \tag{17.88}$$

with (see Eq. (17.49))

$$G' = \frac{G\Gamma_{21}cL}{\mathcal{L}} = \frac{\omega_{21}Nd_{21}^2L}{2\epsilon_0\hbar\mathcal{L}}, \tag{17.89}$$

which must be coupled with the five atomic equations (17.62)–(17.66), giving the complete set of dynamical equations

$$\dot{\alpha} = -\kappa\alpha + iG'r_{21}, \tag{17.90}$$
$$\dot{\rho}_{11} = i(\alpha^*r_{21} - \alpha r_{21}^*) + \gamma_{21}\rho_{22} - \Lambda\rho_{11}, \tag{17.91}$$
$$\dot{\rho}_{33} = i\beta(r_{23} - r_{23}^*) + \gamma_{23}\rho_{22}, \tag{17.92}$$
$$\dot{r}_{21} = i[\alpha(\rho_{11} - \rho_{22}) + \beta r_{31}] - \Gamma_{21}r_{21}, \tag{17.93}$$
$$\dot{r}_{23} = i[\beta(\rho_{33} - \rho_{22}) + \alpha r_{31}^*] - \Gamma_{23}r_{23}, \tag{17.94}$$
$$\dot{r}_{31} = i(\beta r_{21} - \alpha r_{23}^*) - \Gamma_{31}r_{31}. \tag{17.95}$$

This system of nonlinear equations admits the trivial stationary solution with

$$\alpha = r_{21} = r_{31} = 0, \tag{17.96}$$

and with r_{23}, ρ_{11}, ρ_{22} and ρ_{33} given by Eqs. (17.72), (17.76) and (17.77). The linear-stability analysis for this stationary solution is performed in Section 22.2. The result is that the trivial (nonlasing) stationary solution becomes unstable when the following condition is satisfied:

$$A_{\mathrm{LWI}} > 1, \qquad A_{\mathrm{LWI}} = \frac{G'}{\kappa\Gamma_{21}}\sigma_{\mathrm{LWI}} = \frac{GL}{T}\sigma_{\mathrm{LWI}}, \tag{17.97}$$

with

$$\sigma_{LWI} = \frac{4\left[(\Lambda + \gamma_{23})^2 - \gamma_{21}^2\right]\Lambda\beta^2}{\left[(\Lambda + \gamma_{21} + \gamma_{23})\Lambda + 4\beta^2\right]\left[(\gamma_{21} + \gamma_{23})\gamma_{23}\Lambda + (\gamma_{21} + 2\Lambda)4\beta^2\right]}. \qquad (17.98)$$

We note that the condition $A_{LWI} > 1$ has the same form as that for two-level atoms in the resonant case ($\Delta = 0$). Also the definition of the parameter A_{LWI} has the same form (see Eqs. (9.22), (4.34) and (17.89)). The parameter σ_{LWI} plays a similar role to the parameter σ in the two-level case (see Eq. (4.14)), since it is positive when $\Lambda + \gamma_{23} > \gamma_{21}$, i.e. when there is gain (see Eq. (17.84)); in this case the gain does not require population inversion. In the three-level case the parameter σ has a much more complex structure and depends on the atomic relaxation rates and on the intensity of the field β. In Fig. 17.4 the parameters are selected in such a way that the trivial stationary solution is unstable, and the system evolves towards a lasing stationary solution.

PART II

DYNAMICAL PHENOMENA, INSTABILITIES, CHAOS

Introduction to Part II

In Part I of this book we studied almost exclusively stationary solutions of the models; we also mentioned in which parametric ranges they are stable or unstable, but without proving it. In this part we not only describe the technique used to assess the stability of stationary solutions but also, especially, study the temporal phenomena which originate from the instabilities themselves. As a matter of fact, the appearance of spontaneous pulsations in the form of undamped trains of pulses was first observed in masers even before the advent of the laser in 1960. However, a strong interest in temporal optical instabilities arose much later, in the early 1980s, in resonance, as it were, with the general interest in the field of chaos in nonlinear dynamical systems.

In Part II of the book we focus on the dynamical phenomena in nonlinear optical systems, of course taking the general viewpoint of nonlinear dynamical systems. In Chapter 18 we describe first the technique of linear-stability analysis and then we discuss the most relevant and general instability-related dynamical aspects in nonlinear dissipative systems such as, for example, attractors, bifurcations and routes to chaos. This provides the language to describe the results discussed in the following chapters of Part II. For reasons of space, we do not introduce technicalities such as, for instance, Lyapunov exponents.

In order to identify the instabilities, in general we prefer to use a procedure that is more direct than the Routh–Hurwitz stability criterion described in Appendix A. A more straightforward technique is to calculate the stability boundaries in the space of the parameters of the system, as described in Section 18.1 and Appendix B.

In Chapter 19 we discuss the prominent transient dynamical features in class-A and class-B lasers, as is usual in textbooks on laser physics, whereas in the following chapters of Part II we discuss temporal instabilities and their consequences. In Chapter 20 we outline a general scenario that encompasses and structures all temporal instabilities that arise in the framework of two-level systems in ring cavities. A characteristic trait of such an approach is that single-mode and multimode instabilities are treated in parallel, showing their intrinsic and profound interconnections. This notably facilitates the description of instabilities, spontaneous oscillations and optical chaos in lasers (Chapter 22 deals with homogeneously broadened lasers and Chapter 23 with inhomogeneously broadened lasers) and in passive optically bistable systems (Chapter 24). The final chapter, Chapter 25, deals with temporal instabilities in the same systems as were described in Chapter 13, i.e. a laser with an injected signal, a laser with a saturable absorber and an optical parametric oscillator. As in Part III, many experimental results are presented and discussed, in close connection with the theoretical predictions. General discussions of temporal optical instabilities are included in the special issues [211–219] and in the review articles and books [86, 130, 220–233].

18

Some general aspects in nonlinear dissipative dynamical systems

General references for this chapter are [91, 92, 234, 235]. The aim of this chapter is to review some basic notions and techniques that are used in studies of dynamical phenomena in nonlinear dissipative systems in general. In the following chapters they will be applied to the analysis of optical instabilities.

In order to generate chaos or related temporal patterns, one may introduce parameter modulation or external feedback; examples of these procedures can be found in the general references cited in the introduction to Part II. In our book we usually assume that all the parameters are strictly constant in time and that the only feedback arises from the mirrors of the optical resonator; thus, the temporal oscillations predicted by the theory arise spontaneously from the instabilities.

We are interested in systems whose dynamical state can be characterized with the specification of n real variables $x_1, \ldots, x_n$; we give the name of system *phase space* to the n-dimensional manifold of the variables $x_1, \ldots, x_n$ and assume, furthermore, that the point in phase space representing the temporal evolved system state is governed by the set of ordinary differential equations

$$\frac{dx_i}{dt} = f_i(x_1, x_2, \ldots, x_n, \{\beta\}), \qquad i = 1, 2, \ldots, n, \tag{18.1}$$

where f_i are real functions of the variables $x_i, i = 1, 2, \ldots, n$ and of the parameters $\beta_1, \ldots, \beta_m$, which we denote globally with the symbol $\{\beta\}$. If the value of a parameter can be varied with sufficient flexibility during the course of a measurement, we will refer to it as a *control parameter*.

We assume that at least one of the n functions f_i has a nonlinear dependence upon some of its arguments $x_i, i = 1, 2, \ldots, n$ and also assume that the system is dissipative in the sense that, if we follow the evolution of any phase-space region, its volume is on average a decreasing function of time that tends to zero in the long-time limit. Hence, unlike the case of Hamiltonian systems, in which the volume in phase space is constant during the evolution (Liouville's theorem), here the domain of dynamical interest in phase space contracts as time progresses. One can relax the condition that the n variables $x_i, i = 1, 2, \ldots, n$ are real and allow also pairs of complex-conjugate variables. In this case, of course, complex-conjugate variables obey complex-conjugate time-evolution equations. In any case, complex equations can be recast, for example, in terms of real and imaginary parts or of modulus and phase so that, without loss of generality, we can describe the evolution of the system in a real n-dimensional phase space.

18.1 Stationary solutions and their stability

Because the functions f_i are explicitly independent of time, it makes sense to look for stationary solutions determined by the conditions $dx_i/dt = 0$ ($i = 1, 2, \ldots, n$); these solutions obey the set of nonlinear algebraic equations

$$f_i(x_1, x_2, \ldots, x_n, \{\beta\}) = 0, \qquad i = 1, 2, \ldots, n. \tag{18.2}$$

If Eqs. (18.2) have a sufficiently simple structure, the calculation of the stationary states can be carried out analytically; the same holds for the linear-stability analysis.

A stationary solution is stable if the system, after being displaced slightly from the stationary state, returns to its original configuration; it is unstable, instead, if it moves further away from it. To perform the linear-stability analysis we introduce the displacements

$$\delta x_i = x_i - x_{i,s}(\{\beta\}), \qquad i = 1, 2, \ldots, n, \tag{18.3}$$

where $x_{i,s}$ denotes the stationary value of the ith variable, which depends on the values of the parameters $\beta_1, \beta_2, \ldots, \beta_m$. Because the quantities δx_i are small, we can linearize Eq. (18.1) with respect to these displacements. Upon taking into account the defining equation, Eq. (18.2), of the stationary state, the linearized equations acquire the general form

$$\frac{d}{dt}\delta x_i = \sum_{j=1}^{n} \left(\frac{\partial f_i}{\partial x_j}\right)_s \delta x_j, \tag{18.4}$$

where the partial derivatives $\partial f_i/\partial x_j$ are evaluated at the stationary configuration. The solutions of Eq. (18.4) are linear combinations of elementary functions of the form

$$\delta x_i(t) = e^{\lambda t}\delta x_i(0), \qquad i = 1, 2, \ldots, n, \tag{18.5}$$

where the quantities $\delta x_i(0)$ are time-independent. After introducing the ansatz (18.5) into Eqs. (18.4) we obtain the eigenvalue equation

$$(\mathbf{A} - \lambda\mathbf{1})\delta\mathbf{x}(0) = 0, \tag{18.6}$$

where $\delta\mathbf{x}(0)$ is an n-dimensional vector whose components are $\delta x_i(0)$, $\mathbf{1}$ is the identity $n \times n$ matrix and $\mathbf{A}$ is the $n \times n$ matrix with elements $(\partial f_i/\partial x_j)_s$. The eigenvalues are determined by the characteristic equation

$$\det(\mathbf{A} - \lambda\mathbf{1}) = 0. \tag{18.7}$$

This is just a polynomial equation of degree n, with the general form

$$\sum_{k=0}^{n} c_k(\{\beta\})\lambda^k = 0, \tag{18.8}$$

where the coefficients c_k depend upon the stationary configuration and the parameters of the system, and $c_n = 1$. In every case the coefficients c_k are real, so the characteristic equation has real roots or pairs of complex-conjugate roots.

The stationary solution is stable if and only if all the roots of the characteristic equation (18.8) have a negative real part, because in that case all displacements decay to zero (see Eq. (18.5)), and the system returns to its original stationary state.

In order to check the stability, it is not necessary to solve the characteristic equation (18.7), but it is sufficient to analyze the sign of the coefficients c_k (which must be non-negative to ensure stability) and that of some appropriate combinations of these quantities, defined by the Routh–Hurwitz stability criterion (see [236] and page 123 of [91]), which is described in Appendix A. Alternatively, we can calculate directly the stability boundary, which is the hypersurface in parameter space which separates the stable region from the instability domain. We identify the boundary with the condition that one of the eigenvalues has a vanishing real part. For this purpose we set

$$\lambda = -i\omega, \tag{18.9}$$

where ω is a real number, in the eigenvalue equation (18.8). After separating the real from the imaginary parts, we obtain two equations for ω having the general form $P_1(\omega, \{\beta\}) = 0$ and $P_2(\omega, \{\beta\}) = 0$, where P_1 and P_2 are polynomials in the variable ω and with coefficients equal to $+c_k$, $-c_k$ or zero. In particular, we note that $P_2(\omega = 0, \{\beta\}) = 0$, because the coefficients c_k are real, and $P_1(\omega = 0, \{\beta\}) = c_0$. The condition $c_0(\{\beta\}) = 0$ is an equation in parameter space; if this equation has a solution, this then defines a hypersurface that is a stability boundary for $\omega = 0$. The stable side of this boundary corresponds to $c_0 > 0$, if $c_1 > 0$. As one crosses the boundary $c_0(\{\beta\}) = 0$, starting from the stable side, the system may display a steady-state bifurcation or may jump discontinuously into another attractor (see the following sections).

We can also look for solutions of the system

$$P_1(\omega, \{\beta\}) = 0, \qquad P_2(\omega, \{\beta\}) = 0, \tag{18.10}$$

with $\omega \neq 0$. After finding an expression for ω from one of the two equations (18.10) and inserting it into the other, we obtain an equation in parameter space. If this admits a solution, this also defines a hypersurface corresponding to a stability boundary, where the real part of a pair of complex-conjugate eigenvalues changes sign. Upon crossing this boundary starting from the stable side, the system may exhibit a Hopf bifurcation, or may jump discontinuously into a different attractor (see the following sections). An advantage of this procedure is that one obtains, in addition to the boundary, the oscillation frequency corresponding to the boundary itself.

When the system (18.10) allows solutions both for $\omega = 0$ and for $\omega \neq 0$, we obtain two hypersurfaces in parameter space. The effective stability boundary corresponds to the surface which is crossed first as we vary the control parameters starting from the stable region.

We note that, when the stability boundary is approached from the stable side, one can find one or two eigenvalues whose real part becomes progressively smaller. This implies that, if the system is displaced slightly from the stationary state, it requires a longer and longer time to return to the original configuration. This phenomenon, which is associated generally with the approach to an unstable state, is called *critical slowing down*.

The linear-stability analysis is simple in the cases $n = 1$ and $n = 2$. In the first case the system (18.1) reduces to the equation

$$\frac{dx}{dt} = f(x, \{\beta\}).$$ (18.11)

In this case there is only one eigenvalue, given by

$$\lambda = \left(\frac{\partial f}{\partial x}\right)_s (\{\beta\}).$$ (18.12)

Therefore the stationary state is stable when $\lambda < 0$ and unstable when $\lambda > 0$.

On the other hand, for $n = 2$ the characteristic equation has two roots, which may be both real or complex conjugate. By using the Routh–Hurwitz criterion or following the procedure described by Eqs. (18.9) and (18.10) one can decide the stability of the stationary state without calculating the eigenvalues. The stationary state is stable, provided that the two conditions

$$c_1(\{\beta\}) > 0, \qquad c_0(\{\beta\}) > 0$$ (18.13)

are satisfied. The equation $c_0(\{\beta\}) = 0$ defines the hypersurface in parameter space, if any, where one of two real eigenvalues vanishes. The equation $c_1(\{\beta\}) = 0$ defines the hypersurface, if any, where the real part of two complex-conjugate eigenvalues vanishes; the oscillation frequency at the boundary is $\omega^2 = c_0(\{\beta\})$.

The cases $n = 3$, 4 and 5 are discussed in Appendix B to derive the equations which define the boundary in parameter space, where the real part of a pair of complex-conjugate eigenvalues vanishes.

18.2 Attractors and repellers; bistability and multistability

A stable stationary state is said to be an *attractor* because there exists a region of phase space around it with the property that any initial condition selected from this domain approaches the attractor in the long-time limit. This region is called the basin of attraction of the stationary state. Obviously, the definition of stability given in the previous subsection has only a local character, because, if one displaces the system from the stable stationary state, the system returns to the stationary point only if the initial condition lies in the basin of attraction of the stationary state itself.

On the other hand, an unstable stationary solution is said to be a *repeller* because the phase-space trajectories move away from it. It must be kept in mind, however, that in a multi-dimensional problem ($n > 1$) the repulsion may take place only along one or perhaps a few directions in phase space, while attraction develops along other directions. Hence, a point of phase space may actually approach a repeller until the repulsion along the appropriate directions takes over. Because of the unavoidable presence of noise, unstable configurations cannot be physically realized as stationary states.

A system is said to be *bistable* (or *multistable*) if we can find a region in its parameter space such that for every point in this domain the system has two (or more) stable stationary states. During the evolution, the system will approach one or another attractor according to the basin of attraction in which the phase-space point happens to be positioned initially. When the dimension n of the phase space is larger than one, the various basins of attraction of the multistable system may display a complex structure and their boundaries may even be fractal.

18.3 Other kinds of attractors: limit cycles, tori, strange attractors; deterministic chaos; generalized multistability

When $n > 1$ it may happen that the phase-space point does not approach a stationary state in the long-time limit, but instead approaches a different kind of attractor. The simplest object of this type is a limit cycle, which is a closed one-dimensional trajectory. In this case, the long-term evolution of any variable x_i is periodic with a period equal to the time taken by the phase-space point to cover the entire loop. The power spectrum of the function $x_i(t)$ displays a fundamental frequency, which is accompanied, in most cases, by harmonic components.

If $n \geq 3$ the system can display attractors with a different character from that of ordinary stationary states and limit cycles. The simplest of these attractors is a torus, which is a two-dimensional, ring-shaped surface covered by the trajectory during the long-time evolution. The variable x_i in this case usually displays a quasi-periodic evolution and its spectrum is marked by the presence of incommensurate frequencies, in addition to their combination tones. Multidimensional tori can also emerge if the phase space has a number of dimensions larger than 3.

An especially interesting phenomenon that can occur when the dimensionality of the phase space is not smaller than 3 is deterministic chaos. Here the time evolution of the variables $x_i(t)$ exhibits only irregular aperiodic behavior similar to that which arises in the presence of stochastic noise. In this case, however, the equations do not include stochastic noise sources and for that reason this phenomenon is called *deterministic chaos* or deterministic noise. The spectrum is now characterized by a continuous background (i.e. every frequency component is represented during the evolution of the variable $x_i(t)$), which is also similar, qualitatively, to that of ordinary stochastic noise.

The distinguishing feature of deterministic chaos, which is quite unlike stochastic noise, is the sensitivity of the trajectory in phase space to the initial conditions. Sensitivity to the initial conditions implies the existence of a region in phase space such that initially close-by trajectories move exponentially further apart from each other during a sufficiently short period of time. This behavior implies an effective lack of predictability of the evolution of the system, because any small error in the assignment of the initial conditions is amplified exponentially. Note that this divergence of the phase-space trajectories is not in contradiction with the global decrease of the volume in phase space because the divergence produces an expansion of the phase-space region along just one direction (or possibly more, if $n > 3$), whereas contraction ensues for the others in such a way that the volume of the region actually decreases in time.

When performing calculations that are based on a set of nonlinear differential equations, chaos can be identified with the help of a set of real indices called Lyapunov exponents [91, 92, 234, 235]. These indices are descriptive of an entire trajectory and are identical for all the trajectories that lie within the same basin of attraction. The sum of all the Lyapunov exponents is negative for a dissipative system; its modulus gives the rate of contraction of the phase-space volume. There are directions in phase space for which adjacent trajectories do not experience contraction or expansion; these are characterized by the fact that some Lyapunov exponents are equal to zero. One finds a vanishing Lyapunov exponent, for example, in the case of a limit cycle and two in the case of a two-dimensional torus. The presence of directions with exponential expansion, i.e. the presence of chaos, is signaled by the appearance of positive values for at least one of these indices.

The chaotic attractor, or *strange attractor*, can be visualized as a single thread twisted into an infinite number of intertwined loops. It exhibits striking features such as, for example, *self-similarity*, i.e. the property of exhibiting the same structure, in a statistical sense, at every level of magnification. In addition, strange attractors have a fractal dimension [237, 238]. From an experimental point of view, an appropriate indicator of chaos is the fractal dimensionality of the attractor, which can be tested with the help of a number of different techniques; a description of these procedures, however, lies outside the scope of this book.

A necessary condition for the emergence of chaos is that the system be characterized by at least three dynamical variables that evolve over time scales of comparable magnitude. If the time scales are very different from each other, one can perform an adiabatic elimination of the fast variables, so that the number of effective degrees of freedom is reduced. The onset of chaos is favored by the presence of competition between phenomena characterized by different natural frequencies, as one often finds in optical systems, and by the presence of several stationary repellers in phase space. In the latter case, the phase-space point tends to wander from one repeller to the other like the ball in a pin-ball machine, first approaching a stationary state along an attracting direction and then leaving it along the repelling direction. This mechanism, for example, is evident in the case of Lorenz–Haken [239, 240] chaos.

The name of *generalized multistability* was introduced to indicate situations where a system has more than one attractor in phase space for the same values of the parameters, regardless of the nature of the attractors. Thus in the case of generalized multistability one may find coexisting stationary states, limit cycles, chaotic attractors etc., usually with a complex scenario. In the literature this name has been used mainly to denote situations where at least one attractor is of chaotic type.

18.4 Transitions induced by the variation of a control parameter

In our discussion so far we have assumed that the parameters of the system are fixed, and we have analyzed the different kinds of attractors which may develop in phase space. If, as is usually done in experimental investigations, we vary one of the control parameters, e.g. β, in a continuous and slow way, we may find transitions from one attractor to another

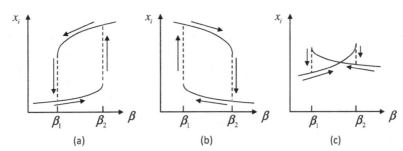

The stable stationary values of the variable x_i are plotted as functions of the control parameter β. The hysteresis cycle can be traced counterclockwise (a) or clockwise (b), or it may have a "butterfly shape" (c).

for particular values of β called critical points, β_c. This is called the *swept-parameter technique*. We now wish to examine a few important examples of this phenomenon.

18.4.1 A steady-state hysteresis cycle

Consider first a system in a stationary state configuration. As we increase the control parameter β to a value β_2, suppose that this stationary state disappears because it coalesces with an unstable stationary state. As a consequence, the system moves discontinuously to another attractor, which we assume to be also a stable stationary state. This behavior occurs, for example, in optical bistability (see Fig. 11.4).

The same behavior may develop if, for $\beta = \beta_2$, the stationary state does not disappear but becomes unstable because a real solution of the eigenvalue equation (18.8) becomes positive for $\beta > \beta_2$. In this case, for $\beta = \beta_2$, one has either a steady-state bifurcation (see below) or the jump to another attractor if two attractors coexist in this region. Let us assume that the other attractor is a stable stationary state as may occur, for example, in a laser with a saturable absorber (see Fig. 13.4).

Once the system has reached the second attractor, we then decrease the control parameter below β_2. Because the second attractor is stable, the system remains in the new configuration, even if $\beta < \beta_2$, until the second attractor disappears at $\beta = \beta_1 < \beta_2$. At this point the system returns to its original state. If we then plot the stationary state of some real variable, say x_1, as a function of β, we obtain a hysteresis cycle, which may have, for example, the form sketched in Fig. 18.1 (counterclockwise, clockwise, or shaped like a butterfly, as we will see in Section 22.6). This behaviour is reminiscent of first order phase transitions in equilibrium systems.

18.4.2 Steady-state bifurcation

We now assume that a stationary state becomes unstable (but does not disappear) when β increases to the value β_c. This instability is induced by a real eigenvalue that becomes positive for $\beta > \beta_c$. We have a *steady-state bifurcation* if, for $\beta \geq \beta_c$, there is at least one other stationary solution that is stable and is such that, for $\beta = \beta_c$, it coincides with the first. In this case, if we plot the value of some real variable x_i, as a function of β, we

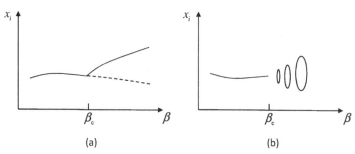

Figure 18.2   (a) Steady-state bifurcation. For $\beta > \beta_c$ the broken-line branch is unstable because a real eigenvalue has become positive. The stable solid-line branch bifurcates at the critical point β_c. (b) Hopf bifurcation. For $\beta > \beta_c$ the stationary branch becomes unstable because a pair of complex-conjugate eigenvalues has a positive real part. At $\beta = \beta_c$ a limit cycle bifurcates from the stationary solution.

find a behavior that is reminiscent of second-order phase transitions in systems at thermal equilibrium (Fig. 18.2(a)). We have met this scenario, for example, in the case of the free-running laser (see Fig. 1.4).

18.4.3 Hopf bifurcation

Now suppose that a stationary state becomes unstable, when β increases to the value β_c, with the instability being induced by a pair of complex-conjugate eigenvalues whose real parts becomes positive for $\beta > \beta_c$. We have a *Hopf bifurcation* if, for $\beta \geq \beta_c$, there exists a stable limit cycle emerging from the stationary solution for $\beta = \beta_c$ and growing as β increases beyond β_c (Fig. 18.2(b)). For $\beta \simeq \beta_c$ the frequency with which the limit cycle is traversed coincides with the imaginary part ω of the pair of complex eigenvalues that are responsible for the instability. In this case, upon crossing the critical value β_c, the system undergoes a transition from a stationary to a time-periodic state with the appearance of oscillations and the spontaneous breaking of time-translational symmetry. For $\beta < \beta_c$ the phase-space point returns to the stable stationary state with a spiraling trajectory, which corresponds to the presence of damped oscillations in the time evolution towards the stationary state. For $\beta > \beta_c$, instead, if the phase-space point lies initially close to the unstable stationary state, it spirals away from it and, in the long-term limit, approaches the limit cycle where the oscillations become self-sustained.

18.4.4 Routes to chaos

There are innumerable examples of systems that, beginning with a stationary state, upon increasing a control parameter, end up to a chaotic state after a sequence of transitions. In some instances, such as for example in the Lorenz–Haken model [239, 240], a system may jump directly from a stationary state to a chaotic behavior (note, however, that for some selected values of the parameters, the Lorenz–Haken model displays also different kinds of instabilities). In many other instances, the transition to chaos involves more than one step, and shows characteristic sequences usually called *routes to chaos*. The best known are the period-doubling route, the quasi-periodicity route and the intermittency route (see Section 22.4 and [241]).

18.4.5 The "potential" case

An especially interesting situation arises when we can identify a function $V(x_1, \ldots, x_n)$ such that the dynamical equations can be cast in the form

$$\frac{dx_i}{dt} = -\frac{\partial V}{\partial x_i}, \qquad i = 1, 2, \ldots, n. \tag{18.14}$$

The function V is called potential for the following reasons. If we consider a classical particle subject to a conservative force, associated with the potential U, and also to a viscous force whose frictional coefficient is denoted by γ, its equations of motion are

$$m\frac{d^2 x_i}{dt^2} + \gamma\frac{dx_i}{dt} = -\frac{\partial U}{\partial x_i}, \tag{18.15}$$

where x_i ($i = 1, 2, 3$) is the ith Cartesian component of the position vector. In the over-damped case the inertial term can be neglected and Eq. (18.15) acquires the form (18.14) with $V = U/\gamma$.

In the case of Eq. (18.14) the system cannot display oscillatory motion such as one would expect in the presence of a limit cycle, or of a chaotic attractor, etc. This result emerges from the fact that the function $V(x_1(t), \ldots, x_n(t))$ is monotonically decreasing in time because

$$\frac{dV}{dt} = \sum_{i=1}^{n} \frac{\partial V}{\partial x_i}\frac{dx_i}{dt} = -\sum_{i=1}^{n} \left(\frac{\partial V}{\partial x_i}\right)^2 \leq 0. \tag{18.16}$$

The phase space can contain only stationary states corresponding to the stationary points of the potential, i.e. the points which satisfy the equations

$$\frac{\partial V}{\partial x_i} = 0, \qquad i = 1, 2, \ldots, n. \tag{18.17}$$

Because of the monotonic behavior of V with time, the stable stationary solutions correspond to the local minima of the function V.

The theory of catastrophes developed by Thom [242] focusses precisely on those situations which are described by Eqs. (18.14). In addition the calculation of the fluctuations becomes very simple in this potential case. In fact, suppose that the evolution of the system in the presence of noise is governed by the stochastic equations

$$\frac{dx_i}{dt} = -\frac{\partial V}{\partial x_i} + \xi_i(t), \qquad i = 1, 2, \ldots, n, \tag{18.18}$$

where the variables ξ_i undergo a stationary, Gaussian stochastic process with

$$\langle \xi_i(t) \rangle = 0, \tag{18.19}$$

$$\langle \xi_i(t), \xi_j(t') \rangle = D\delta(t - t')\delta_{i,j}, \tag{18.20}$$

where $\delta(t)$ denotes the Dirac delta function, so that noise is white and additive, because we assume that the quantity D is independent of the variables x_i. In this case the stationary probability distribution of the variables x_i is given by [91, 234]

$$P_{\mathrm{s}}(x_1, \ldots, x_n) = N \exp\left[-\frac{2}{D}V(x_1, \ldots, x_n)\right], \tag{18.21}$$

where the normalization constant N is determined by the condition

$$\int dx_1, \ldots, dx_n \, P_{\mathrm{s}}(x_1, \ldots, x_n) = 1. \tag{18.22}$$

Equations (18.17) and (18.21) show that V plays the role of a generalized free energy for the system, which is normally quite far removed from thermal equilibrium.

When the number of variables n is 2 or larger, one can write the dynamical equations (18.1) in the form (18.14) only in exceptional cases. In contrast, for $n = 1$ the equation of motion can always be cast into the potential form (18.14) with the definition

$$V = -\int dx \, f(x). \tag{18.23}$$

19

Special limits in the single-mode model

The relaxation rates κ, $\gamma_\perp$ and $\gamma_\parallel$ which appear in the single-mode model are irrelevant for the stationary state. Instead, their values become critical when we consider dynamical, i.e. time-dependent, phenomena. In particular, they control the possibility of eliminating adiabatically some of the variables in play, thus reducing the complexity of the problem. In this chapter we first illustrate, in Section 19.1, the useful laser classification introduced by Arecchi. Next, in Section 19.2, we discuss the dynamical behavior of the solutions of a model obtained from the single-mode model (derived in Chapter 12) by performing the adiabatic elimination of the atomic variables, which procedure is called the *good-cavity limit* in the literature. Section 19.2.1 is devoted to the case of the free-running laser (*class-A laser*) and discusses in particular, at the end, the *cubic single-mode laser model* which is the simplest model capable of describing a laser. On the other hand, in Section 19.2.2 we discuss the case of optical bistability in the good-cavity limit.

If we eliminate adiabatically the atomic polarization only, we obtain, instead, the single-mode rate equations which are discussed in Section 19.3. In particular, Section 19.3.1 concerns the case of class-B lasers and focusses especially on the phenomenon of relaxation oscillations. Special attention is paid to demonstrating the equivalence of the model which describes class-B lasers to the rate-equation laser model derived in Chapter 1 and the semiconductor laser model obtained in Chapter 16. In Section 19.3.2 we describe the generation of a giant pulse in class-B lasers, a phenomenon which is called *active Q-switching*.

The final section, Section 19.4, deals with the model obtained by adiabatically eliminating the field variable in the so-called *bad-cavity limit*.

19.1 Classification of lasers

Arecchi [243] identifies three especially important kinds of laser, which he calls class-A, class-B and class-C lasers, respectively.

(1) Class-A lasers are characterized by the condition

$$\kappa \ll \gamma_\perp, \gamma_\parallel. \tag{19.1}$$

In this case the atomic variables can be adiabatically eliminated. The He–Ne laser, argon laser and quantum-cascade laser belong to this class.

(2) Class-B lasers satisfy the conditions

$$\gamma_\parallel \ll \kappa \ll \gamma_\perp. \tag{19.2}$$

In this case the atomic polarization can be adiabatically eliminated, whereas, as we will see, the population inversion cannot be adiabatically eliminated. The ruby laser, CO_2 laser and semiconductor laser belong to this class.

(3) In the case of class-C lasers the relaxation rates κ, $\gamma_\perp$ and $\gamma_\parallel$ have the same order of magnitude and therefore one must deal with the complete single-mode model.

19.2 Adiabatic elimination of the atomic variables (the good-cavity limit)

If one performs the adiabatic elimination of the atomic variables in the single-mode model (12.43)–(12.45), in the case of optical bistability one obtains Eq. (13.26), where $t'' = \kappa t$. In the laser case (with or without an injected signal) one must replace the bistability parameter $2C$ by the opposite of the pump parameter A. The functional expressions for the normalized atomic polarization and population difference as functions of the field variable x are $P = (1 - i\Delta)x/(1 + \Delta^2 + |x|^2)$ and $D = (1 + \Delta^2)/(1 + \Delta^2 + |x|^2)$, respectively.

19.2.1 The case of class-A lasers. The cubic model for the laser

Let us now focus on the case of the free-running laser, in which the dynamical equation in the good-cavity limit (19.1) reads

$$\frac{dx}{dt''} = -(1 - i\Delta)\left(1 - \frac{A}{1 + \Delta^2 + |x|^2}\right)x, \tag{19.3}$$

where in Eq. (13.26) we have replaced $2C$ by $-A$, and also θ by $-\Delta$ using Eq. (12.57), and we have set $y = 0$. If we now define

$$x = \sqrt{(1 + \Delta^2)\tilde{I}}\, e^{i\varphi}, \qquad a = \frac{A}{1 + \Delta^2}, \tag{19.4}$$

we obtain the equations

$$\frac{d\tilde{I}}{dt''} = 2\left(\frac{a}{1 + \tilde{I}} - 1\right)\tilde{I}, \tag{19.5}$$

$$\frac{d\varphi}{dt''} = \Delta\left(1 - \frac{a}{1 + \tilde{I}}\right). \tag{19.6}$$

Note that the time-evolution equation for the normalized intensity $\tilde{I}$ is self-contained, that the laser threshold corresponds to $a = 1$ and that the nontrivial stationary solution is $\tilde{I} = a - 1$. It is convenient to cast Eq. (19.5) in the form

$$\frac{d\tilde{I}}{dt''} = 2\left[(a - 1)\tilde{I} - \frac{a\tilde{I}^2}{1 + \tilde{I}}\right], \tag{19.7}$$

because one identifies the qualitative features of the time evolution of the intensity starting from a small value of $\tilde{I}$, which simulates an initial fluctuation due to spontaneous emission. In the initial stage of the evolution, the last term on the r.h.s. of Eq. (19.7) is negligible because $\tilde{I}$ is small. If the laser is operating below threshold, the intensity decays to zero exponentially. If, instead, the laser is operating above threshold ($a > 1$) the intensity grows exponentially, because the linear gain a is larger than the loss which arises from the escape of photons from the cavity, until the last term of the r.h.s. of Eq. (19.7), which describes saturation, becomes relevant and forces the system to converge to the stationary state, in which the output intensity is constant.

On the other hand, once the time evolution of $\tilde{I}$ has been calculated, the time evolution of the phase φ is immediately obtained from Eq. (19.6). The laser frequency $\omega_0 - d\varphi/dt''$ evolves with time and, when the laser is operating above threshold, approaches the stationary value ω_0. In order to get rid of the factor of 2 in Eq. (19.5), let us denote by $\tilde{I}'$ the derivative of $\tilde{I}$ with respect to the time variable $2\kappa t$,

$$\tilde{I}' = \left(\frac{a}{1 + \tilde{I}} - 1 \right) \tilde{I}. \tag{19.8}$$

If we linearize Eq. (19.8) around a stationary value $\tilde{I}_s$, we obtain the eigenvalue (see Eq. (18.12))[1]

$$\lambda = \frac{a}{(1 + \tilde{I}_s)^2} - 1. \tag{19.9}$$

Below threshold $\tilde{I}_s = 0$ and therefore $\lambda = a - 1 < 0$, because $a < 1$; hence the trivial stationary solution is stable, whereas it becomes unstable above threshold. The nontrivial stationary solution is $\tilde{I}_s = a - 1$ and therefore $\lambda = (1 - a)/a$. Hence for $a > 1$ the nontrivial stationary solution is the only stable stationary solution.

Since Eq. (19.8) involves just one variable, we can cast it in the potential form (18.14), i.e.

$$\tilde{I}' = -\frac{dV}{d\tilde{I}}, \quad \text{with} \quad V(\tilde{I}) = -\int d\tilde{I} \left(\frac{a\tilde{I}}{1 + \tilde{I}} - \tilde{I} \right) = -a\tilde{I} + \frac{\tilde{I}^2}{2} + a \ln\left(1 + \tilde{I}\right). \tag{19.10}$$

The shape of the potential varies with the parameter a. In Fig. 19.1 we show the potential for three values of a, one below threshold ($a = 0.5$), one at threshold ($a = 1$) and the last above threshold ($a = 2$). For $a < 1$ there is a minimum at $\tilde{I} = 0$, with a parabolic behavior in the neighborhood of the origin. At threshold the minimum is very flat, because it is not parabolic but quartic. Above threshold the solution $\tilde{I}_s = 0$ becomes a maximum, while the potential V presents a minimum for $\tilde{I} = a - 1$. The system approaches the stable stationary state without oscillations, since V is a monotonically decreasing function of time (see Eq. (18.16)).

---

[1] In this section the eigenvalue is normalized with respect to 2κ.

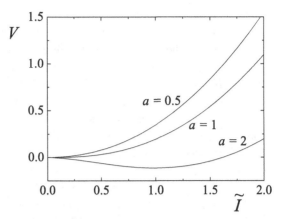

The shape of the generalized potential $V(\tilde{I})$ for three values of a, the first below threshold the second at threshold and the last above threshold.

It is easy to find the exact numerical solution of Eq. (19.8), which reads

$$\frac{\tilde{I}(t)}{\tilde{I}(0)} \left[\frac{a - 1 - \tilde{I}(0)}{a - 1 - \tilde{I}(t)} \right]^a = e^{2\kappa(a-1)t}. \tag{19.11}$$

If $a < 1$ the r.h.s. tends to zero and therefore also $\tilde{I}(t)$ tends to zero for any initial condition. When $a > 1$ we can distinguish two stages. In the initial stage $\tilde{I}(t) \simeq \tilde{I}(0)$ and we can set the argument of the square bracket equal to 1, so that

$$\tilde{I}(t) \simeq \tilde{I}(0)e^{2\kappa(a-1)t}, \tag{19.12}$$

and the intensity grows exponentially. In the final stage we have $\tilde{I}(t) \simeq a - 1$, and if we set $\tilde{I}(t) = a - 1$ in the term which multiplies the square bracket we obtain

$$\tilde{I}(t) = a - 1 - \left[\frac{a-1}{\tilde{I}(0)} \right]^{1/a} [a - 1 - \tilde{I}(0)]e^{-2\kappa(a-1)t/a}. \tag{19.13}$$

The intensity approaches the final value $(a - 1)$ as an exponential with the rate $2\kappa(a - 1)/a$. In both stages the larger a, the faster the laser evolution. This fact emerges with evidence from Fig. 19.2, where one compares the time evolution for $a = 1.1$ and $a = 2$; in both cases the initial value is $\tilde{I}(0) = 10^{-3}$.

A very important remark is that if the laser is near threshold, i.e. $a \simeq 1$, the normalized intensity $\tilde{I}$ is small; hence in Eq. (19.8) one can approximate $1/(1 + \tilde{I})$ by $1 - \tilde{I}$ and write

$$\tilde{I}' = a\tilde{I}(1 - \tilde{I}) - \tilde{I} = (a - 1)\tilde{I} - a\tilde{I}^2 \simeq (a - 1)\tilde{I} - \tilde{I}^2, \tag{19.14}$$

where in the last step we have set $a = 1$ in the quadratic term. This is the simplest dynamical model for the single-mode laser and has often been considered in the literature (see e.g. [7, 8]). The two stationary solutions coincide with those of model (19.8) and the saturation is described by the quadratic term (compare it with Eq. (19.7)). Note that Eq. (19.14) is equivalent to a cubic equation for the normalized electric field x, so in the literature this is usually called the *cubic laser model*.

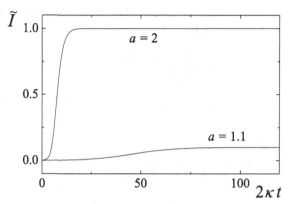

Figure 19.2 Class-A lasers. Time evolution towards the stable stationary state $\tilde{I} = a - 1$ for two values of a.

It is important to remark that the same cubic model holds also for the Fabry–Perot laser. As a matter of fact, if we eliminate the atomic variables adiabatically by inserting Eq. (14.46) into Eq. (14.44), in the case of a free-running laser ($y = 0, \theta = -\Delta$) we obtain the equation

$$\frac{dx}{dt} = -\kappa(1 - i\Delta)\left[1 - \frac{A}{2|x|^2}\left(1 - \frac{1}{\sqrt{1 + 4|x|^2/(1 + \Delta^2)}}\right)\right]x, \qquad (19.15)$$

where the integral in φ is done exactly as in the passage from Eq. (14.47) to Eq. (14.49). Equation (19.15) is the dynamical equation for the class-A Fabry–Perot laser.

If we now assume that the laser is operating only slightly above threshold, i.e. that $A \simeq 1 + \Delta^2$, then we have that $|x|^2 \ll 1$ and we can expand the expression $[1 + 4|x|^2/(1 + \Delta^2)]^{-1/2}$ in Eq. (19.15) up to the second-order term, obtaining

$$\frac{dx}{dt} = -\kappa(1 - i\Delta)\left(1 - a + 3\frac{|x|^2}{1 + \Delta^2}\right)x, \qquad (19.16)$$

where we have used the definition of a given in Eq. (19.4), and we have set $a = 1$ in the cubic term. Finally, by setting $|x|^2 = (1 + \Delta^2)\tilde{I}/3$ (compare this with Eq. (19.4)) and taking into account that $\tilde{I}' = d\tilde{I}/d(2\kappa t)$, we recover Eq. (19.14).

Equation (19.14) can be written in the form $\tilde{I}' = \tilde{I}(a - 1 - \tilde{I})$, which is identical to the logistic Verhulst equation, which is prototypical for population dynamics [91, 234, 235].

19.2.2 The good-cavity limit for optical bistability

Let us now turn our attention to Eq. (13.26), which describes optical bistability in the good-cavity limit (19.1). If we linearize it around a stationary solution as indicated in Eq. (18.4), we obtain the pair of complex-conjugate equations

$$\frac{d\delta x}{dt''} = -(1 + i\theta)\delta x - \frac{2C(1 - i\Delta)}{(1 + \Delta^2 + X)^2}\left[(1 + \Delta^2)\delta x - x^2\delta x^*\right], \qquad (19.17)$$

$$\frac{d\delta x^*}{dt''} = -(1 - i\theta)\delta x^* - \frac{2C(1 + i\Delta)}{(1 + \Delta^2 + X)^2}\left[(1 + \Delta^2)\delta x^* - (x^*)^2\delta x\right], \qquad (19.18)$$

where x denotes the stationary value and $X = |x|^2$. From them we obtain the 2×2 matrix $\mathbf{A}$ which appears in Eq. (18.6) and arrive at the characteristic equation (18.8) with

$$c_1 = 2 + 4C\frac{1 + \Delta^2}{(1 + \Delta^2 + X)^2}, \tag{19.19}$$

$$c_0 = 1 + \theta^2 + 4C\frac{(1 - \Delta\theta)(1 + \Delta^2)}{(1 + \Delta^2 + X)^2} + 4C^2\frac{(1 + \Delta^2)(1 + \Delta^2 - X)}{(1 + \Delta^2 + X)^3}. \tag{19.20}$$

The stability condition $c_1 > 0$ is always satisfied. Insofar as the other stability condition $c_0 > 0$ is concerned, we note that

$$c_0 = \frac{dY}{dX}, \tag{19.21}$$

and therefore, as anticipated in Chapter 11, the stationary states in the positive-slope parts of the stationary curve $X(Y)$ are stable, whereas those on the negative-slope segment are unstable. The fact that the constant c_0 of the eigenvalue equation (18.8) is proportional to dY/dX remains true also in the complete single-mode model without any adiabatic elimination, and therefore this result is valid in general.

Let us now focus on the case of absorptive optical bistability $\Delta = \theta = 0$. If we subdivide the complex variable x into real and imaginary parts by writing $x = x' + ix''$, we can cast Eq. (13.26) into the potential form (see Section 18.4.5)

$$\frac{dx'}{dt''} = -\frac{\partial V(x', x'')}{\partial x'}, \qquad \frac{dx''}{dt''} = -\frac{\partial V(x', x'')}{\partial x''}, \tag{19.22}$$

with the following definition for the potential:

$$V(x', x'') = -yx' + \frac{1}{2}(x'^2 + x''^2) + C\ln(1 + x'^2 + x''^2). \tag{19.23}$$

We note that, if the imaginary part x'' vanishes at the initial time, it remains equal to zero at all times, and therefore we consider only the section of the potential curve for x'' equal to zero. Referring to Fig. 11.4, we see that for $y < y_m$ and $y > y_M$, i.e. in the monostable regions, the potential displays only one minimum, whereas in the bistable domain $y_m < y < y_M$ it has two minima corresponding to the two stable stationary solutions (see Fig. 19.3, curves $y = 7$ and $y = 12$). When y is close to y_m the left minimum is the absolute minimum (curve $y = 9$). As y is increased the two minima attain the same level, and finally the right minimum becomes the absolute minimum (curve $y = 10$).

It is interesting to note that, when the noise arises from thermal fluctuations or from fluctuations of the input field y, the noise obeys the conditions (18.19) and (18.20) [86], so the stationary probability distribution has the form (18.21). From this equation we see that the probability distribution presents peaks corresponding to the stable stationary solutions, and minima corresponding to the unstable stationary solutions. This implies that for the values of y for which there are two stable stationary solutions the stationary probability distribution is *bimodal*, i.e. it shows two peaks, whereas it has only one peak in the monostable cases.

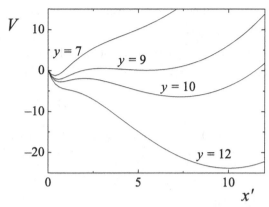

Figure 19.3 The shape of the potential function $V(x', x'' = 0)$ for $C = 10$ and the indicated values of y.

19.3 Adiabatic elimination of the atomic polarization: the single-mode rate-equation model

Under the conditions defined by Eq. (19.2), we can, first of all, eliminate the atomic polarization adiabatically from Eqs. (12.59)–(12.61), obtaining

$$\frac{dx}{dt} = \kappa[y - (1 + i\theta)x + a(1 - i\Delta)Dx], \tag{19.24}$$

$$\frac{dD}{dt} = -\gamma_\parallel \left[D\frac{|x|^2}{1 + \Delta^2} + D - 1 \right], \tag{19.25}$$

where a is defined by Eq. (19.4). The functional expression for the atomic polarization P as a function of x and D is $P = (1 - i\Delta)Dx/(1 + \Delta^2)$. Next, since $\gamma_\parallel \ll \kappa$, when y is different from zero (for a laser with an injected signal or optical bistability) we can proceed to eliminate adiabatically also the field variable x, obtaining

$$\frac{dD}{dt} = -\gamma_\parallel \left[D\frac{y^2}{(1 - aD)^2 + (\theta + \Delta aD)^2} + D - 1 \right], \tag{19.26}$$

where we have assumed that y is real. We note, on the other hand, that, in the case of a free-running laser ($y = 0$, $\theta = -\Delta$), if we set $dx/dt = 0$ in Eq. (19.24) we do not obtain a functional expression for the field variable x as a function of the inversion variable D, and therefore the field x cannot be eliminated adiabatically.

19.3.1 Class-B lasers and relaxation oscillations

Let us now focus on the case of the free-running class-B laser. If we express the variable x in the form (19.4), use Eq. (1.42) and indicate with a prime the derivative with respect to the normalized time[2] $2\kappa t$ we obtain from Eq. (19.24) with $y = 0, \theta = -\Delta$ the set of equations

$$\tilde{I}' = (aD - 1)\tilde{I}, \tag{19.27}$$

$$D' = -\gamma \left[D\left(1 + \tilde{I}\right) - 1 \right], \tag{19.28}$$

$$\varphi' = -\frac{\Delta}{2}(aD - 1). \tag{19.29}$$

Equations (19.27) and (19.28) give a self-contained set of two coupled equations for $\tilde{I}$ and D. If we now introduce the variable

$$n = aD - 1, \tag{19.30}$$

the model assumes the form

$$\tilde{I}' = \tilde{I}n, \tag{19.31}$$

$$n' = \gamma \left[a - (1 + n)\left(1 + \tilde{I}\right) \right]. \tag{19.32}$$

Before analyzing these equations, let us connect them with the rate-equation laser model derived in Chapter 1 and with the semiconductor laser model derived in Chapter 16. We note, first of all, that the variable D defined by Eq. (1.33) coincides with the D, defined in Eq. (4.33), which appears in Eqs. (19.27) and (19.28). Also the variable $\tilde{I}$ defined by Eq. (1.33) coincides with the $\tilde{I}$ which appears in Eqs. (19.27) and (19.28), as one can verify using Eqs. (8.17), (4.30), (2.39) and (19.4). The time $2kt$ used in Eqs. (1.43) and (1.44) coincides with that used in Eqs. (19.27) and (19.28). On the other hand, Chapter 1 deals with a Fabry–Perot cavity, and this implies that the definitions of the parameters a and κ differ from those used in this chapter, which deals with a ring cavity. In this case $a = A/(1 + \Delta^2)$ with $A = g\bar{L}/T$ (where $\bar{L}$ is the sample length), whereas in a Fabry–Perot cavity $A = 2g\bar{L}/T$ (see Eq. (14.36); note also that in Chapter 1 we implicitly assumed that $n_B = 1$). Therefore Eqs. (19.27) and (19.28) coincide with the equations (1.43) and (1.44), which were derived phenomenologically in the first chapter, and this proves that the photon concept is not necessary to obtain the rate equations for the laser, because Eqs. (19.27) and (19.28) have been derived in a purely semiclassical context.

Let us now come back to the equations (16.55) and (16.56) for the semiconductor laser and let us demonstrate their equivalence to the equations that govern class-B lasers. If we introduce the positions

$$F = \sqrt{\tilde{I}}e^{i\varphi}, \qquad n = \tilde{D} - 1, \qquad \gamma = \frac{\gamma_{nr}}{2\kappa}, \tag{19.33}$$

Eqs. (16.55) and (16.56) become

$$\tilde{I}' = \tilde{I}n, \tag{19.34}$$

$$n' = \gamma \left[\mu - (1 + n)\left(1 + \tilde{I}\right) \right], \tag{19.35}$$

$$\varphi' = -\frac{\alpha}{2}n. \tag{19.36}$$

---

[2] In the case of class-A and class-B lasers, it is convenient to use the normalized time $2\kappa t = 2t''$ (see the beginning of Section 19.2.1) in which time is normalized with respect to the inverse of the decay rate of the intensity.

Clearly these three equations coincide with Eqs. (19.31), (19.32) and (19.29), respectively, if we replace α by Δ and μ by a and, in Eq. (19.29), we use the definition (19.30) for n. We observe, however, that the α parameter of semiconductor lasers, even if it plays the same role as the atomic detuning parameter in gas lasers, has a deeply different physical meaning, as explained in Section 16.5.

Now we analyze the stationary solutions of Eqs. (19.31) and (19.32) and their stability. The trivial stationary solution corresponds to $\tilde{I}_s = 0$ and $n_s = a - 1$; the nontrivial stationary solution corresponds to $\tilde{I}_s = a - 1$ and $n_s = 0$ and exists for $a > 1$. The matrix $\mathbf{A}$ (see Eq. (18.6)) of the linearized set of equations reads

$$\mathbf{A} = \begin{pmatrix} n_s & \tilde{I}_s \\ -\gamma(1 + n_s) & -\gamma(1 + \tilde{I}_s) \end{pmatrix}. \tag{19.37}$$

In the case of the trivial stationary solution the matrix becomes

$$\mathbf{A} = \begin{pmatrix} a - 1 & 0 \\ -\gamma a & -\gamma \end{pmatrix}, \tag{19.38}$$

and the eigenvalues are $\lambda_1 = a - 1$ and $\lambda_2 = -\gamma$. Both are negative for $a < 1$, whereas λ_1 is positive for $a > 1$, hence the trivial stationary solution becomes unstable above threshold. In the case of the nontrivial stationary solution the matrix is

$$\mathbf{A} = \begin{pmatrix} 0 & a - 1 \\ -\gamma & -\gamma a \end{pmatrix}, \tag{19.39}$$

and the eigenvalues of the characteristic equation (18.8) are[3]

$$\lambda_\pm = -\frac{\gamma a}{2} \pm \sqrt{\frac{\gamma^2 a^2}{4} - \gamma(a - 1)}. \tag{19.40}$$

Since for a class-B laser $\gamma \ll 1$ (see Eq. (19.2)), the first term in the square root is negligible unless the laser is operating very close to threshold, and we can write

$$\lambda_\pm \simeq -\frac{\gamma a}{2} \pm i\sqrt{\gamma(a - 1)}. \tag{19.41}$$

The nontrivial stationary solution is always stable because the real part is negative, and the laser returns to the stationary state with damped oscillations that are called *relaxation oscillations*. If we go back to the unnormalized time t, we see that the damping rate and the oscillation frequency are given by

$$-2\kappa \, \mathrm{Re} \, \lambda_\pm = \gamma_\parallel, \tag{19.42}$$

$$\Omega_r = \sqrt{2\kappa\gamma_\parallel(a - 1)}, \tag{19.43}$$

respectively. Since $\gamma_\parallel \ll \kappa$ the damping rate is much smaller than the oscillation frequency, so the oscillations undergo a very slow damping (see Fig. 19.4(a)). The fact that the oscillation frequency scales as $\sqrt{\kappa\gamma_\parallel}$, rather than as $\gamma_\parallel$, is related to the fact that the electric-field variable cannot be eliminated adiabatically, because after the adiabatic elimination the only available rate would be $\gamma_\parallel$.

---

[3] In this section the eigenvalues are normalized with respect to 2κ.

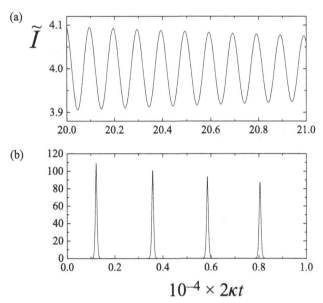

Figure 19.4   The final (a) and initial (b) evolution of the intensity of a class-B laser with $a = 5$ and $\gamma = 10^{-5}$.

The linear-stability analysis describes the laser behavior well only when it is close to the stationary solution. However, in the initial phase of the time evolution the value of the intensity is quite different from the stationary value, and the evolution profile can be very different. Let us assume, for example, that the parameter a is slowly increased up to the threshold value $a = 1$, so that the initial value of n is $n(0) = 0$. If the value of a is abruptly taken above threshold, the initial phase of the time evolution is of the kind shown in Fig. 19.4(b), in which the initial condition for $\tilde{I}$ is the small value $\tilde{I}(0) = 5.55 \times 10^{-11}$ which simulates an initial fluctuation due to spontaneous emission. The laser emits very short and intense pulses, with a period of 2300 temporal units and slow damping. The final stage is shown by Fig. 19.4(a). It is interesting to observe that it is possible to give an analytical description also of the initial phase. Since the variable n always remains very small, in Eq. (19.32) we can neglect n with respect to 1, so that the equation reduces to

$$n' = \gamma \left(a - 1 - \tilde{I}\right). \tag{19.44}$$

One has that Eqs. (19.31) and (19.44) have a Hamiltonian structure equivalent to a Toda oscillator model, and this allows one to solve the equations analytically [244].

19.3.2 Generation of giant pulses by active Q-switching

Another interesting phenomenon that arises in class-B lasers is the generation of an intense short pulse by introducing a fast (ideally instantaneous) modification of the parameters, which is called *Q-switching* [6, 7].

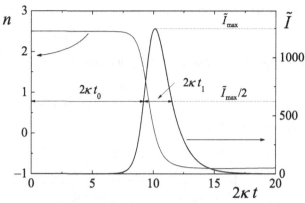

Figure 19.5 Evolution of the variables $\tilde{I}$ and n during a giant pulse generated via the active Q-switching procedure for $\gamma = 10^{-3}$. The initial conditions are $\tilde{I}_i = 10^{-7}$ and $n_i = a - 1 = 2.5$.

The name comes from the fact that in this process the quality factor Q of the cavity is rapidly switched from a low to a high value. More precisely, we speak of active Q-switching to distinguish it from passive Q-switching, which we will discuss in Section 25.2.

The quality factor Q is defined as $Q = \omega_0/(2\kappa)$; hence its value can be varied by varying the cavity losses κ. These can be increased with respect to their normal value by different methods, for instance by misalignment of one mirror, or by means of electro-optical or acousto-optical modulators. As long as the losses are large and the quality factor low, one can pump the laser up to a value of the parameter a larger than 1. Under normal conditions the laser would be above threshold and would emit laser light, but with increased losses the system does not lase and the inversion per atom r_3 assumes the value σ imposed by the pump; hence $D = 1$ and $n = a - 1$ (see Eq. (19.30)). This value of n is larger than the value $n = 0$ in the lasing state. In this way, we have pumped the population inversion r_3 to a value beyond σ_{thr}, which is the maximum value of the inversion in a stationary state (see Fig. 9.3). At a certain time that we take as $t = 0$, the normal values of the cavity losses and quality factor are restored, and the system approaches the lasing state starting from the initial condition $n = a - 1$, $\tilde{I} \approx 0$. This causes a very fast emission of photons, because the excess energy stored in the atomic medium is suddenly released and the process continues even when n decreases below the threshold value $n = 0$, i.e. becomes negative, because of the large number of photons in the cavity, which induces new processes of stimulated emission in a time much faster than that, of order $\gamma_\parallel^{-1}$, required by the pump to repopulate the upper level.

On the other hand, in a time on the order of the inverse of κ, the photons escape from the cavity and the system approaches a configuration in which the internal field vanishes and n takes a negative value n_f. This *giant pulse* [7, 8] is described in Fig. 19.5, which is obtained by solving Eqs. (19.31) and (19.32). Only on the much longer time scale $\gamma_\parallel^{-1}$ does the system finally approach the stationary state via the usual relaxation oscillations.

It is possible to obtain a detailed description of the giant pulse, if we observe that in Eq. (19.32) we can neglect the terms $\gamma(a - 1 - n)$ because initially $n = a - 1$ and later, when the value of n becomes different from $a - 1$, the term $-\gamma \tilde{I}(1 + n)$ dominates since

$\tilde{I}$ is very large. Therefore Eq. (19.32) becomes

$$n' = -\gamma \tilde{I}(1+n). \tag{19.45}$$

From Eqs. (19.31) and (19.45) we obtain the equation

$$\frac{d\tilde{I}}{dn} = -\frac{n}{\gamma(1+n)}, \tag{19.46}$$

from which we identify the constant of motion

$$C = n - \ln(1+n) + \gamma \tilde{I}. \tag{19.47}$$

During the pulse the system passes from the initial state $n_i = a - 1$, $\tilde{I}_i \ll 1$ to the peak value $\tilde{I}_{max}$ which occurs for $n = 0$, and therefore

$$C = a - 1 - \ln a = \gamma \tilde{I}_{max}, \tag{19.48}$$

from which we obtain

$$\tilde{I}_{max} = \frac{a - 1 - \ln a}{\gamma}. \tag{19.49}$$

The intensity $\tilde{I}_{max}$ scales as γ^{-1} times the stationary intensity. With the parameters of Fig. 19.5 $a = 3.5$, $\gamma = 10^{-3}$ and $I_i = 10^{-7}$ we obtain $\tilde{I}_{max} \approx 1247$, in excellent accord with Fig. 19.5. Another important quantity is the pulse buildup time t_0, defined as the time taken to pass from the initial value $\tilde{I}_i$ to the value $\tilde{I}_{max}/2$. In order to evaluate this time we take into account that in the initial stage of the time evolution we can set $n \approx a - 1$ in Eq. (19.31), obtaining

$$2\kappa t \approx \frac{1}{a - 1} \ln \left(\tilde{I}/\tilde{I}_i \right). \tag{19.50}$$

By extrapolating this equation to an intensity as large as $I_{max}/2$, we obtain

$$2\kappa t_0 \approx \frac{1}{a - 1} \ln \left(\tilde{I}_{max}/(2\tilde{I}_i) \right). \tag{19.51}$$

With the parametric values of Fig. 19.5 we have $2\kappa t_0 \approx 9$, which is again in good accord with the numerical simulation.

The pulse width can be estimated taking into account that at the pulse peak $n = 0$ and therefore, in a sufficiently small time interval around the pulse peak, we can assume that $|n|$ remains much smaller than unity and approximate the constant of motion C with

$$C \approx \frac{n^2}{2} + \gamma \tilde{I}. \tag{19.52}$$

Since we also have $C = \gamma \tilde{I}_{max}$, we obtain

$$\tilde{I} = \tilde{I}_{max} - \frac{n^2}{2\gamma}, \quad n = \pm\sqrt{2\gamma \left(\tilde{I}_{max} - \tilde{I} \right)}. \tag{19.53}$$

On the other hand, for small n Eq. (19.45) can be approximated by

$$n' = -\gamma \tilde{I}, \tag{19.54}$$

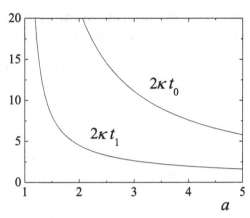

Figure 19.6 The behavior of the buildup time (upper line) and of the duration of the pulse (lower line) as a function of a.

from which it follows that

$$\kappa \, dt = \frac{dn}{n^2 - 2\gamma \tilde{I}_{\max}} \tag{19.55}$$

The pulse width (FWHM) is obtained by integrating from $n = +\sqrt{\gamma \tilde{I}_{\max}}$ to $n = -\sqrt{\gamma \tilde{I}_{\max}}$, i.e. the two values associated with $\tilde{I} = \tilde{I}_{\max}/2$ according to Eq. (19.53). Taking into account Eq. (19.49), we have

$$2\kappa t_1 \approx \frac{2.5}{\sqrt{a - 1 - \ln a}}. \tag{19.56}$$

The curves of the buildup time $2\kappa t_0$ and of the pulse width $2\kappa t_1$ as a function of a are shown in Fig. 19.6. For the parameters of Fig. 19.5 we have $2\kappa t_1 \approx 2$.

It is interesting to calculate the variation of the maximum intensity of the pulse as a function of the quality factor Q after the switching. At first sight, it seems that the intensity is an increasing function of Q. However, we must take into account that $\tilde{I}$ is the intracavity intensity and that the output intensity decreases when Q increases for fixed ω_0, because the photon escape rate κ decreases. Hence there is an optimal value of Q that maximizes the output intensity. This value can be calculated by taking into account that $\tilde{I}_{\max}^{\text{out}}$ is proportional to the intracavity intensity $\tilde{I}_{\max}$ given by Eq. (19.49) and inversely proportional to Q. On the other hand, we know that γ and a (which is inversely proportional to κ) are proportional to Q. On setting $a = bQ$ we have

$$\tilde{I}_{\max}^{\text{out}}(Q) \propto [bQ - 1 - \ln(bQ)]Q^{-2}. \tag{19.57}$$

The output intensity turns out to be maximal when

$$bQ = 1 + 2\ln(bQ), \tag{19.58}$$

and this equation has the approximate solution $bQ = a = 3.5$. From Eq. (19.56) we see that $1/t_1^2$ has the same functional dependence on Q as $\tilde{I}_{\max}^{\text{out}}$ given by Eq. (19.57), therefore the pulse duration is minimal for the same value $a = 3.5$.

19.4 Adiabatic elimination of the electric field (the bad-cavity limit)

Let us now consider the limit in which the electric-field relaxation rate is much larger than the atomic relaxation rates, i.e.

$$\kappa \gg \gamma_\perp, \, \gamma_\parallel. \tag{19.59}$$

This, the opposite of the good-cavity limit, is called the *bad-cavity limit*. It has been analyzed mainly for optical bistability in a quantum-mechanical context [86], and we limit ourselves to the semiclassical model obtained by adiabatically eliminating the field variable from Eqs. (12.43)–(12.45). The functional expression for x as a function of the atomic variables is

$$x = \frac{y - AP}{1 + i\theta}, \tag{19.60}$$

while the dynamical equations for P and D read

$$\frac{dP}{dt} = \gamma_\perp \left[\frac{y - Ap}{1 + i\theta} D - (1 + i\Delta)P \right], \tag{19.61}$$

$$\frac{dD}{dt} = -\gamma_\parallel \left[\frac{y(\text{Re } P - \theta \, \text{Im } P) - A|P|^2}{1 + \theta^2} - D + 1 \right], \tag{19.62}$$

where we have taken y real as usual. The nomenclature "good-cavity limit" and "bad-cavity limit" must be used with care, because the two limits compare the cavity damping rate with the atomic decay rates, and therefore the quality of the cavity is not the only element which determines whether we are in one or in the other limit.

20

The linear-stability analysis of the Maxwell–Bloch equations

In this chapter we outline the general scenario for optical temporal instabilities, starting from the standard Maxwell–Bloch equations. With the exception of Section 22.7, in Part II we will always consider the ring-cavity configuration because only in the case of the ultrathin-medium case considered in Section 22.7 is the treatment of instabilities in Fabry–Perot cavities not substantially more complicated than that in the ring-cavity configuration.

Some authors have introduced special approaches in order to study temporal optical instabilities, exploiting the concepts of gain and of dispersion, which are natural in optical systems and therefore help our intuition. In particular, a hint of the mechanism that produces instabilities in optical systems is offered by the so-called "weak-sideband" approach or "gain-feedback approach" [245–249], whose validity can be justified rigorously with the help of the linear-stability analysis [250]. The central point of this method is the response of the system to a weak probe passing through a collection of two-level atoms driven by a strong field [251–253]. If the strong signal with carrier frequency ω_0 saturates the medium, we can find regions of the frequency axis, which are far removed from the carrier frequency of the signal, where the weak probe can experience gain instead of absorption even if the medium has no population inversion.

In this book we follow the approach of [250], which is rigorously rooted in the linear-stability analysis and, in addition, establishes a general, useful and, in a sense, surprising connection between single-mode and multimode instabilities. We classify as single-mode a temporal instability that involves the resonant mode only, and classify as multimode a temporal instability that involves only off-resonance cavity modes. The discussion in this chapter is mainly focussed on this connection.

In Section 20.1 the multimodal equations derived in Section 12.2 are reformulated by introducing a modal expansion also for the atomic variables. In this way one obtains a complex hierarchy of equations that couple the modal amplitudes for the field and for the atomic variables. Such equations get dramatically simplified when they are linearized around a single-mode stationary solution, because they decouple into blocks of equations, each of which is composed of five equations for five coupled variables. At this stage we can introduce the basic distinction between single-mode and multimode instabilities.

Section 20.2 outlines the general features of multimodal instabilities and describes them in terms of an appropriate gain function. On the other hand, single-mode instabilities are discussed in Section 20.3 in terms of the same gain function as used for multimode instabilities plus an appropriate frequency-dispersion function. This approach allows us to disclose, in Section 20.4, the profound connections between the two classes of instabilities. Examples of connected pairs of single-mode and multimode instabilities will be discussed in the remainder of Part II, especially in Chapters 22, 23 and 24.

In Section 20.5 we focus our attention on the totally resonant configuration, in which the three frequencies ω_0, ω_a and ω_c coincide. In this case each independent set of five linearized equations decouples in turn into a set of three equations and a set of two equations, introducing a further important simplification. In particular, this feature allows us to introduce a further distinction between amplitude and phase instabilities, of which we will meet examples in the following chapters.

20.1 Coupled multimodal equations for field and atomic variables. Single-mode and multimode instabilities

Let us now return to Eqs. (12.35)–(12.37) and introduce the expansions

$$
\begin{pmatrix} F(z,\tau) \\ P(z,\tau) \\ D(z,\tau) \end{pmatrix} = \sum_{n=-\infty}^{+\infty} e^{ik_n z} e^{-i\alpha_n \tau} \begin{pmatrix} \tilde{f}_n(\tau) \\ \tilde{p}_n(\tau) \\ \tilde{d}_n(\tau) \end{pmatrix},
\tag{20.1}
$$

with

$$
k_n = \frac{2\pi n}{L}, \qquad n = 0, \pm 1, \pm 2, \ldots,
\tag{20.2}
$$

while α_n is given by Eq. (12.17). With respect to the treatment of Chapter 12 we note the following differences:

(1) in the expansion for F there is, in addition to the spatial exponential $\exp(ik_n z)$, also the temporal exponential $\exp(-i\alpha_n \tau)$, so that $f_n(\tau) = e^{-i\alpha_n \tau} \tilde{f}_n(\tau)$; and
(2) also the atomic variables P and D are expanded in modal terms.

The expansions for the complex conjugates of F and P are obtained by performing the complex conjugation on the r.h.s. of Eq. (20.1). By inserting the expansions (20.1) into Eqs. (12.35)–(12.37), we obtain the following equations for the modal amplitudes [228, 250]:

$$
\frac{d\tilde{f}_n}{d\tau} = -\kappa \left[(1+i\theta)\tilde{f}_n - A\tilde{p}_n - y\delta_{n,0} \right],
\tag{20.3}
$$

$$
\frac{d\tilde{p}_n}{d\tau} = i\alpha_n \tilde{p}_n + \gamma_\perp \left[\sum_{n'} \tilde{f}_{n'} \tilde{d}_{n-n'} - (1+i\Delta)\tilde{p}_n \right],
\tag{20.4}
$$

$$
\frac{d\tilde{d}_n}{d\tau} = i\alpha_n \tilde{d}_n - \gamma_\parallel \left[\frac{1}{2} \sum_{n'} \left(\tilde{f}_{n'} \tilde{p}^*_{n'-n} + \tilde{f}^*_{n'} \tilde{p}_{n+n'} \right) + \tilde{d}_n - \delta_{n,0} \right].
\tag{20.5}
$$

The equations for $\tilde{f}^*_n$ and $\tilde{p}^*_n$ are given by the complex conjugate of Eq. (20.3) and Eq. (20.4), respectively; note that $\tilde{d}_{-n} = \tilde{d}^*_n$ because D is a real quantity.

The $n = 0$ single-mode stationary solution is given by

$$
\tilde{f}_{n,s} = \tilde{p}_{n,s} = \tilde{d}_{n,s} = 0 \quad \text{for} \quad n \neq 0,
\tag{20.6}
$$

and

$$\tilde{p}_{0,s} = \frac{1 - i\Delta}{1 + \Delta^2 + X} x, \qquad \tilde{d}_{0,s} = \frac{1 + \Delta^2}{1 + \Delta^2 + X}, \qquad (20.7)$$

where we define

$$x = \tilde{f}_{0,s}, \qquad X = |x|^2, \qquad (20.8)$$

and the stationary equation is given by Eq. (13.1) for a laser with an injected signal and optical bistability (in the latter case the pump parameter A must be replaced by $-2C$). In the case of the free-running laser one must set $\theta = -\Delta$, and the stationary equation is given by $X = A - 1 - \Delta^2$, while the phase of x remains arbitrary.

Outside the stationary state, Eqs. (20.3)–(20.5) constitute a very complex hierarchy of equations for the modal amplitudes. In contrast with the coupled equations (12.39), (12.36) and (12.37) with (12.40), they do not lend themselves easily to numerical resolution, because there is no straightforward procedure, to the best of our knowledge, to truncate the hierarchy. In the case of Eqs. (12.39), (12.36) and (12.37) with (12.40), instead, by considering a reasonable number of terms in the modal expansion of F, one obtains a fairly good approximation, which can be improved by increasing the number of terms. However, Eqs. (20.3)–(20.5) get dramatically simplified if we linearize them with respect to the deviations from the stationary values

$$\delta \tilde{f}_n = \tilde{f}_n - \tilde{f}_{n,s}, \qquad \delta \tilde{p}_n = \tilde{p}_n - \tilde{p}_{n,s}, \qquad \delta \tilde{d}_n = \tilde{d}_n - \tilde{d}_{n,s}. \qquad (20.9)$$

The linearized equations read [228, 250]

$$\frac{d\delta \tilde{f}_n}{d\tau} = -\kappa \left[(1 + i\theta)\delta \tilde{f}_n - A\,\delta \tilde{p}_n \right], \qquad (20.10)$$

$$\frac{d\delta \tilde{f}^*_{-n}}{d\tau} = -\kappa \left[(1 - i\theta)\delta \tilde{f}^*_{-n} - A\,\delta \tilde{p}^*_{-n} \right], \qquad (20.11)$$

$$\frac{d\delta \tilde{p}_n}{d\tau} = i\alpha_n \tilde{p}_n + \gamma_\perp \left[\tilde{d}_{0,s}\,\delta \tilde{f}_n + x\,\delta \tilde{d}_n - (1 + i\Delta)\delta \tilde{p}_n \right], \qquad (20.12)$$

$$\frac{d\delta \tilde{p}^*_{-n}}{d\tau} = i\alpha_n \delta \tilde{p}^*_{-n} + \gamma_\perp \left[\tilde{d}_{0,s}\,\delta \tilde{f}^*_{-n} + x^*\,\delta \tilde{d}_n - (1 - i\Delta)\delta \tilde{p}^*_{-n} \right], \qquad (20.13)$$

$$\frac{d\delta \tilde{d}_n}{d\tau} = (i\alpha_n - \gamma_\parallel)\delta \tilde{d}_n - \frac{\gamma_\parallel}{2} \left(\tilde{p}^*_{0,s}\,\delta \tilde{f}_n + x\,\delta \tilde{p}^*_{-n} + \tilde{p}_{0,s}\,\delta \tilde{f}^*_{-n} + x^*\,\delta \tilde{p}_n \right). \quad (20.14)$$

We have written Eqs. (20.11) and (20.13) for the deviations $\delta \tilde{f}^*_{-n}$ and $\delta \tilde{p}^*_{-n}$ instead of $\delta \tilde{f}^*_n$ and $\delta \tilde{p}^*_n$ in order to point out the remarkable fact that the infinite set of linearized equations splits into an infinite set of decoupled blocks, each containing five coupled linear equations.

In particular, the block for $n = 0$ reads

$$\frac{d\delta\tilde{f}_0}{d\tau} = -\kappa \left[(1 + i\theta)\delta\tilde{f}_0 - A\,\delta\tilde{p}_0\right], \tag{20.15}$$

$$\frac{d\delta\tilde{f}_0^*}{d\tau} = -\kappa \left[(1 - i\theta)\delta\tilde{f}_0^* - A\,\delta\tilde{p}_0^*\right], \tag{20.16}$$

$$\frac{d\delta\tilde{p}_0}{d\tau} = \gamma_\perp \left[\tilde{d}_{0,s}\,\delta\tilde{f}_0 + x\,\delta\tilde{d}_0 - (1 + i\Delta)\delta\tilde{p}_0\right], \tag{20.17}$$

$$\frac{d\delta\tilde{p}_0^*}{d\tau} = \gamma_\perp \left[\tilde{d}_{0,s}\,\delta\tilde{f}_0^* + x^*\,\delta\tilde{d}_0 - (1 + i\Delta)\delta\tilde{p}_0^*\right], \tag{20.18}$$

$$\frac{d\delta\tilde{d}_0}{d\tau} = -\gamma_\parallel \left[\frac{1}{2}\left(\tilde{p}_{0,s}^*\,\delta\tilde{f}_0 + x\,\delta\tilde{p}_0^* + \tilde{p}_{0,s}\,\delta\tilde{f}_0^* + x^*\,\delta\tilde{p}_0\right) + \delta\tilde{d}_0\right], \tag{20.19}$$

and corresponds to the set of linearized equations for the single-mode model (12.43)–(12.45).

Next, we apply the exponential ansatz (18.5) to the five linearized equations (20.10)–(20.14), where δx_i ($i = 1, 2, \ldots, 5$) denotes the five deviations, and the exponential factor $e^{\lambda t}$ is *not* complex conjugated in $\delta\tilde{f}_{-n}^*$ and $\delta\tilde{p}_{-n}^*$. In this way we arrive at the fifth-degree characteristic equation (compare this with Eq. (18.8))

$$\sum_{k=0}^{5} c_k^{(n)}(\{\beta\}, \alpha_n)\tilde{\lambda}^k = 0, \quad \tilde{\lambda} = \frac{\lambda}{\gamma_\perp}, \tag{20.20}$$

where $\{\beta\}$ denotes the parameters of the system. In the case of the free-running laser, we will use X instead of A as a parameter, so the parameters are X, Δ, $\tilde{\kappa} = \kappa/\gamma_\perp$ and $\tilde{\gamma} = \gamma_\parallel/\gamma_\perp$. In the case of optical bistability and a laser with an injected signal it is convenient to use X as a parameter instead of the normalized input intensity Y which appears in Eq. (13.1); hence the parameters are X, A ($2C$ in the case of optical bistability), Δ, θ, $\tilde{\kappa}$ and $\tilde{\gamma}$.

The explicit form of the coefficients $c_k^{(n)}$ is given in Appendix C in Eqs. (C.1)–(C.6). Note that they depend on X, the modulus squared of $f_{0,s}$, but not on the phase of $f_{0,s}$. The stationary solution is stable if and only if each of the five roots of Eq. (20.20) has a non-positive real part, for all values of $n = 0, \pm 1, \ldots$. If one of the five roots becomes positive for $n = 0$, while every other root for $n \neq 0$ has a negative real part, the instability is said to be of the *single-mode* type, because it affects only the resonant mode ω_c, which appears in the single-mode model (12.43)–(12.45). If, instead, only modes with $n \neq 0$ become unstable, the instability is said to be of the *multimode* type.

The coefficients $c_k^{(0)}$ of the characteristic equation which governs single-mode instabilities are given by Eqs. (C.1)–(C.6) in Appendix C with $\tilde{\alpha}_n = \alpha_n/\gamma_\perp = 0$. In this connection, the following remarks concerning the constant coefficient $c_0^{(0)}$ of the characteristic equation are important.

(a) In the cases of a laser with an injected signal and optical bistability, by using Eq. (13.1) (with A replaced by $-2C$ in the case of optical bistability) one has that

$$c_0^{(0)} = \tilde{\kappa}^2\tilde{\gamma}(1 + \Delta^2 + X)\frac{dY}{dX}. \tag{20.21}$$

If we take into account the stability condition $c_0^{(0)} \geq 0$, we see that Eq. (20.21) provides a general demonstration of the fact that the segments of the stationary curve $X(Y)$ with negative slope are unstable.

(b) In the case of the free-running laser, by setting $\theta = -\Delta$ and $X = A - 1 - \Delta^2$, one can easily verify that $c_0^{(0)} = 0$, which implies that one of the roots of the characteristic equation is equal to zero. The existence of a zero eigenvalue arises from the fact that the phase of the electric field is arbitrary in the absence of a driving external field to impose the phase. If one starts with any value of the phase and then displaces the phase from this value, there is no mechanism to restore the previous value. We note that also in the laser models considered in Chapter 19 the phase leads to a zero root of the eigenvalue equation. As a matter of fact, if one linearizes also the phase equation (19.6) or (19.29) one obtains an additional root in the characteristic equation with respect to those considered in Chapter 19, and this root is equal to zero.

It should be obvious that the emergence of multimode instabilities requires the atoms to interact with more cavity modes than just the resonant mode at frequency ω_c, and this in turn requires that the atomic gain or absorption profile must be sufficiently wide. For low values of the cavity field this statement implies the inequality

$$\gamma_\perp \gtrsim 2\pi \frac{c}{\Lambda} , \tag{20.22}$$

and, more generally,

$$\gamma_\perp \sqrt{1 + X} \gtrsim 2\pi \frac{c}{\Lambda} , \tag{20.23}$$

if we take into account power broadening. Therefore, multimode instabilities may develop only in the limit of broad atomic lines or when the roundtrip length of the cavity is sufficiently large. Note that condition (20.22) is just the opposite of the condition (9.23) which defines the single-mode limit. It is easy to verify that Eq. (20.22) implies the good-cavity condition $\kappa \ll \gamma_\perp$ because in the low-transmission limit we know that $T \ll 1$. As a matter of fact we have

$$\tilde{\kappa} = \frac{\kappa}{\gamma_\perp} = \frac{T}{2\pi} \frac{2\pi c}{\Lambda \gamma_\perp} \ll 1. \tag{20.24}$$

If, in addition, $\gamma_\parallel$ is comparable in magnitude to $\gamma_\perp$, than we also have that $\kappa \ll \gamma_\parallel$, so the good-cavity condition (19.1) is satisfied and one can eliminate the atomic variables adiabatically. In this chapter, in the treatment of multimode instabilities, we assume that $\kappa \ll \gamma_\perp$, while the case $\gamma_\parallel \ll \gamma_\perp$ will be discussed in Chapter 22.

From what we have said, single-mode instabilities are governed by the characteristic equation (20.20) for $n = 0$, whereas multimode instabilities are governed by Eq. (20.20) for $n \neq 0$. On this basis, one would expect that single-mode and multimode instabilities are only loosely connected. In the following sections we demonstrate, instead, that they are intrinsically connected by general and far-reaching links.

In the case of single-mode instabilities giving rise to spontaneous oscillations, the field intensity oscillates in the same way throughout the cavity because in the single-mode model the field is uniform along the cavity. Multimode instabilities, instead, always correspond to

Hopf instabilities and favor the formation of pulses that propagate along the cavity. In order to understand how this phenomenon develops, we consider the simplest case of multimode oscillatory behavior in which the amplitudes $\tilde{f}_n$ approach stationary values for sufficiently long times. Under these conditions Eq. (20.1) yields

$$F(z, \tau) = \sum_n e^{ik_n z} e^{-i\alpha_n \tau} \tilde{f}_n(\infty), \qquad (20.25)$$

where $\tilde{f}_n(\infty)$ are the asymptotic values of $\tilde{f}_n(\tau)$. If we now transform back to the original space and time variables z and t, then, with the help of Eqs. (12.1), (12.17), (8.10) and (8.12), we obtain, along the atomic sample, the pattern

$$F(z, t) = \sum_n e^{-i\alpha_n(t - z n_B/c)} \tilde{f}_n(\infty). \qquad (20.26)$$

This result describes the appearance of a superposition of waves traveling along the atomic sample and, therefore, it has the structure of a pulse. Obviously, the output pulsations are periodic with a period equal to the cavity roundtrip time Λ/c. Note that this is no longer true when the modal amplitudes do not approach stationary values; in this case one can find, for example, quasi-periodic or chaotic behaviour.

20.2 Multimode instabilities and their features

Since $\kappa \ll \alpha_n$ for $T \ll 1$, in the good-cavity limit we can eliminate the atomic variables adiabatically by setting the time derivatives in Eqs. (20.12)–(20.14) equal to zero. The resulting algebraic equations for the atomic-fluctuation variables can be solved with lengthy but simple manipulations involving the use of Eqs. (20.6) and (20.7). The result of this calculation is the following closed set of two linear differential equations for $\delta \tilde{f}_n$ and $\delta \tilde{f}_{-n}^*$ [228, 250] (see Appendix D for a sketch of the derivation and the definition of the relevant symbols):

$$\frac{d\delta \tilde{f}_n}{d\tau} = -\tilde{\kappa}(1 + i\theta)\delta \tilde{f}_n$$
$$+ \tilde{\kappa} A \left[T_1(\tilde{\alpha}_n, X, \Delta, \tilde{\gamma})\delta \tilde{f}_n + T_2(\tilde{\alpha}_n, X, \Delta, \tilde{\gamma})x^2 \delta \tilde{f}_{-n}^* \right], \qquad (20.27)$$

$$\frac{d\delta \tilde{f}_{-n}^*}{d\tau} = -\tilde{\kappa}(1 - i\theta)\delta \tilde{f}_{-n}^*$$
$$+ \tilde{\kappa} A \left[T_2(\tilde{\alpha}_n, X, -\Delta, \tilde{\gamma})x^{*2} \delta \tilde{f}_n + T_1(\tilde{\alpha}_n, X, -\Delta, \tilde{\gamma})\delta \tilde{f}_{-n}^* \right], \qquad (20.28)$$

where we have set $\tilde{\alpha}_n = \alpha_n/\gamma_\perp$. The functions T_1 and T_2 are defined in Appendix D. Next we introduce the ansatz

$$\begin{pmatrix} \delta \tilde{f}_n(\tau) \\ \delta \tilde{f}_n^*(\tau) \end{pmatrix} = e^{\lambda \tau} \begin{pmatrix} \tilde{f}_n^{(0)} \\ \tilde{f}_{-n}^{(0)} \end{pmatrix}, \qquad (20.29)$$

and derive the characteristic equation (with $\tilde{\lambda} = \lambda/\gamma_\perp$)

$$\left[\tilde{\lambda}/\tilde{\kappa} + 1 + i\theta - AT_1(\Delta)\right]\left[\tilde{\lambda}/\tilde{\kappa} + 1 - i\theta - AT_1(-\Delta)\right] = A^2 T_2(\Delta)T_2(-\Delta)X^2. \quad (20.30)$$

We now introduce the auxiliary functions

$$T' = [T_1(\Delta) + T_1(-\Delta)]/2, \qquad T'' = [T_1(\Delta) - T_1(-\Delta)]/(2i), \quad (20.31)$$

so that the characteristic equation takes the form

$$\left(\tilde{\lambda}/\tilde{\kappa} + 1 - AT'\right)^2 = A^2 T_2(\Delta)T_2(-\Delta)X^2 - \left(\theta - AT''\right)^2. \quad (20.32)$$

We now let

$$\mathcal{F}(\tilde{\alpha}_n, X, \Delta, \tilde{\gamma}, A, \theta) = A^2 T_2(\Delta)T_2(-\Delta)X^2 - \left(\theta - AT''\right)^2, \quad (20.33)$$

and obtain

$$\tilde{\lambda}_\pm/\tilde{\kappa} = -1 + AT'(\tilde{\alpha}_n, X, \Delta, \tilde{\gamma}) \pm \mathcal{F}^{1/2}(\tilde{\alpha}_n, X, \Delta, \tilde{\gamma}, A, \theta). \quad (20.34)$$

The stationary state is stable if and only if the real parts both of $\tilde{\lambda}_+$ and of $\tilde{\lambda}_-$ are negative for all $n = 0, \pm1, \ldots$. In terms of the two functions $\mathcal{G}_\pm$, defined by [228, 250]

$$\mathcal{G}_\pm(\tilde{\alpha}_n, X, \Delta, \tilde{\gamma}, A, \theta) = \mathrm{Re}\left[AT'(\tilde{\alpha}_n, X, \Delta, \tilde{\gamma}) \pm \mathcal{F}^{1/2}(\tilde{\alpha}_n, X, \Delta, \tilde{\gamma}, A, \theta)\right], \quad (20.35)$$

the multimode instability condition takes the form

$$\mathcal{G}_+(\tilde{\alpha}_n, X, \Delta, \tilde{\gamma}, A, \theta) > 1 \quad (20.36)$$

or

$$\mathcal{G}_-(\tilde{\alpha}_n, X, \Delta, \tilde{\gamma}, A, \theta) > 1 \quad (20.37)$$

for at least one value of $n \neq 0$ or, more precisely, for at least one pair of values $\pm n$, because both $\mathcal{G}_+$ and $\mathcal{G}_-$ are even functions of $\tilde{\alpha}_n$. Equations (20.36) and (20.37) can be interpreted as "gain-larger-than-loss" conditions for the cavity modes.

After fixing the independent parameters of the system, the stability analysis of the stationary solution runs as follows. One plots $\mathcal{G}_+$ and $\mathcal{G}_-$ as functions of $\tilde{\alpha}_n$, considered as a continuous variable, and selects the intervals, if any, where $\mathcal{G}_+(\tilde{\alpha}_n)$ or $\mathcal{G}_-(\tilde{\alpha}_n)$ is greater than unity. The stationary state is unstable if at least one of the discrete values of $\tilde{\alpha}_n = 2n\pi c/(\Lambda\gamma_\perp)$ falls within the selected intervals. This condition can always be satisfied by varying the cavity length (provided that either $\mathcal{G}_+$ or $\mathcal{G}_-$ is greater than unity for some range of $\tilde{\alpha}_n$ values). This is the reason why the existence of a multimode instability is usually linked to the existence of intervals of the $\tilde{\alpha}_n$ axis such that $\mathcal{G}_+(\tilde{\alpha}_n) > 1$ or $\mathcal{G}_-(\tilde{\alpha}_n) > 1$; we will continue to follow this criterion. If we now focus our attention on the plane of the control parameters $\tilde{\alpha}_n$ and X and, for each value of X, identify the intervals where $\mathcal{G}_+(\tilde{\alpha}_n) > 1$ or $\mathcal{G}_-(\tilde{\alpha}_n) > 1$, the resulting region of the plane is the "instability domain" corresponding to a given selection of the remaining parameters. In view of the symmetry of this domain with respect to the transformation $\tilde{\alpha}_n \to -\tilde{\alpha}_n$, we can limit our considerations to the half-plane $\tilde{\alpha}_n > 0$.

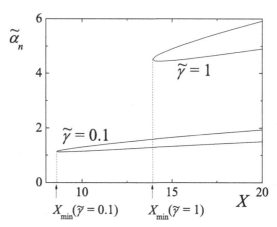

Figure 20.1 A free-running laser on resonance ($\Delta = 0$) for $\gamma_\parallel = \gamma_\perp$ and $\gamma_\parallel = 0.1\gamma_\perp$. The instability domain for the multimode instability is constituted by the "tongues" delimited by $\tilde{\alpha}_{\max}$ and $\tilde{\alpha}_{\min}$.

For the sake of simplicity, in the following we assume that there is a single instability domain for a given value of X, when it exists, characterized by upper and lower bounds $\tilde{\alpha}_{\max}(X) > 0$ and $\tilde{\alpha}_{\min}(X) > 0$. Let us now consider the example of the free-running laser on resonance ($\Delta = \theta = 0$), in which the instability domain can be calculated analytically. The expression for the function $\mathcal{G}_+(\tilde{\alpha}_n)$ is

$$\mathcal{G}_+(\tilde{\alpha}_n) = \frac{\tilde{\alpha}_n^2(1 + \tilde{\gamma}X) + \tilde{\gamma}^2(1 - X^2)}{\tilde{\alpha}_n^4 + (1 - 2\tilde{\gamma}X + \tilde{\gamma}^2)\tilde{\alpha}_n^2 + \tilde{\gamma}^2(1 + X)^2} , \qquad (20.38)$$

where we have used Eqs. (20.35), (20.31), (20.33), (D.11) and (D.12), and we have taken into account that for $\Delta = 0$ one has $A = X + 1$ (see Eq. (9.24) with $|x_{\bar{j}}|^2 = X$). The positive solutions of the equation $\mathcal{G}_+(\tilde{\alpha}_n) = 1$ are

$$\tilde{\alpha}_{\substack{\max \\ \min}} = \sqrt{\frac{\tilde{\gamma}}{2}} \left[3X - \tilde{\gamma} \pm \sqrt{X^2 - 2(4 + 3\tilde{\gamma})X + \tilde{\gamma}^2}\right]^{1/2} , \qquad (20.39)$$

which are real for $X > X_{\min} = 4 + 3\tilde{\gamma} + 2\sqrt{2(2 + 3\tilde{\gamma} + \tilde{\gamma}^2)}$. These functions are plotted in Fig. 20.1 for $\tilde{\gamma} = 1$ and for $\tilde{\gamma} = 0.1$. On the other hand, $\mathcal{G}_-(\tilde{\alpha}_n) = 1/(1 + \tilde{\alpha}_n^2)$ turns out to be always smaller than unity in this case and therefore does not give rise to any instability.

It is evident from Eq. (20.39) that the boundary values $\tilde{\alpha}_{\max}$ and $\tilde{\alpha}_{\min}$ depend, in general, not only on the variable X but also on the other parameters of the system.

Let us finally illustrate the intrinsic connection between multimode instabilities and four-wave mixing, which has been exploited in several papers (see e.g. [245–247]). This connection arises from the circumstance that the growth of two sidemodes, which are symmetrical with respect to the resonant mode, can be interpreted as a four-wave mixing process whereby two pump photons of frequency ω_0 convert into two photons with frequencies $\omega_0 + \delta$ and $\omega_0 - \delta$, where, in our case, δ is the frequency separation between the unstable mode and the resonance frequency.

A fully formal way of materializing the connection is to note the analogy of Eqs. (20.27) and (20.28), which govern multimode instabilities, and Eqs. (7.36) and (7.37), which describe cross-phase modulation and four-wave mixing. In Eqs. (20.27) and (20.28) the quantity T_2 plays the role of the four-wave-mixing coupling coefficient, whereas T_1 corresponds not only to phase modulation but also to gain/absorption, because of its real part. The coefficient $T_1(\alpha_n)$ governs the gain experienced by each mode, while the coefficient $T_2(\alpha_n)$ generates the necessary coupling between modes n and $-n$ which alters the gain experienced by each mode. The final expression for the effective gain, which depends not only upon the atomic parameters but also on the cavity detuning, is given by Eq. (20.35).

20.3 Single-mode instabilities and their features

Let us now turn our attention to the single-mode instabilities [228, 250] and let us consider the instability boundary where $\lambda = -i\omega$ with ω real, i.e. $\text{Re}\,\lambda = 0$. If in the linearized Eqs. (20.15)–(20.19) we assume, as usual, that the fluctuations $\delta f_0^{(0)}$, $\delta f_0^{(0)*}$, $\delta \tilde{p}_0^{(0)}$, $\delta \tilde{p}_0^{(0)*}$ and $\delta \tilde{d}_0^{(0)}$ are proportional to $e^{\lambda t}$, and set $\lambda = -i\omega$, we see that the atomic equations coincide with the multimodal linearized equations (20.10)–(20.14) in which one eliminates the atomic variables adiabatically, if one replaces $\tilde{\alpha}_n$ by ω. Therefore, by following the same procedure as in the previous section, we obtain the following two linear homogeneous equations for $\delta f_0^{(0)}$ and $\delta f_0^{(0)*}$:

$$-i\frac{\tilde{\omega}}{\tilde{\kappa}}\delta f_0^{(0)} = [AT_1(\tilde{\omega}, X, \Delta, \tilde{\gamma}) - 1 - i\theta]\delta f_0^{(0)} + AT_2(\tilde{\omega}, X, \Delta, \tilde{\gamma})x^2\,\delta f_0^{(0)*}, \quad (20.40)$$

$$-i\frac{\tilde{\omega}}{\tilde{\kappa}}\delta f_0^{(0)*} = [AT_1(\tilde{\omega}, X, -\Delta, \tilde{\gamma}) - 1 + i\theta]\delta f_0^{(0)*} + AT_2(\tilde{\omega}, X, -\Delta, \tilde{\gamma})x^{*2}\,\delta f_0^{(0)},$$

$$\cdot\,(20.41)$$

where $\tilde{\omega}$ denotes the scaled frequency $\omega/\gamma_\perp$. If we now derive the characteristic equation associated with Eqs. (20.40) and (20.41) by following the same procedure as in the multimode case, we can easily arrive at the following result:

$$-i\frac{\tilde{\omega}}{\tilde{\kappa}} = -1 + AT'(\tilde{\omega}, X, \Delta, \tilde{\gamma}) \pm \mathcal{F}^{1/2}(\tilde{\omega}, X, \Delta, \tilde{\gamma}, A, \theta), \quad (20.42)$$

whose real and imaginary parts are given by equations of the type [228, 250]

$$1 = \mathcal{G}_\pm(\tilde{\omega}, X, \Delta, \tilde{\gamma}, A, \theta), \quad (20.43)$$

$$-\frac{\tilde{\omega}}{\tilde{\kappa}} = \mathcal{D}_\pm(\tilde{\omega}, X, \Delta, \tilde{\gamma}, A, \theta). \quad (20.44)$$

The functions $\mathcal{G}_\pm$ are defined as in Eq. (20.35) with the simple replacement of $\tilde{\alpha}_n$ by $\tilde{\omega}$, and $\mathcal{D}_\pm$ are given by

$$\mathcal{D}_\pm(\tilde{\omega}, X, \Delta, \tilde{\gamma}, A, \theta) = \text{Im}[AT'(\tilde{\omega}, X, \Delta, \tilde{\gamma}) \pm \mathcal{F}^{1/2}(\tilde{\omega}, X, \Delta, \tilde{\gamma}, A, \theta)]. \quad (20.45)$$

The plus and minus signs must be selected concurrently in Eqs. (20.43) and (20.44). The functions $\mathcal{D}_\pm$ are dispersion functions and are odd in the variable $\tilde{\omega}$.

Since the parameter $\tilde{\kappa}$ does not appear in the functions $\mathcal{G}$ and $\mathcal{D}$, it is convenient to discuss single-mode instabilities in the plane of the parameters X and $\tilde{\kappa}$, assuming that all the other parameters are kept fixed. The stability boundary in the plane of the variables X and $\tilde{\kappa}$ can be obtained by eliminating $\tilde{\omega}$ in Eqs. (20.43) and (20.44). Since $\mathcal{D}_\pm$ are odd functions, Eq. (20.44) is always satisfied if we set $\tilde{\omega} = 0$ (which means $\lambda = 0$), but we focus on the boundary for oscillatory instabilities for which $\tilde{\omega}$ is different from zero.

Let us consider first Eq. (20.43), which is satisfied by $\tilde{\omega} = \tilde{\alpha}_{\max}(X)$ and $\tilde{\omega} = \tilde{\alpha}_{\min}(X)$ (see the previous section, comparing Eq. (20.43) with Eqs. (20.36) and (20.37)). If we replace one of these two values of $\tilde{\omega}$ in Eq. (20.44) and solve this equation with respect to $\tilde{\kappa}$, we obtain

$$
\tilde{\kappa}_{\substack{\max \\ \min}} = -\frac{\tilde{\alpha}_{\substack{\max \\ \min}}(X)}{\mathcal{D}_+\left(\tilde{\alpha}_{\substack{\max \\ \min}}(X), X\right)},
\tag{20.46}
$$

where we have omitted the indication in $\mathcal{D}_+$ of the parameters different from X, because they are held fixed. In addition to Eq. (20.46), we have an identical equation with $\mathcal{D}_+$ replaced by $\mathcal{D}_-$.

The boundary of the single-mode instability in the plane of the variables X and $\tilde{\kappa}$ is formed by the two lines $\tilde{\kappa}_{\max}(X)$ and $\tilde{\kappa}_{\min}(X)$. In the remainder of this chapter we assume for definiteness that the lines $\alpha_{\max}(X)$ and $\alpha_{\min}(X)$ exist only for the $+$ choice in $\mathcal{G}$, meaning that we disregard the $-$ choice. The essential condition for the existence of the single-mode instability is that the functions $\mathcal{D}_+(\alpha_{\max}(X), X)$ and $\mathcal{D}_+(\alpha_{\min}(X), X)$ are negative, because the parameter $\tilde{\kappa}$ is positive. When this condition is satisfied, there is indeed a single-mode instability of the oscillatory type.

In Section 22.6 we will consider the case that the lines $\tilde{\alpha}_{\max}(X)$ and $\tilde{\alpha}_{\min}(X)$ exist only for the $-$ choice in $\mathcal{G}$.

Considering again the case of the free-running laser with $\Delta = \theta = 0$, the expression for the function $\mathcal{D}_+$ is

$$
\mathcal{D}_+(\omega) = \tilde{\omega}\frac{\tilde{\omega}^2 - 2\tilde{\gamma}X + \tilde{\gamma}^2(1 - X)}{\tilde{\omega}^4 + (1 - 2\tilde{\gamma}X + \tilde{\gamma}^2)\tilde{\omega}^2 + \tilde{\gamma}^2(1 + X)^2}.
\tag{20.47}
$$

By using Eqs. (20.46), (20.39) and (20.47) we obtain

$$
\tilde{\kappa}_{\substack{\max \\ \min}} = \frac{1}{2}\left[X - 2 - \tilde{\gamma} \pm \sqrt{X^2 - 2(4 + 3\tilde{\gamma})X + \tilde{\gamma}^2}\right].
\tag{20.48}
$$

If we insert the above expressions into the identity $(\tilde{\kappa} - \tilde{\kappa}_{\max})(\tilde{\kappa} - \tilde{\kappa}_{\min}) = 0$ we obtain a simple expression for the instability boundary in terms of the intensity X,

$$
X = \frac{(\tilde{\kappa} + 1)(\tilde{\kappa} + \tilde{\gamma} + 1)}{\tilde{\kappa} - \tilde{\gamma} - 1},
\tag{20.49}
$$

which just represents the inverse of the function $\tilde{\kappa}(X)$ defined by Eq. (20.48). The instability boundary in the $(\tilde{\kappa}, X)$ plane is shown in Fig. 20.2 for $\tilde{\gamma} = 1$ and $\tilde{\gamma} = 0.1$.

We have seen that the stability boundary is defined by Eq. (20.46) in which we consider the boundary in the $(\tilde{\kappa}, X)$ plane by keeping the other parameters of the system fixed. Needless to say, it is possible to consider the stability boundary in the plane of two arbitrary parameters of the system by keeping the remaining parameters fixed.

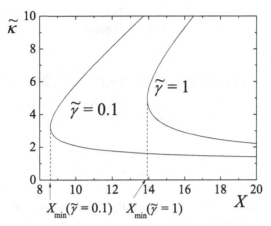

Figure 20.2 A free-running laser on resonance ($\Delta = 0$) for $\gamma_\parallel = \gamma_\perp$ and $\gamma_\parallel = 0.1\gamma_\perp$. The single-mode stationary solution is unstable to the right of the curves.

We observe that, in the cases in which the stability boundary of the single-mode instability can be calculated analytically, as e.g. for the free-running laser on resonance that we have just considered, the most straightforward procedure to calculate this boundary is not the one described in this chapter, but the one described in Section 18.1. On the other hand, the method outlined in Section 20.3 has the virtue of disclosing the far-reaching connections between single-mode and multimode instabilities. An additional advantage is that it provides a simple and general numerical procedure to obtain the stability boundary, which applies also when the field equation is not a differential but an integro-differential equation, so that the eigenvalue equation is not algebraic, as it is in the case of inhomogeneous broadening considered in Chapter 15 or in the case of a Fabry–Perot cavity considered in Sections 14.2 and 14.3.

We conclude this section with two remarks that compare the treatment of single-mode instabilities given in this book with that followed in the literature.

The first remark is that Eqs. (20.43) and (20.44) are formally identical to the consistency relations which are assumed in the weak-sideband approach [245–249], and express the balance between gain and losses for the sidebands on the one hand, and the condition that the phase delay of the sideband signal over a cavity roundtrip is a multiple of 2π on the other. In our approach, Eqs. (20.43) and (20.44) follow naturally from the linear-stability analysis on focussing on the stability boundary.

The second remark is that, both in the weak-sideband approach and in [250], one fixes all the parametric values but one, say X, and one solves first (in general numerically) Eq. (20.44) with respect to $\tilde{\omega}$, thus finding a function $\tilde{\omega}(X)$. Next, one inserts this function into Eq. (20.43) and solves, again numerically in general, Eq. (20.43) with respect to X, thereby finding the boundary expressed as X as a function of the remaining parameters. In the approach of this section, instead one solves first (in general numerically) Eq. (20.43) with respect to $\tilde{\omega}$, thus finding $\tilde{\omega}$ as a function of the parameters of the system except $\tilde{\kappa}$, which does not appear in Eq. (20.43). In a second step, one inserts this function into

Eq. (20.44) and, without any further calculation, obtains the stability boundary expressed as $\tilde{\kappa}$ as a function of the remaining parameters.

20.4 The general connection between single-mode and multimode instabilities

The key point exhibited by the foregoing analysis [228, 250] is that both single-mode and multimode instabilities are governed by the same gain function $\mathcal{G}_+$. In fact, the only differences between Eqs. (20.36) and (20.37) and Eq. (20.43) are the replacement of $\tilde{\alpha}_n$ by $\tilde{\omega}$ and of the inequality by the equality sign. Thus, we can conclude at once that the existence of a single-mode instability implies the existence of a multimode instability. Consider, in fact, a set of fixed values of the parameters; if we can find a value $\tilde{\omega}$ corresponding to this selection such that, for example, $\mathcal{G}_+(\tilde{\omega}) = 1$, then, by continuity, it must be true that $\mathcal{G}_+(\tilde{\alpha}_n) \geq 1$ either on the left or in the right neighborhood of $\tilde{\omega}$. Note, however, that the existence of a multimode instability does not necessarily imply the existence of a single-mode instability because the latter also requires that the condition (20.44) be satisfied. This in turn requires that $\mathcal{D}_+(\tilde{\omega}, X)$ is negative when $\tilde{\omega} = \tilde{\alpha}_{\max}(X)$ and $\tilde{\omega} = \tilde{\alpha}_{\min}(X)$.

This analysis immediately clarifies two significant points. First, it explains why single-mode and multimode instabilities display a particular frequency range in the framework of each particular model. Second, the procedure described in the previous two sections, by which we construct the instability domain for the multimode instability on the one hand, and the boundary of the single-mode instability on the other, ensures that the instability threshold $X_{\min}$ in the variable X for the corresponding single-mode and multimode instabilities is the same, as is manifest also in Figs. 20.1 and 20.2.

Note also that, while multimode instabilities arise in the good-cavity limit, singlemode instabilities arise when κ is on the order of or even larger than $\gamma_\perp$, as happens, for example, in the case of a free-running laser (see Fig. 20.2).

It is also important to observe that the connection between single-mode and multimode instabilities holds also beyond the framework of the Maxwell–Bloch equations considered in Chapter 12, for example

- when the atomic line is inhomogeneously broadened and
- in the case of a laser with a saturable absorber.

20.5 The resonant case, amplitude and phase instabilities

A considerable simplification follows in the resonant case $\omega_0 = \omega_a = \omega_c$, i.e. $\Delta = \theta = 0$. For the driven systems (optical bistability and a laser with an injected signal) the stationary values of F, P, f_0 and p_0 are real and positive if, as we have already assumed, the input field y is real and positive. In the case of the free-running laser the same happens on appropriately choosing the arbitrary phase of the laser field. In turn this implies that amplitude and phase fluctuations around the stationary state basically coincide with the fluctuations of the real and imaginary parts, respectively. In fact, if we let for example

$$F(z, t) = F'(z, t) + i F''(z, t) = \rho(z, t)\exp[i\phi(z, t)], \qquad (20.50)$$

and we denote by $\delta F'(z, t)$, $\delta F''(z, t)$, $\delta\rho$ and $\delta\phi$ the displacements of the corresponding variables from their stationary values, we can immediately obtain the following relations:

$$\delta F'(z, t) = \delta\rho(z, t), \qquad (20.51)$$

$$\delta F''(z, t) = \rho_s(z)\delta\phi(z, t), \qquad (20.52)$$

where ρ_s denotes the stationary value of ρ and we have taken into account that $\phi_s = 0$. In addition, the linearized equations (20.10)–(20.14) assume a block-diagonal form if we introduce the variables

$$\delta\tilde{f}_n' = \frac{\delta\tilde{f}_n + \delta\tilde{f}_{-n}^*}{2}, \qquad \delta\tilde{p}_n' = \frac{\delta\tilde{p}_n + \delta\tilde{p}_{-n}^*}{2}, \qquad \delta\tilde{d}_n' = \delta\tilde{d}_n,$$

$$\delta\tilde{f}_n'' = \frac{\delta\tilde{f}_n - \delta\tilde{f}_{-n}^*}{2i}, \qquad \delta\tilde{p}_n'' = \frac{\delta\tilde{p}_n - \delta\tilde{p}_{-n}^*}{2i}, \qquad (20.53)$$

so that, using Eq. (20.1), one has

$$\delta F'(z, \tau) = \sum_n e^{ik_n z} e^{-i\alpha_n \tau}\, \delta\tilde{f}_n'(\tau),$$

$$\delta F''(z, \tau) = \sum_n e^{ik_n z} e^{-i\alpha_n \tau}\, \delta\tilde{f}_n''(\tau), \qquad (20.54)$$

and similar expressions hold for $\delta P'$ and $\delta P''$. In fact, if $\Delta = \theta = 0$, we obtain

$$\frac{d\delta\tilde{f}_n'}{d\tau} = -\kappa\left(\delta\tilde{f}_n' - A\,\delta\tilde{p}_n'\right), \qquad (20.55)$$

$$\frac{d\delta\tilde{p}_n'}{d\tau} = i\alpha_n\,\delta\tilde{p}_n' + \gamma_\perp\left(\tilde{d}_{0,s}\,\delta\tilde{f}_n' + x\,\delta\tilde{d}_n' - \delta\tilde{p}_n'\right), \qquad (20.56)$$

$$\frac{d\delta\tilde{d}_n'}{d\tau} = i\alpha_n\,\delta\tilde{d}_n' - \gamma_\parallel\left(\tilde{p}_{0,s}\,\delta\tilde{f}_n' + x\,\delta\tilde{p}_n' + \delta\tilde{d}_n'\right), \qquad (20.57)$$

$$\frac{d\delta\tilde{f}_n''}{d\tau} = -\kappa\left(\delta\tilde{f}_n'' - A\,\delta\tilde{p}_n''\right), \qquad (20.58)$$

$$\frac{d\delta\tilde{p}_n''}{d\tau} = i\alpha_n\,\delta\tilde{p}_n'' + \gamma_\perp\left(\tilde{d}_{0,s}\,\delta\tilde{f}_n'' - \delta\tilde{p}_n''\right), \qquad (20.59)$$

where $\tilde{d}_{0,s}$ and $\tilde{p}_{0,s}$ are defined by Eq. (20.7).

It is obvious that Eqs. (20.55)–(20.57) form a closed set for the variables $\delta\tilde{f}_n'$, $\delta\tilde{p}_n'$ and $\delta\tilde{d}_n'$, just as Eqs. (20.58) and (20.59) do with respect to $\delta\tilde{f}_n''$ and $\delta\tilde{p}_n''$. The first three equations govern the fluctuations of the real parts of F, P and D, i.e. the amplitude fluctuations, while the last two govern the phase fluctuations. Taken together, the two sets of equations yield a pair of characteristic equations, a cubic and a quadratic, respectively, with the forms

$$\sum_{k=0}^{3} a_k^{(n)} \lambda^k = 0, \tag{20.60}$$

$$\sum_{k=0}^{2} b_k^{(n)} \lambda^k = 0. \tag{20.61}$$

The explicit expressions for the coefficients $a_k^{(n)}$ and $b_k^{(n)}$ are given in Appendix E. For obvious reasons of nomenclature, the instabilities that emerge from solutions of Eq. (20.60) are called amplitude instabilities, while the others are called phase instabilities. In the case of multimode instabilities under resonance conditions Eqs. (20.27) and (20.28) become

$$\frac{d\delta \tilde{f}_n'}{d\tau} = -\tilde{\kappa}[1 - A(T_1 + T_2 X)]\delta \tilde{f}_n', \tag{20.62}$$

$$\frac{d\delta \tilde{f}_n''}{d\tau} = -\tilde{\kappa}[1 - A(T_1 - T_2 X)]\delta \tilde{f}_n''. \tag{20.63}$$

Therefore the gain and dispersion functions which appear in Eqs. (20.36), (20.37), (20.43) and (20.44) are given by

$$\mathcal{G}_{\pm}(\tilde{\omega}) = A \operatorname{Re}(T_1 \pm T_2 X), \tag{20.64}$$

$$\mathcal{D}_{\pm}(\tilde{\omega}) = A \operatorname{Im}(T_1 \pm T_2 X), \tag{20.65}$$

which is in agreement with Eqs. (20.35) and (20.45), since we can immediately verify that for $\theta = 0$ and $\Delta = 0$ we have $T' = T_1$, $T'' = 0$ (see Eq. (20.31)) and $\mathcal{F} = (AT_2 X)^2$ (see Eq. (20.33)). Therefore amplitude instabilities are related to $\mathcal{G}_+$ and $\mathcal{D}_+$ and phase instabilities are related to $\mathcal{G}_-$ and $\mathcal{D}_-$.

It is now important to signal that in the following we will use the term *phase instabilities* to describe also another kind of multimode instability that arises in the framework of free-running lasers under nonresonant conditions ($\Delta \neq 0$). In order to understand this point let us recall first of all that in the case of the free-running laser the characteristic equation (20.20) for $n = 0$ has always a solution $\lambda = 0$ as a consequence of the phase invariance of the laser equations. This implies that the gain function $\mathcal{G}_-$ is equal to 1 for $\tilde{\alpha}_n = 0$, as one has from Eqs. (20.34) and (20.35). Therefore we will use the term phase instabilities also for those multimodal instabilities which arise under nonresonant conditions when the function $\mathcal{G}_-$ displays an interval of $\tilde{\alpha}_n$ where $\mathcal{G}_-(\tilde{\alpha}_n) > 1$.

21

Adiabatic elimination in the complete Maxwell–Bloch equations

In Section 19.3 we discussed the model which is derived from the single-mode model by adiabatically eliminating the atomic polarization. Similarly, in Section 19.2 we analyzed the model which is obtained from the single-mode model by adiabatic elimination of both atomic variables. The aim of this chapter is to discuss the same adiabatic eliminations, but in the framework of the Maxwell–Bloch equations for the ring cavity (12.35)–(12.37) or, equivalently, of the full multimode model of Eqs. (12.39), (12.36) and (12.37). In the literature, one often finds that a rate-equation model is obtained by a naïve adiabatic elimination of the atomic polarization, which is performed by dropping the time derivative of the atomic polarization in Eq. (12.36), and that a field equation is obtained by the naïve adiabatic elimination of the atomic variables, in which one drops the time derivatives in Eqs. (12.36) and (12.37). In this chapter we discuss both the equations obtained via naïve adiabatic eliminations and equations derived by refined procedures of adiabatic elimination of the atomic polarization and adiabatic elimination of the atomic variables.

In Section 21.1 we illustrate the naïve adiabatic elimination of the atomic polarization, whereas Section 21.2 is devoted to the refined adiabatic elimination and to its comparison with the naïve rate-equation approximation. Finally, in Section 21.3 we discuss the adiabatic elimination of both atomic variables, both in the naïve and in the refined formulation. An important feature is that the refined adiabatic eliminations are capable of reproducing the multimode instabilities, which on the contrary disappear altogether if one performs the naïve adiabatic elimination of both atomic variables.

21.1 The rate-equation approximation

Let us start from Eqs. (12.35)–(12.37) and set the time derivative in the polarization equation (12.36) equal to zero. We obtain the following functional expression for P:

$$P = \frac{1 - i\Delta}{1 + \Delta^2} FD. \tag{21.1}$$

We obtain also two equations for F and D,

$$c\frac{L}{\Lambda}\frac{\partial F}{\partial z} + \frac{\partial F}{\partial \tau} = -\kappa\left[(1 + i\theta)F - y - A\frac{1 - i\Delta}{1 + \Delta^2}FD\right], \tag{21.2}$$

$$\frac{\partial D}{\partial \tau} = -\gamma_\parallel\left[D\left(\frac{|F|^2}{1 + \Delta^2} + 1\right) - 1\right], \tag{21.3}$$

which are the full rate equations, obtained by what we call "naïve adiabatic elimination of the atomic polarization". As one can see by looking at Eqs. (12.39), this procedure is valid when the condition

$$\gamma_\perp \gg \gamma_\parallel, \ |\alpha_n| \tag{21.4}$$

is satisfied for all values of n for which the amplitude f_n is non-negligible. Clearly this condition is very restrictive; for example, it is much more restrictive than (20.22). Note that condition (21.4) for $n = 1$ implies $\gamma_\perp \gg \kappa$. In terms of the modal amplitudes f_n, Eq. (21.2) can be rewritten as follows:

$$\frac{df_n}{d\tau} = -i\alpha_n f_n - \kappa \left[(1 + i\theta) f_n - y\delta_{n,0} - \frac{1 - i\Delta}{1 + \Delta^2} \frac{A}{L} \int_0^L dz \, e^{-ik_n z} F(z, \tau) D(z, \tau) \right], \tag{21.5}$$

and can be coupled to Eq. (21.3) where $F(z, \tau)$ is expressed in terms of the amplitudes f_n by Eq. (12.40).

We note finally that, strictly speaking, the name *rate equations* for Eqs. (21.2) and (21.3) is fully appropriate only for the case of the free-running laser, in which $y = 0$ and $\theta = -\Delta$.

As a matter of fact, in this case, by using again the expression $F(z, \tau) = \rho(z, \tau)\exp(i\phi(z, \tau))$ and introducing the normalized intensity

$$\tilde{I}(z, \tau) = |F(z, \tau)|^2, \tag{21.6}$$

one obtains from Eqs. (21.2) and (21.3) the three equations

$$\frac{L}{\Lambda} \frac{\partial \tilde{I}}{\partial z} + \frac{\partial \tilde{I}}{\partial \tau} = -2\kappa \tilde{I} \left(1 - \frac{A}{1 + \Delta^2} D \right), \tag{21.7}$$

$$\frac{\partial D}{\partial \tau} = -\gamma_\parallel \left[D \left(\frac{\tilde{I}}{1 + \Delta^2} + 1 \right) - 1 \right], \tag{21.8}$$

$$\frac{L}{\Lambda} \frac{\partial \phi}{\partial z} + \frac{\partial \phi}{\partial \tau} = \kappa \Delta \left(1 - \frac{A}{1 + \Delta^2} D \right). \tag{21.9}$$

In particular, Eqs. (21.7) and (21.8) constitute a closed set of equations that, in the literature, are called rate equations. However, in the following we will call Eqs. (21.2) and (21.3) rate equations also for the cases of the laser with an injected signal and optical bistability.

21.2 Adiabatic elimination of the atomic polarization and comparison with the rate-equation approximation

Let us now illustrate what we call "refined adiabatic elimination" of the atomic polarization. We start from Eqs. (20.4) and we set $d\tilde{p}_n/d\tau = 0$ assuming that

$$\gamma_\perp \gg \gamma_\parallel, \ \kappa, \tag{21.10}$$

which is much less restrictive than (21.4). We obtain

$$\tilde{p}_n(\tau) = \frac{1}{1 + i(\Delta - \tilde{\alpha}_n)} \sum_{n'} \tilde{f}_{n'}(\tau) \tilde{d}_{n-n'}(\tau)$$

$$= \frac{1}{1 + i(\Delta - \tilde{\alpha}_n)} \frac{e^{i\alpha_n \tau}}{L} \int_0^L dz \, e^{-ik_n z} F(z, \tau) D(z, \tau), \qquad (21.11)$$

where $\tilde{\alpha}_n = \alpha_n / \gamma_\perp$ as usual and we have used Eqs. (20.1) and (20.2). By inserting this expression into Eq. (20.3) we obtain

$$\frac{d\tilde{f}_n}{d\tau} = -\kappa \left[(1 + i\theta) \tilde{f}_n - y\delta_{n,0} - \frac{A}{1 + i(\Delta - \tilde{\alpha}_n)} \frac{e^{i\alpha_n \tau}}{L} \int_0^L dz \, e^{-ik_n z} F(z, \tau) D(z, \tau) \right]$$

$$(21.12)$$

and hence, taking into account that $\tilde{f}_n(\tau) = e^{i\alpha_n \tau} f_n(\tau)$, we arrive at the equation

$$\frac{df_n}{d\tau} = -i\alpha_n f_n - \kappa \left[(1 + i\theta) f_n - y\delta_{n,0} - \frac{1}{1 + i(\Delta - \tilde{\alpha}_n)} \frac{A}{L} \int_0^L dz \, e^{-ik_n z} F(z, \tau) D(z, \tau) \right],$$

$$(21.13)$$

which, together with Eq. (12.37), forms a closed system of equations for f_n and D if we take into account the expression for F in terms of f_n provided by Eq. (12.40) and the following expression for P:

$$P(z, \tau) = \sum_n \frac{e^{ik_n z}}{1 + i(\Delta - \tilde{\alpha}_n)} \frac{1}{L} \int_0^L dz \, e^{-ik_n z} F(z, \tau) D(z, \tau), \qquad (21.14)$$

that follows from Eqs. (20.1) and (21.11).

The connection between the equations derived by the refined adiabatic elimination of the atomic polarization and those obtained by the naïve adiabatic elimination becomes immediately clear if in Eqs. (21.13) and (21.14) we drop the term $\tilde{\alpha}_n$ in the denominator. As a matter of fact, Eq. (21.14) reduces to Eq. (21.1) and Eq. (21.13), using Eq. (12.40), leads to Eq. (21.2), i.e. we recover the rate-equation model. As we said, the difference introduced by the term $\tilde{\alpha}_n$ is important, because the refined adiabatic elimination is valid under the conditions (21.10) which are much less restrictive than the conditions (21.4) for the rate-equation model.

21.3 Adiabatic elimination of the atomic variables

As in the previous section, let us start from the naïve adiabatic elimination. If we set the time derivatives equal to zero in the atomic equations (12.36) and (12.37), we obtain the expression $P = (1 - i\Delta)F/(1 + \Delta^2 + |F|^2)$ for P, which, once inserted into Eq. (12.35), leads to the field equation

$$c \frac{L}{\Lambda} \frac{\partial F}{\partial z} + \frac{dF}{d\tau} = -\kappa \left[(1 + i\theta)F - y - A \frac{1 - i\Delta}{1 + \Delta^2 + |F|^2} F \right]. \qquad (21.15)$$

This procedure requires the very restrictive conditions

$$\gamma_\perp,\ \gamma_\parallel \gg |\alpha_n| \tag{21.16}$$

for all values of n for which the amplitude f_n is non-negligible.

Let us now turn to the refined adiabatic elimination of the atomic variables. In this context, let us introduce the quantities

$$\begin{pmatrix} \underline{F}(z,\tau) \\ \underline{P}(z,\tau) \\ \underline{D}(z,\tau) \end{pmatrix} = \sum_n e^{ik_n z} \begin{pmatrix} \tilde{f}_n(\tau) \\ \tilde{p}_n(\tau) \\ \tilde{d}_n(\tau) \end{pmatrix}, \tag{21.17}$$

where the slowly varying amplitudes $\tilde{f}_n$, $\tilde{p}_n$, $\tilde{d}_n$ coincide with those defined in Eq. (20.1). By multiplying Eqs. (20.3)–(20.5) by $e^{ik_n z}$ and summing, we see that the fields defined by Eq. (21.17) obey the equations

$$\frac{\partial \underline{F}}{\partial \tau} = -\kappa[(1+i\theta)\underline{F} - y - A\underline{P}], \tag{21.18}$$

$$\frac{\partial \underline{P}}{\partial \tau} = c\frac{L}{\Lambda}\frac{\partial \underline{P}}{\partial z} + \gamma_\perp[\underline{F}\,\underline{D} - (1+i\Delta)\underline{P}], \tag{21.19}$$

$$\frac{\partial \underline{D}}{\partial \tau} = c\frac{L}{\Lambda}\frac{\partial \underline{D}}{\partial z} - \gamma_\parallel\left[\frac{1}{2}(\underline{F}\,\underline{P}^* + \underline{F}^*\,\underline{P}) + \tilde{D} - 1\right]. \tag{21.20}$$

Let us focus on Eqs. (21.19) and (21.20) and let us set the derivatives with respect to time equal to zero. This step requires the condition

$$\gamma_\perp,\ \gamma_\parallel \gg \kappa, \tag{21.21}$$

which is far less restrictive than (21.16). From Eq. (21.20) we obtain the equation for $\underline{D}$,

$$c\frac{L}{\Lambda}\frac{\partial \underline{D}}{\partial z} = \gamma_\parallel\left[\frac{1}{2}(\underline{F}(z,\tau)\underline{P}^*(z,\tau) + \text{c.c.}) + \underline{D}(z,\tau) - 1\right], \tag{21.22}$$

while from Eq. (21.19) we derive

$$\tilde{p}_n = \frac{1}{1+i(\Delta-\tilde{\alpha}_n)}\frac{1}{L}\int_0^L dz\, e^{-ik_n z}\underline{F}(z,\tau)\underline{D}(z,\tau). \tag{21.23}$$

By inserting Eq. (21.23) into Eq. (20.3) we obtain the equations for $\tilde{f}_n$:

$$\frac{d\tilde{f}_n}{d\tau} = -\kappa\left[(1+i\theta)\tilde{f}_n - y\delta_{n,0} - \frac{A}{1+i(\Delta-\tilde{\alpha}_n)}\frac{1}{L}\int_0^L dz\, e^{-ik_n z}\underline{F}(z,\tau)\underline{D}(z,\tau)\right]. \tag{21.24}$$

Now Eqs. (21.24), coupled with Eq. (21.22), constitute a closed set of equations, if we take into account the expression for $\underline{F}$ and $\underline{P}$ in terms of $\tilde{f}_n$ and $\tilde{p}_n$ in Eq. (21.17) and use Eq. (21.23), so

$$\underline{P}(z,\tau) = \sum_n \frac{e^{ik_n z}}{1+i(\Delta-\tilde{\alpha}_n)}\frac{1}{L}\int_0^L dz'\, e^{-ik_n z'}\underline{F}(z',\tau)\underline{D}(z',\tau), \tag{21.25}$$

The quantities $F(z', \tau)$ and $D(z', \tau)$ are expressed in terms of $\underline{F}(z, \tau)$ and $\underline{D}(z, \tau)$ as follows:

$$F(z', \tau) = \sum_n e^{ik_n z'} e^{-i\alpha_n \tau} \frac{1}{L} \int_0^L dz\, e^{-ik_n z} \underline{F}(z, \tau), \qquad (21.26)$$

$$D(z', \tau) = \sum_n e^{ik_n z'} e^{-i\alpha_n \tau} \frac{1}{L} \int_0^L dz\, e^{-ik_n z} \underline{D}(z, \tau). \qquad (21.27)$$

A very important point is that Eqs. (21.24) and (21.22) are perfectly in agreement with the linearized equations (20.27) and (20.28), which were derived by adiabatically eliminating the atomic variables from the full set of linearized equations (20.10)–(20.14). As a matter of fact, if one linearizes Eqs. (21.24) and (21.22) around the stationary state taking into account Eq. (21.17) and eliminates the variables $\tilde{d}_n$, one arrives at exactly the same set of Eqs. (20.27) and (20.28). Therefore multimode instabilities are perfectly preserved in the framework of Eqs. (21.24) and (21.22). The same is true for the set of equations (21.13) and (12.37) which is valid in the refined adiabatic elimination of the atomic polarization.

In contrast, if one starts from Eq. (21.15) obtained by the naïve adiabatic elimination procedure, which amounts to dropping not only the derivatives with respect to time but also the derivatives with respect to space in Eqs. (21.19) and (21.20), then the parameter $\tilde{\alpha}_n$ is replaced by zero in Eqs. (20.3)–(20.5), and therefore multimode instabilities disappear altogether. As a matter of fact, when $\tilde{\alpha}_n$ is replaced by 0 it becomes impossible to satisfy the inequalities (20.36) and (20.37).

22

Dynamical aspects in the laser

While the stationary behavior of the laser was described in Part I, now we focus on its dynamical aspects. Some of them have been illustrated already in Chapter 19, but only in the single-mode regime and for class-A and class-B lasers, in which the only instability which arises concerns the trivial stationary solution and leads to the transition from the nonlasing to the lasing state. Chapter 20, on the other hand, provided a general picture of single-mode and multimode instabilities in active as well as in passive systems, and of the structural relations which link single-mode and multimode instabilities. In this chapter we focus on the instabilities which arise in the laser and lead to spontaneous temporal oscillations and chaos.

We start in Section 22.1 with the linear-stability analysis of the trivial stationary solution in the general multimode case for the standard laser. The results of this analysis have been anticipated in previous chapters, but here you will find their derivation. The same problem is considered in Section 22.2 for the laser without inversion, but limited to the single-mode case under fully resonant conditions as in Section 17.5.

Of fundamental importance is the analogy between the single-mode laser model and the Lorenz model, which is prototypical for the general field of chaos. This matter is discussed in Section 22.3, and is immediately followed by the treatment of the resonant single-mode laser instability in Section 22.4.

On the other hand, multimodal instabilities in the ring laser are discussed in Sections 22.5 and 22.6. The first is devoted to the amplitude instability, completing the picture given in Section 20.2, and the second to the phase instability. The multimode amplitude instability is related to a classic dynamical phenomenon in lasers, mode-locking, which topic is illustrated in Chapter 23.

Finally, Section 22.7 is devoted to the matter of amplitude instabilities again, but in the case of a Fabry–Perot laser containing an ultrathin medium discussed in Section 14.5. We emphasize the elements which, in the Fabry–Perot case, enrich the picture with respect to the simpler scenario of the multimode amplitude instability in the ring laser.

22.1 Linear-stability analysis of the trivial stationary solution in the standard laser

Equations (20.10)–(20.14) govern the linear-stability analysis of the single-mode stationary solutions. Here we consider the case of a free-running laser and the trivial stationary solution, whereas in Chapter 20 we analyzed the stability of the nontrivial stationary solution. For the trivial stationary solution Eq. (20.6) remains valid as it is, whereas Eqs. (20.7) and (20.8) reduce to

$$x = \tilde{f}_{0,s} = 0, \qquad \tilde{p}_{0,s} = 0, \qquad \tilde{d}_{0,s} = 1. \tag{22.1}$$

Hence the linearized equations become

$$\frac{d\delta\tilde{f}_n}{d\tau} = -\kappa\left[(1 - i\Delta)\delta\tilde{f}_n - A\,\delta\tilde{p}_n\right], \tag{22.2}$$

$$\frac{d\delta\tilde{f}^*_{-n}}{d\tau} = -\kappa\left[(1 + i\Delta)\delta\tilde{f}^*_{-n} - A\,\delta\tilde{p}^*_{-n}\right], \tag{22.3}$$

$$\frac{d\delta\tilde{p}_n}{d\tau} = i\alpha_n\tilde{p}_n + \gamma_\perp\left[\delta\tilde{f}_n - (1 + i\Delta)\delta\tilde{p}_n\right], \tag{22.4}$$

$$\frac{d\delta\tilde{p}^*_{-n}}{d\tau} = i\alpha_n\,\delta\tilde{p}^*_{-n} + \gamma_\perp\left[\delta\tilde{f}^*_{-n} - (1 - i\Delta)\delta\tilde{p}^*_{-n}\right], \tag{22.5}$$

$$\frac{d\delta\tilde{d}_n}{d\tau} = (i\alpha_n - \gamma_\parallel)\delta\tilde{d}_n. \tag{22.6}$$

Note that Eqs. (22.2) and (22.4) constitute a self-contained set of equations for $\delta\tilde{f}_n$ and $\delta\tilde{p}_n$, whereas Eqs. (22.3) and (22.5) constitute another self-contained set of equations for the variables $\delta\tilde{f}^*_{-n}$ and $\delta\tilde{p}^*_{-n}$, and Eq. (22.6) is a self-contained equation for $\delta\tilde{d}_n$.

Let us now focus on the single-mode model, in which the only mode in play is $n = 0$. Equations (22.2) and (22.4) read

$$\frac{d\delta\tilde{f}_0}{d\tau} = -\kappa\left[(1 - i\Delta)\delta\tilde{f}_0 - A\,\delta\tilde{p}_0\right], \tag{22.7}$$

$$\frac{d\delta\tilde{p}_0}{d\tau} = \gamma_\perp\left[\delta\tilde{f}_0 - (1 + i\Delta)\delta\tilde{p}_0\right]. \tag{22.8}$$

Equations (22.3) and (22.5) for $n = 0$ are the complex conjugates of Eqs. (22.7) and (22.8), respectively. The quintic characteristic equation (20.20) factorizes into two quadratic equations and one linear equation for λ. The linear equation $\lambda + \gamma_\parallel = 0$ does not lead to any instability. Therefore it suffices to consider the characteristic equation associated with the self-contained set of equations (22.7) and (22.8), which reads

$$\lambda^2 + [\kappa + \gamma_\perp + i\Delta(\gamma_\perp - \kappa)]\lambda - \kappa\gamma_\perp(A - 1 - \Delta^2) = 0. \tag{22.9}$$

This equation has complex coefficients; however, we can, as usual, look for the boundary of the stability domain in the parameter space by setting $\lambda = -i\omega$ with ω real. By separating real and imaginary parts we obtain the two equations

$$\omega^2 - \Delta(\gamma_\perp - \kappa)\omega + \kappa\gamma_\perp(A - 1 - \Delta^2) = 0, \tag{22.10}$$

$$(\kappa + \gamma_\perp)\omega = 0. \tag{22.11}$$

From Eq. (22.11) we obtain that $\omega = 0$, so the instability leads to the bifurcation of a stationary state, and Eq. (22.10) gives $A = 1 + \Delta^2$, which corresponds to the lasing threshold. Hence the trivial stationary solution is stable below threshold and unstable above threshold.

Next, let us now return to the general case of Eqs. (22.2) and (22.4) and let us assume that $\kappa \ll \gamma_\perp$, so that the polarization fluctuations $\delta \tilde{p}_n$ can be eliminated adiabatically, giving from Eq. (22.4)

$$\delta \tilde{p}_n = \frac{\delta \tilde{f}_n}{1 + i(\Delta - \tilde{\alpha}_n)}, \tag{22.12}$$

where, as usual, we set $\tilde{\alpha}_n = \alpha_n/\gamma_\perp$. If we insert Eq. (22.12) into Eq. (22.2) we obtain

$$\frac{d\delta \tilde{f}_n}{d\tau} = -\kappa \left[1 - i\Delta - \frac{A}{1 + i(\Delta - \tilde{\alpha}_n)} \right] \delta \tilde{f}_n, \tag{22.13}$$

so the relevant eigenvalue is $\lambda/\kappa = -1 + i\Delta + A/[1 + i(\Delta - \tilde{\alpha}_n)]$. Taking into account that $\Delta = (\omega_a - \omega_0)/\gamma_\perp$ and that $\omega_0 = \omega_c$ in the limit $\kappa \ll \gamma_\perp$, we have that

$$\Delta - \tilde{\alpha}_n = \frac{\omega_a - (\omega_c + \alpha_n)}{\gamma_\perp} \equiv \Delta_n. \tag{22.14}$$

Note that, in the limit $\kappa \ll \gamma_\perp$, Δ_j defined by Eq. (9.14) coincides with Δ_n defined by Eq. (22.14), since $\kappa' = \kappa$ in the low-transmission limit and $(j - \bar{j}) = n$. Therefore we can write

$$\frac{\mathrm{Re}\,\lambda}{\kappa} = \mathcal{G}(\tilde{\alpha}_n, A) - 1 \quad \text{with} \quad \mathcal{G}(\tilde{\alpha}_n, A) = \frac{A}{1 + \Delta_n^2}. \tag{22.15}$$

Since we assumed that ω_c is the cavity frequency nearest to ω_a, we have that the gain is maximum for $n = 0$, i.e. for the resonant mode.

Note that, if one uses the rate equations (21.2) and (21.3) instead of performing the correct adiabatic elimination of the polarization, one obtains a mode-independent gain $\mathcal{G} = A$ instead of (22.15). Note also that the gain $\mathcal{G}$ in Eqs. (22.15) refers to the trivial stationary solution, whereas the gain $\mathcal{G}$ which appears in Eqs. (20.36) and (20.37) refers, in the case of the free-running laser, to the nontrivial stationary solution.

22.2 Linear-stability analysis of the trivial stationary solution in the laser without inversion

Let us start from the set of dynamical equations (17.90)–(17.95) which governs the laser without inversion for the Λ level scheme shown in Fig. 17.1. We remind the reader that the laser frequency, the cavity frequency in play and the atomic transition frequency between levels 2 and 1 are assumed to be exactly resonant in these equations.

Let us focus on the trivial stationary solution, which is described by Eq. (17.96) and in the following text line. We linearize the dynamical equations around the trivial stationary solution by setting, for each variable x_n, $x_n(t) = x_{ns} + \delta x_n(t)$, with x_{ns} being the stationary value of the variable x_n in the trivial stationary solution and δx_n the deviation of the variable from its stationary value, and by keeping only the terms which are linear in the deviations. It turns out that the linearized equations for $\delta \alpha$, δr_{21} and δr_{32} form the self-contained set of equations

$$\dot{\delta\alpha} = -\kappa\, \delta\alpha + i\, G'\, \delta r_{21}, \tag{22.16}$$

$$\dot{\delta r}_{21} = i(\rho_{11,s} - \rho_{22,s})\delta\alpha + i\beta\, \delta r_{31} - \Gamma_{21}\, \delta r_{21}, \tag{22.17}$$

$$\dot{\delta r}_{31} = i(\beta\, \delta r_{21} - r^*_{23,s}\, \delta\alpha) - \Gamma_{31}\, \delta r_{31}. \tag{22.18}$$

By introducing the usual exponential ansatz $\delta x_n(t) = \delta x_n(0)\exp(\lambda t)$ we obtain the homogeneous linear set of equations

$$\begin{pmatrix} \lambda + \kappa & -iG' & 0 \\ -i(\rho_{11,s} - \rho_{22,s}) & \lambda + \Gamma_{21} & -i\beta \\ ir^*_{23,s} & -i\beta & \lambda + \Gamma_{31} \end{pmatrix} \begin{pmatrix} \delta\alpha(0) \\ \delta r_{21}(0) \\ \delta r_{31}(0) \end{pmatrix} = 0, \tag{22.19}$$

which leads to the cubic characteristic equation

$$\lambda^3 + c_2\lambda^2 + c_1\lambda + c_0 = 0, \tag{22.20}$$

with

$$c_2 = \kappa + \Gamma_{21} + \Gamma_{31}, \tag{22.21}$$

$$c_1 = \Gamma_{21}\Gamma_{31} + \beta^2 + \kappa(\Gamma_{21} + \Gamma_{31}) + G'(\rho_{11,s} - \rho_{22,s}), \tag{22.22}$$

$$c_0 = \kappa(\Gamma_{21}\Gamma_{31} + \beta^2) + G'\Gamma_{31}(\rho_{11,s} - \rho_{22,s}) - G'\beta^2(\rho_{33,s} - \rho_{22,s})/\Gamma_{23}, \tag{22.23}$$

where we have used Eq. (17.72) for $r_{23,s}$. The coefficients c_1 and c_2 are clearly positive because we assume $\Lambda < \gamma_{21}$ (see Eqs. (17.76)–(17.78)), so we must analyze the sign of c_0. The first and second terms are positive, whereas the last contribution is negative. By taking into account Eqs. (17.76) and (17.77) for the $\rho_{ii,s}$ terms and Eq. (17.61) for the Γ_{ij} terms, the instability condition $c_0 < 0$ can be cast in the form

$$G' > \kappa\Gamma_{21} \frac{[(\Lambda + \gamma_{21} + \gamma_{23})\Lambda + 4\beta^2][(\gamma_{21} + \gamma_{23})\gamma_{23}\Lambda + (\gamma_{21} + 2\Lambda)4\beta^2]}{4[(\Lambda + \gamma_{23})^2 - \gamma_{21}^2]\Lambda\beta^2}. \tag{22.24}$$

Now, if we introduce the definition of the pump parameter A_{LWI} for the laser without inversion given by Eqs. (17.97) and (17.98), the instability condition (22.24) reads $A_{\mathrm{LWI}} > 1$, as stated in Eq. (17.97). Hence the lasing threshold corresponds to the case that the pump parameter is equal to unity, exactly as for the standard laser, homogeneously or inhomogeneously broadened, in the resonant configuration.

22.3 Class-C lasers: the analogy with the Lorenz model and optical chaos

Let us now consider the free-running laser for general values of the temporal relaxation rates κ, $\gamma_\perp$ and $\gamma_\parallel$, not restricted by the conditions (19.1) and (19.2) which define class-A and class-B lasers, respectively. This is called a class-C laser.

Let us start from the single-mode model governed by Eqs. (12.43)–(12.45) assuming resonance, i.e. $\Delta = \theta = 0$ and $y = 0$. Hence the equations reduce to

$$\dot{x} = -\kappa(x - AP), \tag{22.25}$$

$$\dot{P} = \gamma_\perp(xD - P), \tag{22.26}$$

$$\dot{D} = -\gamma_\parallel(xP + D - 1), \tag{22.27}$$

where we have taken x and P real. Therefore, in the nontrivial stationary solution the phase of x is no longer arbitrary and can take only the two values 0 and π, so that $x_s = \pm\sqrt{A - 1}$. Let us now introduce the normalized time $t' = \gamma_\perp t$ and the variables

$$x_L = \tilde{\gamma}^{1/2}x, \qquad y_L = \tilde{\gamma}^{1/2}AP, \qquad z_L = (1 - D)A, \tag{22.28}$$

where $\tilde{\gamma} = \gamma_\parallel/\gamma_\perp$, and let us define the parameters

$$\rho = A, \qquad \sigma = \kappa/\gamma_\perp, \qquad \beta = \tilde{\gamma}, \tag{22.29}$$

so that Eqs. (22.25)–(22.27) take the forms

$$\frac{dx_L}{dt'} = \sigma(y_L - x_L), \tag{22.30}$$

$$\frac{dy_L}{dt'} = x_L(\rho - z_L) - y_L, \tag{22.31}$$

$$\frac{dz_L}{dt'} = x_L y_L - \beta z_L. \tag{22.32}$$

This set of equations was introduced and analyzed in 1963 by Lorenz [239] as a model for atmospheric convection, and is called *the Lorenz model* in the literature.

As we will discuss in the following section, some solutions of the Lorenz model display a typical chaotic behavior and this model became a paradigm in the field of chaos. The equations of the Lorenz model were derived from the equations describing fluid circulation in a shallow fluid layer heated from below. For $\rho < 1$ the fluid is motionless, and the value $\rho = 1$ marks the onset of convection (Rayleigh–Bénard instability [91, 234, 235]). The Lorenz model predicts that beyond a second threshold for the parameter ρ the two nontrivial stationary solutions become unstable because of an oscillatory instability and the system can approach a chaotic regime.

The analogy between the single-mode laser and the Lorenz model was demonstrated by Haken [240]. While the Lorenz model does not represent a realistic picture for atmospheric convection because it is derived by exceedingly crude approximations, the single-mode model provides a realistic picture for a free-running laser in a ring cavity. This circumstance primed a very lively interest in *optical chaos* or *optical turbulence* in the late 1970s and in the 1980s.

In the following section we discuss the scenario predicted by the resonant single-mode laser model. We finally observe that in the laser case the parameter $\tilde{\gamma}$ is always smaller than 2, whereas in the Lorenz model the parameter ρ can take any positive value, even values that are unphysical for the laser.

22.4 The resonant single-mode laser instability

Let us start from Eqs. (22.25)–(22.27), and linearize these equations with respect to the nontrivial stationary solution $x_s = \pm\sqrt{A-1}$, $P_s = x_s/(1+x_s^2) = x_s/A$, $D_s = 1/A$. The matrix $\mathbf{A}$ of the linearized system (see Eq. (18.6)) is

$$\mathbf{A} = \begin{pmatrix} -\kappa & \kappa A & 0 \\ \gamma_\perp/A & -\gamma_\perp & \gamma_\perp x_s \\ -\gamma_\parallel x_s/A & -\gamma_\parallel x_s & -\gamma_\parallel \end{pmatrix}, \tag{22.33}$$

where we use A instead of X as a parameter. Hence the characteristic equation (18.8) is cubic, with

$$c_2 = \kappa + \gamma_\perp + \gamma_\parallel, \qquad c_1 = \gamma_\parallel(\gamma_\perp A + \kappa), \qquad c_0 = 2\kappa\gamma_\perp\gamma_\parallel(A-1). \tag{22.34}$$

By following the procedure described in Appendix B, we find that the boundary of the oscillatory instability is given by

$$A = 1 + \frac{(\tilde{\kappa}+1)(\tilde{\kappa}+\tilde{\gamma}+1)}{\tilde{\kappa}-\tilde{\gamma}-1} = \frac{\tilde{\kappa}(\tilde{\kappa}+\tilde{\gamma}+3)}{\tilde{\kappa}-\tilde{\gamma}-1}, \tag{22.35}$$

with $\tilde{\kappa} = \kappa/\gamma_\perp$. This result agrees with Eq. (20.49) since $X = x_s^2 = A - 1$. The nontrivial stationary solution becomes unstable when the pump parameter A exceeds the value specified by the r.h.s. of Eq. (22.35), which is often called the *second threshold of the laser*. Note that the second threshold (22.35) exists only when $\kappa > \gamma_\perp + \gamma_\parallel$, which is basically a bad-cavity condition, and requires values of A much larger than unity, i.e. much larger than the first threshold. A favorable condition to lower the second threshold is that $\gamma_\parallel \ll \gamma_\perp$, in which case its expression reduces to

$$A = \frac{\tilde{\kappa}(\tilde{\kappa}+3)}{\tilde{\kappa}-1}. \tag{22.36}$$

One easily verifies that this expression is minimum for $\tilde{\kappa} = 3$, corresponding to which we have that $A = 9$. According to the Routh–Hurwitz stability criterion (see Appendix A), the nontrivial stationary solution is unstable for $c_0 - c_1 c_2 > 0$, i.e. for A larger than the second threshold. Our procedure based on the calculation of the stability boundary has the additional bonus of providing the oscillation frequency corresponding to the boundary. From Eq. (B.3) we have that such a frequency is given by

$$\omega^2 = \frac{c_0}{c_2} = \frac{2\kappa\gamma_\perp\gamma_\parallel(A-1)}{\kappa+\gamma_\perp+\gamma_\parallel} = \frac{2\gamma_\perp\gamma_\parallel(A-1)}{1+(\gamma_\perp+\gamma_\parallel)/\kappa} \simeq \gamma_\perp\gamma_\parallel(A-1) = \Omega^2, \tag{22.37}$$

where we have taken into account that $(\gamma_\perp + \gamma_\parallel)/\kappa < 3$ since $\gamma_\parallel < 2\gamma_\perp$ and $\tilde{\kappa} > 1$, so $2/[1 + (\gamma_\perp + \gamma_\parallel)/\kappa]$ is of order unity. Also, we have taken into account that $A - 1 = x_s^2 = \Omega^2/(\gamma_\perp\gamma_\parallel)$, where Ω is the Rabi frequency (see Eqs. (8.19) and (4.30)). This means that at the instability threshold the relaxation oscillations of the laser fluctuations around the stationary state have a frequency close to the Rabi frequency of the laser field. Therefore the instability itself can be interpreted as a consequence of the resonance between these two frequencies of the system.

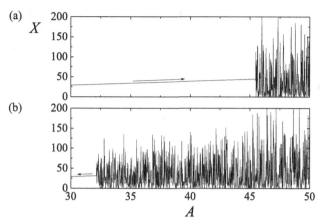

Figure 22.1 The behavior of the output intensity $X = x^2$ as a function of the parameter A when the latter is slowly varied from 30 to 50 (a) and from 50 to 30 (b).

In order to see the dynamical behavior of the laser beyond the second threshold let us integrate the time-evolution equations numerically. Let us fix the parametric values $\tilde{\kappa} = 1.4253$ and $\tilde{\gamma} = 0.2778$, which are appropriate to describe an ammonia laser, the only laser for which an experimental observation of the Lorenz–Haken chaos has been reported, to the best of our knowledge. With such values the laser dynamics becomes chaotic as soon as the second threshold $A = 45.45$ (see Eq. (22.35)) is surpassed. To be precise, let us first gradually increase the pump parameter well beyond the second threshold, so that it reaches and stays in the chaotic regime, and then let us gradually decrease it back to values below the second threshold. This is called the *swept-parameter technique*.

The remarkable feature is that, when A returns to values below the second threshold, the system does not come back to a regime in which the output intensity is stationary, but the output remains chaotic until roughly the value $A = 32.5$, below which the intensity returns to be stationary. Therefore there is hysteresis (i.e. a bistability range) between a stationary solution and a chaotic solution, as shown in Fig. 22.1 in which the pump parameter varies slowly from 30 to 50 (a) or from 50 to 30 (b). This implies that a chaotic solution exists even below the second threshold, in the interval $32.5 < A < 45.45$.

In this range, if the system lies in one of the two stationary solutions $x_s = \pm\sqrt{A - 1}$ and is slightly displaced from it (*soft mode excitation*), the system returns to such a stationary solution, which is stable. If, instead, the displacement is large enough (*hard mode excitation*), or the initial condition is far enough from the two stationary solutions, then the system may evolve to the chaotic solution. For instance, the system reaches the chaotic state shown in Figs. 22.2(a)–(c), which corresponds to $A = 40$, if the initial condition is close to the unstable trivial stationary solution $x_s = 0$. Figures 22.2(a)–(c) show the temporal behavior of the field variable x (a), the projection of the phase-space trajectory onto the plane $P = 0$ (b) and the optical spectrum, i.e. the modulus squared of the Fourier transform of the amplitude x, (c), respectively.

In Fig. 22.2(a) we observe that the field jumps from positive to negative values and vice versa without any regularity. From the phase-space trajectory in Fig. 22.2(b) we see that the

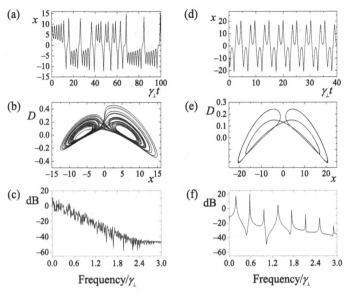

Figure 22.2 The temporal evolution of the electric field, phase-space trajectory in the subspace $P = 0$ and optical spectrum for a chaotic regime with $A = 40$ (a)–(c) and for a periodic regime with $A = 112.5$ (d)–(f).

phase-space point covers some loops around one of the two nontrivial stationary solutions, then passes to the other side, covering some loops around the other nontrivial stationary solutions and so on, and the passages to the other side occur at unpredictable times. This behavior gives rise to the typical butterfly shape of the Lorenz attractor. The spectrum, on a logarithmic scale, exhibits a linear decreasing trend without evident frequency peaks. The phase-space trajectories display the exponential divergence which characterizes the deterministic chaotic behavior.

When the value of the pump parameter grows in the unstable region one finds windows within which the dynamical behavior becomes periodic. One of these windows is illustrated in Figs. 22.2(d)–(f), for $A = 112.5$. Finally, one observes that when A is increased to values larger than 146.5 one always finds periodic behavior. Such a transition from a chaotic to a periodic regime occurs via an inverted period-doubling route, i.e. one starts from the chaotic solution (infinite period) and upon passing through solutions with decreasing period, one arrives at period 1, a simple periodic regime. The period 4 is illustrated in Figs. 22.3(a)–(c), period 2 in Figs. 22.3(d)–(f), and period 1 in Figs. 22.3(g)–(i). In each of the three cases the first plot shows the intensity instead of the field, in order to indicate the different peak intensities. This implies that in the case of period n the peaks repeat with period $2n$. The phase-space trajectories are shown in the subspace $D = 0$, and close themselves after n loops. In the power spectra there are $n - 1$ secondary peaks between the principal peaks, i.e. those of Fig. 22.3(i).

The period-doubling route to chaos is one of the three most common routes by which a generic nonlinear dynamical system can reach a chaotic regime. The second route is the quasi-periodic route, or Ruelle–Takens–Newhouse scenario [235], which displays a sequence of three bifurcations when a control parameter is varied. The first bifurcation brings the system from a fixed point (stationary state) to a periodic orbit, whereas after the

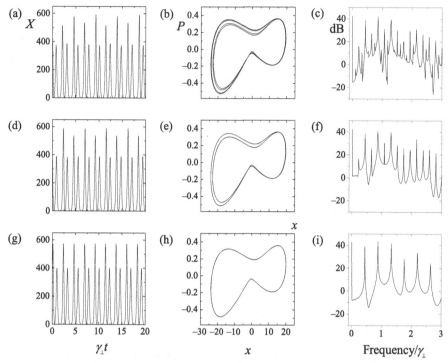

Figure 22.3 The behavior of the electric field's normalized intensity $X = x^2$ as a function of time, phase-space trajectory in the subspace $D = 0$ and power spectrum for $A = 140$ (a)–(c), $A = 142$ (d)–(f) and $A = 147$ (g)–(i).

second bifurcation the attractor becomes a torus because the second frequency generated is incommensurate with the first and this evolution is quasi-periodic. Finally, after the third bifurcation a third incommensurate frequency arises and one passes to a hypertorus. This last attractor is usually unstable and the dynamics becomes chaotic.

The third route is called the intermittency route or Pomeau–Manneville scenario [235]. In this case there is one bifurcation only, after which the periodic oscillations get interrupted, at time intervals of random length, by turbulent phases. When the value of the control parameter is increased, the turbulent phases arise more and more often and have a longer and longer duration, until the chaotic behavior becomes permanent. Three types of intermittency have been identified.

Coming back to the period-doubling route to chaos, Feigenbaum [254] has demonstrated that, independently of the dynamical system which is considered, this route is characterized by a universal property. If a_n is the value of the control parameter a at which the system bifurcates from period 2^n to period 2^{n+1}, then the ratio

$$\delta_n = \frac{a_n - a_{n-1}}{a_{n+1} - a_n} \tag{22.38}$$

tends, for $n \to \infty$, to the limit value $\delta = 4.669\ldots$.

The experimental observation of the Lorenz–Haken chaos is made difficult by the circumstance that the second threshold of the laser is at least nine times larger than the first and, in addition, the laser must satisfy the bad-cavity condition $\kappa > \gamma_\perp + \gamma_\parallel$. The only experimental observation which has been reported was obtained by Weiss and coworkers using an ammonia (NH_3) laser, which operates in the far infrared [255]. The experimental data display a good qualitative agreement with the Lorenz–Haken scenario. However, the ammonia laser cannot be accurately described by a two-level model. In fact, a more realistic description requires a three-level scheme as a consequence of the coherence introduced by the external pump mechanism, which is itself a laser in this case. There are, however, regions of parameter space where the three-level model of the far-infrared laser displays a dynamical behavior that is similar to the one predicted by the Lorenz–Haken model [256].

22.5 The multimode amplitude instability

The multimode amplitude instability of the laser is usually called the Risken–Nummedal–Graham–Haken instability [124, 257]. Even though this has already been discussed in part in Section 20.2 (see Eqs. (20.38) and (20.39)), here we show an alternative derivation of such results, accompanied by an extensive discussion of the scenario in which this instability arises. One reason for the introduction of an alternative derivation is that the derivation in Chapter 20 utilizes a modal expansion for the atomic variables, which is not possible, in general, in the case of Fabry–Perot lasers. Another reason is that in the resonant case $\Delta = 0$ the alternative procedure allows one to obtain the eigenvalues in a more straightforward way.

Let us start from Eqs. (12.35)–(12.37). In the case of a free-running laser we set $y = 0$ and $\theta = -\Delta$ (see Eq. (12.57)). In addition we assume resonance, i.e. that $\Delta = 0$, and, for the sake of simplicity, we treat F and P as real variables. In conclusion the set of dynamical equations reads

$$c\frac{L}{\Lambda}\frac{\partial F}{\partial z} + \frac{\partial F}{\partial \tau} = -\kappa(F - AP), \tag{22.39}$$

$$\frac{\partial P}{\partial \tau} = \gamma_\perp(FD - P), \tag{22.40}$$

$$\frac{\partial D}{\partial \tau} = -\gamma_\parallel(FP + D - 1). \tag{22.41}$$

As we did in the single-mode configuration, let us now linearize such equations around the stationary solution by setting

$$F(z, \tau) = x_s + \delta F(z)e^{\lambda\tau}, \quad P(z, \tau) = \frac{x_s}{A} + \delta P(z)e^{\lambda\tau}, \quad D(z, \tau) = \frac{1}{A} + \delta D(z)e^{\lambda\tau}, \tag{22.42}$$

with $x_s = \pm\sqrt{A - 1}$. We obtain the linearized equations

$$c\frac{L}{\Lambda}\frac{\partial \delta F}{\partial z} + \lambda\,\delta F = -\kappa\,\delta F + \kappa A\,\delta P, \tag{22.43}$$

$$\lambda\,\delta P = \gamma_\perp(x_s\,\delta D + \delta F/A) - \gamma_\perp\,\delta P, \tag{22.44}$$

$$\lambda\,\delta D = -\gamma_\parallel x_s(\delta P + \delta F/A) - \gamma_\parallel\,\delta D. \tag{22.45}$$

By combining Eqs. (22.44) and (22.45) we get

$$\delta P = \frac{\gamma_\perp}{A} \frac{\lambda + \gamma_\| - \gamma_\| x_s^2}{(\lambda + \gamma_\|)(\lambda + \gamma_\perp) + \gamma_\| \gamma_\perp x_s^2} \delta F. \tag{22.46}$$

By inserting the last equation into Eq. (22.43), this linear equation can be solved at once, and, by imposing the periodic boundary condition $\delta F(L) = \delta F(0)$, one arrives at the equation

$$\exp\left\{-\frac{\Lambda}{c}\left[\lambda + \kappa\left(1 - \gamma_\perp \frac{\lambda + \gamma_\|(2 - A)}{(\lambda + \gamma_\|)(\lambda + \gamma_\perp) + \gamma_\| \gamma_\perp (A - 1)}\right)\right]\right\} = 1, \tag{22.47}$$

where, again, we use A instead of x_s^2 as a parameter. If we take into account that $1 = e^{-2i\pi n}$, with $n = 0, \pm 1, \pm 2, \ldots$, and take definition (12.17) into account, we arrive at the eigenvalue equation[1]

$$\lambda = -i\alpha_n - \kappa\left[1 - \gamma_\perp \frac{\lambda + \gamma_\|(2 - A)}{(\lambda + \gamma_\|)(\lambda + \gamma_\perp) + \gamma_\| \gamma_\perp (A - 1)}\right]. \tag{22.48}$$

Since $\kappa/\alpha_n \propto T/n \ll 1$, we can calculate the eigenvalues which can give rise to an instability by solving Eq. (22.48) iteratively:

$$\lambda \simeq -i\alpha_n - \kappa\left[1 - \gamma_\perp \frac{\gamma_\|(2 - A) - i\alpha_n}{(\gamma_\| - i\alpha_n)(\gamma_\perp - i\alpha_n) + \gamma_\| \gamma_\perp (A - 1)}\right]. \tag{22.49}$$

This result is clearly correct when $\kappa \ll \gamma_\perp, \gamma_\|$. In Appendix F we describe a rigorous procedure to determine the stability boundary associated with the eigenvalue equation (22.48), from which it becomes clear that the only condition for the validity of Eq. (22.49) is that $\kappa \ll \gamma_\perp$, so the validity persists e.g. for class-B lasers.

From Eq. (22.49) we see that, if we define a gain $\mathcal{G}$ such that $(\mathrm{Re}\,\lambda)/\kappa = \mathcal{G} - 1$ as we did in Section 20.2, the gain is given by

$$\mathcal{G}(\tilde{\alpha}_n) = \gamma_\perp \,\mathrm{Re}\, \frac{\gamma_\|(2 - A) - i\alpha_n}{(\gamma_\| - i\alpha_n)(\gamma_\perp - i\alpha_n) + \gamma_\| \gamma_\perp (A - 1)}$$

$$= \frac{\tilde{\alpha}_n^2[\tilde{\gamma}(A - 1) + 1] + \tilde{\gamma}^2 A(2 - A)}{\tilde{\alpha}_n^4 + [(1 + \tilde{\gamma})^2 - 2\tilde{\gamma} A]\tilde{\alpha}_n^2 + \tilde{\gamma}^2 A^2}, \tag{22.50}$$

where we define $\tilde{\alpha}_n = \alpha_n/\gamma_\perp$ and $\tilde{\gamma} = \gamma_\|/\gamma_\perp$. This expression coincides with that given in Eq. (20.38) if we consider that $A = X + 1$. Therefore the stability boundary $\mathcal{G}(\tilde{\alpha}_n) = 1$ is given by $\tilde{\alpha}_n = \tilde{\alpha}_{\max}$, $\tilde{\alpha}_n = \tilde{\alpha}_{\min}$, where $\tilde{\alpha}_{\max}$ and $\tilde{\alpha}_{\min}$, are given by Eq. (20.39). When expressing $\alpha_{\max}$ and $\alpha_{\min}$ in terms of A instead of X, one finds that in the limit $\gamma_\| \ll \gamma_\perp$ which is valid, e.g., for class-B lasers, they take the form

$$\alpha_{\max \atop \min} = \left\{\frac{\gamma_\| \gamma_\perp}{2}\left[3(A - 1) \pm \sqrt{(A - 1)(A - 9)}\right]\right\}^{1/2}, \tag{22.51}$$

from which we see at once that the multimode amplitude instability, in the case of class-B lasers, requires that $A > 9$ exactly like the resonant single-mode instability. For $A \to \infty$ we have that $\alpha_{\max} \to \sqrt{2A\gamma_\perp \gamma_\|}$ and $\alpha_{\min} \to \sqrt{A\gamma_\perp \gamma_\|}$ (see Fig. 22.4, where we plot

---

[1] The most straightforward procedure to derive the eigenvalue equation (22.48) is to linearize Eqs. (12.39) and use Eqs. (22.46) and (12.40).

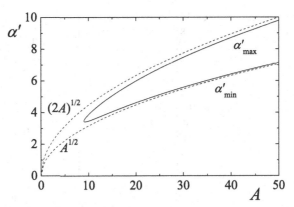

Figure 22.4 The instability domain of the multimode amplitude laser instability lies between the curves α'_{max} and α'_{min}. We define $\alpha' = \alpha/\sqrt{\gamma_\parallel \gamma_\perp}$ and we assume that $\gamma_\parallel \ll \gamma_\perp$.

$\alpha' = \alpha/\sqrt{\gamma_\perp \gamma_\parallel}$). In this limit we can set $A \approx A - 1 = X$, so that $\sqrt{A\gamma_\perp \gamma_\parallel} = \Omega$ (see Eqs. (8.19) and (4.30)). Hence the multimode instability arises when there is a cavity mode whose frequency differs from that of the resonant mode by a quantity on the order of the Rabi frequency Ω and the instability itself can be seen as an amplification of the Rabi oscillations. Hence the Rabi frequency plays the central role both in the resonant single-mode instability and in the multimode amplitude instability. However, in the multimode case there is not the bad-cavity condition $\kappa > \gamma_\perp + \gamma_\parallel$, but the opposite condition $\kappa \ll \gamma_\perp$. We also note that the asymptotic behavior of α_{max} and α_{min} for large A holds also when $\gamma_\parallel$ has the same order of magnitude as $\gamma_\perp$, e.g. in the case of class-A lasers.

The stability analysis shows that, for a given value of A, if the frequency separation α_1 lies in the instability domain $\alpha_{min} < \alpha_1 < \alpha_{max}$, only the two symmetrical sidemodes of the resonant cavity mode are unstable. At first sight one would expect this configuration to give rise to a dynamical behavior in which only the resonant mode and the two sidemodes are active, so that the field intensity oscillates with a frequency equal to the beat note between the sidemodes and the central frequency. Actually the numerical integration reveals that, due to the nonlinear interaction among the cavity modes, beyond the instability threshold other modes get excited in addition to the unstable modes. This implies that the output intensity displays regular pulses, with a period equal to the roundtrip cavity time Λ/c, rather than sinusoidal oscillations. Figure 22.5 shows the long-time evolution of the electric-field envelope F (a), the normalized intensity $X = F^2$ (b) and the power spectrum (c) for selected values of the parameters. In the power spectrum we see the beat notes of several longitudinal modes, even though their intensities decrease rapidly for large frequencies, indicating that the contribution of sidemodes of order n is less and less important for larger n. The form of the pulse is determined by the special phase relation which links the modes. As a matter of fact, the numerical solutions show that the modes realize spontaneously a *mode-locked* configuration in which, if we express the modal amplitudes f_n in terms of modulus and phase in Eq. (20.26), i.e.

$$f_n(\infty) = \rho_n e^{i\phi_n}, \qquad (22.52)$$

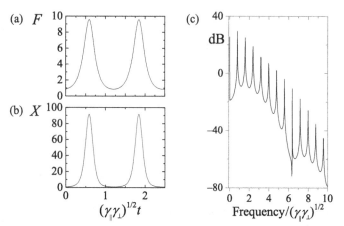

Figure 22.5 The temporal behavior of the electric-field envelope F (a), the normalized intensity $X = F^2$ (b) and the power spectrum in the presence of a multimode instability for $A = 20$, $\alpha_1 = 5\sqrt{\gamma_\parallel \gamma_\perp}$, $\gamma_\perp = 100\gamma_\parallel$ and $\kappa = 3\gamma_\parallel$ (c).

the phases ϕ_n obey the relation

$$\phi_n = n\phi, \tag{22.53}$$

i.e. they are all multiples of the same phase ϕ, which condition characterizes the pheno-menon of mode-locking. Therefore, on using Eq. (20.26) we have that

$$F(z, t) = \sum_n e^{-i\alpha_n(t - n_B z/c)} \rho_n e^{in\phi} = \rho_0 + \sum_{n \neq 0} \rho_n e^{-in[\alpha_1(t - n_B z/c) - \phi]}$$

$$= \rho_0 + 2\sum_{n > 0} \rho_n \cos\left\{ n\left[\alpha_1\left(t - \frac{n_B z}{c} \right) - \phi \right] \right\}, \tag{22.54}$$

where, by shifting the time origin, the ϕ in the exponent may be erased. In Chapter 23 we will briefly discuss the phenomenon of mode-locking, showing that it can lead to pulsations in which each pulse has a duration much shorter than the time interval between subsequent pulses. Figure 22.5(b) shows an example of this behavior.

Let us assume that the value of the pump parameter A is equal to the second threshold $A = 9$, which is valid for $\gamma_\parallel \ll \gamma_\perp$. The instability can develop only if one of the modes is such that α_n is equal to the common value of α_{max} and α_{min} for $A = 9$, i.e. if (see Eq. (22.51))

$$2\pi \frac{c}{\Lambda} n = \sqrt{12\gamma_\perp \gamma_\parallel}. \tag{22.55}$$

This can be seen as a condition on the cavity length

$$\Lambda = \Lambda_c = \frac{2\pi c}{\sqrt{12\gamma_\perp \gamma_\parallel}} n. \tag{22.56}$$

The minimum value of Λ is obtained when mode 1 (and -1) is the unstable mode. The critical value (22.55) of the optical length of the cavity is usually too high for one to observe the Risken–Nummedal–Graham–Haken instability. Only with the advent of the fiber laser, which allows one to realize very long cavities, has it become possible to overcome this problem. In the case of the experiment which was conducted at the Pirelli Labs by Fontana and collaborators [258], the laser parameters were $\gamma_\perp = 2\pi \times 10^{12}$ s^{-1} and $\gamma_\parallel = 100$ s^{-1}, and the cavity was completely filled by a fiber with background refractive index $n_B = 1.5$. For these values of the parameters one obtains a critical length $L_c = \Lambda_c/n_B \approx 14.5$ m and the cavity used in the experiment was approximately 20 m long. An apparent contradiction with theory was that the instability was observed rather close to threshold. However, this feature can be explained by taking into account that in the experiment the effective pump parameter was the parameter w associated with the optical pumping, see Section 10.3, which is different from the normalized pump parameter A [259]. The laser utilized was a fiber erbium-doped laser. The Er atom can be modeled as a three-level laser, so we can use Eq. (10.28) for the parameter σ. Since $A = gL/T$ is proportional to σ because g is proportional to σ (see Eq. (4.34)), we can set

$$A = G\frac{w - \gamma_{int}}{w + \gamma_{int}}, \tag{22.57}$$

where we have called G the proportionality constant which links A and σ. By inverting Eq. (22.57) we can express the experimental pump parameter w in terms of the theoretical pump parameter A,

$$w = \gamma_{int}\frac{G + A}{G - A}. \tag{22.58}$$

The key point is that in the Er case the G coefficient is large, on the order of 100, whereas the value of A required in order to reach the instability is on the order of 10. Therefore we can write

$$w = \gamma_{int}\frac{1 + A/G}{1 - A/G} \approx \gamma_{int}\left(1 + \frac{A}{G}\right)^2 \approx \gamma_{int}\left(1 + 2\frac{A}{G}\right). \tag{22.59}$$

Since the first laser threshold is $A_{thr}^{(1)} = 1$ and the second threshold is $A_{thr}^{(2)} \approx 9$, the ratio of the two thresholds in terms of the experimental pump parameter w is

$$\frac{w_{thr}^{(2)}}{w_{thr}^{(1)}} = \frac{1 + 2A_{thr}^{(2)}/G}{1 + 2A_{thr}^{(1)}/G} = \frac{1 + 18/G}{1 + 2/G} \approx 1 + \frac{16}{G}, \tag{22.60}$$

which turns out to be only slightly larger than unity. For further reading see [260, 261].

22.6 The multimode phase instability

In the previous section we analyzed the resonant configuration of the free-running laser and discussed at length the multimode amplitude instability which arises from the gain $\mathcal{G}_+$ associated with the amplitude eigenvalue λ_+ (see Eqs. (20.34) and (20.35)). In this section we focus on the detuned configuration and on the multimode phase instability which arises from the gain $\mathcal{G}_-$ associated with the eigenvalue λ_- which, in the case of a free-running laser, can be called the phase eigenvalue even in the detuned case, as explained at the end of Section 20.5.

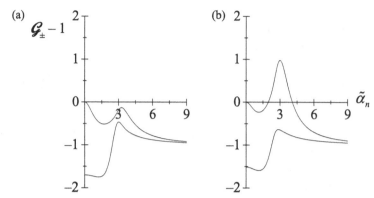

Figure 22.6 For increasing values of the detuning parameter Δ, the phase eigenvalue eventually develops a positive real part. In both illustrations $A = 10$, $\tilde{\alpha}_1 = 3$ and $\bar{\gamma} = 0.8$. The single-mode stationary state $j = \bar{j}$ is stable in (a), where $\Delta = 0.7$, and unstable in (b), where $\Delta = 1.2$ (phase instability).

We linearize [88] (see also [262, 263]) the dynamical equations around the stationary solution defined by Eqs. (20.6)–(20.8), with $x = x_{\bar{j}}$ given by Eq. (9.24). As usual in the case of multimode instabilities, we assume that $\kappa \ll \gamma_\perp$, so that we can neglect the κ which appears in Eq. (9.24) and Δ amounts to $(\omega_a - \omega_c)/\gamma_\perp$.

Let us discuss Fig. 22.6, which shows the real part of the two eigenvalues λ_+ and λ_- (i.e. the quantities $\mathcal{G}_+ - 1$ and $\mathcal{G}_- - 1$, see Eqs. (20.34)–(20.37)) as a function of $\tilde{\alpha}_n$ treated as a continuous variable. The effect of increasing the detuning parameter Δ, i.e the detuning between the center of the atomic line and the cavity frequency ω_c (see Eq. (9.24)) is seen by comparing Figs. 22.6(a) and (b). In both cases the gain of the pump parameter of the system is much smaller than what would be required for the appearance of the amplitude instability. The stationary state is stable for the chosen value of Δ in Fig. 22.6(a), but becomes unstable for the larger detuning used in Fig. 22.6(b), because one of the eigenvalues develops a positive real part for the sideband frequency $\tilde{\alpha}_n = \tilde{\alpha}_1$. Note that the instability is caused by the eigenvalue λ_-, which vanishes for $\tilde{\alpha}_n = 0$ and characterizes a phase instability.

In the analysis of the phase instability, we must discuss the stability of other stationary solutions, in addition to that associated with the cavity mode of frequency $\omega_c = \omega_{\bar{j}}$. Such stationary solutions are described by Eqs. (9.14) and (9.15) and Fig. 9.2. Figure 9.2(a) shows that by increasing the detuning parameter Δ one can pass from a stationary solution to other stationary solutions, whereas the situation is ambiguous in the case of Fig. 9.2(b), because there are intervals of Δ where two different stationary solutions coexist, and therefore it is necessary to determine the stability of the stationary solutions in play in order to clarify the scenario.

For example, let us fix our attention on the stationary solution with frequency $\omega_{(\bar{j}+1)}$. As we see from Eqs. (9.14) and (9.15), this solution is obtained from the previous stationary solution by replacing $\Delta = \Delta_{\bar{j}}$ by $\Delta_{(\bar{j}+1)} = \Delta - (2\pi c/\Lambda)/\gamma_\perp$ (remember that we assume $\kappa \ll \gamma_\perp$). In general, all the results obtained for the mode $j = \bar{j}$ remain valid for the mode $j = (\bar{j} + 1)$, provided that, in any graph that describes the results as a function of the parameter Δ, one operates a translation to the right by the amount $(2\pi c/\Lambda)/\gamma_\perp = \tilde{\alpha}_1$. This is evident in Fig. 9.2. We will use this translation property to determine the stability domain

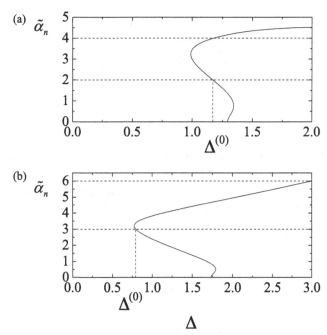

Figure 22.7 The instability boundary for the stationary state $j = \bar{j}$ as a function of the detuning parameter Δ. The stationary state is stable to the left of the solid line. The horizontal dashed lines mark the positions of the first off-resonance cavity modes. (a) $A = 6$, $\bar{\gamma} = 2$, $\tilde{\alpha}_1 = 2$; (b) $A = 10$, $\bar{\gamma} = 0.8$, $\tilde{\alpha}_1 = 3$.

of the stationary solution with frequency $\omega_{(\bar{j}+1)}$ from the stability domain of the stationary state with frequency $\omega_{(\bar{j})}$.

The instability boundary of the stationary solution of frequency $\omega_{\bar{j}}$, defined by the condition $\mathcal{G}_-(\alpha_n) = 1$, is shown in Fig. 22.7 [88] for two different sets of values of the parameters. By considering Fig. 22.7, we can now discuss the mode–mode competition between mode $j = \bar{j}$ (i.e. $n = 0$) and mode $j = \bar{j} + 1$ (i.e. $n = 1$) when we perform a detuning scan, in which we vary the parameter $\Delta = (\omega_{\mathrm{a}} - \omega_{\mathrm{c}})/\gamma_\perp$ from zero to $\tilde{\alpha}_1$ and vice versa. Beyond this point the pattern repeats itself. The two possible situations are shown schematically in Figs. 22.8(a) and (b), where we display the domain of stability of the two relevant states as Δ is varied.

Figure 22.8(a) corresponds to Fig. 22.7(a). The stability domain of the solution $j = \bar{j}$ covers the interval $0 < \Delta < \Delta^{(0)}$, where $\Delta^{(0)}$ is indicated on Fig. 22.7(a). Actually, the full stability range for the solution $j = \bar{j}$ extends symmetrically to the left of the point $\Delta = 0$, but in Fig. 22.7(a) we indicate only the part in the interval $0 < \Delta < \tilde{\alpha}_1$. On the other hand, the stability domain of the solution $j = \bar{j} + 1$ is obtained from the stability domain of the solution $j = \bar{j}$ by shifting it to the right by $\tilde{\alpha}_1$, and in Fig. 22.8(a) we indicate only the part included in the interval $0 < \Delta < \tilde{\alpha}_1$. Figure 22.8(b) corresponds in the same way to Fig. 22.7(b). In both cases the stability diagram is symmetrical with respect to the middle point $\tilde{\alpha}_1/2$, i.e. to the mid point of the free spectral range normalized to $\gamma_\perp$.

In the case of Fig. 22.8(a) the two stationary solutions coexist and are stable in the interval $\Delta^{(1)} < \Delta < \Delta^{(0)}$. Hence the system exhibits bistable behavior with a discontinuous

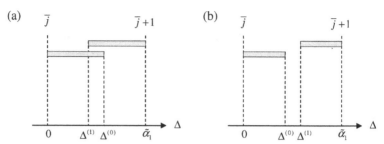

Figure 22.8 The horizontal shaded areas labeled $\bar{j}$ and $\bar{j}+1$ represent the stability domains of the stationary intensity solutions in which the laser oscillates with the modes $j = \bar{j}$ and $j = \bar{j}+1$, respectively, for different values of the detuning parameter $\Delta = (\omega_a - \omega_0)/\gamma_\perp$. (a) A detuning scan will produce discontinuous transitions and hysteresis at $\Delta^{(0)}$ and $\Delta^{(1)}$, respectively. (b) The same as (a) except that the detuning scan will produce persistent oscillations in the region where neither of the two stationary solutions is stable.

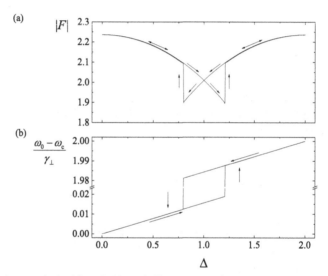

Figure 22.9 A simulated detuning scan obtained from the Maxwell–Bloch equations for the same values of the parameters as in Fig. 22.7(a). The detuning Δ is adiabatically swept from 0 to 2 and back in 2×10^7 time units. (a) The variation of the output field. (b) The variation of the frequency of the output field.

transition from the state in which the laser oscillates with the mode $j = \bar{j}$ to the state in which the laser oscillates with the mode $j = \bar{j}+1$ for $\Delta = \Delta^{(0)}$ and the opposite transition for $\Delta = \Delta^{(1)}$. This behavior is shown in Figs. 22.9(a) and (b), with the hysteresis cycles of the output field (a) and of the oscillation frequency (b). Note that the cycle of the field is butterfly-shaped [88].

On the other hand, Fig. 22.8(b) shows schematically the second possibility. Here the stability domains of the two nontrivial stationary solutions $j = \bar{j}$ and $j = \bar{j}+1$ do not overlap and in the range $\Delta^{(0)} < \Delta < \Delta^{(1)}$ no stable stationary solution exists, since also the trivial stationary solution is unstable. As a consequence, the system shows the emergence of oscillations in which both modes coexist. The output intensity exhibits periodic oscillations

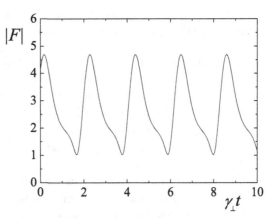

Figure 22.10 The long-time evolution of the modulus of the output field obtained by solving the Maxwell–Bloch equations for the same values of the parameters as in Fig. 22.7(b) and $\Delta = 1.2$. The system evolves into a self-oscillatory state, as one would expect from the results of the linear-stability analysis.

with the fundamental frequency equal to the beat note between the frequencies of the two competing modes. This behavior is displayed in Fig. 22.10, which shows the long-time pulsations of the modulus of the output field, which are obtained by using Eqs. (12.39), (12.36), (12.37) and (12.40) [88].

In general, the situation illustrated in Fig. 22.8(a) is favored by systems with a small value of the ratio $\gamma_\parallel/\gamma_\perp$, while the case of Fig. 22.8(b) is more typical of laser systems for which $\gamma_\parallel/\gamma_\perp$ is on the order of unity. Experimental verifications of the phase instability were reported by Tredicce and collaborators in [88, 264]; here we see evidence of the hysteretic behavior of the type shown in Fig. 22.9. We underline that the phase instability of the laser requires values of the pump parameter A substantially lower than those required by the amplitude instability.

22.7 An ultrathin medium: the multimode amplitude instability in the Fabry–Perot laser

In this section we focus on the case of an ultrathin medium studied in Section 14.5. Let us first consider the ring-cavity configuration described by the difference-differential equations (14.108) with $y = 0$, (14.106) and (14.107). Since all these equations have real coefficients, we can assume for simplicity that the variables F and P are real. Similarly to Eq. (22.42), we set

$$F(t) = x_s + \delta F\, e^{\lambda t}, \quad P(t) = \frac{x_s}{A} + \delta P\, e^{\lambda t}, \quad D(t) = \frac{1}{A} + \delta D\, e^{\lambda t}, \tag{22.61}$$

and we obtain the equation

$$\delta F = (1 - T)\delta F\, e^{-\lambda t_R} + \frac{AT}{2}(1 + e^{-\lambda t_R})\delta P. \tag{22.62}$$

The expression for δP is still given by Eq. (22.46). By inserting it into Eq. (22.62) and taking into account that $A = 1 + x_s^2$ we get the equation

$$1 = (1 - T)e^{-\lambda t_R} + \frac{T}{2}(1 + e^{-\lambda t_R})\gamma_\perp \frac{\gamma_\parallel(2 - A) + \lambda}{(\lambda + \gamma_\parallel)(\lambda + \gamma_\perp) + \gamma_\parallel\gamma_\perp(A - 1)}. \tag{22.63}$$

In the low-transmission limit $T \ll 1$ we can solve the equation perturbatively by setting $\lambda = \lambda^{(0)} + T\lambda^{(1)} + \cdots$. At zeroth order in T we obtain immediately $\lambda^{(0)} = -i\alpha_n$, where $\alpha_n = 2\pi n/t_R = 2\pi c n/\mathcal{L}$ because the roundtrip time for a ring cavity is $t_R = \mathcal{L}/c$. At first order, using Eq. (12.38) with $\Lambda = \mathcal{L}$ and $t_R = \mathcal{L}/c$, we obtain a result identical to Eq. (22.49).

Therefore the result for the multimode stability analysis with an ultrathin medium ($L \ll \lambda$), for the ring-cavity configuration, is identical to that obtained in the case $L \gg \lambda$, and the boundary of the instability domain is still defined by Eqs. (20.39) (see also Fig. 20.1). Also the spontaneous pulsations which arise from the instability practically coincide with those obtained from the numerical resolution of the Maxwell–Bloch equations (see Section 22.5).

Let us now turn our attention to the case of a Fabry–Perot cavity containing an ultrathin medium, on the basis of the map (14.103) with $y = 0$. We see that the variables $\mathcal{F}$ and $\tilde{P}$ can be assumed real only in the cases $\varphi = 0$ (mod π) and $\varphi = \pi/2$ (mod π). The first corresponds to the configuration in which the nonlinear medium is positioned exactly at an antinode of the standing-wave pattern. The second corresponds to the configuration of a nonlinear medium located at a node, in which the medium practically does not interact with the field and therefore this second case is uninteresting. For $\varphi = 0$ (mod π) the map reduces to

$$\mathcal{F}(t) = R\mathcal{F}(t - t_R) + \frac{AT}{2}\left[\tilde{P}(t) + \tilde{P}(t - t_R) + \tilde{P}(t - \bar{t}) + \tilde{P}(t - t_R + \bar{t})\right], \tag{22.64}$$

where, according to Eq. (14.104), $\bar{t} = 2\bar{z}/c = r t_R$ and

$$r = \frac{\bar{z}}{L} \tag{22.65}$$

is the relative position of the ultrathin nonlinear medium inside the cavity. In this case we obtain, by setting $\mathcal{F}(t) = \mathcal{F}_s + \delta\mathcal{F}\,e^{\lambda t}$, $P(t) = \mathcal{F}_s/(1 + \mathcal{F}_s^2) + \delta P\,e^{\lambda t}$ and $D(t) = 1/(1 + \mathcal{F}_s^2) + \delta D\,e^{\lambda t}$, instead of Eq. (22.62), the equation

$$\delta\mathcal{F} = (1 - T)\delta\mathcal{F}\,e^{-\lambda t_R} + \frac{AT}{2}\left[1 + e^{-\lambda t_R} + e^{-\lambda\bar{t}} + e^{-\lambda(t_R - \bar{t})}\right]\delta\tilde{P}. \tag{22.66}$$

Again, we can use Eq. (22.46) with $x_s = \mathcal{F}_s$ and solve the equation perturbatively in T, obtaining

$$\begin{aligned}
\lambda &= -i\alpha_n - \kappa\left[1 - \gamma_\perp\cos^2(\pi n r)\frac{\gamma_\parallel(1 - \mathcal{F}_s^2) - i\alpha_n}{(\gamma_\parallel - i\alpha_n)(\gamma_\perp - i\alpha_n) + \gamma_\parallel\gamma_\perp\mathcal{F}_s^2}\right] \\
&= -i\alpha_n - \kappa\left[1 - \gamma_\perp\cos^2(\pi n r)\frac{\gamma_\parallel(2 - A/A_{\text{thr}}) - i\alpha_n}{(\gamma_\parallel - i\alpha_n)(\gamma_\perp - i\alpha_n) + \gamma_\parallel\gamma_\perp(A/A_{\text{thr}} - 1)}\right],
\end{aligned} \tag{22.67}$$

where α_n is given by Eq. (14.33) and we have taken into account that $A/A_{\text{thr}} = 1 + \mathcal{F}_s^2$ (see Eq. (14.113) with $\phi = 0$). Amplitude and phase instabilities are still decoupled as in

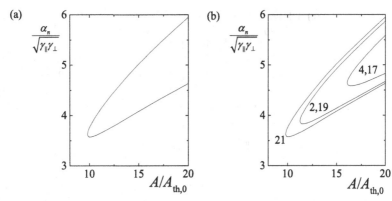

Figure 22.11 (a) The instability domain of the stationary solution $n = 0$ for a Fabry–Perot cavity with $\tilde{\gamma} = 0.13$, $\varphi = 0$ and $r = 1/2$. For a given A the solution is unstable against mode n if n is even and $\alpha_n / \sqrt{\gamma_\parallel \gamma_\perp}$ lies between the two curves. (b) The same for $r = 10/21$. Now different instability domains are associated with different sidemodes. The figure shows the instability domains associated with mode 21, which has the minimum threshold, with the pair 2 and 19, and with the pair 4 and 17. $A_{\text{th},0}$ is the threshold value of A for the stationary solution in which only mode $n = 0$ is active.

the ring cavity, but now the shape of the instability domains depends on the modal index n through the coefficients $\cos^2(\pi n r)$, which range from 0 to 1. For $\cos^2(\pi n r) = 1$ the instability domain coincides with that of the ring cavity when it is plotted as a function of A/A_{thr} because Eq. (22.67) coincides with Eq. (22.49) since $A_{\text{thr}} = 1$ for the ring cavity. Such an instability domain turns out to move rapidly towards larger A and $\tilde{\alpha}_n$ as $\cos^2(\pi n r)$ departs from 1 [148].

This means that a variety of new Risken–Nummedal–Graham–Haken (RNGH)-like instability scenarios are possible in the Fabry–Perot cavity, depending on r. In the simplest case $r = 1/2$, i.e. when the active medium is placed exactly at the center of the cavity, the coefficients $\cos^2(\pi n r)$ can assume only two values: 1, for even n; and 0, for odd n. In the latter case there is no instability. Thus, the RNGH instability domain of Fig. 22.11(a) is the same as for a ring cavity (see Fig. 22.4), but in the Fabry–Perot cavity the resonant stationary solution can be destabilized only by even modes.

In Fig. 22.11(b) the instability domains associated with different sidemodes are shown for a medium slightly displaced from the cavity center, i.e. for $r = 10/21$. In this case only for modes of order $n = 21$ and multiples we recover the same result as for a ring cavity. For other values of n, the instability domains coincide when $n = m$ and $n = 21 - m$, with m being an even integer. In Fig. 22.11(b) we show those for $m = 2$ and $m = 4$, because for larger m the instability threshold is very high.

The circumstance that for a given pump value the resonant mode can be destabilized only by two particular sidemodes, of very different order, has consequences on the laser dynamics after the onset of the instability. Let us consider for instance $A = 12$ and $\alpha_n = 4\sqrt{\gamma_\parallel \gamma_\perp}$. Apart from mode $n = 21$, the resonant mode can be destabilized by modes $n = 2$ and $n = 19$. If the instability is caused by mode $n = 2$ the resulting dynamics, shown in Fig. 22.12(a), is that of two modes beating at the frequency $\tilde{\alpha}_2$. If, instead, the instability is

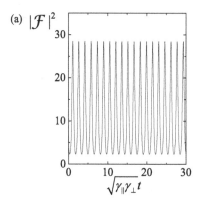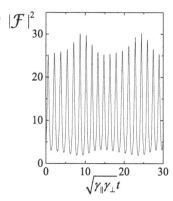

Figure 22.12 The time evolution of the field intensity for the same parameters as in Fig. 22.11(b), with $A = 12$ and (a) $\alpha_2 = 4\sqrt{\gamma_\parallel \gamma_\perp}$ and (b) $\alpha_{19} = 4\sqrt{\gamma_\parallel \gamma_\perp}$.

caused by mode $n = 19$, i.e. the cavity is 9.5 times shorter, then, after an initial transient similar to the previous one, the sidemodes of order $n = 2$ are also excited, and this produces the slow modulation visible in Fig. 22.12(b), at a frequency 9.5 times smaller than the faster one.

Finally, let us analyze the case of generic φ, where the map for a Fabry–Perot laser reads

$$\mathcal{F}(t) = R\mathcal{F}(t - t_R) + \frac{AT}{2} \left[\tilde{P}(t) + \tilde{P}(t - t_R) + e^{2i\varphi} \tilde{P}(t - \bar{t}) + e^{-2i\varphi} \tilde{P}(t - t_R + \bar{t}) \right].$$
(22.68)

In this case the variables $\mathcal{F}$ and $\tilde{P}$ are in general complex. We look for single-mode solutions of this equation and of Eqs. (14.94) and (14.95) with stationary intensity. They have the form

$$\mathcal{F}(t) = \mathcal{F}_s e^{-i\Delta\omega t}, \qquad \tilde{P}(t) = \tilde{P}_s e^{-i\Delta\omega t}, \qquad D = D_s.$$
(22.69)

The equations for $\mathcal{F}_s$, $\tilde{P}_s$ and D_s are

$$\mathcal{F}_s = R\mathcal{F}_s e^{i\Delta\omega t_R} + \frac{AT}{2} \left[1 + e^{i\Delta\omega t_R} + e^{2i\varphi + i\Delta\omega\bar{t}} + e^{-2i\varphi - i\Delta\omega\bar{t} + i\Delta\omega t_R} \right] \tilde{P}_s,$$
(22.70)

$$-i\,\Delta\omega\,\tilde{P}_s = \gamma_\perp \left(\mathcal{F}_s D_s - \tilde{P}_s \right),$$
(22.71)

$$0 = \frac{1}{2} \left(\mathcal{F}_s^* \tilde{P}_s + \mathcal{F}_s \tilde{P}_s^* \right) + D_s - 1.$$
(22.72)

From Eqs. (22.71) and (22.72) we obtain the stationary polarization

$$\tilde{P}_s = \frac{1 + i\,\Delta\omega/\gamma_\perp}{1 + \Delta\omega^2/\gamma_\perp^2 + |\mathcal{F}_s|^2} \mathcal{F}_s,$$
(22.73)

and insert it into Eq. (22.70), obtaining

$$1 = (1 - T)e^{i\Delta\omega t_R} + \frac{AT}{2} \frac{1 + i\,\Delta\omega/\gamma_\perp}{1 + \Delta\omega^2/\gamma_\perp^2 + |\mathcal{F}_s|^2}$$
$$\times \left[1 + e^{i\Delta\omega t_R} + e^{2i\varphi + i\Delta\omega\bar{t}} + e^{-2i\varphi - i\Delta\omega\bar{t} + i\Delta\omega t_R} \right].$$
(22.74)

Again, we solve the equation perturbatively in T. By writing the frequency $\Delta\omega$ as $\Delta\omega_n = \alpha_n + T\,\Delta\omega_n^{(1)} + \cdots$ we get, to first order in T,

$$1 = i\,\Delta\omega_n^{(1)}\, t_R + 2A\,\frac{1 + i\alpha_n/\gamma_\perp}{1 + \alpha_n^2/\gamma_\perp^2 + |\mathcal{F}_{s,n}|^2}\,\cos^2(\varphi + \pi nr). \tag{22.75}$$

By equating separately to zero the real and imaginary parts we obtain

$$1 = \frac{2A}{1 + \alpha_n^2/\gamma_\perp^2 + |\mathcal{F}_{s,n}|^2}\,\cos^2(\varphi + \pi nr), \tag{22.76}$$

$$\Delta\omega_n^{(1)} = -\frac{\alpha_n}{t_R\gamma_\perp}. \tag{22.77}$$

Equation (22.77) gives the first-order displacement of the actual frequency $\omega_n = \omega_c + \Delta\omega_n$ from the resonance frequency ω_c,

$$\Delta\omega_n \simeq \alpha_n\left(1 - \frac{T}{t_R\gamma_\perp}\right) \simeq \frac{\alpha_n}{1 + T/(\tau_R\gamma_\perp)} = \frac{\alpha_n}{1 + \tilde{\kappa}}. \tag{22.78}$$

Note that, if we start from Eq. (9.8) with κ' replaced by κ in the low-transmission limit and observe that $\omega_{0j} = \omega_n$ with $j = \bar{j} + n$, then, since we are considering the resonant case $\omega_a = \omega_c$ and $\tilde{\kappa} = \kappa/\gamma_\perp$, we obtain $\omega_n = \omega_c + \alpha_n/(1 + \tilde{\kappa})$ in agreement with Eq. (22.78), and hence the latter equation coincides with the mode-pulling formula.

From Eq. (22.76) we obtain the stationary intensity $|\mathcal{F}_{s,n}|^2$ for the nth mode,

$$|\mathcal{F}_{s,n}|^2 = 2A\cos^2(\varphi + \pi nr) - 1 - \alpha_n^2/\gamma_\perp^2. \tag{22.79}$$

The lasing threshold for that mode is

$$2A_{th,n} = \frac{1 + \alpha_n^2/\gamma_\perp^2}{\cos^2(\varphi + \pi nr)}. \tag{22.80}$$

The calculation of the relevant eigenvalues when $\varphi \neq 0$ is not straightforward because amplitude and phase instabilities no longer decouple, and we refer the reader to [148] for this point.

In the expression (22.80) for the threshold of the stationary-intensity solution with $n \neq 0$, the numerator displays the parabolic increase of the threshold with frequency typical of two-level atoms, but the denominator, when $\varphi \neq 0$, can compensate for it and make the threshold for a mode with index $n \neq 0$ smaller than that for $n = 0$.

Let us consider for instance a Fabry–Perot cavity with $\varphi = -\pi/5$ and $r = 0.1125$. The instability domain of the stationary solution $n = 0$ with respect to mode $n = 1$ is shown in Fig. 22.13(a) [148]. A new instability domain besides the RNGH-like one appears for smaller values of the pump parameter and frequency offset. The stationary solution $n = 0$ is unstable with respect to mode $n = 1$ already at threshold when the frequency offset $\alpha_1/\sqrt{\gamma_\parallel\gamma_\perp}$ is not too large. Depending on the cavity length, we have two different scenarios: For shorter cavities ($\alpha_1 > 3.873\sqrt{\gamma_\parallel\gamma_\perp}$) we have the RNGH instability; for longer cavities ($1.765\sqrt{\gamma_\parallel\gamma_\perp} < \alpha_1 < 2.2\sqrt{\gamma_\parallel\gamma_\perp}$) we meet a novel instability, which arises much closer to threshold, and causes a switch from mode $n = 1$ to mode $n = 0$ as A is increased, as shown in Fig. 22.13(b). The relevant point is that the switch is not abrupt because there is a range of values of A for which none of the modes is stable and the output

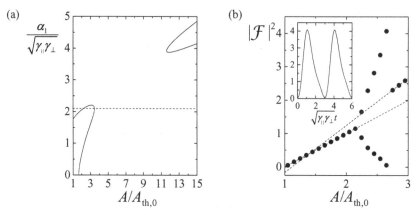

Figure 22.13 (a) The instability domain of the stationary solution $n = 0$ with respect to mode $n = 1$ in a Fabry–Perot cavity with $\varphi = -\pi/5$ and $r = 0.1125$. (b) The output intensity for different A and $\alpha_1 = (2\pi/3)\sqrt{\gamma_\| \gamma_\perp}$, which corresponds to the dashed line of (a). The inset shows the pulsations for $A = 2.64 A_{th,0}$.

is oscillatory. In that case the circles in Fig. 22.13(b) represent the minima and maxima of the oscillations, which are shown in the inset of the figure. After the laser has switched to mode $n = 1$, the opposite switch is observed if A is decreased. The dashed lines represent the stationary intensities of the two single-mode solutions $n = 0$ and $n = 1$. Over a large range of values of A the laser emits the stationary light corresponding to mode $n = 0$, although the intensity of the stationary light corresponding to mode $n = 1$ would be larger. This scenario presents some similarities to that of the multimode phase instability described in Figs. 22.8(b) and 22.10.

An experimental observation of the RNGH instability in a Fabry–Perot quantum cascade laser with ultrathin medium has been achieved by Capasso and his team [265].

23

Single-mode and multimode operation in inhomogeneously broadened lasers

The instabilities in homogeneously broadened lasers, which we described in the previous chapter, have a paradigmatic importance because they arise in models which allow one to perform the linear-stability analysis in a simple analytic form. The treatment of the corresponding instabilities in the case of inhomogeneously broadened lasers is substantially more complex in general. However, it is very interesting because, as we shall see, the instabilities arise in parametric domains that are much more accessible from the experimental viewpoint.

A related issue is that of the possibility of multimodal emission in a laser. It is common knowledge that in order to attain multimode emission it is necessary to have a sort of inhomogeneity in the gain medium, because in a completely homogeneous configuration all modes interact with all atoms, and the arising competition among modes implies that only one of them survives.

There are basically two main kinds of inhomogeneity, the first is of spatial nature and the other is of spectral nature. The first kind is realized in Fabry–Perot lasers, because different cavity modes interact with atoms at different spatial positions in the cavity thanks to the standing-wave configuration of the field and to the consequent phenomenon of spatial hole-burning (see Section 14.2 and Fig. 14.1). This circumstance is capable of weakening the mode–mode competition and allowing multimode operation. In unidirectional ring lasers, on the other hand, there is an "inhomogeneous" configuration in the case of inhomogeneous broadening, because different groups of atoms have different transition frequencies and therefore interact with different modes thanks to the spectral hole-burning (see Section 15.2 and Fig. 15.3). Again, this weakens the mode–mode competition and allows multimode operation.

However, these are qualitative considerations, and the only reliable and quantitative method to determine the possibility of multimodal operation is the linear-stability analysis of the single-mode stationary solutions. On the basis of this procedure, we saw in the previous section that multimode operation in homogeneously broadened lasers is not forbidden, but can arise thanks to the coherence of the atom–field interaction in the case of the amplitude instability described in Section 22.5 or thanks to the mechanism of the phase instability described in Section 22.6. It is difficult, however, to have more than one pair of symmetrical modes become unstable, because to have many unstable modes the cavity must be extremely long. In the case of the amplitude instability there is the additional severe requirement that the laser must be at least nine times above threshold. In this section we demonstrate that such severe limitations, concerning the cavity length and the amount of pumping necessary to reach the instability threshold, can disappear in the case of inhomogeneously broadened lasers.

Also with respect to the single-mode instability described in Section 22.4 the situation improves substantially with inhomogeneous broadening, because the instability threshold in the pump parameter, which coincides with the instability threshold of the amplitude instability, is substantially lowered, and one can obtain the onset of spontaneous oscillations even slightly above threshold for appropriate amounts of inhomogeneous broadening.

In Section 23.1 we calculate the instability boundary in parallel for the multimode and the single-mode instability, focussing on the fully resonant case $\theta = \Delta = 0$, which is the same as for the Risken–Nummedal–Graham–Haken multimode instability [124, 257] and the Lorenz–Haken single-mode instability [239, 240] in the homogeneous case. We limit our analysis to small values of the parameter $\gamma_\parallel/\gamma_\perp$, which situation is common in a large class of lasers, and allows us to perform an analytical treatment. We find a substantial reduction of the instability threshold and, in the multimode case, the relevant possibility of having a large number of modes simultaneously unstable, and therefore a markedly multimodal regime, without any need for a long cavity. On the other hand, Section 23.2 is devoted to the phenomenon of *mode-locking*, in which one can produce a regular train of short pulses.

23.1 Multimode and single-mode instabilities

In the following analysis we will not study the general case, which is considered, however, in [266] (see also [130, 232]). First of all, we assume resonance, i.e. $\omega_a = \omega_c$, which means that $\theta = \Delta = 0$, and assume Lorentzian line broadening, which allows one to perform analytical calculations. Therefore, using Eq. (15.26), the stationary equation reads

$$A_{\mathrm{IB}} = \frac{\xi(\xi + \tilde{\sigma}_{\mathrm{D}})}{1 + \tilde{\sigma}_{\mathrm{D}}}, \tag{23.1}$$

where A_{IB} is defined by Eq. (15.27), ξ by Eq. (15.14), $\tilde{\sigma}_{\mathrm{D}} = \sigma_{\mathrm{D}}/\gamma_\perp$ and σ_{D} is the inhomogeneous linewidth. By solving Eq. (23.1) with respect to ξ we obtain

$$\xi = \frac{1}{2}\left[\sqrt{\tilde{\sigma}_{\mathrm{D}}^2 + 4(1 + \tilde{\sigma}_{\mathrm{D}})A_{\mathrm{IB}}} - \tilde{\sigma}_{\mathrm{D}}\right]. \tag{23.2}$$

The linear-stability analysis of the single-mode stationary solution corresponding to Eq. (23.2), Eqs. (15.10) and (15.11) with $\Delta = 0$ and $W(\bar{\Delta})$ given by Eq. (15.2), follows the lines of Chapter 20.

Let us start with the multimode instability analyzed in Section 20.2 in the homogeneous case. We just have to take into account that in Eqs. (20.27) and (20.28) in the functions T_1 and T_2 the atomic detuning Δ must be replaced by $\Delta + \bar{\Delta} = \bar{\Delta}$, the two functions must be multiplied by the frequency distribution $W(\bar{\Delta})$ and they must be integrated in $\bar{\Delta}$ over the entire real axis. Therefore, the generalized equations for the fluctuations $\delta \bar{f}_n$ and $\delta \bar{f}_{-n}^*$ with inhomogeneous broadening read

$$\frac{d\delta \tilde{f}_n}{d\tau} = -\tilde{\kappa}(1 + i\theta)\delta \tilde{f}_n$$

$$+ \tilde{\kappa} A \int_{-\infty}^{+\infty} d\bar{\Delta}\, W(\bar{\Delta}) \left[T_1(\tilde{\alpha}_n, X, \bar{\Delta}, \tilde{\gamma})\delta \tilde{f}_n + T_2(\tilde{\alpha}_n, X, \bar{\Delta}, \tilde{\gamma})x^2\,\delta \tilde{f}_{-n}^* \right], \quad (23.3)$$

$$\frac{\delta \tilde{f}_{-n}^*}{d\tau} = -\tilde{\kappa}(1 - i\theta)\delta \tilde{f}_{-n}^*$$

$$+ \tilde{\kappa} A \int_{-\infty}^{+\infty} d\bar{\Delta}\, W(\bar{\Delta}) \left[T_2(\tilde{\alpha}_n, X, -\bar{\Delta}, \tilde{\gamma})x^{*2}\,\delta \tilde{f}_n + T_1(\tilde{\alpha}_n, X, -\bar{\Delta}, \tilde{\gamma})\delta \tilde{f}_{-n}^* \right].$$

$$(23.4)$$

The resonant case $\Delta = \theta = 0$ that we are considering allows us to apply the same distinction between amplitude and phase instabilities as in Section 20.5. This is because the functions T_1 and T_2, as a glance at Eqs. (D.11) and (D.12) confirms, can be decomposed into the sum of two terms with even and odd parity with respect to Δ, which is now replaced by $\bar{\Delta}$. Since the frequency distribution is an even function of $\bar{\Delta}$, the integrals containing the odd components of T_1 and T_2 vanish. Hence in Eqs. (23.3) and (23.4) we can replace T_1 and T_2 by their even components T_1^e and T_2^e which, by definition, are such that $T_{1,2}^e(-\bar{\Delta}) = T_{1,2}^e(\bar{\Delta})$.[1] Moreover, we can set $x^2 = x^{*2} = X$, assuming the stationary value x to be real without loss of generality. In this way the two equations are perfectly symmetrical and by defining the new fluctuations $\delta \tilde{f}_n^\pm = \delta \tilde{f}_n \pm \delta \tilde{f}_{-n}^*$ (compare this with Eq. (20.53)) we can uncouple them, obtaining

$$\frac{d\delta \tilde{f}_n^\pm}{d\tau} = \tilde{\kappa} A \int_{-\infty}^{+\infty} d\bar{\Delta}\, W(\bar{\Delta}) \left[T_1^e(\tilde{\alpha}_n, X, \bar{\Delta}, \tilde{\gamma}) \pm T_2^e(\tilde{\alpha}_n, X, \bar{\Delta}, \tilde{\gamma})X \right]\delta \tilde{f}_n^\pm - \tilde{\kappa}\,\delta \tilde{f}_n^\pm.$$

$$(23.5)$$

With the usual exponential ansatz we have

$$\frac{\tilde{\lambda}_\pm}{\tilde{\kappa}} = \mathcal{G}_\pm + i\mathcal{D}_\pm - 1, \qquad (23.6)$$

with

$$\mathcal{G}_\pm = A \int_{-\infty}^{+\infty} d\bar{\Delta}\, W(\bar{\Delta})\mathrm{Re}\,(T_1^e \pm T_2^e X), \quad \mathcal{D}_\pm = A \int_{-\infty}^{+\infty} d\bar{\Delta}\, W(\bar{\Delta})\mathrm{Im}\,(T_1^e \pm T_2^e X).$$

$$(23.7)$$

Note that the $+$ and $-$ signs are associated, respectively, with amplitude and phase fluctuations and that Eqs. (23.5) and (23.7) generalize the homogeneous case, Eqs. (20.62)–(20.65). As in the homogeneous limit, the phase eigenvalue equation does not give rise to any instability, because either it does not lead to any instability boundary, or the instability boundary of the phase instability is entirely contained within the unstable domain of the amplitude instability. Therefore we focus on the amplitude equation and drop the $+$ subscript. From

---

[1] Note that T_1^e coincides with T' defined in Eq. (20.31).

Eqs. (D.11), (D.12) and (D.3) we obtain

$$T_1^e + T_2^e X = \frac{1 - i\tilde{\alpha}_n}{\xi^2 + \bar{\Delta}^2} \frac{1 + 2\Gamma X + \bar{\Delta}^2}{(1 - i\tilde{\alpha}_n)(1 - i\tilde{\alpha}_n - 2\Gamma X) + \bar{\Delta}^2}, \tag{23.8}$$

with $\xi^2 = 1 + X$ (see Eq. (15.14)) and $2\Gamma = \tilde{\gamma}/(i\tilde{\alpha}_n - \tilde{\gamma})$ (see Eq. (D.2)). In the case of Lorentzian broadening (15.2) we can write

$$\mathcal{G} + i\mathcal{D} = \frac{\tilde{\sigma}_D}{\pi} A(1 - i\tilde{\alpha}_n) \int_{-\infty}^{+\infty} d\bar{\Delta} \frac{1 + 2\Gamma X + \bar{\Delta}^2}{(\tilde{\sigma}_D^2 + \bar{\Delta}^2)(\xi^2 + \bar{\Delta}^2)(\chi^2 + \bar{\Delta}^2)}, \tag{23.9}$$

having defined

$$\chi^2 = (1 - i\tilde{\alpha}_n)(1 - i\tilde{\alpha}_n - 2\Gamma X). \tag{23.10}$$

The integral in Eq. (23.9) can be decomposed into the sum of elementary integrals as follows:

$$
\begin{aligned}
\mathcal{G} + i\mathcal{D} &= \frac{\tilde{\sigma}_D}{\pi} A \frac{1 - i\tilde{\alpha}_n}{\xi^2 - \tilde{\sigma}_D^2} \int_{-\infty}^{+\infty} d\bar{\Delta} \left(\frac{1 + 2\Gamma X - \tilde{\sigma}_D^2}{\tilde{\sigma}_D^2 + \bar{\Delta}^2} - \frac{1 + 2\Gamma X - \xi^2}{\xi^2 + \bar{\Delta}^2} \right) \frac{1}{\chi^2 + \bar{\Delta}^2} \\
&= \frac{\tilde{\sigma}_D}{\pi} A \frac{1 - i\tilde{\alpha}_n}{\xi^2 - \tilde{\sigma}_D^2} \left[\frac{1 + 2\Gamma X - \tilde{\sigma}_D^2}{\chi^2 - \tilde{\sigma}_D^2} \int_{-\infty}^{+\infty} d\bar{\Delta} \left(\frac{1}{\tilde{\sigma}_D^2 + \bar{\Delta}^2} - \frac{1}{\chi^2 + \bar{\Delta}^2} \right) \right. \\
&\qquad\qquad \left. - \frac{1 + 2\Gamma X - \xi^2}{\chi^2 - \xi^2} \int_{-\infty}^{+\infty} d\bar{\Delta} \left(\frac{1}{\xi^2 + \bar{\Delta}^2} - \frac{1}{\chi^2 + \bar{\Delta}^2} \right) \right] \\
&= \tilde{\sigma}_D A \frac{1 - i\tilde{\alpha}_n}{\xi^2 - \tilde{\sigma}_D^2} \left[\frac{1 + 2\Gamma X - \tilde{\sigma}_D^2}{\chi^2 - \tilde{\sigma}_D^2} \left(\frac{1}{\tilde{\sigma}_D} - \frac{1}{\chi} \right) - \frac{1 + 2\Gamma X - \xi^2}{\chi^2 - \xi^2} \left(\frac{1}{\xi} - \frac{1}{\chi} \right) \right].
\end{aligned}
\tag{23.11}
$$

By taking Eqs. (23.1) and (15.27) into account we obtain finally

$$\mathcal{G} + i\mathcal{D} = \frac{1 - i\tilde{\alpha}_n}{\xi - \tilde{\sigma}_D} \frac{1}{\chi} \left(\xi \frac{1 + 2\Gamma X - \tilde{\sigma}_D^2}{\chi + \tilde{\sigma}_D} - \tilde{\sigma}_D \frac{1 + 2\Gamma X - \xi^2}{\chi + \xi} \right). \tag{23.12}$$

It is important to observe that Eqs. (23.6) and (23.12) provide the generalization of Eq. (22.49) to the case of inhomogeneous broadening and that the only condition for their validity is $\tilde{\kappa} \ll 1$ ($\kappa \ll \gamma_\perp$) as in the case of Eq. (22.49). Following the lines described in Section 20.2, the boundary of the multimode instability is obtained by solving the equation $\mathcal{G} = 1$ with respect to $\tilde{\alpha}_n$, which gives two solutions, $\tilde{\alpha}_{max}(\xi)$ and $\tilde{\alpha}_{min}(\xi)$. These two curves constitute the instability boundary in the $(\xi, \tilde{\alpha}_n)$ plane for fixed values of $\tilde{\gamma}$ and $\tilde{\sigma}_D$.

Let us now turn our attention to the single-mode instability. As described in Section 20.3, the functions $\mathcal{G}$ and $\mathcal{D}$ allow for obtaining also, and in a straightforward way, the boundary of the single-mode instability, when it exists. As a matter of fact it turns out that, if we call ω the oscillation frequency at the boundary of the single-mode instability, one has that $\tilde{\omega} = \omega/\gamma_\perp = \tilde{\alpha}_{max}$ or $\tilde{\omega} = \tilde{\alpha}_{min}$, and the boundary of the single-mode instability in the plane of the variables $(\xi, \tilde{\kappa})$ is formed by the curves $\tilde{\kappa} = -\tilde{\alpha}_{max}(\xi)/\mathcal{D}(\tilde{\alpha}_{max}(\xi))$ and $\tilde{\kappa} = -\tilde{\alpha}_{min}(\xi)/\mathcal{D}(\tilde{\alpha}_{min}(\xi))$ for fixed values of $\tilde{\gamma}$ and $\tilde{\sigma}_D$. The single-mode instability exists only on condition that the functions $\mathcal{D}(\tilde{\alpha}_{max}(\xi))$ and $\mathcal{D}(\tilde{\alpha}_{min}(\xi))$ are negative. Note that, in contrast with the multimode instability, the single-mode instability is not subject to the condition $\tilde{\kappa} \ll 1$.

From the simple-looking expression (23.12) it is not actually easy to obtain $\mathcal{G}$ and $\mathcal{D}$ in the general case, because the parameter χ is complex. For this reason, in the following we limit our analysis to the condition $\tilde{\gamma} \ll 1$, which is anyway common to the greatest majority of lasers, in the following two cases:

(i) $\tilde{\alpha}_1 \gtrsim 1$ and
(ii) $\tilde{\alpha}_1$ of order $\tilde{\gamma}^{1/2}$.

In this treatment we follow basically the paper [267] insofar as the multimode instability is concerned.

23.1.1 The case $\tilde{\gamma} \ll 1, \tilde{\alpha}_1 \gtrsim 1$

In this limit Γ is of order $\tilde{\gamma}$ and according to Eq. (23.10) we can set $\chi \simeq 1 - i\tilde{\alpha}_n$. This allows us to write

$$\mathcal{G}(\tilde{\alpha}_n, \xi, \tilde{\sigma}_D) = \frac{1}{\xi - \tilde{\sigma}_D} \left[\xi \frac{(1 + \tilde{\sigma}_D)^2 (1 - \tilde{\sigma}_D)}{(1 + \tilde{\sigma}_D)^2 + \tilde{\alpha}_n^2} - \tilde{\sigma}_D \frac{(1 + \xi)^2 (1 - \xi)}{(1 + \xi)^2 + \tilde{\alpha}_n^2} \right], \quad (23.13)$$

$$\mathcal{D}(\tilde{\alpha}_n, \xi, \tilde{\sigma}_D) = \frac{\tilde{\alpha}_n}{\xi - \tilde{\sigma}_D} \left[\xi \frac{1 - \tilde{\sigma}_D^2}{(1 + \tilde{\sigma}_D)^2 + \tilde{\alpha}_n^2} - \tilde{\sigma}_D \frac{1 - \xi^2}{(1 + \xi)^2 + \tilde{\alpha}_n^2} \right]. \quad (23.14)$$

After some calculations one finds that the condition $\mathcal{G} = 1$, which defines the multimode instability boundary, is satisfied when $\tilde{\alpha}_n = \tilde{\alpha}_{\min}$ and $\tilde{\alpha}_n = \tilde{\alpha}_{\max}$ with

$$\tilde{\alpha}_{\min} = 0, \qquad \tilde{\alpha}_{\max}^2 = \xi^2(\tilde{\sigma}_D - 1) + \xi(\tilde{\sigma}_D - 1)(\tilde{\sigma}_D + 2) - (\tilde{\sigma}_D + 1)^2 \equiv \mathcal{A}(\xi, \tilde{\sigma}_D). \tag{23.15}$$

Obviously the zero value of $\tilde{\alpha}_{\min}$ is incompatible with our initial assumption that $\tilde{\alpha}_1 \gtrsim 1$, hence we keep only the upper boundary $\tilde{\alpha}_{\max}$. The instability exists only if $\mathcal{A} > 0$, which implies that $\tilde{\sigma}_D > 1$ and $\xi > \xi_{\min}$ with

$$\xi_{\min} = \frac{\tilde{\sigma}_D + 2}{2} \left[\sqrt{1 + \frac{4(\tilde{\sigma}_D + 1)^2}{(\tilde{\sigma}_D - 1)(\tilde{\sigma}_D + 2)^2}} - 1 \right]. \tag{23.16}$$

Using Eq. (23.2), we plot in Fig. 23.1 the instability boundary $\tilde{\alpha}_{\max}$ as a function of the pump parameter A_{IB} instead of ξ, for various values of $\tilde{\sigma}_D$. The unstable domain lies to the right of each curve. Obviously we must disregard the part of the boundary curve where $\tilde{\alpha}_n \ll 1$.

We observe that the results described by Eqs. (23.13) and (23.15) were first obtained by Khanin [230] by following a somewhat different procedure, in which the conditions $\tilde{\gamma} \ll 1$ and $\tilde{\alpha}_1 \gtrsim 1$ justified neglecting the fluctuations of population inversion.

Let us now compare the scenario of the multimode instability in inhomogeneously broadened lasers with $\tilde{\kappa} \ll 1$, $\tilde{\gamma} \ll 1$ and $\tilde{\alpha}_1 \gtrsim 1$ with that in homogeneously broadened lasers with $\tilde{\kappa} \ll 1$ and $\tilde{\gamma} \ll 1$.

The first major difference concerns the instability threshold. We have seen from Eq. (22.51) that in the homogeneous configuration the instability threshold ($A = 9$) is

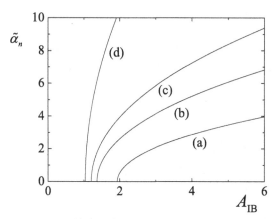

The boundary of multimode instability in the limit $\tilde{\gamma} \ll 1$ and $\tilde{\alpha}_n$ of order 1 for (a) $\tilde{\sigma}_D = 2$, (b) $\tilde{\sigma}_D = 3$, (c) $\tilde{\sigma}_D = 4$ and (d) $\tilde{\sigma}_D = 10$.

nine times larger than the lasing threshold ($A = 1$). On the other hand, from Eq. (23.16) in the inhomogeneous limit $\tilde{\sigma}_D \gg 1$ we find $\xi_{\min} \simeq 1 + 4/\tilde{\sigma}_D^2$. Furthermore, for $\tilde{\sigma}_D \gg 1$ we have from Eqs. (15.29) and (15.14) that $\xi \simeq A_{IB}$, hence we conclude that the instability starts from the value

$$A_{IB,\min} \simeq 1 + \frac{4}{\tilde{\sigma}_D^2}, \tag{23.17}$$

which is close to the lasing threshold $A_{IB} = 1$.

Let us also remark that, when the parameter $\tilde{\sigma}_D$ becomes large, the boundary curve of the multimode instability in the inhomogeneously broadened case displays further features, in addition to the instability threshold, which are drastically different from those of the boundary curves in the Risken–Nummedal–Graham–Haken instability in the homogeneously broadened case.

As a matter of fact, as we saw in Section 22.5, in the homogeneous configuration the boundary curves $\alpha_{\max}$ and $\alpha_{\min}$ are on the order of $\sqrt{\gamma_\parallel \gamma_\perp A}$ when the pump parameter A is large. On the other hand, in the inhomogeneous configuration from Eq. (23.15) we have for $\tilde{\sigma}_D$ large that $\alpha_{\max} \simeq \gamma_\perp \tilde{\sigma}_D \sqrt{\xi - 1}$. Since for $\tilde{\sigma}_D \gg 1$ we have that $\xi \simeq A_{IB}$, for $A_{IB} \gg 1$ we can conclude that $\alpha_{\max} \simeq \gamma_\perp \tilde{\sigma}_D \sqrt{A_{IB}}$.

Therefore, if we compare the cases of homogeneous broadening and strong inhomogeneous broadening,[2] we find a double benefit in the latter: In the expression for $\alpha_{\max}$, $\gamma_\parallel$ is replaced by $\gamma_\perp$ and, next, $\gamma_\perp$ is multiplied by $\tilde{\sigma}_D \gg 1$, which amounts to replacing the homogeneous linewidth $\gamma_\perp$ by the inhomogeneous linewidth σ_D (see Eq. (15.2)). This means that the frequency band of the instability domain is dramatically enlarged from a narrow band to a wide band, i.e. the number of modes that become unstable can be large. A related benefit is that the need for a very long cavity, which constitutes a major problem in the homogeneous configuration, disappears in the inhomogeneous case. If we look at Eq. (22.56), we see that the minimum cavity length Λ_c scales as the inverse of $\sqrt{\gamma_\perp \gamma_\parallel}$. On

---

[2] The linear-stability analysis in the limit $\tilde{\sigma}_D \rightarrow \infty$ has been discussed in [152, 232, 268].

the basis of the previous considerations, we can say that, in the case of the inhomogeneous configuration with $\tilde{\sigma}_D \gg 1$, this value is reduced by a factor $\tilde{\sigma}_D/\tilde{\gamma}^{1/2} \gg 1$.

The previous considerations led to the conclusion that inhomogeneously broadened lasers with large inhomogeneous broadening can be easily operated in a strongly multi-modal regime for values of the pump parameter A_{1B} close to the threshold for single-mode operation, without the problem of a long-cavity condition. In the case of homogeneously broadened lasers, in contrast, for the amplitude instability the multimode instability thres-hold is high and the problem of the long-cavity condition is severe. These problems are reduced for the phase instability in homogeneously broadened lasers, as we saw in Sec-tion 22.6, but the situation is much better in the case of the multimode amplitude instability in strongly inhomogeneously broadened lasers.

Let us now turn our attention to the single-mode instability, still under the conditions $\tilde{\gamma} \ll 1$ and $\tilde{\alpha}_1 \gtrsim 1$. Indeed, in this case no single-mode instability is associated with the multimode instability because $\mathcal{D}(\tilde{\alpha}_{max})$ is positive.

23.1.2 The case $\tilde{\gamma} \ll 1, \tilde{\alpha}_1$ of order $\tilde{\gamma}^{1/2}$

Now the free spectral range α_1 has the same order of magnitude as the Rabi frequency of the single-mode solution, a necessary condition for the appearance of the Risken–Nummedal–Graham–Haken (multimode) and Lorenz–Haken (single-mode) instabilities in the homogeneous limit. Here we generalize those instabilities to arbitrary values of $\tilde{\sigma}_D$.

To this end we set $\tilde{\alpha}_n = \alpha'_n \tilde{\gamma}^{1/2}$ with $\alpha'_n = \alpha_n/(\gamma_\perp \gamma_\parallel)^{1/2}$ of order 1 and expand the r.h.s. of Eq. (23.12) in power series of $\tilde{\gamma}^{1/2}$ up to order $\tilde{\gamma}$. We find that the terms of order 1 and $\tilde{\gamma}$ are real, while the term of order $\tilde{\gamma}^{1/2}$ is purely imaginary. Therefore the gain function $\mathcal{G}$ can be written as

$$\mathcal{G} = 1 + \frac{\mathcal{A}(\xi, \tilde{\sigma}_D)\alpha'^4_n + \mathcal{B}(\xi, \tilde{\sigma}_D)\alpha'^2_n + \mathcal{C}(\xi, \tilde{\sigma}_D)}{(1+\xi)^2(1+\tilde{\sigma}_D)^2\alpha'^2_n}\tilde{\gamma}, \qquad (23.18)$$

with $\mathcal{A}(\xi, \tilde{\sigma}_D)$ given by Eq. (23.15) and

$$\mathcal{B}(\xi, \tilde{\sigma}_D) = 3(\xi^2 - 1)\left[(\xi + \tilde{\sigma}_D + 1)^2 - \xi\tilde{\sigma}_D\right], \qquad (23.19)$$

$$\mathcal{C}(\xi, \tilde{\sigma}_D) = -\xi(\xi - 1)(\xi + 1)^3(\xi + \tilde{\sigma}_D)(\tilde{\sigma}_D + 2), \qquad (23.20)$$

while the dispersion function $\mathcal{D}$ is

$$\mathcal{D} = \frac{(1 + \xi + \tilde{\sigma}_D - \xi\tilde{\sigma}_D)\alpha'^2_n - 2(1 + \xi + \tilde{\sigma}_D)(\xi^2 - 1)}{(1+\xi)(1+\tilde{\sigma}_D)\alpha'_n}\tilde{\gamma}^{1/2}. \qquad (23.21)$$

One can easily verify that in the homogeneous limit $\tilde{\sigma}_D = 0$ Eqs. (23.18) and (23.21) reduce to expressions that coincide with those obtained from Eqs. (20.38) and (20.47), respectively, if one sets $\tilde{\alpha}^2_n = \alpha'^2_n\tilde{\gamma}$ in Eq. (20.38) and $\tilde{\omega}^2 = \alpha'^2\tilde{\gamma}$ in Eq. (20.47) and expands in powers of $\tilde{\gamma}$.

Note that, on the basis of the assumption that $\tilde{\alpha}_1$ is of order $\tilde{\gamma}^{1/2}$, in Eqs. (23.18) and (23.21) one cannot take $\alpha'_n \ll 1$.

The condition $\mathcal{G} = 1$ for multimode instability is equivalent to the condition that the numerator in the fraction of Eq. (23.18) vanishes. This gives two roots $(\alpha'_{min})^2$ and $(\alpha'_{max})^2$

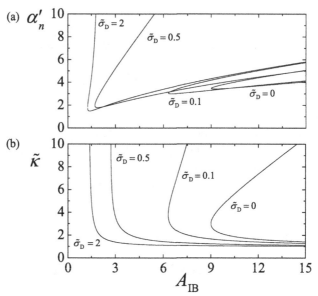

Figure 23.2 The boundaries of (a) multimode and (b) single-mode instability in the limit $\tilde{\gamma} \ll 1$ and $\tilde{\alpha}_n$ of order $\tilde{\gamma}^{1/2}$ for $\tilde{\sigma}_D = 0$ (the homogeneous limit), $\tilde{\sigma}_D = 0.1$, $\tilde{\sigma}_D = 0.5$ and $\tilde{\sigma}_D = 2$.

given by

$$\left(\alpha'_{\substack{max \\ min}}\right)^2 = \frac{-\mathcal{B} \mp \sqrt{\mathcal{B}^2 - 4\mathcal{A}\mathcal{C}}}{2\mathcal{A}}. \tag{23.22}$$

Since $\mathcal{B} > 0$ and $\mathcal{C} < 0$, it is easy to verify that the instability threshold ξ_{th} ($A_{IB,th}$), which is determined by the equation

$$\mathcal{A} = \frac{\mathcal{B}^2}{4\mathcal{C}}, \tag{23.23}$$

occurs for $\mathcal{A} < 0$ (see Eq. (23.15)) and, therefore, when $\tilde{\sigma}_D > 1$ (i.e. when ξ_{min} does exist), we have $\xi_{th} < \xi_{min}$ ($A_{IB,th} < A_{IB,min}$), where ξ_{min} is given by Eq. (23.16). It is easy to verify that $(\alpha'_{min})^2$ is positive for $A_{IB} > A_{IB,th}$ whereas, for $\tilde{\sigma}_D > 1$, α'_{max} exists only for $\mathcal{A} < 0$, i.e. $A_{IB,th} < A_{IB} < A_{IB,min}$, and the curve of α'_{max} has a vertical asymptote as $A_{IB} \to A_{IB,min}^-$, since $\mathcal{A} \to 0$ (see Eq. (23.22)).

By taking into account Eqs. (23.15), (23.19) and (23.20) one can find that Eq. (23.23) leads to a fifth-order equation for ξ, the solution of which gives the value of the instability threshold ξ_{th} as a function of $\tilde{\sigma}_D$.

The boundary for the single-mode instability in the ($A_{IB}, \tilde{\kappa}$) plane is given by $\tilde{\kappa}_{min} = -\alpha'_{min}\tilde{\gamma}^{1/2}/\mathcal{D}(\alpha'_{min})$ and $\tilde{\kappa}_{max} = -\alpha'_{max}\tilde{\gamma}^{1/2}/\mathcal{D}(\alpha'_{max})$. Figure 23.2 shows the boundaries for the multimode (a) and single-mode (b) instabilities for various values of the inhomogeneous broadening. The instability threshold is greatly reduced with respect to the homogeneous case even for moderate values of $\tilde{\sigma}_D$. It can be shown that $A_{IB,th} \simeq 9 - 36\tilde{\sigma}_D$ for $\tilde{\sigma}_D \ll 1$. On the other hand, the "bad-cavity" condition $\tilde{\kappa} > 1$ for the single-mode instability in the homogeneous case persists in the inhomogeneous configuration.

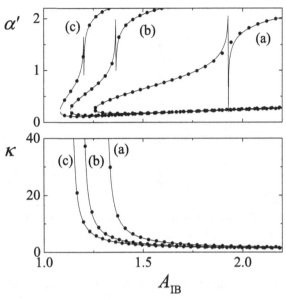

Figure 23.3 Complete boundaries of (top) multimode and (bottom) single-mode instability for (a) $\tilde{\sigma}_D = 2$, (b) $\tilde{\sigma}_D = 3$ and (c) $\tilde{\sigma}_D = 4$, with $\tilde{\gamma} = 10^{-4}$. The solid curves are obtained from the analytical results, the dots from the numerical results (see the text).

The substantial decrease of the instability threshold both in the multimode instability and in the single-mode instability holds also when $\tilde{\gamma}$ is of order unity, see [130, 152, 232].

Note that the vertical asymptote in the expression (23.22) for α'_{max} for $\xi = \xi_{min}$ ($A_{IB} = A_{IB,min}$) is an artifact of the approximation which leads to Eq. (23.18), because the upper boundary of the multimode instability in Fig. 23.2(a) for $A_{IB} < A_{IB,min}$ connects to that shown in Fig. 23.1 when $\tilde{\sigma}_D > 1$, i.e. when $A_{IB,min}$ exists. The latter starts just at $A_{IB} = A_{IB,min}$.

The complete picture is shown in the top panel of Fig. 23.3, where, for three different values of $\tilde{\sigma}_D > 1$, the full curve is obtained from Eq. (23.22) to plot $\alpha'_{min}(A_{IB})$ for $A_{IB} > A_{IB,th}$ and to plot $\alpha'_{max}(A_{IB})$ for $A_{IB,th} < A_{IB} < A_{IB,min}$, and from Eq. (23.15) to plot $\alpha'_{max}(A_{IB})$ for $A_{IB} > A_{IB,min}$. The circles indicate the numerical results for the boundary of the multimode instability obtained from the exact expression (23.12). We see that, apart from a neighborhood of $A_{IB,min}$, the analytical expressions that we derived for the boundary provide an excellent approximation.

On the other hand, in the bottom panel of Fig. 23.3 we observe, for the same values of $\tilde{\sigma}_D > 1$, the instability boundary of the single-mode instability. The solid curve is obtained from the analytical results (23.22) and (23.21) by taking into account that the boundary in the $(A_{IB}, \tilde{\kappa})$ plane corresponds to $\tilde{\kappa} = -\alpha'_{min}(A_{IB})\tilde{\gamma}^{1/2}/\mathcal{D}(\alpha'_{min}(A_{IB}), A_{IB})$. Again, the circles correspond to the boundary which is obtained numerically by using Eq. (23.12) and, again, the agreement between analytical and numerical results is excellent.

From Fig. 23.2 we see that when $\tilde{\sigma}_D < 1$ both the boundary of the multimode instability and that of the single-mode instability present an instability threshold $A_{IB,th}$, which is the same for both cases. On the other hand, from Fig. 23.3 we see that, for $\tilde{\sigma}_D > 1$, only the

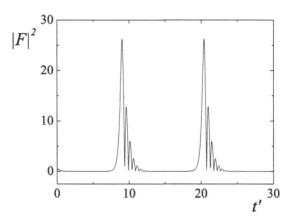

Figure 23.4 The time evolution of the field intensity for $A_{IB} = 8$, $\tilde{\sigma}_D = 4$, $\tilde{\kappa} = 4$ and $\tilde{\gamma} = 0.05$ obtained by numerically solving Eqs. (15.7)–(15.9) for inhomogeneously broadened lasers.

multimode instability presents an instability threshold, whereas the single-mode instability presents a vertical asymptote, which in this case is real rather than being the artifact of an approximation, and is positioned at a value $A_{IB} = A_{IB,as}$ such that $A_{IB,th} < A_{IB,as} < A_{IB,min}$, where $A_{IB,th}$ denotes the threshold of the multimode instability. This feature arises from the fact that $\mathcal{D}(\alpha'_{min}(A_{IB}), A_{IB})$ vanishes for $A_{IB} = A_{IB,as}$. For $A_{IB,th} < A_{IB} < A_{IB,as}$ $\mathcal{D}$ is positive and therefore there is no boundary for the single-mode instability. The position of the asymptote can be obtained by coupling the two equations $\mathcal{G}(\alpha'^2, \xi, \tilde{\sigma}_D) = 1$ and $\mathcal{D}(\alpha'^2, \xi, \tilde{\sigma}_D) = 0$ and by eliminating α'^2 between them. This leads again to a fifth-order equation for ξ, the solution of which provides the value of the asymptote ξ_{as} as a function of $\tilde{\sigma}_D$.

From Fig. 23.3 we see that the multimode instability is associated with a single instability domain, but for $\tilde{\sigma}_D > 1$ this domain is subdivided into two parts, one on the left and the other on the right of the line $A_{IB} = A_{IB,min}$. In the two parts the instability displays quite different features. On the left, the instability represents the generalization of the Risken–Nummedal–Graham–Haken instability of the homogeneous case, and the instability threshold is much lower. The frequency of the unstable modes is on the order of the Rabi frequency of the intracavity field, and to reach the instability it is necessary that the cavity length is large. On the other hand, on the right the frequency of the unstable modes can be much larger than the Rabi frequency, the instability domain extends to large values of $\tilde{\alpha}_n$ and one can have a large number of unstable modes, i.e. a markedly multimodal operation, without any need for a long cavity. Inhomogeneously broadened lasers have not been investigated experimentally in a specific search for the type of multimode instabilities discussed in this section. However, in the early years of operation of gas lasers, multimodal operation was observed frequently.

On the other hand, the circumstance that the instability threshold $A_{IB,th}$ (or the asymptote $A_{IB,as}$ when $\tilde{\sigma}_D > 1$) is substantially lower than the instability threshold of the homogeneous case has allowed various experimental observations of the single-mode instability in the inhomogeneous case to be made, see [269, 270] and references quoted in [130], and has primed a noteworthy literature on single-mode instabilities in inhomogeneously

(a)

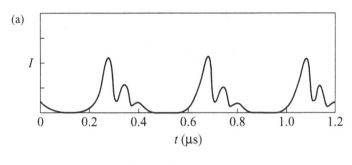

(b)

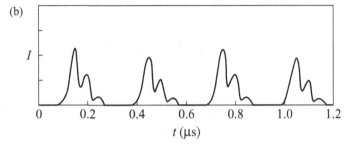

Figure 23.5 Experimental plots of output pulsations (arbitrary units) of the oscillatory instability with time in a 3.51-μm xenon laser (Copyright (1978) by IEEE. Reprinted, with permission, from ref [269]).

broadened lasers. Figure 23.4 shows a numerical example of the pulsing solution that arises in the single-mode model for $\tilde{\gamma} \ll 1$. As shown by Fig. 23.5, the shape of these pulses is qualitatively similar to that observed in some experiments by Casperson in xenon lasers [269].

23.2 Mode-locking

When a laser operates in a multimodal regime such as that described e.g. by Eqs. (20.3)–(20.5), one speaks of *mode-locking* when the set of modes is equispaced not only in frequency but also in phase. For example, in the case described by Eq. (22.52) the phases of the mode amplitudes f_n obey the relation (22.53), i.e. they are multiples of the same phase ϕ. As we saw in Section 22.5, this condition is realized in the case of the Risken–Nummedal–Graham–Haken amplitude instability, thanks to the circumstance that the electric field is coherent and all modes interact with the same atoms. More precisely, this corresponds to a self-mode-locking, because it is spontaneously realized by the laser when the multimode instability arises. In the case of an inhomogeneously broadened laser, in contrast, the modes oscillate independently of one another, because modes with different frequencies interact with different atomic packets, and hence the phases ϕ_n are no longer constant. With appropriate procedures, however, it is possible to obtain that the modal phases are not only constant but also equal to integer multiples of a phase ϕ, thus realizing a mode-locked configuration [6]. If we assume that we have N consecutive modes, all with unitary

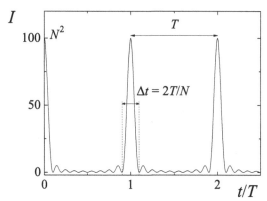

Figure 23.6 A mode-locked train of pulses according to Eq. (23.26) with $N = 10$ and $T = 2\pi/\alpha_1$.

amplitude, we can write

$$F(t) = \sum_{n=0}^{N-1} e^{-in(\alpha_1 t + \phi)} \longrightarrow \sum_{n=0}^{N-1} e^{-in\alpha_1 t}, \qquad (23.24)$$

where we have conveniently redefined the time origin in such a way that ϕ disappears from the equation. Therefore we obtain

$$F(t) = \frac{1 - e^{-iN\alpha_1 t}}{1 - e^{-i\alpha_1 t}} = \frac{e^{-iN\alpha_1 t/2}}{e^{-i\alpha_1 t/2}} \frac{e^{iN\alpha_1 t/2} - e^{-iN\alpha_1 t/2}}{e^{i\alpha_1 t/2} - e^{-i\alpha_1 t/2}} = \frac{e^{-iN\alpha_1 t/2}}{e^{-i\alpha_1 t/2}} \frac{\sin(N\alpha_1 t/2)}{\sin(\alpha_1 t/2)}, \quad (23.25)$$

so the emitted intensity displays the temporal behavior

$$I(t) = |F(t)|^2 = \frac{\sin^2(N\alpha_1 t/2)}{\sin^2(\alpha_1 t/2)}, \qquad (23.26)$$

with an expression analogous to the interference term in a diffraction grating. The temporal evolution is therefore identical to the spatial configuration produced by the grating. In particular we have the following (see Fig. 23.6).

(i) The intensity is maximal and proportional to N^2 when $\alpha_1 t/2 = n\pi$ with n integer, i.e. the emission is in the form of pulses with period $T = 2\pi/\alpha_1 = \Lambda/c$.

(ii) The average intensity is proportional to N, because the interference terms have zero average in time.

(iii) The intensity vanishes, for the first time after the nth peak, for $\alpha_1 t/2 = n\pi + \pi/N$, i.e. for $t = nT + T/N$, because for this value of t the numerator in Eq. (23.26) vanishes while the denominator does not vanish. Hence the peak duration, defined as the time interval between the two instants at which the intensity vanishes before and after the peak, is given by $\Delta t = 2T/N$.

From these remarks it follows that, when the number N of mode-locked modes increases, the peak intensity increases and their duration decreases. The pulse is more localized in time and space the less localized it is in Fourier space, i.e. the more numerous its spectral components. The formation of a comb of regular pulses arises from the circumstance that the relative phases of the modes are constant in time and obey condition (22.53). For this reason one speaks of mode-locking, which means that the modal phases are coupled to one another.

In homogeneously broadened lasers the mode-locking is granted by the coherent nature of the process which leads to multimodal emission. In the case of inhomogeneous broadening, in contrast, the emission from the various modes is in general uncorrelated, such that the mode-locking is not spontaneous, but must be induced by an external action. A commonly used procedure consists in modulating the resonator losses at a frequency equal to the free spectral range α_1. In this way the laser has gain larger than losses only in periodic intervals of time. The emission pattern which best matches this operating condition is that in which the intensity is maximal in the time slots where the losses are low and minimal in the temporal slots where the losses are high. In this way the mode-locked solution prevails over the other possible configurations in which the phases are random, through a sort of Darwinian mechanism of "survival of the fittest". Actually what happens is that the possible pulse tails which might develop due to the random fluctuations of the phases get suppressed by the high losses, and in such a way the ideal situation of constant relative phases is recovered. Further details about mode-locking procedures can be found in [6].

24

Dynamical aspects in optical bistability

Although the discovery of oscillatory instabilities (both single-mode and multimode) in free-running lasers was achieved in the 1960s, investigations on oscillatory instabilities in optical bistability did not take place until the 1970s and 1980s.

The first section of this chapter does not deal with oscillatory instabilities, but instead concerns an interesting dynamical phenomenon in optical bistability that arises when the input field approaches a boundary of the bistable domain, and is called *critical slowing down*. This has been investigated both theoretically and experimentally.

The remainder of this chapter is entirely devoted to oscillatory instabilities, following a historical order. With an inverse approach with respect to the laser case, we discuss multimode instabilities first. In Section 24.2.1 we start with the description of the instability which arises under exactly resonant conditions. While this is the counterpart in optical bistability of the resonant multimode laser instability discussed in Section 22.5, a striking difference from the laser case is that in the passive configuration there is no corresponding single-mode instability. This difference disappears, however, when we consider the generalization of the multimode instability of optical bistability to the detuned configuration. Finally we show that in the detuned case it has been possible to achieve a convincing experimental observation of the multimode instability in a long microwave cavity.

After some relevant miscellaneous considerations in Section 24.2.2, in Section 24.2.3 we address the subject of the multimode Ikeda instability which leads to period doubling and chaos and aroused a lot of interest in optical chaos. We emphasize also the intrinsic connections with the standard multimode instability discussed in Section 24.2.1, showing, in particular, that the Ikeda instability requires conditions far from the low-transmission limit, i.e. that αL is not small.

In Section 24.3 we discuss single-mode instabilities in optical bistability. In Section 24.3.1 we illustrate a configuration for the single-mode instability that includes also the Gaussian shape of the electric field in the beam section, and leads to periodic, close-to-sinusoidal self-pulsations. This behavior has been observed experimentally in careful experiments by Kimble and collaborators. In Section 24.3.2 we turn our attention to the pure plane-wave model, in which we find scenarios of period doubling and chaos.

24.1 Critical slowing down

The stationary hysteresis cycle of optical bistability can be experimentally observed by sweeping the incident field back and forth in a slow (adiabatic) fashion, so that the system can continuously adapt itself to each new value of the input field.

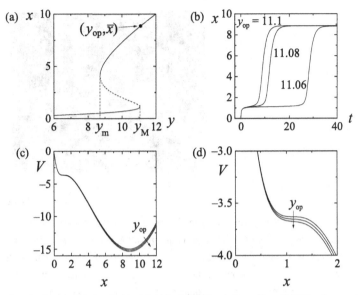

Figure 24.1 (a) The hysteresis cycle of transmitted vs. incident field for $C = 10$ and $\Delta = \theta = 0$. (b) The time evolution of the transmitted field, showing "lethargy" and critical slowing down for three values of y_{op} approaching $y_M = 11.056$. Time is expressed in units of κ^{-1}. (c) The shape of the potential $V(x)$ for the same values of y_{op} increasing in the direction of the arrow. (d) Enlargement of the previous figure around the inflection point of V.

In this section, we consider the transient approach of the system to the stationary state. This approach is governed by the characteristic time constants κ, $\gamma_\perp$ and $\gamma_\parallel$. In Section 18.1 we mentioned the general phenomenon of *critical slowing down*, which arises when we approach a stability boundary in parameter space. If we slightly displace the state of the system from the stationary value, the time the system takes to return to the stationary state becomes longer and longer as the stability boundary is approached. This phenomenon occurs, in particular, when the value of the input field approaches the upper bistability threshold y_M from below or the lower bistability threshold y_m from above (see Figs. 11.4 and 24.1(a)).

Here we focus on an interesting effect that is intrinsically connected to the critical slowing down, which occurs on the other side of the bistability thresholds [271]. Let us consider for definiteness Eq. (13.26), which governs optical bistability in the low-transmission, single-mode and good-cavity limits and let us focus on the resonant configuration $\Delta = \theta = 0$. To be precise, let us think of the following experiment. Let us assume that initially the system lies in the stationary state with zero input field, so that $x(t = 0) = 0$. At this point, we abruptly switch the incident light to some operative value y_{op} larger than the upper bistability threshold y_M, so that the transmitted field approaches the stationary value $\bar{x}$ in the high-transmission branch corresponding to the value y_{op} of the input field.

Equation (13.26) with $\Delta = \theta = 0$ can be solved numerically or analytically [271, 272] and gives the picture shown in Fig. 24.1(b). Clearly, the approach to the high-transmission stationary state exhibits a kind of "lethargy", and the time the system takes to reach it

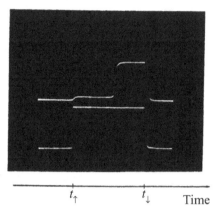

$t_\uparrow$　　　　$t_\downarrow$　　Time

Figure 24.2   An experimental observation of critical slowing down in microwave absorptive optical bistability in ammonia. Upper trace: transmitted power; lower trace: incident power. At $t = t_\uparrow$ the incident field is switched on to a value slightly larger than the bistability threshold y_M. At $t = t_\downarrow$ the incident power is switched off. In this experiment $\gamma_\parallel^{-1}$ is on the order of a microsecond, and the cavity buildup time is $\approx$60 ns. The gas pressure is on the order of 1 mTorr and the mirror transmissivity is 0.01. Reprinted from [273] with kind permission from Springer Science and Business Media.

becomes longer and longer as $y_{op} \to y_M$. This phenomenon can be intuitively understood on the basis of the "potential" picture provided by Eqs. (19.22) and (19.23) and by Fig. 19.3. As a matter of fact, when y_{op} is only slightly larger than y_M the potential V presents a flat part that the system takes a long time to cover and becomes flatter and flatter as y_{op} approaches y_M (Figs. 24.1(c) and (d)), and hence the time of approach to the stationary state becomes longer and longer.

This critical slowing down has been observed experimentally [273–275]. Figure 24.2 shows an experimental observation by the group led by Gozzini at microwave frequencies.

The same kind of scenario is met if, considering an initial condition such that the system is in the high-transmission branch, the value of the input field is abruptly reduced to a value y_{op} slightly smaller than the lower bistability threshold y_m.

The analysis of this section shows that due to the critical slowing down the time of approach to the stationary state can be much longer than the relaxation times of the system, i.e. the inverses of the rates κ, $\gamma_\perp$ and $\gamma_\parallel$. This implies that, in order to observe the stationary hysteresis cycle, the incident field must be swept back and forth in a time 100–1000 times longer than the largest characteristic time in play. If the incident field intensity is swept more rapidly, the cycle turns out to be rounded and the transitions from the lower to the upper branch and vice versa are no longer discontinuous. This effect has been discussed extensively in [232].

If one adds thermal noise, the lethargic stage of critical slowing down produces the effect of transient bimodality, i.e. the onset of a bimodal (two-peaked) probability distribution during the transient approach to the stationary state. This phenomenon was theoretically predicted [276] and experimentally observed [277].

24.2 Multimode instabilities in optical bistability

24.2.1 The Bonifacio–Lugiato instability

The resonant case

Let us start from the resonant configuration $\Delta = \theta = 0$, following the treatment of Section 20.2. In this subsection we outline the resonant multimode amplitude instability of optical bistability predicted in [278], which we call the Bonifacio–Lugiato (BL) instability in the following. With respect to the laser case treated in Section 22.5, the difference is not only that the pump parameter A must be replaced by $-2C$, where C is the bistability parameter but, especially, that in the laser case the pump parameter A and the output intensity X are linked by the stationary equation $X = A - 1$, whereas in optical bistability C and X are independent parameters. As a matter of fact, the stationary equation (11.34) involves also the input intensity Y, which is independent of C. As we stipulated in Chapter 20, we use X instead of Y as a parameter, and therefore X and C are independent parameters. By using the equations of Section 22.5, one finds the following expression for the gain $\mathcal{G}_+$:

$$\mathcal{G}(\tilde{\alpha}_n) = -\frac{2C}{1+X} \frac{\tilde{\alpha}_n^2(1+\tilde{\gamma}X) + \tilde{\gamma}^2(1-X^2)}{\tilde{\alpha}_n^4 + (1 - 2\tilde{\gamma}X + \tilde{\gamma}^2)\tilde{\alpha}_n^2 + \tilde{\gamma}^2(1+X)^2}. \tag{24.1}$$

Note that if we replace $-2C$ by A and take into account that $A = X + 1$, Eq. (24.1) reduces to Eq. (20.38) of the laser case. As usual, $\tilde{\alpha}_n = \alpha_n/\gamma_\perp$ and $\tilde{\gamma} = \gamma_\parallel/\gamma_\perp$. Correspondingly, the expressions of the instability boundaries $\tilde{\alpha}_{max}$ and $\tilde{\alpha}_{min}$ in the case of optical bistability become

$$\tilde{\alpha}_{\substack{max \\ min}}(X) = \left\{ X\tilde{\gamma} - C\frac{1 + \tilde{\gamma}X}{1+X} - \frac{1+\tilde{\gamma}^2}{2} \right.$$

$$\left. \pm \left[C^2\left(\frac{1+\tilde{\gamma}X}{1+X}\right)^2 - X\tilde{\gamma}(1+\tilde{\gamma})^2 + \frac{(1-\tilde{\gamma}^2)^2}{4} + C(1-\tilde{\gamma}^2)\frac{1-\tilde{\gamma}X}{1+X} \right]^{1/2} \right\}^{1/2}. \tag{24.2}$$

The instability domain included between the curves $\tilde{\alpha}_{max}(X)$ and $\tilde{\alpha}_{min}(X)$ is shown in Fig. 24.3 for $C = 20$ and $\tilde{\gamma} = 1$. In this case, as we see from Fig. 24.4, the segment of the high-transmission branch such that $X < (C/2)^2$ is unstable against the onset of symmetrical nonresonant cavity modes.

Let us fix our attention on α_{max}, let us assume that C is large and, in order to obtain an estimate of the order of magnitude of α_{max}, let us consider the right boundary of the instability domain where $X \propto C^2$, so that $X \gg C$. From Eq. (24.2) we see that $\alpha_{max} \simeq (\gamma_\perp\gamma_\parallel X)^{1/2} = (\gamma_\perp\gamma_\parallel)^{1/2}x = \Omega$, where Ω is the Rabi frequency of the output field (see Eqs. (8.19) and (5.23)). Hence also in the case of optical bistability the multimode instability is intimately linked to the Rabi frequency of the intracavity field, exactly like for the

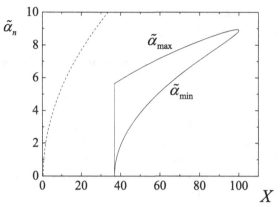

Figure 24.3 The instability region in the plane of the variables X and $\tilde{\alpha}_n$ for $C = 20$ and $\check{\gamma} = 1$. The dashed line is a graph of the function $\tilde{\alpha}_n = \sqrt{3X - 1}$.

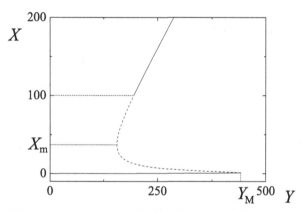

Figure 24.4 Purely absorptive bistability. A stationary curve with indication of the stable (solid lines) and unstable (broken line) states for $C = 20$ and $\check{\gamma} = 1$. In the broken segment of the part with positive slope the points are unstable, provided that at least one cavity frequency $\tilde{\alpha}_n$ lies in the range $\tilde{\alpha}_{\text{min}} < \tilde{\alpha}_n < \tilde{\alpha}_{\text{max}}$, where $\tilde{\alpha}_{\text{min}}$ and $\tilde{\alpha}_{\text{max}}$ are shown in Fig. 24.3.

laser as we saw in Chapter 22. This indicates that coherence plays a major role in such instabilities.

By solving Eqs. (12.39), (12.36) and (12.37) with A replaced by $-2C$ and $\Delta = \theta = 0$, we obtain the time evolution of the output intensity in the presence of the multimode instability [279]. The spontaneous oscillations which originate from the instability have a period essentially equal to the cavity roundtrip time Λ/c and form an envelope that evolves much more slowly than the roundtrip time. To be precise, from Eqs. (20.1), (20.2), (12.1), (12.17), (8.10) and (8.12) we have

$$F(z, t) = \sum_n e^{-i\tilde{\alpha}_n(t - z n_B/c)} \tilde{f}_n\left(t + \frac{z}{L}\Delta t\right). \tag{24.3}$$

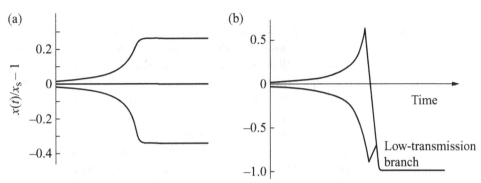

Figure 24.5 The envelope of the time evolution of the transmitted field for $C = 20$ and $\tilde{\gamma} = 1$ obtained by numerically solving the time-evolution equations. The transmitted field oscillates between the upper and the lower lines with a period roughly equal to the cavity roundtrip time Λ/c. (a) $x = 9$, $\tilde{\alpha}_1 = 8$. In the long–time limit the system approaches a steady self-pulsing regime in which the envelope is perfectly flat. (b) $x = 6.15$, $\tilde{\alpha}_1 = 3.6$. For long times, the oscillations vanish and the system precipitates to the low-transmission branch.

In particular, by setting $z = L$ we obtain

$$F(L, t) = \sum_n e^{-i\alpha_n(t - Ln_B/c)} \tilde{f}_n(t + \Delta t). \tag{24.4}$$

On the other hand, from Eqs. (8.17), (8.13) and (8.10), assuming that $L' \ll L$ as in Fig. 14.4, we have that

$$x(t) = F(L, t - \Delta t). \tag{24.5}$$

Hence by combining Eqs. (24.4) and (24.5) and using Eqs. (8.10) and (8.12) we obtain

$$x(t) = \sum_n e^{-i\alpha_n(t - \Lambda/c)} \tilde{f}_n(t). \tag{24.6}$$

If we now take into account only the resonant mode $n = 0$ and the two unstable sidemodes $n = \pm 1$, and assume that

$$\tilde{f}_1(t) = \tilde{f}^*_{-1}(t) = \rho_1(t)e^{i\psi(t)}, \tag{24.7}$$

we have

$$x(t) = f_0(t) + 2\rho_1(t)\cos[\alpha_n(t - \Lambda/c) - \psi(t)], \tag{24.8}$$

so that the envelope is given by

$$x_{\mathrm{env}}(t) = f_0(t) \pm 2\rho_1(t). \tag{24.9}$$

The numerical calculation of the time evolution [86, 279] is done by solving Eqs. (12.39), (12.36) and (12.37), which include the time τ. In this connection we observe that, in the calculation of the output field x, i.e. of $F(L, t)$ as a function of time, the real time t is simply retarded by Δt with respect to τ (see Eqs. (12.1) with $z = L$).

Figure 24.5 shows the two kinds of time evolution one finds from an initial condition only slightly displaced from the unstable stationary state in the higher-transmission branch, for the same values of the parameters C and $\tilde{\gamma}$ as in Figs. 24.3 and 24.4. What is shown is the envelope of the spontaneous oscillations of the output field $x(t)$ (or, more precisely, its variation relative to the unstable stationary value x_s in the upper branch) which emerge from the instability. The oscillations are first exponentially amplified, during the stage of the time evolution that is governed by the linearized equations. In the case of Fig. 24.5(b), in the long term the oscillations vanish because the system precipitates to the lower-transmission state, which corresponds to the same value of the input field and is stable. On the other hand, in the case of Fig. 24.5(a) the system approaches a steady self-pulsing regime, in which the envelope is flat. In general, the shape of the single pulses (of period equal to the cavity roundtrip time) depends on the number of unstable modes and, when only the two modes adjacent to the resonant one are unstable, the pulses are basically sinusoidal. In this self-pulsing regime the system behaves as an all-optical device that converts the continuous-wave input light into pulsed light.

An important point is finally that in the case of the resonant multimode instability of optical bistability there is no corresponding single-mode instability, in contrast with what happens in the laser case. We can check this by following the analysis of Section 20.3. The function $\mathcal{D}_+(\tilde{\omega})$ has the expression

$$\mathcal{D}_+(\tilde{\omega}) = -\frac{2C}{1+X}\tilde{\omega}\frac{\tilde{\omega}^2 - 2\tilde{\gamma}X + \tilde{\gamma}^2(1-X)}{\tilde{\omega}^4 + (1 - 2\tilde{\gamma}X + \tilde{\gamma}^2)\tilde{\omega}^2 + \tilde{\gamma}^2(1+X)^2}. \tag{24.10}$$

The condition $\mathcal{D}_+(\tilde{\omega}) < 0$ is satisfied for

$$\tilde{\omega} > \left[\tilde{\gamma}X(2+\tilde{\gamma}) - \tilde{\gamma}^2\right]^{1/2}, \tag{24.11}$$

and, as illustrated in Fig. 24.3, is not satisfied for $\tilde{\omega} = \tilde{\alpha}_n = \tilde{\alpha}_{\max}, \tilde{\alpha}_{\min}$.

The detuned configuration, theory and experiment

Let us now turn our attention to the detuned case, in which at least one of the two parameters Δ and θ is different from zero. As shown by the analysis of [86, 280], in this case the instability can develop both for $\Delta\theta > 0$ and for $\Delta\theta < 0$ when C is large enough. The instability domain may be much larger than the bistability domain, or can even exist in the absence of bistability.

The multimode instability of optical bistability has been observed experimentally under detuned conditions by Segard and Macke [281, 282] with the help of a folded 182-m-long Fabry–Perot cavity, operating in the microwave regime and filled with hydrogen cyanide ($HC^{15}N$). The cavity free spectral range was roughly 50 times larger than the unsaturated width of the molecular transition, but the power broadening was very pronounced because $|F|$ was of the order of 40. These conditions favored the appearance of a multimode instability with simple sinusoidal oscillations (see the inset of Fig. 24.6). In Fig. 24.6 the input power is slowly swept forward and backward, and the oscillations appear in the upper branch in the backward scan. On the other hand, in Fig. 24.7 we show the observed spectrum of the transmitted light with three main peaks labeled ((0), (1) and (2)) corresponding to the frequency of the pump and to the frequencies of the unstable modes. The experiments also show the expected correlation between the oscillation frequency and the Rabi frequency of

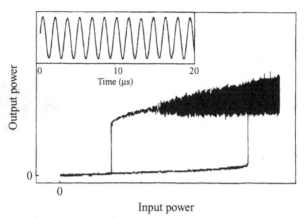

Figure 24.6   The bistability cycle experimentally observed for $\omega_a = \omega_c$ and $\Delta\gamma_\perp/(2\pi) = 317$ kHz, at pressure $p \simeq 0.53$ mTorr. Inset: sine-wave self-oscillations obtained when the incident power is fixed at its maximum. Reprinted figure with permission from [282]. Copyright (1989) by the American Physical Society.

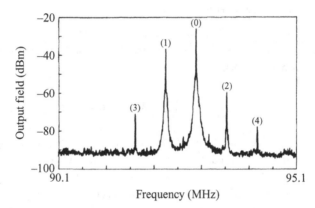

Figure 24.7   The spectrum of the field transmitted by the cavity for $p = 1.33$ mTorr and $\Delta\gamma_\perp/(2\pi) = 270$ kHz. Vertical scale: 10 dB/division. The frequency increases from left to right (500 kHz/division). The different components correspond to field frequencies equal to (0) $\nu_0 = \omega_0/(2\pi)$, (1) $\nu_0 - \nu_{sp}$, (2) $\nu_0 + \nu_{sp}$, (3) $\nu_0 - 2\nu_{sp}$ and (4) $\nu_0 + 2\nu_{sp}$, where $\nu_{sp} = 645$ kHz is the frequency of the spontaneous oscillations in the output intensity. Reprinted figure with permission from [282]. Copyright (1989) by the American Physical Society.

the intracavity field. Note that in the figures of this section the frequency is $\nu = \omega/(2\pi)$, and this holds also for the Rabi frequency.

It is important to mention that the experimental situation is far from the conditions of the low-transmission limit, therefore the numerical simulations of the experiments [282] have been based on the use of the original Maxwell–Bloch equations (4.35)–(4.37), with the appropriate boundary conditions, instead of Eqs. (12.35)–(12.37). As shown by the example of Fig. 24.8, the numerical results are in reasonable agreement with the experimental findings. In these simulations we use a ring-cavity model for a cavity with $L' \ll L$ (see

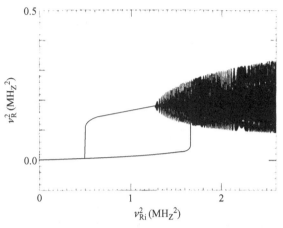

Figure 24.8   Numerical simulation of the experimental Fig 24.6. The parametric values are the same as in that figure. To be precise, $\alpha L = 146$, $R = 0.69$, $c/\mathcal{L} = 830$ kHz and $\gamma_\perp/(2\pi) = \gamma_\parallel/(2\pi) = 22.75$ kHz/mTorr. The input power is slowly swept forward (lower branch) and backward (upper branch). The square of the intracavity Rabi frequency ν_R is plotted as a function of the square of the input Rabi frequency ν_{Ri}. Reprinted figure with permission from [282]. Copyright (1989) by the American Physical Society.

e.g. Fig. 14.4). Simulations for these experiments using a Fabry–Perot model have been performed in [283, 284].

24.2.2 Miscellaneous considerations

While the literature on optical instabilities in ring cavities is abundant, there is a much more limited number of publications on instabilities in Fabry–Perot cavities. Let us consider, for example, [144], which discusses multimode instabilities in absorptive optical bistability in a standing-wave cavity. The time-evolution equations which govern the system are linearized with respect to the single-mode stationary solution and the usual exponential ansatz (18.5) is introduced. The procedure basically generalizes to the Fabry–Perot case that of Section 22.5 for the laser in a ring cavity. The fluctuations of the atomic variables are eliminated and one obtains two coupled linear differential equations in the longitudinal variable z for the fluctuations of the forward and backward propagating fields. These equations are solved taking into account the appropriate boundary conditions of the cavity, which include the position of the atomic sample in the cavity itself. In this way one arrives at the characteristic equation, which is perturbatively solved, similarly to what we did in the case of Eq. (22.48). The results are related also to the Ikeda instability, which is discussed in the following subsection.

The other main consideration of this subsection concerns the simplified form of the ring-cavity model obtained in Section 14.4. In the case $n_B = 1$, $L' \ll \mathcal{L}$ we have derived the model of Eqs. (14.85)–(14.87), for which the boundary condition (14.84) does not include any retardation in time and the location of the atomic sample is described by the function $\chi(z)$ defined by Eq. (14.7). The absence of retardation carries an important advantage, but here we must observe that in the framework of the linearized treatment, which is necessary in order to determine the stability of the stationary solutions, the presence of the function $\chi(z)$ represents a problem because, if one expands the atomic variables $P(z,t)$ and $D(z,t)$ over the modes $(1/\mathcal{L})e^{i2\pi nz/\mathcal{L}}$, one no longer obtains equations that can be treated analytically. This implies that the original model with a boundary condition with retardation is advantageous in the linearized framework not only because it allows one to include the possible presence of the background refractive index n_B, but also because it allows one to carry out the linear-stability analysis analytically. In the framework of the numerical solution of the full model, in contrast, the presence of the function $\chi(z)$ is not a problem, of course.

24.2.3 The Ikeda "chaotic" instability and its relations to the standard multimode instability

One of the most interesting among the oscillatory instabilities in optical bistability goes under the name of Ikeda [145, 146], who proposed its existence and the subsequent emergence of chaos. He started from the ring-cavity model with boundary conditions formulated in [85] and generalized to the detuned configuration. By performing the naïve adiabatic elimination of the atomic polarization (see Eq. (21.4)) he succeeded in casting the resulting rate equations into the form of difference-differential equations and discovered a period-doubling route to chaos that starts from oscillations of period equal to twice the cavity roundtrip time. In a further step, he performed the full (naïve) adiabatic elimination of both atomic variables (see Eq. (21.16)), transforming the Maxwell–Bloch equations with boundary conditions into a system of finite-difference equations (map) for the electric field, with a time step equal to the cavity roundtrip time. Condition (21.16) requires an extremely long cavity and greatly enhances the role of the delay introduced by the propagation of light through the empty section of the resonator. In practice, the experimental observation of such an instability requires the use of a hybrid electro-optical device [285].

In this section we focus on the case of full adiabatic elimination of all atomic variables. Unlike the standard multimode instability of optical bistability (see Sections 24.2.1 and 24.2.2), the Ikeda instability requires a large value for αL and disappears in the low-transmission limit. The following treatment, which is based on [286], clearly illustrates the connections between the Ikeda instability and the standard multimode instability. Since we are not in the uniform-field limit, we start from the general Maxwell–Bloch equations (4.35)–(4.37) with $n_B = 1$ and g replaced by $-\alpha$, and perform the naïve adiabatic elimination of the atomic variables by setting $\partial P/\partial t = 0$ and $\partial D/\partial t = 0$. As usual we set $F = \rho \exp(i\phi)$ and obtain the equations

$$\frac{\partial \rho}{\partial z} + \frac{1}{c}\frac{\partial \rho}{\partial t} = -\alpha \frac{\rho}{1 + \Delta^2 + \rho^2}, \qquad (24.12)$$

$$\frac{\partial \phi}{\partial z} + \frac{1}{c}\frac{\partial \phi}{\partial t} = \alpha \frac{\Delta}{1 + \Delta^2 + \rho^2}. \qquad (24.13)$$

With the definitions $\tilde{I} = \rho^2/(1 + \Delta^2)$ and $\bar{\alpha} = \alpha/(1 + \Delta^2)$ the equations read

$$\frac{\partial \tilde{I}}{\partial z} + \frac{1}{c}\frac{\partial \tilde{I}}{\partial t} = -2\bar{\alpha}\frac{\tilde{I}}{1 + \tilde{I}}, \tag{24.14}$$

$$\frac{\partial \phi}{\partial z} + \frac{1}{c}\frac{\partial \phi}{\partial t} = \bar{\alpha}\frac{\Delta}{1 + \tilde{I}}. \tag{24.15}$$

Next, we linearize these equations around a stationary state. By defining

$$\delta\tilde{I}(z, t) = \tilde{I}(z, t) - \tilde{I}_s(z), \quad \delta\phi(z, t) = \phi(z, t) - \phi_s(z) \tag{24.16}$$

and making the usual exponential ansatz $\delta\tilde{I}(z, t) = \delta\tilde{I}(z)e^{\lambda t}$, $\delta\phi(z, t) = \delta\phi(z)e^{\lambda t}$ we obtain the linearized equations

$$\frac{d\delta\tilde{I}}{dz} + \frac{\lambda}{c}\delta\tilde{I} = -2\bar{\alpha}\frac{\delta\tilde{I}}{(1 + \tilde{I}_s)^2}, \tag{24.17}$$

$$\frac{d\delta\phi}{dz} + \frac{\lambda}{c}\delta\phi = -\bar{\alpha}\Delta\frac{\delta\tilde{I}}{(1 + \tilde{I}_s)^2}. \tag{24.18}$$

By multiplying Eq. (24.17) by $\Delta/2$ and subtracting it from Eq. (24.18) we obtain

$$\frac{d}{dz}\left(\delta\phi - \frac{\Delta}{2}\delta\tilde{I}\right) + \frac{\lambda}{c}\left(\delta\phi - \frac{\Delta}{2}\delta\tilde{I}\right) = 0, \tag{24.19}$$

from which it follows immediately that

$$\delta\phi(z) = \delta\phi(0)e^{-\lambda z/c} + \frac{\Delta}{2}\left[\delta\tilde{I}(z) - \delta\tilde{I}(0)e^{-\lambda z/c}\right]. \tag{24.20}$$

Hence, we need just calculate the solution $\delta\tilde{I}(z)$ of Eq. (24.17). This can be done taking into account the stationary equation for $\tilde{I}_s(z)$,

$$\frac{d\tilde{I}_s(z)}{dz} = -2\bar{\alpha}\frac{\tilde{I}_s(z)}{1 + \tilde{I}_s(z)}, \tag{24.21}$$

which follows from Eq. (24.14) with $\partial\tilde{I}/\partial t = 0$. By inserting Eq. (24.21) into (24.17) we obtain the equation

$$\frac{d\delta\tilde{I}}{dz} + \frac{\lambda}{c}\delta\tilde{I} = \frac{1}{\tilde{I}_s(1 + \tilde{I}_s)}\frac{d\tilde{I}_s}{dz}\delta\tilde{I}, \tag{24.22}$$

whose solution is

$$\delta\tilde{I}(z) = \frac{\tilde{I}_s(z)}{\tilde{I}_s(0)}\frac{1 + \tilde{I}_s(0)}{1 + \tilde{I}_s(z)}\delta\tilde{I}(0)e^{-\lambda z/c}. \tag{24.23}$$

Let us now turn to the boundary conditions associated with Eqs. (24.14) and (24.15). If we start from the boundary condition (8.36), which holds for both laser and optical bistability, and take into account that the stationary field $F_s(z)$ itself obeys this condition, we obtain the following boundary condition for the deviation $\delta F(z, t) = F(z, t) - F_s(z)$:

$$\delta F(0, t) = Re^{-i\delta_0}\delta F(L, t - \Delta t), \quad \Delta t = \frac{\mathcal{L} - L}{c}, \tag{24.24}$$

which, in terms of $\delta \tilde{I}$ and $\delta \phi$, can be rewritten as

$$\frac{\delta \tilde{I}(0, t)}{2\tilde{I}_s(L)} = \left[\frac{\tilde{I}_s(0)}{\tilde{I}_s(L)}\right]^{1/2} R \left[\frac{\cos \Phi}{2\tilde{I}_s(L)} \delta \tilde{I}(L, t - \Delta t) - \sin \Phi \, \delta \phi(L, t - \Delta t)\right], \quad (24.25)$$

$$\delta \phi(0, t) = \left[\frac{\tilde{I}_s(L)}{\tilde{I}_s(0)}\right]^{1/2} R \left[\frac{\sin \Phi}{2\tilde{I}_s(L)} \delta \tilde{I}(L, t - \Delta t) + \cos \Phi \, \delta \phi(L, t - \Delta t)\right], \quad (24.26)$$

where

$$\Phi = \phi_s(L) - \phi_s(0) - \delta_0. \quad (24.27)$$

We insert Eqs. (24.20) and (24.23) into Eqs. (24.25) and (24.26), thereby obtaining a homogeneous system of two equations for $\delta \tilde{I}(0)$ and $\delta \phi(0)$. The condition that this system admits nontrivial solutions gives the eigenvalue equation. On setting

$$B = R \left[\frac{\tilde{I}_s(L)}{\tilde{I}_s(0)} \frac{1 + \tilde{I}_s(0)}{1 + \tilde{I}_s(L)}\right]^{1/2}, \quad (24.28)$$

$$S = \frac{\Delta \left[\tilde{I}_s(0) - \tilde{I}_s(L)\right] \sin \Phi + \left[2 + \tilde{I}_s(0) + \tilde{I}_s(L)\right] \cos \Phi}{2 \left[1 + \tilde{I}_s(0)\right]^{1/2} \left[1 + \tilde{I}_s(L)\right]^{1/2}}, \quad (24.29)$$

the equation reads

$$B^2 e^{-2\lambda \mathcal{L}/c} - 2S B e^{-\lambda \mathcal{L}/c} + 1 = 0. \quad (24.30)$$

On solving Eq. (24.30) we see that the eigenvalues are given by

$$\lambda_{n,\pm} = -\frac{c}{\mathcal{L}} \ln \left(\frac{S \pm (S^2 - 1)^{1/2}}{B}\right) + i\alpha_n, \quad n = 0, \pm 1, \ldots, \quad (24.31)$$

where α_n is defined by Eq. (12.17) with $\Lambda = \mathcal{L}$ because $n_B = 1$.

It can be shown [286] that Eq. (24.31) coincides with the expression for the eigenvalues given in [287, 288] obtained from the treatment of Ikeda [145].

As usual in the naïve adiabatic elimination, all the eigenvalues $\lambda_{n,+}$ have the same real part and the same holds for the eigenvalues $\lambda_{n,-}$. Let us now focus on the case $S < -1$, in which the argument of the logarithm in Eq. (24.31) is real and negative. We assume also that $B < 1$, so that $-S - B > 0$. Under these conditions, one easily sees that the real part of the eigenvalue $\lambda_{n,+}$ is positive (and therefore the stationary state is unstable) when $S > -1 - (1 - B)^2/(2B)$. This condition can be satisfied only for Δ and θ different from zero and leads to instabilities in parts of the curve $X(Y)$ with positive slope. Under suitable further conditions [145] several segments of the curve $X(Y)$ are unstable, and one can find ranges of values of the incident intensity for which no stable stationary state exists. In this situation the system shows a self-pulsing behavior, which can be either a regular self-pulsing with a period roughly equal to an even multiple of the roundtrip cavity time and square-wave pulses (Fig. 24.9(b)), or a chaotic self-pulsing (Fig. 24.9(c)). As it has been shown in [287], this period-doubling bifurcation sequence to chaos agrees with the general theory of Feigenbaum [254].

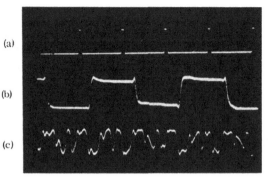

Figure 24.9   An observation of chaotic dynamics in a hybrid device. (a) Time calibration; one pulse every transit time, equal to 40 ms. (b) The output intensity versus time in the periodic domain. (c) The intensity versus time in the chaotic regime. Reprinted figure with permission from [285]. Copyright (1981) by the American Physical Society.

The fact that in Ikeda instability the period is twice the cavity roundtrip time can be easily understood from Equation (24.31). Since for $S < -1$ the argument of the logarithm in λ_+ is real and negative, we have

$$\text{Im } \lambda_{n,+} = \frac{c}{\mathcal{L}}\pi + \alpha_n = 2\pi \left(2\frac{\mathcal{L}}{c}\right)^{-1}(1+2n), \tag{24.32}$$

which gives a period $2\mathcal{L}/c$.

In the low-transmission limit the Ikeda instability disappears. As a matter of fact, the real part of the eigenvalues is the same for all n, and in the low-transmission limit for $n = 0$ the real part of the eigenvalues is negative in the parts of the stationary curve with positive slope.

This is the only section of the book in which we perform a linear-stability analysis outside the low-transmission limit. A general linear-stability analysis of the Maxwell–Bloch equations is discussed in [289] (see also references quoted therein).

24.3 Single-mode instabilities in optical bistability

24.3.1 The single-mode instability. Theory and experiment

The single-mode instability emerges only under detuned conditions and may arise both when the stationary curve of transmitted versus incident intensity is single-valued and when it is S-shaped. The frequency of the spontaneous oscillations produced by this instability is approximately equal to $|\omega_c - \omega_0|$, in accord with the interpretation of this phenomenon as the result of a beat between the driving field and the destabilized resonant mode. As in the case of the corresponding multimode instability, the oscillation frequency is close to the Rabi frequency of the intracavity field.

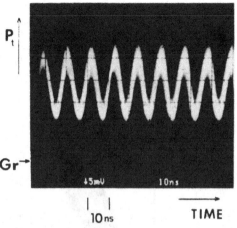

Figure 24.10   An oscilloscope trace of the output intensity as a function of time when the system is self-oscillating for fixed values of C, Δ, θ and the input intensity. The pulsations are quite stable over periods of minutes. Reprinted figure with permission from [293]. Copyright (1989) by the American Physical Society.

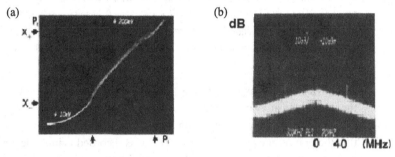

Figure 24.11   (a) Input–output characteristics (transmitted vs. incident power) as recorded on an oscilloscope; the fuzzy region limited by the arrows indicates the presence of an instability; the symbols X_+ and X_- denote the limits of the instability in the transmitted intensity. (b) Spectrum analyzer trace of the current for fixed input intensity. The parameters are $C = 50 \pm 10, \theta = -2.5 \pm 1, \Delta = 2.5 \pm 0.5, \kappa/\gamma_\perp = 0.31$ and $\gamma_\parallel/\gamma_\perp = 1.6$. Reprinted figure with permission from [293]. Copyright (1989) by the American Physical Society.

The single-mode instability of optical bistability was predicted in plane-wave models in [290] in the case of a Kerr medium, and in [291] for two-level atoms. Some specific theoretical results will be discussed in the following subsection. This instability was observed experimentally by Kimble and his collaborators [292, 293] using a ring or a Fabry–Perot cavity crossed at right angles by several atomic beams of sodium. The experimental setting ensured the validity of the requirements for a two-level system, the homogeneously broadened character of the atomic transition and the single-mode condition. Special care was taken to monitor the profile of the electric-field intensity in the directions transverse with respect to the propagation direction, even under unstable conditions. The observed oscillations are of the sinusoidal type (Fig. 24.10); the spectrum of the output intensity (Fig. 24.11(b)) shows a coherent spike, whose frequency shifts as one detunes the cavity.

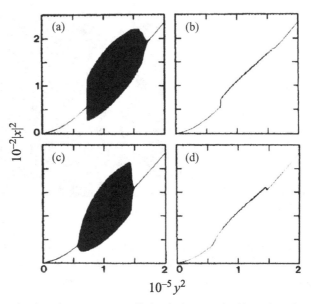

$10^{-2}|x|^2$

$10^{-5}y^2$

Figure 24.12 Numerical simulations that show the spontaneous oscillations in the transmitted intensity as the input intensity is swept slowly for $C = 55$, $\Delta = 2.5$, $\theta = -2.5$, $\kappa/\gamma_\perp = 0.31$ and $\gamma_\parallel/\gamma_\perp = 1.6$. Plots (a) and (b) correspond to a forward scan, plots (c) and (d) to a backward scan. Plots (b) and (d) show the average behavior of the oscillations. Reprinted figure with permission from [293]. Copyright (1989) by the American Physical Society.

The cavity length (cavity detuning) and the frequency of the input field may be fixed and the input intensity varied to locate the instability boundary as a function of the parameters (Fig. 24.11(a)); the corresponding oscillation frequency is shown in Fig. 24.11(b). Note that the output intensity in Fig. 24.11(a) was recorded using a detector with a narrow frequency response, so in the unstable interval between X_- and X_+ it is not possible to see the full oscillations but only a partially integrated version. By changing the cavity length and the input frequency for a fixed value of the bistability parameter C, one can explore the instability boundary in the (Δ, θ) plane; the process is then repeated for different values of C.

These experimental observations have been matched with the theoretical predictions in a quantitative comparison without adjustable parameters [293]. While the plane-wave model exhibits considerable disagreement with the data, the predictions of the single-mode model with a Gaussian transverse profile turn out to be in good accord with the experimental findings on the instability boundaries and the frequency of the oscillations. This model will be described in Section 29.2. As an example, we show in Fig. 24.12 the numerical simulations corresponding to Fig. 24.11(a).

24.3.2 Period doubling and chaos in the plane-wave model

Oscillatory instabilities in the plane-wave single-mode model of optical bistability with two-level atoms were first analyzed in the framework of the rate-equation approximation,

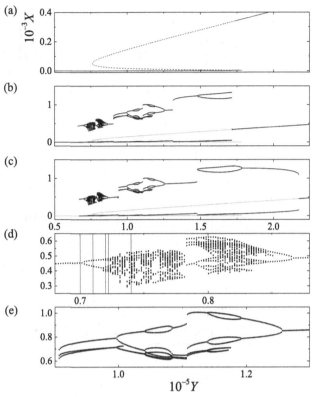

Figure 24.13

Stationary and dynamical features of optical bistability for $C = 400$, $\Delta = -1$, $\theta = 20$, $\tilde{\kappa} = 0.25$ and $\tilde{\gamma} = 0.9$. (a) A plot of stationary output vs. input intensity, showing that the broken part of the upper branch is Hopf unstable. In plots (b) and (c) the stationary curve, represented by a gray line, is superposed on the bifurcation diagrams obtained by plotting the maxima and the minima of the oscillations for 1000 values of Y ranging from 235 000 to 50 000 in the backward scan (b) and from 67 000 to 235 000 in the forward scan (c). For each value of Y the Maxwell–Bloch equations are integrated with the final conditions of the previous run as initial conditions. The integration time is 500 in units of $\gamma_{\perp}^{-1}$, but only the extrema found in the last 20 time units are plotted, in order to discard transients. Plots (d) and (e) show a magnification of the chaotic and periodic attractors. In (d) the vertical lines indicate the values of Y used in Fig. 24.14.

i.e. within the model obtained by adiabatically eliminating the atomic polarization and given by Eqs. (19.24) and (19.25) with a defined by Eq. (19.4) and A replaced by $-2C$ [291]. For extremely large values of C it was shown that the system exhibits period doubling and chaos, similarly to what was found in a model that describes optical bistability in a cavity containing a Kerr medium [290]. Later on it was shown [293] that, by considering the full plane-wave single-mode model of optical bistability without any adiabatic elimination, i.e. Eqs. (12.43)–(12.45) with A replaced by $-2C$, one can find the same scenario for values of C accessible to experiments.

Let us consider, for example, the values $C = 400$, $\Delta = -1$, $\theta = 20$, $\tilde{\kappa} = 0.25$ and $\tilde{\gamma} = 0.9$. Figure 24.13(a) shows the stationary hysteresis curve, with indication of the

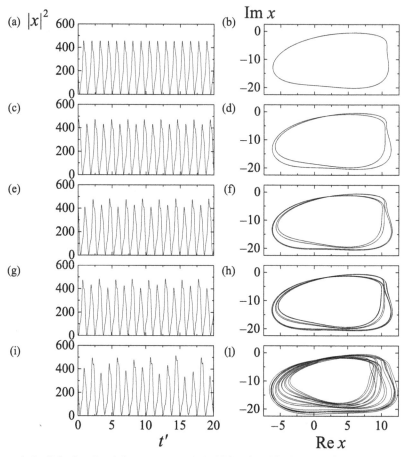

Figure 24.14 The time evolution (left column) and phase-space trajectories (right column) for the same parameters as in Fig. 24.13 and the values of Y corresponding to the vertical lines of Fig. 24.13(d): (a) and (b) $Y = 0.7 \times 10^5$, (c) and (d) $Y = 0.71 \times 10^5$, (e) and (f) $Y = 0.72 \times 10^5$, (g) and (h) $Y = 0.7222 \times 10^5$, and (i) and (l) $Y = 0.74 \times 10^5$.

stable and unstable portions. In this case, the characteristic equation is quintic (whereas in [291] it was cubic) with the coefficients defined by Eqs. (C.1)–(C.6). The boundary of the oscillatory instability is defined by Eq. (B.15).

The integration of the full set of Maxwell–Bloch equations reveals the existence of three separated attractors, as shown in the bifurcation diagrams of Figs. 24.13(b) and (c), where the maxima and the minima of the field intensity after elimination of the transient are displayed as the input intensity Y varies from 235 000 to 50 000 in (b) and from 67 000 to 235 000 in (c).

In the backward scan (b) the stationary solution jumps to a period-2 oscillating solution as the boundary of the Hopf instability is crossed. We then observe a continuous transition to a period-1 solution, followed by a jump to another period-1 solution with smaller oscillations. As Y is further decreased this solution undergoes a sequence of period-doubling bifurcations followed by an inverse sequence. In this region period-3 and period-6 solutions are also visible. Finally the period-3 solution becomes unstable in favor of a period-1 solution with smaller oscillations, which displays a full period-doubling transition to chaos followed by an inverse transition to a period-1 solution. Periodic windows are visible inside the chaotic attractor. The dynamical solution persists even when the left turning point of the hysteresis cycle is crossed, until the system jumps to the lower stationary branch.

The forward scan of Fig. 24.13(c) starts before the oscillating solution destabilizes. The chaotic attractor and its associated bifurcations are apparently identical to those of the backward scan. Instead, the subsequent transitions between attractors display hysteresis. For instance, in the transition from the second to the third attractor, bistability of two period-1 solutions with different amplitudes is found, and in the transition from the third attractor to the upper-branch stationary solution there is a wide range of input intensity beyond the Hopf boundary where the stationary solution coexists with period-2 and period-1 oscillating solutions.

Figure 24.14 exhibits the spontaneous temporal oscillations of the output field, which emerge from the instability, and the projection of the phase-space trajectory onto the $(\mathrm{Re}\, x, \mathrm{Im}\, x)$ plane. In frames (a) and (b) we have simple period-1 oscillations, whose nonlinear character is, however, visible in the asymmetric shape of the waveform; in frames (c) and (d) we observe period-2 oscillations, which are characterized by the fact that the phase-space trajectory closes after two loops; in frames (e) and (f) and (g) and (h) we see period-4 and period-8 oscillations, respectively; and in frames (i) and (l) we observe chaotic oscillations. Notice that this period-doubling transition to chaos occurs in an interval of input intensity where only the lower branch of the stationary solution exists (Fig. 24.13(a)).

25

Self-pulsing in other optical systems

This chapter concludes the discussion about temporal instabilities and self-oscillations, considering the instabilities that arise in the systems analyzed in Chapter 13, with the exception of optical bistability, which has been discussed separately in Chapter 11.

Section 25.1 focuses on the laser with an injected signal. As already mentioned in Section 13.1, in this system self-pulsing instabilities are rather ubiquitous, and on exploring the parameter space one meets an extremely rich variety of oscillatory behaviors. For example, under extremely high-gain conditions, breathing, spiking and chaos have been identified over various ranges of the driving field strength [294]. In this section we show some results that arise under parametric conditions that are more accessible experimentally. We illustrate especially the coexistence, for the same parametric values, of different oscillatory behaviors that are reached by starting from different initial conditions.

On the other hand, Section 25.2 deals with a laser with a saturable absorber. In this system even the full scenario of stationary intensity solutions in the single-mode model is very complex. For the sake of simplicity, in this section we focus on a simpler rate-equation model that is obtained from the single-mode model of Eqs. (13.13)–(13.17) by adiabatically eliminating the atomic polarization of both active and passive medium. As in Chapter 13, we assume that the atomic transition frequencies are equal and also coincide with a cavity frequency. In this way all the variables in the model can be assumed real, and the linearization around the stationary solution described in Section 13.2 leads to a cubic eigenvalue equation. The spontaneous oscillations which arise from this instability are called *repetitive passive Q-switching* in the literature.

Section 25.3 considers the stability of the stationary solutions of the degenerate optical parametric oscillator described in Section 13.4 and outlines the scenario of period doubling and chaos that arises under appropriate parametric conditions.

25.1 A laser with an injected signal. Frequency locking and coexisting attractors

Let us start from Eqs. (12.43)–(12.45). The stationary solutions of this single-mode model have already been discussed in Section 13.1, and the linear-stability analysis for such solutions in Sections 20.1 and 20.3. In Figs. 13.1 and 13.2 we showed that it is easy to find segments of the stationary curve with positive slope that are unstable against the onset of spontaneous oscillations. We have observed that when the frequency of the free-running laser ω_L given by Eq. (13.2) differs from the frequency of the input field ω_0, and the values of the parameters are such that the free-running laser (corresponding to $Y = 0$) is above threshold, a segment of the stationary curve which starts from the origin is unstable. On the other hand, when the input intensity is increased enough it becomes capable of forcing the laser to oscillate with the input frequency, so the stationary state becomes stable.

The minimum value of the input intensity Y for which this situation prevails is called the *injection-locking threshold*. For the values of the input intensity from 0 to the locking threshold the output intensity displays spontaneous oscillations. For small driving intensities the oscillations are sinusoidal with a frequency equal to the beat note between the laser frequency and the driving frequency, and usually the oscillation profile becomes more complex when the input intensity is increased.

The injection-locking threshold is assocated with a Hopf bifurcation of the stationary intensity. It is easy to determine its position in the two limits of class-A and class-B lasers where $\tilde\kappa \ll 1$, $\tilde\gamma \sim 1$ and $\tilde\kappa \ll 1$, $\tilde\gamma \ll \tilde\kappa$, respectively [295]. The Hopf bifurcation is a single-mode instability that is determined by a pair of complex conjugate eigenvalues whose real part crosses zero. The eigenvalues are solutions of the fifth-order characteristic equation

$$c_5^{(0)}\tilde\lambda^5 + c_4^{(0)}\tilde\lambda^4 + c_3^{(0)}\tilde\lambda^3 + c_2^{(0)}\tilde\lambda^2 + c_1^{(0)}\tilde\lambda + c_0^{(0)} = 0, \tag{25.1}$$

where $\tilde\lambda = \lambda/\gamma_\perp$ and the coefficients $c_i^{(0)}$, $i = 1, \dots, 5$, are obtained from the coefficients $c_i^{(n)}$, $i = 1, \dots, 5$, of Appendix C on setting $\tilde\alpha_n = 0$. Assuming that the relevant eigenvalues (those with a real part that can become positive) are of order $\tilde\kappa$, we can perform an approximate stability analysis at the leading order in $\tilde\kappa$.

For class-A lasers the leading order is $\tilde\kappa^2$ and the reduced characteristic equation reads

$$\tilde\lambda^2 + 2\tilde\kappa[1 - (1 + \Delta^2)AB^2]\tilde\lambda + \tilde\kappa^2\frac{dY}{dX} = 0, \tag{25.2}$$

where A is the pump parameter, $B = (1 + \Delta^2 + X)^{-1}$ and we used Eq. (20.21). We focus on the positive-slope branches of the stationary state, for which the last term is positive, and the boundary of the Hopf instability is given by the condition that the coefficient of the first-order term in $\tilde\lambda$ vanishes (see the end of Section 18.1), that is $(1 + \Delta^2)AB^2 = 1$. This yields the critical value of the intensity

$$X_H = \sqrt{1 + \Delta^2}\left(\sqrt{A} - \sqrt{1 + \Delta^2}\right). \tag{25.3}$$

The injected intensity at the locking point can be obtained by inserting this expression into Eq. (13.1).

For class-B lasers we can neglect all terms of order $\tilde\gamma$. The reduced characteristic equation at the leading order in $\tilde\kappa$, which is now $\tilde\kappa^3$, reads

$$\tilde\lambda^3 - 2\tilde\kappa(AB - 1)\tilde\lambda^2 + \tilde\kappa^2\left[(1 - AB)^2 + (\theta + \Delta AB)^2\right]\tilde\lambda = 0. \tag{25.4}$$

Apart from the trivial solution $\lambda = 0$, the equation admits the pair of solutions

$$\lambda_\pm = +\tilde\kappa\left[\frac{A}{1 + \Delta^2 + X} - 1 \pm i\left(\theta + \frac{\Delta A}{1 + \Delta^2 + X}\right)\right]. \tag{25.5}$$

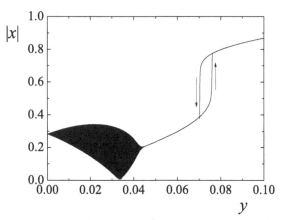

Figure 25.1 The temporal evolution of the output field amplitude $|x|$ for a slowly varying input field $y(t') = y_0 + Vt'$ with $t' = \gamma_\perp t$ and $V = 2 \times 10^{-7}$. The forward and backward scans are overlapped in this figure. The parameters are the same as in Fig. 13.2(b).

Clearly the real part of these eigenvalues is zero at the critical intensity

$$X_H = A - \left(1 + \Delta^2\right), \tag{25.6}$$

whereas the imaginary part is Im $\lambda_\pm = \pm\tilde{\kappa}(\theta + \Delta)$. If we compare these results with Eqs. (13.2) and (13.5) evaluated in the limit $\tilde{\kappa} \ll 1$ ($\kappa \ll \gamma_\perp$), we see that X_H is just the stationary intensity of the laser without injection while Im $\lambda_\pm$ is the frequency mismatch of the master and slave lasers (i.e. of the input frequency and the free-running laser's frequency).

Let us consider examples of dynamical behaviors that can be observed below the injection-locking point [295, 296]. In Fig. 25.1 we see the outcome of a slow scan (forward and backward) of the input field y (real and positive for definiteness) for the same values of the parameters as in Fig. 13.2(b). In the range $0 < y \lesssim 0.042$ one observes sponta-neous oscillations (Fig. 25.1 shows the envelope of the oscillations), whereas for $y \gtrsim 0.042$ the stationary solution is stable and the output field amplitude $|x|$ traces essentially the stationary curve because of the very small scan rate.

As long as the pulsations are produced by the same attractor, the observed patterns are the same for both forward and backward scans. It is easy, however, to find values of the input field for which this is not the case; switching between different domains and hysteretic behavior can then be observed. An example of richer dynamical behavior is shown in Fig. 25.2. Both in forward (b) and in backward (c) scans one can observe regions of chaotic oscillations and either continuous or discontinuous changes in the amplitude of self-pulsing. The transition from one pattern of pulsations to another is a consequence of the existence of different domains of attraction, which often overlap (i.e. coexist) over certain ranges of values of the driving field, exactly as we have already found in Fig. 24.13. A priori, the presence of hysteresis might be ascribed to lag effects due to the scan; actually, this is not the case for the scans shown here.

Self-pulsing in a laser with an injected signal has been observed in [297, 298].

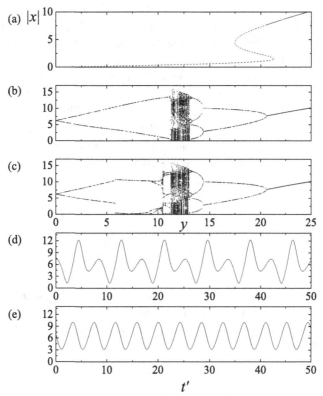

(a) The stationary curve for a laser with an injected signal with $A = 40$, $\Delta = 1$ and $\theta = 2$. The broken part of the curve is unstable. Parts (b) and (c) show the maxima and minima of the oscillations of $|x|$ corresponding to a slowly varying driving field of the type $y(t') = y_0 + Vt'$ during (b) a forward scan with $V = 2.5 \times 10^{-4}$ and (c) a backward scan with the same velocity; $\tilde{\kappa} = 0.5$ and $\tilde{\gamma} = 0.05$. Parts (d) and (e) show the time traces for (d) $y = 14$ and (e) $y = 15$.

25.2 A laser with a saturable absorber. Repetitive passive Q-switching

Let us start from Eqs. (13.13)–(13.17) assuming that the atomic transition frequencies in the active and in the passive medium coincide with a cavity frequency, i.e. that $\theta = 0$, $\Delta = 0$ and $\bar{\Delta} = 0$.

As we mentioned in Section 13.2, even in this resonant case the scenario of a laser with a saturable absorber may be quite complex. This circumstance arises from the fact that the trivial (i.e. zero-intensity) stationary solution can become unstable before reaching the threshold $A = 1 + 2C$ [129, 130, 232]. When the instability threshold of the trivial stationary solution is lower than the laser threshold, one may have the bifurcation of stationary intensity states or of periodic states in which the output intensity oscillates in time [130, 232]. The stationary intensity states are drastically different from the usual stationary state which appears at the laser threshold. As a matter of fact, even under complete resonance conditions (hence, there is no mode-pulling) the frequency in these stationary intensity solutions is different from the cavity frequency [129, 130, 232]. These stationary intensity solutions require that $\gamma_\perp > \bar{\gamma}_\perp$.

The calculation of the time-dependent solutions which arise from instabilities of the nontrivial stationary states has generated an extensive literature (see e.g. [130, 232, 299, 300]). In this section we will analyze the stability of both trivial and nontrivial stationary solutions under drastically simpler conditions. To be precise, we assume that $\gamma_\perp, \bar{\gamma}_\perp \gg \kappa$, $\gamma_\parallel, \bar{\gamma}_\parallel$, so that we can adiabatically eliminate the atomic polarization variables of both the active and the passive medium. By following the usual procedure described in Section 10.1 we obtain the equations [233, 301]

$$\dot{x} = \kappa(AD - 2C\bar{D} - 1)x, \tag{25.7}$$

$$\dot{D} = -\gamma_\parallel[D(1 + x^2) - 1], \tag{25.8}$$

$$\dot{\bar{D}} = -\bar{\gamma}_\parallel[\bar{D}(1 + sx^2) - 1], \tag{25.9}$$

where dot means derivative with respect to t as usual, and for the sake of simplicity we have taken x real. By introducing the symbols $X = x^2$, $\gamma = \gamma_\parallel/(2\kappa)$ and $\bar{\gamma} = \bar{\gamma}_\parallel/(2\kappa)$ and denoting with a prime the derivative with respect to $2\kappa t$ we cast the equations in the forms

$$X' = (AD - 2C\bar{D} - 1)X, \tag{25.10}$$

$$D' = -\gamma[D(1 + X) - 1], \tag{25.11}$$

$$\bar{D}' = -\bar{\gamma}[\bar{D}(1 + sX) - 1]. \tag{25.12}$$

The rate equations (25.10)–(25.12) of course cannot describe effects that arise from the coherent atom–field interaction; for example, the oscillatory instability we will describe below has nothing to do with the Rabi frequency of the intracavity field.

We linearize Eqs. (25.10)–(25.12) around a stationary solution $X_s, D_s, \bar{D}_s$ as prescribed in Section 18.1 and arrive at a matrix $\mathbf{A}$ (see Eqs. (18.6) and (18.7)) given by

$$\mathbf{A} = \begin{pmatrix} AD_s - 2C\bar{D}_s - 1 & AX_s & -2CX_s \\ -\gamma D_s & -\gamma(1 + X_s) & 0 \\ -\bar{\gamma}s\bar{D}_s & 0 & -\bar{\gamma}(1 + sX_s) \end{pmatrix}. \tag{25.13}$$

In the case of the trivial stationary solution $X_s = 0, D_s = 1, \bar{D}_s = 1$ one obtains a factorized characteristic equation that gives directly the three roots $\lambda_1 = A - 2C - 1, \lambda_2 = -\gamma$ and $\lambda_3 = -\bar{\gamma}$ (in this section λ is normalized with respect to 2κ). Hence the trivial stationary solution is stable below threshold ($A < 1 + 2C$) and unstable above threshold.

In the case of the nontrivial stationary solution $X_s, D_s = (1 + X_s)^{-1}, \bar{D}_s = (1 + sX_s)^{-1}$ the eigenvalues of $\mathbf{A}$ are given by the cubic equation

$$\lambda^3 + c_2\lambda^2 + c_1\lambda + c_0 = 0, \tag{25.14}$$

where the coefficients are given by

$$c_2 = \gamma(1 + X_s) + \bar{\gamma}(1 + sX_s),$$

$$c_1 = \gamma\bar{\gamma}(1 + X_s)(1 + sX_s) + X_s\left(\gamma\frac{A}{1 + X_s} - \bar{\gamma}\frac{2Cs}{1 + sX_s}\right), \quad (25.15)$$

$$c_0 = \gamma\bar{\gamma}X_s\left(A\frac{1 + sX_s}{1 + X_s} - 2Cs\frac{1 + X_s}{1 + sX_s}\right).$$

The boundary of the stationary instability is given by $c_0 = 0$, i.e.

$$\frac{c_0}{\gamma\bar{\gamma}} = X_s(1 + sX_s)\left[\frac{A}{1 + X_s} - 2Cs\frac{1 + X_s}{(1 + sX_s)^2}\right] = X_s(1 + sX_s)\frac{dA}{dX_s} = 0, \quad (25.16)$$

where the function $A(X_s)$ is given by

$$A = (1 + X_s)\left(1 + \frac{2C}{1 + sX_s}\right), \quad (25.17)$$

as obtained by dividing Eq. (13.21) by x and by solving it with respect to A with X replaced by X_s. Therefore the boundary of the static instability coincides with the turning point of the stationary curve, and the segment of the stationary curve with negative slope is unstable because $c_0 < 0$.

On the other hand, as prescribed by Eq. (B.4) the boundary of the oscillatory instability is given by the equation $c_1 c_2 = c_0$, which, using Eq. (25.15), reads explicitly

$$\bar{\gamma}^2 2Cs X_s - \gamma^2 AX_s - \gamma\bar{\gamma}(1 + X_s)(1 + sX_s)[\gamma(1 + X_s) + \bar{\gamma}(1 + sX_s)] = 0. \quad (25.18)$$

In general, the procedure is to insert Eq. (25.17) into Eq. (25.18) and solve numerically the resulting quartic equation in X_s, thus finding the boundary point X_H where an oscillatory solution bifurcates from the nontrivial stationary solution. The corresponding value A_H of A is obtained by setting $X_s = X_H$ in Eq. (25.17). The most complex of the Routh–Hurwitz instability conditions (see Appendix A) reads $c_1 c_2 < c_0$ and indicates that the unstable side of the boundary is $X_s < X_H$, $A < A_H$. The analysis becomes drastically simpler in the limit

$$\gamma = \frac{\gamma_\parallel}{2\kappa} \ll 1, \qquad \bar{\gamma} = \frac{\bar{\gamma}_\parallel}{2\kappa} \ll 1, \quad (25.19)$$

because the third term in the l.h.s. of Eq. (25.18) becomes negligible and we obtain the simple result

$$A_H = \left(\frac{\bar{\gamma}}{\gamma}\right)^2 2Cs. \quad (25.20)$$

Then the value of X_H is obtained by inserting the value (25.20) of A_H into the expression for X_+ given by Eq. (13.22).

It must be noted that this simple expression for A_H is accurate only for γ, $\bar{\gamma} \sim 10^{-3}$, 10^{-4}. Nevertheless, it provides the important information that, for a given stationary curve, i.e. fixed values of $2C$ and s, the position of the Hopf bifurcation point depends mainly on the quantity $(\bar{\gamma}/\gamma)^2$. In particular, the Hopf instability threshold A_H is larger than the laser

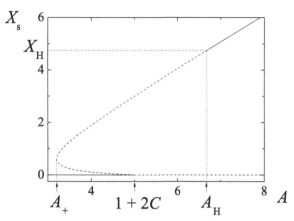

Figure 25.3 The stationary intensity as a function of the pump parameter A in a laser with saturable absorber with $2C = 4$ and $s = 5$. The Hopf instability threshold (A_H, X_H) is indicated for $\gamma = 0.01$ and $\bar{\gamma} = 0.008$. The broken part of this curve is unstable.

threshold $A = 1 + 2C$ when

$$\frac{\bar{\gamma}}{\gamma} > \sqrt{\frac{1 + 2C}{2Cs}}, \tag{25.21}$$

and, in the bistable regime $s > 1 + 1/(2C)$, it is larger than the turning point A_+ of the stationary curve (see Eq. (13.23) and Fig. 13.4) when

$$\frac{\bar{\gamma}}{\gamma} > \frac{\sqrt{s-1} + \sqrt{2C}}{s\sqrt{2C}}. \tag{25.22}$$

For the sake of definiteness, let us consider the parametric values $2C = 4$ and $s = 5$, for which the bistable stationary curve is shown in Fig. 25.3 and the r.h.s. of Eqs. (25.21) and (25.22) are equal, respectively, to 0.5 and 0.4. By choosing $\gamma = 0.01$ and $\bar{\gamma} = 0.8\gamma$ both conditions are satisfied.

Figure 25.4 shows the regular pulsations that appear for four values of A in the range $1 + 2C < A < A_H$; the height of the pulses and their period decreases as A approaches the Hopf bifurcation.

In order to describe the basic features of these oscillations let us note first that, since we assumed the conditions $\gamma_\|, \bar{\gamma}_\| \ll \kappa \ll \gamma_\perp, \bar{\gamma}_\perp$ (see Eq. (25.19)), if we compare with Eq. (19.2) we can say that these conditions correspond to a class-B laser with a saturable absorber. The oscillations in Fig. 25.4 are pulses well separated in time.

In Section 19.3.2 we described the generation of a giant pulse by a procedure called active Q-switching. In the case of a laser with a saturable absorber the pulsations emerge spontaneously from the Hopf instability, instead of an active procedure that we apply to the laser, and consist in a sequence of repeated pulses. For all these reasons the phenomenon described by Fig. 25.4 is called *repetitive passive Q-switching* [301, 302].

The mechanism which produces the passive Q-switching can be described in the following way. It arises from the interplay of a slow response of the population difference in the amplifier and in the absorber and a fast response of their field (see Eq. (25.19)). Under these

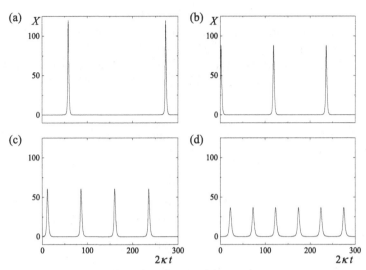

Figure 25.4 Repetitive passive Q-switching. Regular pulses emitted by the laser for (a) $A = 5.1$, (b) $A = 5.5$, (c) $A = 6$ and (d) $A = 6.5$. The other parameters are as in Fig. 25.3.

conditions, when the pump parameter is such that the unsaturated gain overcomes the unsaturated losses, the field builds up rapidly without appreciable changes in the populations. In a second stage, the emission is largely enhanced by the fact that the absorber gets bleached (saturated). Finally, when also the amplifier gets saturated so that the gain becomes smaller than the losses, the field begins to decrease and finally decays to (practically) zero. At this point, also the populations are back to their original values, hence the process repeats itself again and again.

A final remark in this section is that the two-level model which we used in Sections 13.2 and 25.2 represents a rather idealized description of the system. More refined models have been proposed and analyzed in the literature (see [135, 303–305] and references cited therein).

25.3 A degenerate optical parametric oscillator, period doubling and chaos

We start from Eqs. (13.51) and (13.52) with $y_1^{(0)} = 0$, which describe a degenerate optical parametric oscillator. In Section 13.4 we have already discussed the stationary solutions and shown, in particular, that when $\theta_1\theta_2 > 1$ the stationary curve exhibits bistability between a trivial stationary solution and a nontrivial stationary solution. In this section we perform the linear-stability analysis and discuss the oscillatory behaviors which arise from the oscillatory instability [139].

The linear-stability analysis of the trivial stationary solution (13.54) is left to the reader as an exercise; this is easy to do by following the same line of reasoning as for the case of the trivial stationary solution of the laser described in Section 22.1. The result is that the trivial stationary solution is stable for $0 \leq Y < 1 + \theta_1^2$, where Y is defined in Eq. (13.53), i.e. below threshold. In the following we analyze the stability of the nontrivial stationary solutions.

As usual, we start by expressing the variables x_1 and x_2 as the sum of the corresponding stationary values plus fluctuation variables

$$x_i(t) = x_{is} + \delta x_i(t), \quad i = 1, 2. \tag{25.23}$$

Then we linearize Eqs. (13.51) and (13.52) with respect to the fluctuation variables, obtaining

$$\delta \dot{x}_1 = -\kappa_1 [(1 + i\theta_1)\delta x_1 - x_{2s} \, \delta x_1^* - x_{1s}^* \, \delta x_2], \tag{25.24}$$

$$\delta \dot{x}_1^* = -\kappa_1 [(1 - i\theta_1)\delta x_1^* - x_{2s}^* \, \delta x_1 - x_{1s} \, \delta x_2^*], \tag{25.25}$$

$$\delta \dot{x}_2 = -\kappa_2 [(1 + i\theta_2)\delta x_2 + 2x_{1s} \, \delta x_1], \tag{25.26}$$

$$\delta \dot{x}_2^* = -\kappa_2 [(1 - i\theta_2)\delta x_2^* + 2x_{1s}^* \, \delta x_1^*]. \tag{25.27}$$

As usual, we introduce the exponential ansatz

$$\begin{pmatrix} \delta x_1(t) \\ \delta x_1^*(t) \\ \delta x_2(t) \\ \delta x_2^*(t) \end{pmatrix} = \exp(\lambda t) \begin{pmatrix} \delta x_1(0) \\ \delta x_1^*(0) \\ \delta x_2(0) \\ \delta x_2^*(0) \end{pmatrix}, \tag{25.28}$$

which leads to the quartic characteristic equation

$$\lambda^4 + c_3 \lambda^3 + c_2 \lambda^2 + c_1 \lambda + c_0 = 0, \tag{25.29}$$

with

$$\begin{aligned} c_3 &= 2(\kappa_1 + \kappa_2), \\ c_2 &= \kappa_2 \left[4\kappa_1 + (4\kappa_1 X_{1s} + \kappa_2) \left(1 + \theta_2^2 \right) \right], \\ c_1 &= 2\kappa_1 \kappa_2 \left(1 + \theta_2^2 \right) [\kappa_2 + 2X_{1s}(\kappa_1 + \kappa_2)], \\ c_0 &= 4X_{1s}\kappa_1^2 \kappa_2^2 \left(1 + \theta_2^2 \right) \left[1 - \theta_1 \theta_2 + X_{1s} \left(1 + \theta_2^2 \right) \right], \end{aligned} \tag{25.30}$$

where X_1 is defined by Eq. (13.53). The boundary of the stationary instability in parameter space is given by the equation $c_0 = 0$ which, by using Eq. (13.53), can be cast in the form

$$c_0 = 2\kappa_1^2 \kappa_2^2 X_{1s} (1 + \theta_2^2) \left. \frac{dY}{dX_1} \right|_s. \tag{25.31}$$

Hence the boundary of the stationary instability coincides with the turning point A of the stationary curve in Fig. 13.5(b), and the instability condition $c_0 < 0$ states, as usual, that the part of the stationary curve with negative slope is unstable.

According to Eq. (B.9) in Appendix B for the quartic case, the boundary of the oscillatory (Hopf) instability in parameter space is defined by the equation $c_1^2 - c_3(c_1 c_2 - c_0 c_3) = 0$, which, using Eqs. (13.53) and (13.56), leads to the result

$$X_{1s} = -\frac{\kappa_2 \left[(2\kappa_1 + \kappa_2)^2 + \kappa_2^2 \theta_2^2 \right]}{2(\kappa_1 + \kappa_2)^2 \left[2\kappa_1 (1 + \theta_1 \theta_2) + \kappa_2 \left(1 + \theta_2^2 \right) \right]}, \tag{25.32}$$

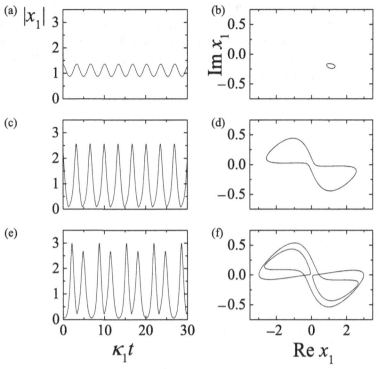

Figure 25.5 Spontaneous oscillations for $\theta_1 = 3, \theta_2 = -1$ and $\kappa_1 = \kappa_2$. Parts (a), (c) and (e) show the time evolution of $|x_1|$; parts (b), (d) and (f) show the projection of the phase-space trajectory onto the (Re x_1, Im x_1) plane. In (a) and (b) $X_{1s} = 0.63$, in (c) and (d) $X_{1s} = 1$ and in (e) and (f) $X_{1s} = 1.2$.

where the r.h.s. is positive when

$$\theta_1 \theta_2 < - \left(1 + \frac{1 + \theta_2^2}{2\kappa_1/\kappa_2} \right). \tag{25.33}$$

The most complex of the Routh–Hurwitz instability conditions (see Appendix A) is $c_1^2 - c_3 (c_1 c_2 - c_0 c_3) < 0$, so the unstable segment of the stationary curve corresponds to the points where X_1 is larger than the r.h.s. of Eq. (25.32). This is indicated in Fig. 13.5(a) as a broken part.

Note that condition (25.33) implies that the oscillatory instability arises only when both modes are detuned and the product $\theta_1 \theta_2$ is negative, whereas the bistability condition (13.57) requires, again, that both modes are detuned but $\theta_1 \theta_2$ must be positive. Hence in the case of the degenerate optical parametric oscillator the two phenomena of bistability and spontaneous oscillations are mutually exclusive.

In Fig. 25.5 we see the self-pulsing oscillations for the values of the parameters indicated in the caption, which correspond to those of Fig. 13.5(a), and increasing values of X_{1s}. As we have from Eq. (25.32), the instability threshold is $X_{1s} = X_H = 0.625$ (see Fig. 13.5(a)). Slightly beyond this threshold (Figs. 25.5(a) and (b)), the system undergoes small amplitude oscillations and the projection of the phase-space trajectory onto the (Re x_1, Im x_1) plane is a small limit cycle that bifurcates smoothly from the stationary solution at the instability threshold. Increasing the pump intensity Y and therefore X_{1s}, the oscillation amplitude grows (Figs. 25.5(c) and (d)). One finds period-2 (Figs. 25.5(e) and (f)), period-4, and chaotic oscillations [139].

PART III

TRANSVERSE OPTICAL PATTERNS

Introduction to Part III

The field of optical pattern formation concerns the spatial and spatio-temporal phenomena that arise in the structure of the electromagnetic field in the planes orthogonal with respect to the direction of propagation. Most theoretical treatments of the interaction between matter and radiation introduce the plane-wave approximation, i.e. assume that the electric field is uniform in each transverse plane. In this way, the time-evolution equations depend only on one spatial variable, i.e. the longitudinal variable z which corresponds to the direction of propagation. By dropping the plane-wave approximation, one opens the door to the fascinating world of pattern formation. In the paraxial approximation this corresponds to keeping, in the time-evolution equation of the electric field, the term with the transverse Laplacian which describes diffraction of radiation; this term couples the different points of the transverse (x, y) plane, as is necessary for pattern formation. With the exception of the discussion of spatial Kerr solitons in Section 7.3, in the analysis of Maxwell–Bloch equations, or of other nonlinear optical models, in Parts I and II of this book we have always adopted the plane-wave approximation, whereas now in Part III we study the transverse effects.

The interaction of light with linear inhomogeneous media can give rise to spatial structures of interesting and remarkable complexity. However, the field of optical pattern formation concerns mainly the interaction with nonlinear media, where the phenomena emerge spontaneously as a consequence of an instability; another name that is commonly used to designate optical pattern formation is "transverse nonlinear optics" or "optical morphogenesis". Historically, the interest in optical pattern formation emerged as a natural evolution of the previous development of the field of optical instabilities and chaos, when the main attention shifted gradually from purely temporal effects to spatio-temporal phenomena. The evolution was made possible also by the spectacular increase of the computational capabilities of available computers. For both of the fields of optical instabilities and optical pattern formation, continuous inspiration arose from the formulation of general disciplines such as Haken's synergetics [91, 92] and Prigogine's theory of dissipative structures [234, 235].

The existence of transverse effects has been well known since the earliest days of laser physics. In order to obtain the simple Gaussian transverse structure which is desired for most applications, one introduces apertures in the laser cavity. Otherwise, the system spontaneously generates more or less complex configurations. These phenomena were, however, mostly considered undesirable or difficult to control. A basic understanding, though, was achieved with the analysis of diffractive effects upon propagation of Gaussian beams with phenomena such as self-focussing and self-defocussing.

Despite the existence of an important early literature on this subject (see Refs. 8–22 in [306]), systematic investigations on optical pattern formation were initiated only in the

1980s. A great impulse was given by the attention devoted to the case of nonlinear materials contained in optical cavities in the low-transmission and single-longitudinal-mode limits, because such systems are described by sets of partial differential equations with two spatial variables plus time, exactly as in the case of two-dimensional patterns in hydrodynamics and nonlinear chemical reactions [307–309]. On the other hand, the study of systems with a single feedback mirror [310, 311] produced the best compromise between simplicity of theoretical treatment and accessibility to experimental realization.

The simplest models for studying optical pattern formation assume translational symmetry in the transverse plane. This implies that, if there are mirrors, they must be plane mirrors; and, if there is a driving field, it must have a plane-wave configuration. This is the most fundamental setting, because it allows one to study the spontaneous onset of a pattern from a homogeneous state with breaking of the translational symmetry, as a consequence of an instability arising from the interplay of nonlinearity and diffraction. In this case one can perform analytically both the calculation of the homogeneous stationary solution and its linear-stability analysis. When there is a cavity, the modes of the empty cavity correspond to the plane waves tilted with respect to the propagation direction, with a continuous frequency spectrum.

The assumption of translational symmetry implies that the system is infinitely extended in the transverse directions, and this is conveniently formalized by using periodic boundary conditions. Clearly, this kind of model is strongly idealized, because in practice one always has a beam confined to a certain region of the transverse plane. A confinement can be introduced in different ways, e.g. by including in the model the transverse shape of the medium or, if an input beam is present, by considering a Gaussian or a flat-top transverse profile. In this case, everything must be calculated numerically. When the transverse dimensions of the confinement region are much larger than the length which characterizes the spatial modulation of the pattern, one can recover the results of the idealized model qualitatively.

In the cavity case, one can obtain the transverse confinement by considering spherical instead of planar mirrors. In this case the system has only rotational symmetry; the modes of the empty cavity correspond to Gauss–Laguerre functions, with a discrete frequency spectrum. An important advantage of this configuration is that the number of modes in play can be controlled by the geometrical parameters of the cavity. For example, by reducing the Fresnel number one can cut off Gauss–Laguerre modes of high order. By varying the radius of curvature of the spherical mirrors and their distance, one controls the frequency spacing between adjacent transverse modes.

The overwhelming diverse phenomena one meets in optical pattern formation display several similarities with those, for example, in hydrodynamics and nonlinear chemical reactions, where diffusion and not diffraction couples to the nonlinearity to govern the arising patterns. A formal analogy between laser equations and hydrodynamics is described in Section 27.5. In definite domains of the parameter space, the exact dynamical equations can be approximated by Ginzburg–Landau or Swift–Hohenberg equations, or similar equations; pattern formation in a wide variety of fields has been described by these types of equations. For the sake of keeping our book to a reasonable size, we do not discuss such equations.

Undoubtedly, hydrodynamics and nonlinear chemical reactions have a much longer tradition than optics has in the study of pattern formation. However, radiation–matter interaction is fundamental in physics and chemistry, and this is already a strong motivation for studying pattern formation in optics. In addition, optics presents two special features that are interesting and stimulating. First, optical systems are very fast and have a large frequency bandwidth; hence they lend themselves naturally to applications, e.g. in telecommunications and information technology. The most relevant example of a useful application of optical structures is provided by solitonic transmission in optical fibers. The investigations of transverse nonlinear optics offer, in principle, the possibility of an approach to parallel optical information processing, by encoding information in the transverse structure of the electric field.

The second special feature is that optical systems are macroscopic or mesoscopic, yet they are capable of displaying interesting quantum effects, even at room temperature. This has led to investigations on the quantum aspects in transverse optical structures, which in turn have contributed in an important way to the birth of a new field called quantum imaging [67].

Chapter 26 studies the propagation of the electric field in vacuum, including transverse effects, and calculates the fundamental Gaussian beams and the higher-order Gauss–Hermite or Gauss–Laguerre beams. This classic piece of analysis is essential to describe the modes of cavities with spherical mirrors, both in the case of Fabry–Perot cavities and in the case of ring cavities.

The remaining chapters deal with optical pattern formation arising from the nonlinear interaction of radiation with matter, with the illustration of key concepts necessary to understand the nature of the investigations in the field of optical pattern formation and their developments. Chapters 27 and 28 treat the case of cavities with planar mirrors and Chapter 29 the configuration of cavities with spherical mirrors. Outstanding aspects relating to applications concerning the temporal version of cavity solitons are illustrated in Section 28.2.

The final chapter, Chapter 30, treats the aspects which are most interesting for the possibility of practical applications of the field of optical pattern formation. We describe the formation and manipulation of cavity solitons in cavities with planar mirrors. Especially interesting is the case of broad-area vertical-cavity surface-emitting lasers (VCSELs) driven by an external coherent field.

Reviews of the field of optical pattern formation can be found in [229,306,312,314–323]; a special issue can be found in [324]. Transverse effects in cold atomic systems are discussed in [325].

In this chapter we describe the case of a free electric field that does not interact with any material, so we start from Eq. (3.30) with the atomic polarization P_0 set equal to zero. In addition, we focus on the stationary solutions, hence we drop also the term with the time derivative, and the transverse Laplacian plays the crucial role in determining the field configuration.

In Section 26.1 we show that the field equation admits stationary solutions in which the field has a Gaussian configuration with a beam radius (equal to the halfwidth of the Gaussian) that varies as a function of the longitudinal coordinate z, and that the variation is governed by a parameter, called the *Rayleigh length*, which characterizes the diffraction of the beam. The Gaussian solution corresponds to the fundamental mode of the radiation field, and the plane-wave solutions constitute a limit case of the Gaussian solutions obtained when the beam radius tends to infinity.

On the other hand, in Section 26.2 we derive the higher-order modes, in the form of Gauss–Hermite modes, given by the product of the fundamental Gaussian mode and Hermite polynomials, and Gauss–Laguerre modes, given by the product of the fundamental Gaussian mode and Laguerre polynomials. The Gauss–Hermite modes are appropriate to treat problems with square symmetry; the Gauss–Laguerre modes are appropriate to treat problems with axial symmetry. In both cases the modes depend on one independent parameter, the Rayleigh length, and in both cases the modes constitute an orthonormal and complete set of functions of the transverse variables x and y for any arbitrary value of the longitudinal variable z.

Sections 26.3–26.5 are devoted to the case of Gaussian modes in a cavity. To be precise, in Section 26.3 we focus on the case of a Fabry–Perot cavity with spherical mirrors and calculate the Rayleigh length which characterizes the set of cavity modes as a function of the parameters of the cavity, i.e. the radii of curvature of the spherical mirrors and the distance between the mirrors. In Section 26.4 we derive the same result in a more straightforward way by using the so-called $ABCD$ matrix approach, which is then utilized in Section 26.5 to calculate the modes of a ring cavity including spherical mirrors.

In Section 26.6 we evaluate the frequencies of the cavity modes both for Fabry–Perot cavities (Section 26.6.1) and for ring cavities (Section 26.6.2) with spherical mirrors. The frequencies depend on three integer indices: one is the longitudinal index j, which is the same as that of the plane-wave theory (see Chapter 8); the other two are transverse indices, which are different for Gauss–Hermite and Gauss–Laguerre modes.

26.1 Gaussian-shaped beams

Let us start from Eq. (3.30) without the terms with the atomic polarization and the time derivative, i.e.

$$\nabla_\perp^2 E_0 + 2ik_0 \frac{\partial E_0}{\partial z} = 0. \tag{26.1}$$

In order to arrive at the fundamental Gaussian mode let us consider a spherical wave centered at the origin $\exp(ik_0 r_{3D})/r_{3D}$, where $r_{3D} = \sqrt{x^2 + y^2 + z^2}$ is the three-dimensional distance from the origin. In the paraxial approximation we have

$$r_{3D} = \sqrt{x^2 + y^2 + z^2} \simeq z\left(1 + \frac{x^2 + y^2}{2z^2}\right) = z + \frac{x^2 + y^2}{2z}, \tag{26.2}$$

where we have assumed that $z > 0$, so the spherical wave becomes

$$\frac{e^{ik_0 r_{3D}}}{r_{3D}} \simeq \frac{e^{ik_0(x^2+y^2)/(2z)}}{z} e^{ik_0 z} = E_0(\mathbf{x})e^{ik_0 z}, \tag{26.3}$$

and, if we multiply $E_0(\mathbf{x})$ by an arbitrary constant z_0, we can write

$$E_0(\mathbf{x}) = \frac{e^{i\tilde{r}^2/\eta}}{\eta}, \tag{26.4}$$

where

$$\tilde{x} = \frac{x}{w_0}, \quad \tilde{y} = \frac{y}{w_0}, \quad \tilde{r} = \frac{r}{w_0} = \frac{\sqrt{x^2 + y^2}}{w_0}, \quad \eta = \frac{z}{z_0}, \tag{26.5}$$

and w_0 is linked to z_0 by the relation[1]

$$z_0 = \frac{k_0 w_0^2}{2}. \tag{26.6}$$

In terms of the scaled dimensionless variables $\tilde{r}$ and η, Eq. (26.1) can be rewritten as follows:

$$\tilde{\nabla}_\perp^2 E_0 + 4i \frac{\partial E_0}{\partial \eta} = 0, \quad \tilde{\nabla}_\perp^2 = \frac{\partial^2}{\partial \tilde{x}^2} + \frac{\partial^2}{\partial \tilde{y}^2} = \frac{1}{\tilde{r}} \frac{\partial}{\partial \tilde{r}}\left(\tilde{r} \frac{\partial}{\partial \tilde{r}}\right) + \frac{1}{\tilde{r}^2} \frac{\partial^2}{\partial \varphi^2}. \tag{26.7}$$

One can easily check that Eq. (26.4) is a solution of Eq. (26.7) both for $\eta > 0$ ($z > 0$) and for $\eta < 0$ ($z < 0$). One can also verify that $E_0(\tilde{r}, \eta - c)$, where c is any complex number, is a solution too. By taking $c = i$ we consider the solution

$$E_0(\tilde{r}, \eta) = \frac{1}{\eta - i} \exp\left(\frac{i\tilde{r}^2}{\eta - i}\right) = \frac{1}{\sqrt{1 + \eta^2}} \exp\left[i\frac{\tilde{r}^2 \eta}{1 + \eta^2} - \frac{\tilde{r}^2}{1 + \eta^2} + i \arctan\left(\frac{1}{\eta}\right)\right]$$

$$= \frac{1}{w(\eta)} e^{i\tilde{r}^2/R(\eta)} e^{-\tilde{r}^2/w^2(\eta)} e^{i \arctan(1/\eta)}, \tag{26.8}$$

[1] Note that the scaled coordinates $\tilde{x}$, $\tilde{y}$ and η correspond to the variables $\tilde{x}$, $\tilde{y}$ and $\tilde{z}$ considered in Section 3.1, respectively.

with

$$w(\eta) = \sqrt{1 + \eta^2}, \qquad R(\eta) = \frac{1 + \eta^2}{\eta}. \tag{26.9}$$

This solution corresponds to a beam with a Gaussian profile whose halfwidth is $w_0\sqrt{1 + (z/z_0)^2}$, in the original variables r and z. The halfwidth is minimal at $z = 0$ where its value is w_0, which is called the *beam waist*. The width grows by a factor of $\sqrt{2}$ (i.e. the area doubles) at a distance $z = z_0$ from the origin. Therefore the parameters w_0 and z_0, which are linked by the relation (26.6), provide the spatial scales over which the beam varies appreciably in the radial direction and in the longitudinal direction, respectively. The parameter z_0 is called the *Rayleigh length* (or *Rayleigh range*) or diffraction length.

Let us now show that $R(z) = z_0 R(\eta)$, with $R(\eta)$ given by Eq. (26.9), represents the radius of curvature of the wavefront at position z. As a matter of fact, the phase of the wave is given by

$$\phi(\tilde{r}, \eta) = k_0 z_0 \eta + \arctan\left(\frac{1}{\eta}\right) + \frac{\tilde{r}^2}{R(\eta)}, \tag{26.10}$$

which, in unnormalized units, reads

$$\begin{aligned}\phi(r, z) &= k_0 z + \arctan\left(\frac{z_0}{z}\right) + k_0 \frac{r^2 z}{2(z^2 + z_0^2)} \\ &= k_0 \left[z + \frac{\lambda}{2\pi} \arctan\left(\frac{z_0}{z}\right) + \frac{r^2 z}{2(z^2 + z_0^2)} \right]. \end{aligned} \tag{26.11}$$

For z sufficiently larger than λ we can neglect the second term with respect to the first so that

$$\phi(r, z) = k_0 \left[z + \frac{r^2 z}{2(z^2 + z_0^2)} \right]. \tag{26.12}$$

Let us consider the wavefront defined by the equation $\phi(r, z) = \bar{\phi}$. On the axis $r = 0$ this equation defines the coordinate value $\bar{z} = \bar{\phi}/k_0$. For z different from 0 but small, one has the same value of the phase at a smaller value of z, which we call $\bar{z} - dz$. We can write

$$\bar{\phi} \approx k_0 \bar{z} - k_0\, dz + k_0 \frac{r^2 \bar{z}}{2(\bar{z}^2 + z_0^2)}, \tag{26.13}$$

where in the last term the approximation of z by $\bar{z}$ holds when r is small enough. Therefore dz is given by

$$dz \approx \frac{r^2 \bar{z}}{2(\bar{z}^2 + z_0^2)}. \tag{26.14}$$

On the other hand, if we consider a circle of radius R and we call h the segment of the radius cut by a chord of length $2r$ one has the relation $(R - h)^2 + r^2 = R^2$, i.e. $-2hR + h^2 + r^2 = 0$. If r is small, also h is small, and therefore h^2 is negligible with respect to $2hR$. Hence we have

$$h = \frac{r^2}{2R}. \tag{26.15}$$

By comparing Eq. (26.15) with Eq. (26.14) we can conclude that $R(z) = (z^2 + z_0^2)/z$ is the radius of curvature of the wavefront in unnormalized coordinates, and $R(\eta) = (\eta^2 + 1)/\eta$ is the normalized radius of curvature. Note that by this convention the radius of curvature is negative for $z < 0$.

At $z = 0$ the radius of curvature is infinite and therefore the wavefront is planar, and the phase of the wave is given by $\arctan(\infty) = \pi/2$. For setting the phase at $z = 0$ equal to zero it suffices to multiply the expression (26.8) by $\exp(-i\pi/2)$. As a matter of fact, from the relation[2]

$$\arctan\alpha + \arctan\beta = \arccos\left(\frac{1 - \alpha\beta}{\sqrt{(1 + \alpha^2)(1 + \beta^2)}}\right), \tag{26.16}$$

we have that

$$\arctan\eta + \arctan\left(\frac{1}{\eta}\right) = \frac{\pi}{2}, \tag{26.17}$$

and therefore the general Gaussian solution with phase equal to zero for $z = 0$ reads

$$E_0(\tilde{r}, \eta) = \sqrt{\frac{2}{\pi}}\frac{1}{w(\eta)}e^{i\tilde{r}^2/R(\eta)}e^{-\tilde{r}^2/w^2(\eta)}e^{-i\arctan\eta}, \tag{26.18}$$

where the constant $\sqrt{2/\pi}$ is determined by imposing that the integral of $|E_0|^2$ over the transverse plane is equal to unity (one can easily verify that this condition holds for all values of η). Figure 26.1(a) shows the graph of the function $R(\eta)$ defined by Eq. (26.9), and Fig. 26.1(b) shows the wavefronts and the graph of the function $w(\eta)$. Note that the radius of curvature is a non-monotonic function that first decreases and then increases.

Note that for $z \gg z_0$ the beam halfwidth $w_0 w(\eta = z/z_0)$ is given by $z(w_0/z_0)$, i.e. it is linear in z (see Fig. 26.1(b)), so we have a divergence angle of the beam $\theta_d = w_0 w(\eta = z/z_0)/z = w_0/z_0 = \lambda/(\pi w_0)$, where we have used Eq. (26.6) and λ is the wavelength. This is in agreement with the fact that diffraction produces an angular divergence equal to the

---

[2] From the relation

$$\tan(\theta + \varphi) = \frac{\sin\theta\cos\varphi + \sin\varphi\cos\theta}{\cos\theta\cos\varphi - \sin\theta\sin\varphi} = \frac{\tan\theta + \tan\varphi}{1 - \tan\theta\tan\varphi}$$

it follows that

$$\tan(\arctan\alpha + \arctan\beta) = \frac{\alpha + \beta}{1 - \alpha\beta} \equiv \tan\gamma.$$

Therefore

$$\cos(\arctan\alpha + \arctan\beta) = \pm[1 + \tan^2\gamma]^{-1/2} = \pm\left[1 + \left(\frac{\alpha + \beta}{1 - \alpha\beta}\right)^2\right]^{-1/2}$$

$$= \pm\sqrt{\frac{(1 - \alpha\beta)^2}{(1 + \alpha^2)(1 + \beta^2)}} = \frac{1 - \alpha\beta}{\sqrt{(1 + \alpha^2)(1 + \beta^2)}}.$$

The sign ambiguity has been resolved by considering the case $\alpha = \beta = 0$.

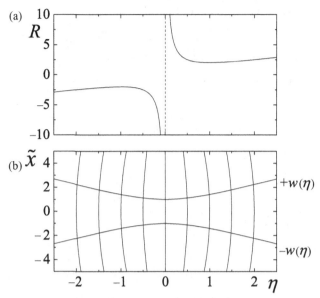

Figure 26.1 (a) A graph of the function $R(\eta) = (1 + \eta^2)/\eta$, where $\eta = z/z_0$ and z_0 is the Rayleigh length. The radius of curvature at $z = z_0\eta$ is $z_0\,R(\eta)$. (b) Wavefronts of a Gaussian beam obtained from Eq. (26.10) by neglecting the term $\arctan(1/\eta)$ and setting $k_0 z_0 = 200$ and $\phi(\tilde{r}, \eta) = \text{constant}$, and a graph of the lines $\pm w(\eta)$ (see Eq. (26.9)). The beam halfwidth at position $z = z_0\eta$ is $w_0 w(\eta)$.

ratio of the wavelength to the beam diameter within a factor of order unity (in our case the beam diameter is equal to $2w_0$) [6].

26.2 Higher-order modes

Equation (26.7) admits, in addition to the Gaussian solutions (26.18), also solutions with a more complex dependence on the transverse coordinates $\tilde{x}$ and $\tilde{y}$. The structure of these higher-order modes is, however, strictly linked to that of the fundamental Gaussian mode, i.e. the parameters $w(\eta)$ and $R(\eta)$, which control the beam halfwidth and the radius of curvature of the wavefront, respectively, maintain their meaning in these modes. Therefore we look for a solution of the form

$$E_0(\tilde{x}, \tilde{y}, \eta) = \chi(\tilde{x}, \tilde{y}, \eta)G(\tilde{x}, \tilde{y}, \eta), \qquad (26.19)$$

where G is the Gaussian given by the r.h.s. of Eq. (26.18). By inserting this equation into Eq. (26.7) we obtain

$$\chi\,\tilde{\nabla}_\perp^2 G + G\,\tilde{\nabla}_\perp^2 \chi + 2(\tilde{\nabla}_\perp \chi)\cdot(\tilde{\nabla}_\perp G) + 4i\,\chi\,\frac{\partial G}{\partial \eta} + 4i\,G\,\frac{\partial \chi}{\partial \eta} = 0. \qquad (26.20)$$

Since G is a solution of Eq. (26.7) and

$$\tilde{\nabla}_\perp G = \frac{2i\,G}{\eta - i}(\tilde{x}\hat{e}_x + \tilde{y}\hat{e}_y),$$

(26.21)

we can write the following equation for the function χ

$$\tilde{\nabla}_\perp^2 \chi + \frac{4i}{\eta - i}\left(\tilde{x}\frac{\partial\chi}{\partial\tilde{x}} + \tilde{y}\frac{\partial\chi}{\partial\tilde{y}}\right) + 4i\frac{\partial\chi}{\partial\eta} = 0.$$

(26.22)

By introducing the change of variables

$$x' = \frac{\sqrt{2}\tilde{x}}{w(\eta)}, \qquad y' = \frac{\sqrt{2}\tilde{y}}{w(\eta)},$$

(26.23)

we arrive finally at the equation

$$\frac{\partial^2\chi}{\partial x'^2} + \frac{\partial^2\chi}{\partial y'^2} - 2\left(x'\frac{\partial\chi}{\partial x'} + y'\frac{\partial\chi}{\partial y'}\right) + 2i(1 + \eta^2)\frac{\partial\chi}{\partial\eta} = 0.$$

(26.24)

There are basically two kinds of solution of this equation. The first type is represented by solutions that can be factorized as a product of a function of x' and a function of y'. Since it turns out that such solutions correspond to Hermite polynomials, one calls them *Gauss–Hermite modes*. This kind of solution is appropriate for problems with square symmetry. For problems with cylindrical symmetry, instead, the solutions which factorize as a product of a function of the radial coordinate and a function of the angular coordinate are more convenient. Since the function which depends on the radial coordinate turns out to correspond to a Laguerre polynomial, one calls them *Gauss–Laguerre modes*.

26.2.1 Gauss–Hermite modes

Let us seek solutions of Eq. (26.24) of the form

$$\chi(x', y', \eta) = f(x', \eta)g(y', \eta).$$

(26.25)

It is easy to verify that the starting equation can be decomposed in the two equations

$$\frac{\partial^2 f}{\partial x'^2} - 2x'\frac{\partial f}{\partial x'} + 2i(1 + \eta^2)\frac{\partial f}{\partial \eta} = 0,$$

(26.26)

$$\frac{\partial^2 g}{\partial y'^2} - 2y'\frac{\partial g}{\partial y'} + 2i(1 + \eta^2)\frac{\partial g}{\partial \eta} = 0.$$

(26.27)

Let us consider the first equation and let us look for solutions of the form

$$f(x', \eta) = \tilde{f}(x')e^{-i\theta(\eta)}.$$

(26.28)

We obtain the equation

$$\frac{d^2\tilde{f}}{dx'^2} - 2x'\frac{d\tilde{f}}{dx'} + 2(1 + \eta^2)\frac{d\theta}{d\eta}\tilde{f} = 0.$$

(26.29)

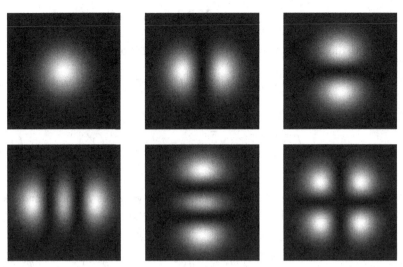

Figure 26.2 Gauss–Hermite modes of order (m, n). From left to right for the top row and then the bottom row: $(0, 0)$, $(1, 0)$, $(0, 1)$, $(2, 0)$, $(0, 2)$ and $(1, 1)$.

If we set $\theta(\eta) = m \arctan \eta$ with $m = 0, 1, 2, \ldots$, Eq. (26.29) becomes

$$\frac{d^2 \tilde{f}}{dx'^2} - 2x' \frac{d\tilde{f}}{dx'} + 2m\tilde{f} = 0, \qquad (26.30)$$

which admits as solutions the Hermite polynomials $H_m(x')$ [326]. The Hermite polynomials of lowest order are

$$H_0(x') = 1, \qquad H_1(x') = 2x', \qquad H_2(x') = 4x'^2 - 2. \qquad (26.31)$$

In order to calculate the polynomials of higher order, one can use the recursive relation

$$H_{n+1}(x') = 2x' H_n(x') - 2n H_{n-1}(x'). \qquad (26.32)$$

The polynomials, multiplied by $e^{-x'^2/2}$, constitute a complete basis in the Hilbert space of square summable functions in the interval $-\infty < x' < +\infty$, with the orthonormality property

$$\int_{-\infty}^{\infty} dx\, e^{-x^2} H_n(x) H_m(x) = 2^n n! \sqrt{\pi} \delta_{m,n}. \qquad (26.33)$$

Equation (26.27) for the function g leads to Hermite polynomials in the same way.

Hence, in conclusion, the Gauss–Hermite modes of order (n, m) are labeled by the two non-negative integer indices n and m and are given by

$$\psi_{mn}(\tilde{x}, \tilde{y}, \eta) = \sqrt{\frac{2}{\pi}} \frac{1}{w(\eta)} \frac{1}{\sqrt{2^{m+n} m! n!}} H_m \left[\frac{\sqrt{2}\tilde{x}}{w(\eta)} \right] H_n \left[\frac{\sqrt{2}\tilde{y}}{w(\eta)} \right]$$
$$\times e^{i\tilde{r}^2/R(\eta)} e^{-\tilde{r}^2/w^2(\eta)} e^{-i(m+n+1)\arctan \eta}, \qquad (26.34)$$

with $\tilde{x} = x/w_0$ and $\tilde{y} = y/w_0$, as we obtain from Eqs. (26.23), (26.5), (26.19), (26.18) and (26.28). Note that $\psi_{00}(\tilde{x}, \tilde{y}, \eta)$ coincides with the fundamental Gaussian mode given by Eq. (26.18). The Gauss–Hermite modes of lowest order are shown in Fig. 26.2.

26.2.2 Gauss–Laguerre modes

In this case we seek solutions of Eq. (26.24) of the form

$$\chi(x', y', \eta) = f(\rho', \eta)e^{il\varphi}, \qquad l = 0, \pm 1, \pm 2, \ldots, \tag{26.35}$$

with

$$\rho' = \sqrt{x'^2 + y'^2} = \frac{\sqrt{2}}{w(\eta)}\tilde{r}, \qquad \varphi = \arctan\left(\frac{y'}{x'}\right) = \arctan\left(\frac{\tilde{y}}{\tilde{x}}\right) = \arctan\left(\frac{y}{x}\right). \tag{26.36}$$

By performing the transformation to the variables ρ' and φ, and taking into account that the transverse Laplacian in polar coordinates is given by

$$\frac{\partial^2}{\partial x'^2} + \frac{\partial^2}{\partial y'^2} = \frac{\partial^2}{\partial \rho'^2} + \frac{1}{\rho'}\frac{\partial}{\partial \rho'} + \frac{1}{\rho'^2}\frac{\partial^2}{\partial \varphi^2}, \tag{26.37}$$

one obtains the following equation for $f(\rho', \eta)$:

$$\frac{\partial^2 f}{\partial \rho'^2} + \frac{1}{\rho'}\frac{\partial f}{\partial \rho'} - \frac{l^2}{\rho'^2}f - 2\rho'\frac{\partial f}{\partial \rho'} + 2i(1 + \eta^2)\frac{\partial f}{\partial \eta} = 0. \tag{26.38}$$

At this point it is convenient to introduce the new radial coordinate $u = \rho'^2$ in such a way that the equation for $f(u, \eta)$ becomes

$$4u\frac{\partial^2 f}{\partial u^2} + 4(1 - u)\frac{\partial f}{\partial u} - \frac{l^2}{u}f + 2i(1 + \eta^2)\frac{\partial f}{\partial \eta} = 0. \tag{26.39}$$

We look for solutions of the form

$$f(u, \eta) = u^{|l|/2}\tilde{f}(u)e^{-i\theta(\eta)}, \tag{26.40}$$

and obtain the following equation for $\tilde{f}$:

$$4u\frac{d^2\tilde{f}}{du^2} + 4(|l| + 1 - u)\frac{d\tilde{f}}{du} - 2|l|\tilde{f} + 2(1 + \eta^2)\frac{d\theta}{d\eta}\tilde{f} = 0. \tag{26.41}$$

If we set $\theta(\eta) = (2p + |l|)\arctan \eta$ with $p = 0, 1, 2, \ldots$, we have finally the equation

$$u\frac{d^2\tilde{f}}{du^2} + (|l| + 1 - u)\frac{d\tilde{f}}{du} + p\tilde{f} = 0, \tag{26.42}$$

which admits as solutions the Laguerre polynomials $L_p^{|l|}(u)$ [326]. The Laguerre polynomials of lowest order are given by

$$L_0^{|l|}(u) = 1, \qquad L_1^{|l|}(u) = -u + |l| + 1,$$

$$L_2^{|l|}(u) = \frac{1}{2}[u^2 - 2(|l| + 2)u + (|l| + 1)(|l| + 2)], \tag{26.43}$$

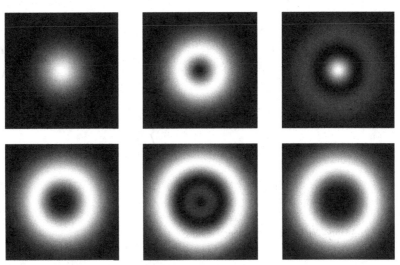

Figure 26.3   Gauss–Laguerre modes of order (p, l). From left to right and for the top row and then the bottom row: $(0, 0)$, $(0, \pm 1)$, $(1, 0)$, $(0, \pm 2)$, $(1, \pm 1)$ and $(0, \pm 3)$.

and the polynomials of higher order can be obtained by using the recursive relation

$$p L_p^{|l|}(u) = (2p + |l| - 1 - u) L_{p-1}^{|l|}(u) - (p + |l| - 1) L_{p-2}^{|l|}(u). \tag{26.44}$$

Once multiplied by $e^{-u/2} u^{|l|/2}$, the Laguerre polynomials are a complete basis in the Hilbert space of square summable functions in the interval $0 < u < \infty$ with the orthonormality property

$$\int_0^\infty du\, e^{-u} u^{|l|} L_p^{|l|}(u) L_q^{|l|}(u) = \frac{(p + |l|)!}{p!} \delta_{p,q}. \tag{26.45}$$

The normalized Gauss–Laguerre modes are labeled by the two indices $p = 0, 1, 2, \ldots$ and $l = 0, \pm 1, \pm 2, \ldots$; in the literature the modes with $p = 0$ are called "doughnut modes" (see Fig. 26.3) and are given by

$$\psi_{pl}(\tilde{r}, \varphi, \eta) = \sqrt{\frac{2}{\pi}} \frac{1}{w(\eta)} \sqrt{\frac{p!}{(p + |l|)!}} \left[\frac{2\tilde{r}^2}{w^2(\eta)} \right]^{|l|/2} L_p^{|l|} \left[\frac{2\tilde{r}^2}{w^2(\eta)} \right] e^{il\varphi}$$
$$\times\, e^{i\tilde{r}^2/R(\eta)} e^{-\tilde{r}^2/w^2(\eta)} e^{-i(2p+|l|+1)\arctan \eta}, \tag{26.46}$$

with $\tilde{r} = r/w_0 = \sqrt{x^2 + y^2}/w_0$, $\varphi = \arctan(y/x)$, where we have used Eqs. (26.18), (26.5), (26.19), (26.40) and (26.35). Also in this case the lowest mode, $\psi_{00}(\tilde{r}, \phi, \eta)$, coincides with the fundamental Gaussian mode (26.18). The Gauss–Laguerre modes of lowest order are represented in Fig. 26.3.

It is important to remark that one can consider, instead of the exponentials $e^{il\varphi}$, the two linear combinations $\cos(l\varphi)$ and $\sin(l\varphi)$, with $l = 0, 1, 2, \ldots$. This provides an alternative choice for the Gauss–Laguerre modes. In this book we will use the exponentials $e^{il\varphi}$ with $l = 0, \pm 1, \pm 2, \ldots$.

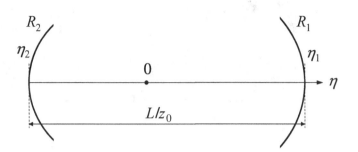

Figure 26.4 A Fabry–Perot cavity with spherical mirrors.

26.3 Gaussian modes in a cavity with spherical mirrors. The case of Fabry–Perot cavities

In the case of resonators with spherical mirrors the cavity modes have the form of Gaussian modes. The key parameter which controls the Gaussian modes and the beam waist w_0, i.e. the Rayleigh length z_0, is determined by the geometrical properties of the cavity.

Let us consider a Fabry–Perot cavity composed of two spherical mirrors, 1 and 2, concave and with radii of curvature R_1 and R_2, respectively, at a distance L from each other. Inside the cavity there is a point of the longitudinal axis z for which the wavefront is planar. We take this point as the origin of the axis, and call η_1 and η_2 the positions of the two mirrors on the η axis such that (see Fig. 26.4)

$$\eta_1 - \eta_2 = \frac{L}{z_0}. \tag{26.47}$$

A Gaussian mode is a mode of this cavity, provided that its radius of curvature at the position of the mirrors coincides with the radius of curvature of the mirrors themselves. Hence we must have

$$-\frac{R_2}{z_0} = \frac{1 + \eta_2^2}{\eta_2}, \qquad \frac{R_1}{z_0} = \frac{1 + \eta_1^2}{\eta_1}, \tag{26.48}$$

where we have taken into account that for negative η the radius of curvature is negative. Using Eqs. (26.47) and (26.48) we can obtain an expression for the Rayleigh length z_0 in terms of the cavity length R and the mirror radii R_1 and R_2. First of all, let us define the two geometrical dimensionless factors

$$g_1 = 1 - \frac{L}{R_1}, \qquad g_2 = 1 - \frac{L}{R_2}. \tag{26.49}$$

With the help of Eqs. (26.47) and (26.48) we can write them in terms of η_1 and η_2 as follows:

$$g_1 = 1 - \frac{L}{z_0}\frac{z_0}{R_1} = 1 - (\eta_1 - \eta_2)\frac{\eta_1}{1+\eta_1^2} = \frac{1+\eta_1\eta_2}{1+\eta_1^2},$$

$$(26.50)$$

$$g_2 = 1 - \frac{L}{z_0}\frac{z_0}{R_2} = 1 + (\eta_1 - \eta_2)\frac{\eta_2}{1+\eta_2^2} = \frac{1+\eta_1\eta_2}{1+\eta_2^2}.$$

Now, by combining the two equations (26.48) and taking into account Eq. (26.47), we obtain

$$\frac{R_1 + R_2}{z_0} = \frac{(1+\eta_1^2)\eta_2 - (1+\eta_2^2)\eta_1}{\eta_1\eta_2} = \frac{(\eta_1 - \eta_2)(\eta_1\eta_2 - 1)}{\eta_1\eta_2} = \frac{L}{z_0}\frac{\eta_1\eta_2 - 1}{\eta_1\eta_2}. \quad (26.51)$$

By solving Eq. (26.51) with respect to $\eta_1\eta_2$ one obtains

$$1 + \eta_1\eta_2 = \frac{R_1 + R_2 - 2L}{R_1 + R_2 - L}, \quad (26.52)$$

and, using Eqs. (26.47), (26.48), (26.50) and (26.52), we can write

$$\frac{L}{z_0} = \eta_1 - \eta_2 = \frac{z_0}{R_1}(1+\eta_1^2) + \frac{z_0}{R_2}(1+\eta_2^2) = z_0(1+\eta_1\eta_2)\left(\frac{1}{g_1 R_1} + \frac{1}{g_2 R_2}\right)$$

$$= z_0 \frac{R_1 + R_2 - 2L}{R_1 + R_2 - L}\frac{g_1 R_1 + g_2 R_2}{g_1 g_2 R_1 R_2}, \quad (26.53)$$

from which we derive an expression for z_0^2,

$$z_0^2 = \frac{g_1 g_2 R_1 R_2 L(R_1 + R_2 - L)}{(g_1 R_1 + g_2 R_2)(R_1 + R_2 - 2L)}. \quad (26.54)$$

Next, by using the definition (26.49) of g_1 and g_2, we obtain

$$R_1 R_2(1 - g_1 g_2) = L(R_1 + R_2 - L), \qquad g_1 R_1 + g_2 R_2 = R_1 + R_2 - 2L, \quad (26.55)$$

so that z_0^2 can be written as

$$z_0^2 = g_1 g_2(1 - g_1 g_2)\left(\frac{R_1 R_2}{g_1 R_1 + g_2 R_2}\right)^2. \quad (26.56)$$

This expression provides an acceptable result for z_0 only if the r.h.s. is positive, i.e.

$$0 \leq g_1 g_2 \leq 1, \quad (26.57)$$

and the stability domain in the (g_1, g_2) plane is shown in Fig. 26.5. Outside the stability domain a Gaussian beam does not reproduce itself after a roundtrip and the cavity is unstable. An especially interesting case is that of a symmetric cavity, i.e. $R_1 = R_2 = R$ and $g_1 = g_2 = g = 1 - L/R$, in which Eq. (26.56) reduces to

$$z_0 = \frac{R}{2}\sqrt{1 - g^2} = \frac{\sqrt{L(2R - L)}}{2}. \quad (26.58)$$

Three special cases of a symmetric cavity are the following.

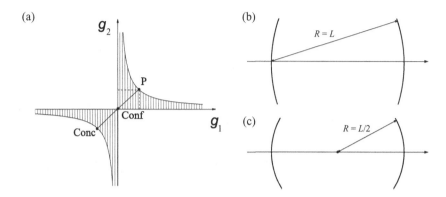

Figure 26.5 (a) The stability diagram of a spherical mirror Fabry–Perot cavity. The hyperbola corresponds to the line $g_1 g_2 = 1$ and the shaded region corresponds to the stability domain of the cavity. The segment $g_1 = g_2$ corresponds to a symmetrical cavity ($R_1 = R_2 = R$). In correspondence to the two broken segments, one of the two mirrors is planar ($R_1 = \infty$ or $R_2 = \infty$). The point P indicates a symmetric planar cavity. The points Conc and Conf indicate a concentric and a confocal cavity like those shown in parts (b) and (c), respectively.

- A planar cavity ($R = \infty$), corresponding to point P in Fig. 26.5(a) and shown in Fig. 8.1. In this case the Rayleigh length and the beam waist are equal to infinity too.
- A confocal cavity ($R = L$), corresponding to the point labeled Conf in Fig. 26.5(a) and shown in Fig. 26.5(b). In this case $z_0 = L/2$.
- A concentric cavity ($R = L/2$), corresponding to the point labeled Conc in Fig. 26.5(a) and shown in Fig. 26.5(c). In this case $z_0 = w_0 = 0$. In this limit case the paraxial approximation, which requires the beam waist to be much larger than the wavelength, does not apply.

26.4 The *ABCD* matrix method

An alternative convenient procedure to obtain the expression for the Rayleigh range in a Gaussian mode is provided by the so-called *ABCD* matrix method [9, 10]. In this section we derive again the expression for z_0 as a function of the cavity parameters, because in the following section we will apply this method to the ring-cavity case. We will describe this method concisely; for the details we refer the reader to [9, 10]. Let us define the quantity

$$q(\eta) = \eta - i, \tag{26.59}$$

which contains all the information about the geometry of the beam because

$$\frac{1}{q(\eta)} = \frac{\eta}{1 + \eta^2} + \frac{i}{1 + \eta^2} = \frac{1}{R(\eta)} + \frac{i}{w^2(\eta)}. \tag{26.60}$$

The condition which ensures that a Gaussian mode is a cavity mode is that the $q(\eta)$ parameter assumes again its initial value after a complete roundtrip of the cavity. The transformations of $q(\eta)$ inside the cavity are described by bilinear transformations of the form

$$q' = \frac{Aq + B}{Cq + D},$$
(26.61)

which correspond to 2×2 matrices

$$\begin{pmatrix} A & B \\ C & D \end{pmatrix}.$$

For cavities that do not include optical elements other than spherical mirrors it is necessary to define only the *ABCD* matrices which describe the translation of the beam and its reflection on the surface of a mirror.

On the basis of the definition (26.59) of q, the translation of the beam along a segment of length d corresponds to the transformation $q' = q + d/z_0$, which is described by the matrix

$$\begin{pmatrix} 1 & d/z_0 \\ 0 & 1 \end{pmatrix}.$$
(26.62)

Insofar as the reflection on a mirror is concerned, let us recall, first of all, that $R(\eta)$ has a sign, being positive for $\eta > 0$ and negative for $\eta < 0$. When a beam with radius of curvature R impinges on a spherical mirror with the same radius of curvature the reflected field will have radius of curvature $-R$, because the direction of propagation (i.e. the η axis) is reversed. Hence using Eq. (26.60) means that the transformation induced by a spherical mirror on the parameter $q(\eta)$ is $1/q' = 1/q - 2z_0/R$, which corresponds to the matrix

$$\begin{pmatrix} 1 & 0 \\ -2z_0/R & 1 \end{pmatrix}.$$
(26.63)

As a result, the total transformation of q in a roundtrip is given by the product of the individual transformation matrices taken in order. Let us apply this formalism to the case of a symmetrical Fabry–Perot cavity. If we start from point $z = 0$, where we know that the wavefront is planar, it suffices to impose that the beam reproduces itself after half a roundtrip, i.e. when we return to $z = 0$. The *ABCD* matrix which describes the back and forth path from $z = 0$ to $z = 0$ is given by the product of three matrices,

$$\begin{pmatrix} A & B \\ C & D \end{pmatrix} = \begin{pmatrix} 1 & L/(2z_0) \\ 0 & 1 \end{pmatrix} \begin{pmatrix} 1 & 0 \\ -2z_0/R & 1 \end{pmatrix} \begin{pmatrix} 1 & L/(2z_0) \\ 0 & 1 \end{pmatrix}$$

$$= \begin{pmatrix} 1 - L/R & [L/(2z_0)](2 - L/R) \\ -2z_0/R & 1 - L/R \end{pmatrix},$$
(26.64)

where we have used Eqs. (26.62) and (26.63). The condition $q' = q$ leads to the equation $Cq^2 + Dq = Aq + B$. Since $A = D$ in Eq. (26.64) and $q = -i$ for $z = 0$, we must have $B = -C$, i.e.

$$\frac{L}{2z_0} \left(2 - \frac{L}{R} \right) = \frac{2z_0}{R},$$
(26.65)

which is equivalent to Eq. (26.58).

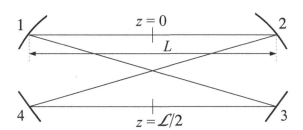

Figure 26.6 A ring cavity with two spherical mirrors (1 and 2) and two planar mirrors (3 and 4). $\mathcal{L}$ is the total length of the cavity.

26.5 Gaussian modes in a cavity with spherical mirrors. The case of ring cavities

Let us consider a ring cavity formed by two mirrors, of which two are spherical (with the same radius of curvature R) and two are planar, as shown in Fig. 26.6 [327]. Let us denote by $\mathcal{L}$ the total cavity length and by L the distance between the two spherical mirrors. The cavity modes will have a different Rayleigh range in the two parts of the cavity between the spherical mirrors; to be precise z_0 is the Rayleigh range for the part of the beam centered at $z = 0$ and z_1 is that for the part of the beam centered at point $\mathcal{L}/2$ (see Fig. 26.6).

Note that the Rayleigh length remains fixed in the transformations, and therefore we obtain a different Rayleigh length according to the point from which we start the roundtrip. In order to determine z_0 let us start from point $z = 0$ and evaluate the transformation that the $q(\eta)$ parameter undergoes after a complete roundtrip in the cavity. The matrix corresponding to such a transformation is

$$
\begin{pmatrix} A & B \\ C & D \end{pmatrix} = \begin{pmatrix} 1 & L/(2z_0) \\ 0 & 1 \end{pmatrix} \begin{pmatrix} 1 & 0 \\ -2z_0/R & 1 \end{pmatrix} \begin{pmatrix} 1 & (\mathcal{L} - L)/z_0 \\ 0 & 1 \end{pmatrix} \begin{pmatrix} 1 & 0 \\ -2z_0/R & 1 \end{pmatrix} \begin{pmatrix} 1 & L/(2z_0) \\ 0 & 1 \end{pmatrix}
$$

$$
= \begin{pmatrix} 1 - 2(R\mathcal{L} + L^2 - \mathcal{L}L)/R^2 & (R - L)(R\mathcal{L} + L^2 - \mathcal{L}L)/(R^2 z_0) \\ 4z_0(\mathcal{L} - L - R)/R^2 & 1 - 2(R\mathcal{L} + L^2 - \mathcal{L}L)/R^2 \end{pmatrix}. \quad (26.66)
$$

Also in this case we have $A = D$ and $q = -i$, and therefore we must impose that $B = -C$, which amounts to

$$
\frac{(R - L)(R\mathcal{L} + L^2 - \mathcal{L}L)}{R^2 z_0} = \frac{4z_0(R + L - \mathcal{L})}{R^2}, \quad (26.67)
$$

and gives the result

$$
z_0^2 = \frac{1}{4} \frac{R - L}{R + L - \mathcal{L}} (R\mathcal{L} + L^2 - \mathcal{L}L). \quad (26.68)
$$

In order to determine z_1 it is not necessary to repeat the calculation, because it suffices to observe that the equation for z_1 can be obtained from that for z_0 by simply replacing L by $\mathcal{L} - L$. Hence from Eq. (26.68) we obtain

$$z_1^2 = \frac{1}{4} \frac{(R + L - \mathcal{L})}{R - L}(R\mathcal{L} + L^2 - \mathcal{L}L). \tag{26.69}$$

Note that the result (26.58) for the Fabry–Perot cavity can be recovered from Eqs. (26.68) and (26.69) by setting $\mathcal{L} = 2L$.

26.6 Mode frequencies

In the previous sections we determined the Rayleigh length of a cavity with spherical mirrors as a function of the geometrical parameters of the cavity itself. The aim of this section is, instead, to determine the cavity frequencies. In order to do that, we must focus on the phase of the Gaussian modes, which is independent of the parameter $q(\eta)$ and is given by

$$\phi(z) = k_0 z - (m + n + 1)\arctan(z/z_0), \tag{26.70}$$

for the Gauss–Hermite mode of order (m, n) (see Eqs. (26.34) and (26.3)) and by

$$\phi(z) = k_0 z - (2p + |l| + 1)\arctan(z/z_0), \tag{26.71}$$

for the Gauss–Laguerre mode of order (p, l) (see Eqs. (26.46) and (26.3)). Note that the phase $\phi(z)$ includes also the phase $k_0 z$ of the carrier wave. We can observe that what counts in the coefficient of the arctan term is the combination $(m + n)$ in the case of Gauss–Hermite modes and $(2p + |l|)$ in the case of Gauss–Laguerre modes. Let us indicate by q these two combinations of indices, so that the phase can be written in general in the form

$$\phi(z) = k_0 z - (q + 1)\arctan(z/z_0). \tag{26.72}$$

The integer q identifies a family of modes with the same frequency (see next subsection), composed of $(q + 1)$ elements. The family $q = 0$ is composed only of the fundamental Gaussian mode. The family $q = 1$ is composed of Gauss–Hermite modes $(1, 0)$ and $(0, 1)$ and Gauss–Laguerre modes $(0, \pm 1)$. The family $q = 2$ is composed of Gauss–Hermite modes $(2, 0)$, $(0, 2)$ and $(1, 1)$ and Gauss–Laguerre modes $(0, \pm 2)$ and $(1, 0)$, etc.

26.6.1 Mode frequencies for a Fabry–Perot cavity

At this point, it is necessary to recall that Eq. (26.1) holds for unidirectional propagation and therefore, in the case of a Fabry–Perot cavity, it holds for the forward-propagating field envelope E_F (see Section 14.1). The same is true for the Gaussian modes which we calculated in the previous sections. The equation which governs the backward-propagating field envelope E_B is obtained from Eq. (26.1) by simply replacing z with $-z$. Since the phase for the forward-propagating field is given by Eq. (26.72), i.e.

$$\phi_F(z) = k_0 z - (q+1)\arctan(z/z_0),\tag{26.73}$$

the phase for the backward-propagating field is given by

$$\phi_B(z) = -\phi_F(z).\tag{26.74}$$

In a Fabry–Perot cavity the phase accumulated in a roundtrip is the sum of the phase accumulated by the forward field in the propagation from the left to the right mirror and of the phase accumulated by the backward field in the propagation from the right to the left mirror, and the cavity frequencies are determined by the condition that this sum is an integer multiple of 2π. From Eqs. (26.73) and (26.74) we obtain

$$
\begin{aligned}
2\pi j &= 2[\phi(\eta_1 z_0) - \phi(\eta_2 z_0)] \\
&= 2\{k_0 \eta_1 z_0 - (q+1)\arctan(\eta_1) - [k_0 \eta_2 z_0 - (q+1)\arctan(\eta_2)]\} \\
&= 2\{k_0(\eta_1 - \eta_2)z_0 - (q+1)[\arctan(\eta_1) - \arctan(\eta_2)]\} \\
&= 2\left[k_0 L - (q+1)\arccos\left(\frac{1 + \eta_1 \eta_2}{\sqrt{(1+\eta_1^2)(1+\eta_2^2)}} \right) \right],
\end{aligned}
\tag{26.75}
$$

where, as in Fig. 26.4, we have indicated by η_1 and η_2 the positions of the two cavity mirrors, and in the last passage we used Eqs. (26.16) and (26.47). From Eq. (26.75) and taking into account Eq. (26.50) we see that the cavity frequency depends on the longitudinal index j and the transverse index q according to

$$\omega_{j,q} = \frac{c}{L}[\pi j + (q+1)\arccos(\pm\sqrt{g_1 g_2})]\tag{26.76}$$

and turns out to be the sum of the longitudinal contribution $j\pi c/L$, which coincides with the frequency in the plane-wave theory (see Eq. (14.17)) and a transverse contribution that is the same for all transverse modes of the qth family. The argument of the arccos term must be taken with the $+$ sign if both g_1 and g_2 are positive and with the $-$ sign if they are both negative. It is interesting to compare the longitudinal frequency spacing

$$\Delta\omega_L = \omega_{j+1,q} - \omega_{j,q} = \frac{c}{L}\pi\tag{26.77}$$

with the transverse frequency spacing

$$\Delta\omega_T = \omega_{j,q+1} - \omega_{j,q} = \frac{c}{L}\arccos(\pm\sqrt{g_1 g_2}).\tag{26.78}$$

We can write

$$\frac{\Delta\omega_T}{\Delta\omega_L} = \frac{1}{\pi}\arccos(\pm\sqrt{g_1 g_2}).\tag{26.79}$$

Taking into account that $g_i = 1 - L/R_i$, $i = 1, 2$, two cases are especially interesting.

The first is the symmetric cavity ($R_1 = R_2 = R$),

$$\frac{\Delta\omega_T}{\Delta\omega_L} = \frac{1}{\pi} \arccos g = \frac{1}{\pi} \arccos\left(1 - \frac{L}{R}\right) = 2\arctan\left(\frac{L}{2z_0}\right)$$

$$= 2\arctan\sqrt{\frac{L}{2R - L}}. \tag{26.80}$$

The last expression can be obtained either directly from the previous expression or, more straightforwardly, by taking into account that in a symmetrical cavity $\eta_1 = -\eta_2 = L/(2z_0)$ (see Eq. (26.47)), so that, by using Eq. (26.75), one obtains $\omega_{j,q} = (c/L)[\pi j + (q + 1)2\arctan(L/2z_0)]$.

The second case is the plane-concave cavity ($R_1 = R$, $R_2 = \infty$),

$$\frac{\Delta\omega_T}{\Delta\omega_L} = \frac{1}{\pi} \arccos\sqrt{1 - \frac{L}{R}}, \tag{26.81}$$

in which one must have $L < R$.

For a symmetrical cavity the ratio of the transverse frequency spacing to the longitudinal one amounts to 0, 1/2 and 1 for a planar cavity ($R = \infty$, Fig. 8.1), a confocal cavity ($R = L$, Fig. 26.5(b)) and a concentric cavity ($R = L/2$, Fig. 26.5(c)), respectively. In the case of a quasi-planar cavity ($R \gg L$, Fig. 26.7(a)) the transverse frequency spacing is much smaller than the longitudinal frequency spacing [6]. In the case of a confocal cavity (Fig. 26.7(b)) the mode (j, q) is frequency degenerate with the modes $(j - 1, q + 2)$, $(j - 2, q + 4)$ and so on. The set of modal frequencies coincides with that of the planar cavities plus all the frequencies which lie halfway between the modal frequencies of the planar cavity. In the case of a concentric cavity (Fig. 26.7(c)) the mode (j, q) is frequency degenerate with the modes $(j - 1, q + 1)$, $(j - 2, q + 2)$ and so on. The set of modal frequencies coincides with that of the planar cavity.

26.6.2 Mode frequencies for a ring cavity

In the case of the symmetrical ring cavity considered in Fig. 26.6 [327] the phase accumulation in a roundtrip is the sum of the phase accumulated in the segment of length L, where the beam has Rayleigh range z_0, and of the phase accumulated in the remaining part of length $\mathcal{L} - L$, where the beam has Rayleigh range z_1. As usual, the phase accumulated in a roundtrip must be equal to an integer multiple of 2π

$$2\pi j = k_0 \mathcal{L} - 2(q + 1)\left[\arctan\left(\frac{L}{2z_0}\right) + \arctan\left(\frac{\mathcal{L} - L}{2z_1}\right)\right], \tag{26.82}$$

from which it follows that

$$\omega_{j,q} = ck_0 = \frac{2c}{\mathcal{L}}\left\{\pi j + (q + 1)\left[\arctan\left(\frac{L}{2z_0}\right) + \arctan\left(\frac{\mathcal{L} - L}{2z_1}\right)\right]\right\}. \tag{26.83}$$

If we set

$$\eta_1 = \frac{L}{2z_0}, \qquad \eta_2 = \frac{L - \mathcal{L}}{2z_1}, \tag{26.84}$$

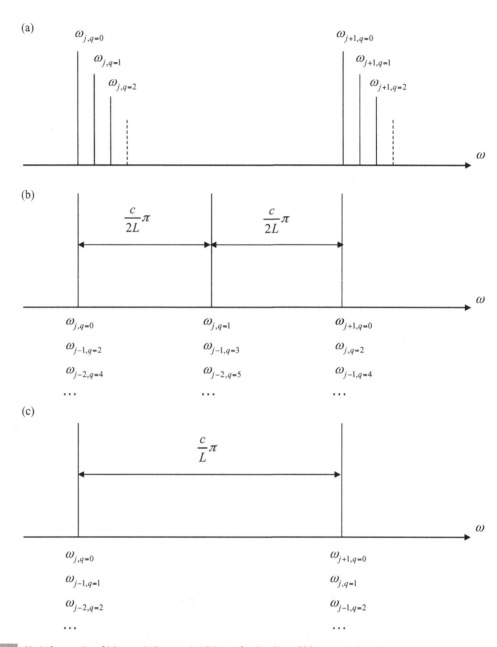

Figure 26.7   Mode frequencies of (a) a quasi-planar cavity, (b) a confocal cavity and (c) a concentric cavity.

and use relation (26.16) we can write an equation similar to (26.75) for the frequency of mode (j, q):

$$\omega_{j,q} = \frac{2c}{\mathcal{L}} \left[\pi j + (q + 1)\arccos \left(\frac{1 + \eta_1 \eta_2}{\sqrt{(1 + \eta_1^2)(1 + \eta_2^2)}} \right) \right]. \tag{26.85}$$

On the other hand, using Eqs. (26.68) and (26.69) for z_0 and z_1, after some lengthy calculations we find

$$1 + \eta_1 \eta_2 = \frac{R - L}{R}(1 + \eta_1^2) = \frac{R - \mathcal{L} + L}{R}(1 + \eta_2^2). \tag{26.86}$$

Hence, similarly to what we did for the Fabry–Perot cavity, we can introduce the geometrical factors

$$g_1 = 1 - \frac{L}{R} = \frac{1 + \eta_1 \eta_2}{1 + \eta_1^2}, \qquad g_2 = 1 - \frac{\mathcal{L} - L}{R} = \frac{1 + \eta_1 \eta_2}{1 + \eta_2^2}, \tag{26.87}$$

which allow us to write Eq. (26.85) as

$$\omega_{j,q} = \frac{2c}{\mathcal{L}}[\pi j + (q + 1)\arccos(\pm\sqrt{g_1 g_2}\,)]. \tag{26.88}$$

The symmetric ring cavity turns out to be equivalent to an asymmetric Fabry–Perot cavity of length $\mathcal{L}/2$ and with mirrors with different radii of curvature $R_1 = R\mathcal{L}/(2L)$ and $R_2 = R\mathcal{L}/(2(\mathcal{L} - L))$.

If $R = \mathcal{L} - L$ or $R = L$ one has $g_1 g_2 = 0$ and $\Delta\omega_\mathrm{T}/\Delta\omega_\mathrm{L} = 1/2$, which is analogous to a confocal Fabry–Perot cavity. If $R = (\mathcal{L} - L)L/\mathcal{L}$ one has $g_1 g_2 = 1$, and the negative sign must be taken in the argument of the arccos term because both g_1 and g_2 are negative; therefore $\Delta\omega_\mathrm{T}/\Delta\omega_\mathrm{L} = 1$ and this ring cavity is analogous to a concentric Fabry–Perot cavity.

General features about optical pattern formation in planar cavities

In this chapter we focus on the simplest context to illustrate the field of transverse optical pattern formation, i.e. cavities with planar mirrors.

In Section 27.1 we describe the models which are analyzed in this chapter and the following chapter. Section 27.2 carries out the central task of explaining the general principles of pattern formation in planar cavities. An outstanding example is provided in Section 27.3 with the discussion of patterns in optical parametric oscillators. The case of single feedback mirrors, which in a sense corresponds to half a planar cavity, plays a relevant role in the domain of optical pattern formation and is discussed in Section 27.4. Section 27.5 formulates the analogy between optical pattern formation and hydrodynamics and illustrates the phenomenon of optical vortices.

27.1 Dynamical models with diffraction

In the field of nonlinear optics the Maxwell–Bloch equations play a role analogous to that of the Navier–Stokes equations in hydrodynamics; they provide the paradigm for the resonant interaction between matter and radiation by focussing on the simplest picture of a collection of two-level atoms. We shall limit our attention to the case of homogeneously broadened atomic systems and ring cavities. In addition, in this and in the following chapter we will consider cavities with planar mirrors.

, Let us start from Eqs. (3.31)–(3.33), adding the damping terms to the atomic equations. By rescaling the variables as indicated in Eqs. (4.30), (4.32) and (4.33) and using the definition (4.34) we arrive at the set of equations

$$\frac{n_B}{c} \frac{\partial F}{\partial t} + \frac{\partial F}{\partial z} = gP + \frac{i}{2k_0} \nabla_\perp^2 F, \tag{27.1}$$

$$\frac{\partial P}{\partial t} = \gamma_\perp [FD - (1 + i\Delta)P], \tag{27.2}$$

$$\frac{\partial D}{\partial t} = -\gamma_\parallel \left[\frac{1}{2}(FP^* + F^*P) + D - \chi(x, y) \right], \tag{27.3}$$

where the function $\chi(x, y)$ describes the transverse shape of the pumped region.

Next, we perform the transformations (12.1), (12.8) and (12.9), implement the low-transmission approximation and the single-mode approximation and arrive at a set of equations like Eqs. (12.43)–(12.45) with the addition, in the field equation (12.43), of a term that describes diffraction. Before writing such equations, we must complete the definition of the low-transmission approximation in order to include the transverse modes in the description. In order to do that, let us start from the expression for the frequencies of the modes of a planar cavity,

$$\omega = \frac{c}{n_\mathrm{B}} \left(k_x^2 + k_y^2 + k_z^2\right)^{1/2}. \tag{27.4}$$

In the paraxial approximation $k_\perp \ll k_z \approx k_0$, with $k_\perp = \sqrt{k_x^2 + k_y^2}$, they reduce to

$$\omega = \frac{c}{n_\mathrm{B}} k_z \left(1 + \frac{k_x^2 + k_y^2}{k_z^2}\right)^{1/2} \approx \omega_\mathrm{c} + \frac{c k_\perp^2}{2 n_\mathrm{B} k_0}, \tag{27.5}$$

with a continuous frequency distribution from ω_c to ∞. Note that the expression $c k_\perp^2 / (2 n_\mathrm{B} k_0)$ is precisely what we obtain by applying the transverse Laplacian to the transverse modes,

$$-\frac{c}{2 n_\mathrm{B} k_0} \nabla_\perp^2 e^{i(k_x x + k_y y)} = \frac{c k_\perp^2}{2 n_\mathrm{B} k_0} e^{i(k_x x + k_y y)}. \tag{27.6}$$

Since we include the transverse modes, the low-transmission approximation must be completed by the requirement that, for the relevant range of transverse wave vectors $k_\perp$, the transverse contribution to the frequency must be much smaller than the free spectral range, i.e.

$$\frac{c k_\perp^2}{2 n_\mathrm{B} k_0} \ll \frac{c}{\Lambda}. \tag{27.7}$$

Therefore the low-transmission approximation is defined as

$$T \ll 1, \qquad gL = \mathcal{O}(T), \qquad \delta_0 = \frac{\omega_\mathrm{c} - \omega_0}{c/\Lambda} = \mathcal{O}(T), \qquad \frac{k_\perp^2 \Lambda}{2 n_\mathrm{B} k_0} = \mathcal{O}(T), \tag{27.8}$$

while the single-mode approximation, which we must now more correctly call the *single-longitudinal-mode approximation*, reads

$$\frac{\gamma_\perp}{c/\Lambda} = \mathcal{O}(T). \tag{27.9}$$

By using such approximations one arrives at a single-longitudinal-mode model [308, 318] that involves the time t and the two spatial variables x and y, and therefore it is much less demanding in terms of CPU time than are models that include all three spatial variables and, in addition, it is structurally analogous to the spatially two-dimensional models analyzed in the fields of hydrodynamics and of nonlinear chemical reactions

$$\frac{\partial F(x, y, t)}{\partial t} = -\kappa \big[(1 + i\theta)F(x, y, t) - F_\mathrm{I}(x, y) - A P(x, y, t) - i\bar{a}\, \nabla_\perp^2 F(x, y, t)\big], \tag{27.10}$$

$$\frac{\partial P(x, y, t)}{\partial t} = \gamma_\perp [F(x, y, t)D(x, y, t) - (1 + i\Delta)P(x, y, t)], \tag{27.11}$$

$$\frac{\partial D(x, y, t)}{\partial t} = -\gamma_\parallel \Big[\frac{1}{2}(F^*(x, y, t)P(x, y, t) + F(x, y, t)P^*(x, y, t))$$

$$+ D(x, y, t) - \chi(x, y) \Big], \tag{27.12}$$

where κ is the cavity damping constant (or cavity linewidth) cT/Λ, A is the pump parameter gL/T and θ is the cavity detuning parameter $(\omega_c - \omega_0)/\kappa$. The parameter $\bar{a}$, which has the dimensions of an area, is defined as

$$\bar{a} = \frac{c}{2n_B k_0 \kappa} = \frac{1}{2n_B} \frac{\lambda \Lambda}{2\pi T}. \tag{27.13}$$

With respect to Eqs. (12.43)–(12.45), we indicate by F and F_I the normalized output and input fields, instead of x and y, respectively, because in Part III x and y denote the transverse coordinates. We have written $F_I(x, y)$ in order to allow in general for a possible spatial dependence of the input field. The photon flux exiting from any transverse region R is proportional to $\int_R dx\, dy |F(x, y)|^2$.

Two special cases of the model (27.10)–(27.12), obtained in appropriate limits, have been analyzed extensively in the literature. Both include the standard adiabatic elimination of atomic variables for $\kappa \ll \gamma_\perp, \gamma_\parallel$, and refer to the case of optical bistability, in which A is replaced by $-2C$ and we have set $\chi = 1$. The first corresponds to the purely absorptive case $\Delta = 0$ [308, 328],

$$\frac{\partial F}{\partial t} = -\kappa \left[(1 + i\theta)F - F_I(x, y) + \frac{2CF}{1 + |F|^2} - i\bar{a}\nabla_\perp^2 F \right], \tag{27.14}$$

while the other corresponds to the opposite purely dispersive limit of Eq. (13.31) [307, 329],

$$\frac{\partial F}{\partial t} = -\kappa [F - F_I(x, y) - i\eta(|F|^2 - \theta)F - i\bar{a}\nabla_\perp^2 F], \tag{27.15}$$

where $\eta = +1$ and $\eta = -1$ correspond to the self-focussing and self-defocussing cases $\Delta < 0$ and $\Delta > 0$, respectively, and we have written θ instead of $\bar{\bar{\theta}}$. Equation (27.15) also describes the case of a pure Kerr medium, i.e. of a refractive nonlinearity (see Section 11.2).

On the other hand, also the case of $\chi^{(2)}$ media has been considered in research on optical pattern formation. Here we write a model that describes a degenerate optical parametric oscillator with plane mirrors [330] and represents a generalization of the model [137, 138] to include the transverse degrees of freedom. In this case the medium in the cavity induces the partial transformation of a pump field F_2 of frequency $\omega_2 = 2\omega_1$, which is injected into the cavity, into a signal field of frequency ω_1. Without transverse effects, the model is given by Eqs. (13.51) and (13.52). Its generalization, which includes transverse effects, reads

$$\frac{\partial F_1}{\partial t} = -\kappa_1 \left[(1 + i\theta_1)F_1 - F_1^* F_2 - i\bar{a}_1 \nabla_\perp^2 F_1 \right], \tag{27.16}$$

$$\frac{\partial F_2}{\partial t} = -\kappa_2 \left[(1 + i\theta_2)F_2 - F_I(x, y) + F_1^2 - i\bar{a}_2 \nabla_\perp^2 F_2 \right], \tag{27.17}$$

where κ_1 and κ_2 are the cavity damping rates of the two fields, respectively; and $\bar{a}_1$ and $\bar{a}_2$ are defined as

$$\bar{a}_1 = \frac{c^2}{2\kappa_1 \omega_1 n^2} = \frac{1}{2n} \frac{\lambda_1 \Lambda}{2\pi T_1}, \qquad \bar{a}_2 = \frac{c^2}{2\kappa_2 \omega_2 n^2} = \frac{1}{2n} \frac{\lambda_2 \Lambda}{2\pi T_2}, \tag{27.18}$$

with n being the refractive index of the medium, which is the same for the two fields, as is necessary for phase-matched degenerate emission (see Section 13.4). The cavity detuning parameters are defined by Eqs. (13.47) and (13.40).

It is useful to note that it is possible to eliminate the parameters κ and a from the dynamical equations (27.10)–(27.12) by appropriate scaling of the time and space variables in such a way that they become dimensionless. Specifically, we can set

$$t'' = \kappa t = \frac{cT}{\Lambda}t, \qquad \bar{x} = \frac{x}{\sqrt{a}}, \qquad \bar{y} = \frac{y}{\sqrt{a}}. \tag{27.19}$$

If we compare Eqs. (27.19) and (27.13) for $\mathcal{L} = L$ (i.e. $\Lambda = n_B L$) with Eqs. (3.19) and (3.26) we see that in the cavity the length scale L of Eq. (3.26) is replaced by L/T, with the same length enhancement by a factor of $1/T$ as noted in the plane-wave theory at the end of Section 12.2. In this connection, we observe that, if we consider the characteristic transverse size $d = \sqrt{a}$ and define the Fresnel number (see Eq. (27.13))

$$\mathcal{F} = \frac{d^2}{\lambda\mathcal{L}} = \frac{1}{4\pi T}\frac{\Lambda}{\mathcal{L}n_B}, \tag{27.20}$$

we have that $\mathcal{F} \gg 1$, since $T \ll 1$. We can say that, in the framework of optical pattern formation, $\mathcal{F}$ plays the role of an aspect ratio and, as in the case of hydrodynamics and nonlinear chemical reactions, the aspect ratio is large. The dynamical equations such as Eqs. (27.10)–(27.12) and (27.14)–(27.17) are solved numerically using periodic boundary conditions, which corresponds to assuming that the integration window is a small portion of the transverse section of the system, sufficiently far from its physical boundaries that the dynamics within it is not influenced by the boundaries themselves.

27.2 Systems with translational symmetry. The mechanisms for pattern formation

Let us now focus on the case when the system has translational and rotational symmetry; this requires in particular that the input field F_I, if it exists, and χ do not depend on (x, y). We review now the basic mechanisms that govern the formation of a spatial pattern from a homogeneous state. Let us indicate by B a generic dynamical variable in the time-evolution equations in play (e.g. B can be F, P, D, F^*, or P^* in the set of equations (27.10)–(27.12)). Because of translational symmetry in the transverse plane, the dynamical equations admit a spatially homogeneous stationary solution (there might be more than one, but we will ignore this in our discussion); let us denote by B_s the stationary value of B. Whether this solution is stable or not is analyzed by linearizing the equations of motion around the stationary solution and checking the eigenvalues of these linearized equations. Whereas an eigenvalue with positive real part indicates instability, a stable solution has only eigenvalues with negative real part. To be precise, one sets $B(x, y, t) = B_s + \delta B(x, y, t)$, where δB represents a small perturbation from the stationary state, and one linearizes the dynamical equations with respect to δB. Because the coefficients are independent of x and y, one focuses on solutions of the form

$$\delta B(x, y, t) = \delta B_k(t)e^{i\mathbf{k}_\perp \cdot \mathbf{x}}, \tag{27.21}$$

where $\mathbf{k}_\perp = (k_x, k_y)$ is the same for all variables B and $\mathbf{x} = (x, y)$. This amounts to assuming that the perturbation is modulated with transverse wavelength $2\pi/k_\perp$, which explains the term *modulational instabilities* used for those instabilities that arise for wave vector $k_\perp \neq 0$.

The ansatz (27.21) corresponds to performing a Fourier analysis in space; in the case of optics the Fourier transform of the intracavity field corresponds to the far-field configuration. With the position (27.21), the linearized equations become ordinary differential equations with constant coefficients, so that one can set

$$\delta B(t) = \overline{\delta B}\, e^{\lambda t}, \tag{27.22}$$

where λ (not to be confused with the wavelength) is the same for all variables B. Ansatz (27.22) leads to a set of linear homogeneous algebraic equations with the form of an eigenvalue equation with eigenvalue λ. This leads to a characteristic equation for λ of order equal to the number n of variables B

$$\sum_{i=0}^{n} c_i(k_\perp^2, \beta)\lambda^i = 0, \tag{27.23}$$

where the coefficients c_i depend on the modulus of the transverse wavevector $\mathbf{k}_\perp$ (as a matter of fact $\nabla_\perp^2 \exp(i\mathbf{k}_\perp \cdot \mathbf{x}) = -k_\perp^2 \exp(i\mathbf{k}_\perp \cdot \mathbf{x})$) and on one control parameter β (which may be, for example, the input field F_I or the pump parameter A), bearing in mind that the other parameters are to be kept fixed.

In the discussion of the characteristic equation (27.23), it is best to focus on the boundary of the stability domain in the $(\beta, k_\perp^2)$ plane, which is defined by the condition $\text{Re}\ \lambda = 0$. On setting $\lambda = -i\omega$ and taking into account that the coefficients c_i are real, by separating the real and imaginary terms one obtains two algebraic equations of the form (see Eq. (18.10))

$$P_1(\omega, k_\perp^2, \beta) = 0, \qquad P_2(\omega, k_\perp^2, \beta) = 0, \tag{27.24}$$

where P_1 and P_2 are polynomials in the variable ω with coefficients equal to c_i, $-c_i$ or 0. In particular $P_2(\omega = 0, k_\perp^2, \beta) = 0$ and $P_1(\omega = 0, k_\perp^2, \beta) = c_0(k_\perp^2, \beta)$.

There are, in general, two kinds of boundaries. The first is characterized by the condition $\omega = 0$ and, in the $(\beta, k_\perp^2)$ plane, corresponds to the line $c_0(k_\perp^2, \beta) = 0$. The second boundary is obtained by eliminating ω between the two equations (27.24); in this case ω is different from zero. Of course, it may happen that one boundary or the other (or both) does not exist. In Fig. 27.1 we draw qualitatively the two boundaries, assuming that for small enough β the homogeneous stationary solution is stable. On increasing β one hits the instability boundary for $\beta = \beta_c$, $k_\perp^2 = k_c^2$ where we assume that k_c is different from 0; these coordinates define the *critical point* in the $(\beta, k_\perp^2)$ plane and correspond to the instability threshold.

In Fig. 27.1(a) the critical point belongs to the boundary with $\omega = 0$, while in Fig. 27.1(b) it lies on the boundary with $\omega \neq 0$. Some general qualitative features of the solution which emerges immediately beyond threshold can be specified when this solution develops with continuity from the homogeneous stationary solution which has become unstable. When this happens and $k_c \neq 0$, in the case of Fig. 27.1(a) one has the onset of a stationary pattern. Those instabilities which lead to the formation of a stationary spatial pattern are usually called *Turing instabilities* [331]. On the other hand, in the case of Fig. 27.1(b), when $k_c \neq 0$ one has a dynamical pattern characterized by the temporal frequency ω.

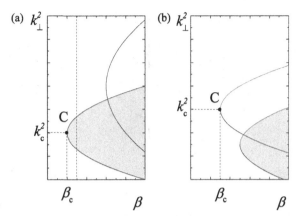

Figure 27.1 The qualitative shape of the instability domain in the plane of the control parameter β and of the modulus squared of the transverse wave vector. The point C of coordinates (β_c, k_c^2) is the critical point. The shaded region is the instability domain with $\omega = 0$.

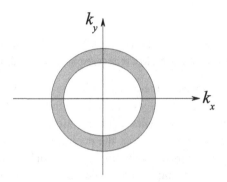

Figure 27.2 The annulus of unstable wave vectors for the value of β corresponding to the dashed line in Fig. 27.1(a).

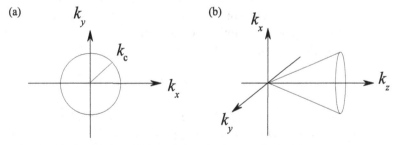

Figure 27.3 (a) The critical circle in the (k_x, k_y) plane. (b) The critical cone in the space (k_x, k_y, k_z).

Let us analyze first the case of Fig. 27.1(a). When β is larger than β_c, there is a range of unstable modes, which tend to grow in time; if we look at the plane of transverse wave vectors (k_x, k_y) there is an annulus of unstable modes (Fig. 27.2) for the value of β indicated by the dashed line in Fig. 27.1(a). At threshold, the annulus reduces to a circle with radius equal to k_c; this is called the critical circle (Fig. 27.3(a)). If we consider the whole wave vector (k_x, k_y, k_z), we have a cone of unstable wave vectors (Fig. 27.3(b)). It may happen

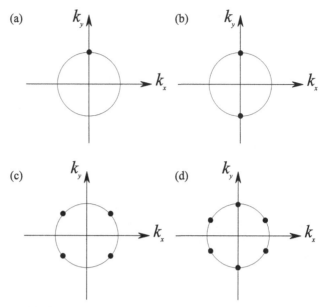

Figure 27.4 The simplest examples of far-field patterns. In case (a) only one mode survives. In case (b) there are two modes with opposite wave vectors, in case (c) four modes evolve and in case (d) one has the formation of hexagons or honeycombs.

that all the waves of the critical cone are emitted so that, in the far field, one has a circle that corresponds to the critical circle. In this case, the near field also has circular symmetry and it is called a *ring pattern* (Fig. 27.5(c)).

Most commonly, however, the nonlinearity of the system causes a spontaneous break-ing of the rotational symmetry such that only some of the modes of the critical circle build up, suppressing the other ones. Some simple examples of possible far fields, which were obtained for various systems slightly above the instability threshold, are shown in Fig. 27.4; for all these cases the far field consists in a number of spatially separated beams, corresponding to the spots over the critical circle. Of course, the orientation of these patterns is arbitrary due to the rotational symmetry. The corresponding near-field pattern is given by a linear superposition of plane waves

$$\delta B(x, y) = \sum_l b_l e^{i\mathbf{k}_l \cdot \mathbf{x}}, \qquad (27.25)$$

where the sum is extended to the spots in the far-field pattern. The exact solution has in general contributions also from spatial harmonics and combinations of the wave-vectors, but these corrections are negligible when close to threshold. Examples of patterns observed in experiments by Grynberg and collaborators [332] in the near and far field are given in Fig. 27.5.

Clearly there is a competition between the different patterns that can evolve from the Turing instability, and the question is which one among the different possibilities is realized for a specific system. The conceptually simplest way to answer is to use a computer and solve the dynamical equations numerically. In two transverse dimensions the problem of numerically solving partial differential equations, in which $\nabla_\perp^2$ has a purely imaginary coefficient, is rather tricky and codes usually rely on the split-step technique [333]. In

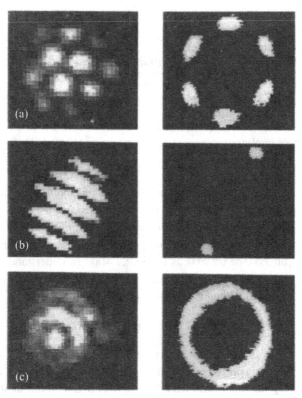

Figure 27.5   Near-field (left) and far-field (right) patterns for (a) hexagons, (b) rolls and (c) rings (from [332], courtesy of G. Grynberg).

general, it is wise to check numerical results by comparing them with those obtained by analytical methods, or by other numerical techniques that directly calculate the stationary solution, bypassing the integration in time.

A technique used to calculate analytically the patterns which emerge beyond the critical point is called nonlinear analysis or bifurcation analysis [334]. Basically, this method treats the quantity $(\beta - \beta_c)^{1/2}$ as a perturbative smallness parameter. For each simple pattern (e.g. rolls, squares, hexagons) one considers the ansatz (27.25) and perturbatively derives from the exact model a set of time-evolution equations for the coefficients. These equations allow one to calculate the possible stationary patterns, and to check their stability with respect to the onset of other patterns (for example, the stability of rolls against the onset of hexagons and vice versa).

In the case of nonlinear optics, however, consideration of the nonlinear interaction process can provide in a very straightforward way an intuitive picture of the scenario which emerges. Let us illustrate this point by invoking three examples.

(a) In the laser case, we must consider Eqs. (27.10)–(27.12) with $F_I = 0$. For the discussion of the spatial instability, it is convenient to take as the reference frequency ω_0 the atomic transition frequency ω_a, in contrast with what we did in Parts I and II of this book. Therefore $\Delta = 0$ by definition, and θ is an independent parameter.

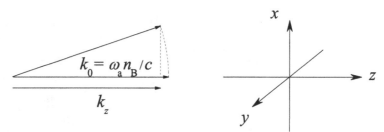

Figure 27.6 A longitudinal wave with $k_0 = \omega_a n_B/c > k_z = \omega_c n_B/c$ is detuned from resonance. By tilting the wave one achieves exact resonance.

The critical point is just the laser threshold. Here there is a situation of maximum competition between transverse modes and it is not surprising that only one mode survives; the near field has exactly the form $\exp(i\mathbf{k}_c \cdot \mathbf{x})$, where $\mathbf{k}_c$ is any vector with modulus k_c. The corresponding far field consists in a single spot, which may be emitted on axis, or off axis as in Fig. 27.4(a). To understand what happens, let us insert the expression $F \propto \exp(i\mathbf{k}_\perp \cdot \mathbf{x})$ in Eq. (27.10). The effect of the term with $\nabla_\perp^2$ is to produce the total detuning of mode $k_\perp$,

$$\theta_{k_\perp} = \theta + \bar{a}k_\perp^2 = \frac{\omega_{k_\perp} - \omega_0}{\kappa}, \tag{27.26}$$

where we used Eq. (27.5) with $\omega = \omega_{k_\perp}$, the definition $\theta = (\omega_c - \omega_0)/\kappa$ and Eq. (27.13). Because in the laser case we have taken ω_0 equal to the atomic transition frequency ω_a, the mode with the largest gain is that with the smallest detuning $\theta_{k_\perp}$. When $\theta \geq 0$ the detuning is minimum for $k_\perp = 0$, hence the axial mode is emitted. If $\theta < 0$, then, instead, the modes with

$$k_\perp = k_c = \sqrt{-\theta/\bar{a}} \tag{27.27}$$

are exactly on resonance, and therefore an off-axis wave onto the critical circle is emitted (Fig. 27.4(a)). An equivalent way to understand the origin of this tilted wave emission is illustrated in Fig. 27.6. This mechanism was pointed out in [335], and has a validity that extends beyond the laser case, as we shall see in the case of the optical parametric oscillator. This mechanism basically coincides with that whereby the spontaneous emission of an atom in a cavity occurs off-axis when the atom–cavity detuning has the appropriate sign [336]. Note that the intensity distribution of the near field of the laser above threshold is uniform also for $k_c \neq 0$. In raising the pump parameter A beyond threshold, however, one meets further instabilities that give rise to intensity patterns of increasing complexity [337, 338].

(b) Let us now consider the case of the degenerate optical parametric oscillator (OPO), which was considered in Section 13.4. The physics of this system is governed by the process of optical parametric down-conversion considered in Chapter 6. Similarly to the laser, the OPO has a threshold. Below threshold, according to the semiclassical equations (27.16) and (27.17) in the stationary state the signal field is zero. Again, the critical point is the threshold of the OPO, and the selection of the critical wave vector k_c is governed by the tilted-wave mechanism applied to the signal field, i.e. by the resonance condition $\theta_1 + \bar{a}_1 k_c^2 = 0$, as will be discussed explicitly in the next

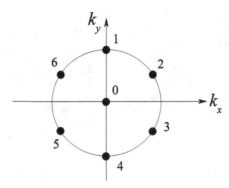

Figure 27.7   The generation of a hexagonal far field (see the text).

section [330]. The basic difference from the laser case is that, since photons are created in pairs, the far field for $\theta_1 < 0$ consists in two spots with opposite wave vectors (case (b) of Fig. 27.4). The two waves interfere with each other and, in the near field, one obtains

$$F_1(\mathbf{x}) = \sigma\, e^{i(\varphi_+ + \mathbf{k}_c \cdot \mathbf{x})} + \sigma\, e^{i(\varphi_- - \mathbf{k}_c \cdot \mathbf{x})} = 2\sigma \cos\left[\mathbf{k}_c \cdot \mathbf{x} + (\varphi_+ - \varphi_-)/2\right] e^{i(\varphi_+ + \varphi_-)/2},$$

(27.28)

so that the intensity distribution $|F_1|^2$ is a roll pattern (stripe pattern), as in Fig. 27.8(b). The name *roll pattern* originates from the hydrodynamical Rayleigh–Bénard instability [91, 234, 235].

(c) Let us now turn to the case of a Kerr medium, i.e. the cavity contains a medium with a cubic nonlinearity (see Chapter 7). In this case the behavior of the system is determined by the four-wave-mixing processes which conserve the total transverse momentum (wave vector) in the simultaneous annihilation of two photons and creation of two photons. In particular, there is a process of annihilation of two photons of the input field ($k_\perp = 0$) and the emission of two symmetrically tilted photons. This directly leads to a far-field configuration with three spots: two opposite spots as in the OPO and a central spot corresponding to the mode with $k_\perp = 0$ (input field).

The near field is given by Eq. (27.28) with the addition of a constant contribution from the mode $k_\perp = 0$, and corresponds again to a roll pattern. The contribution of the homogeneous mode, however, destabilizes the roll pattern according to the following consideration. The first four-wave-mixing process generates, from a pair of photons with $k_\perp = 0$ (Fig. 27.7), two other photons (e.g. 1 and 4) with opposite transverse wave vector. But then a second four-wave-mixing process can create two additional photons (2 and 6) from 0 and 1, or the pair 3 and 5 from 0 and 4, which gives a hexagonal structure. This argument was elucidated by Grynberg for the case of counterpropagating waves [332], but holds also in other situations such as, for example, the cavity model (27.15). Another more formal argument for the onset of hexagonal patterns was offered by Firth: If one considers a term of the form $F|F|^2$ and sets $F = F_s + \delta F$, then one obtains for δF quadratic terms that are characteristic for hexagon formation [339].

As an illustration of steady-state patterns, we give an example of how one meets different patterns on continuously varying a control parameter. We refer to the case of the absorptive model (27.14) and the results were obtained in [328]. In particular, the modulational

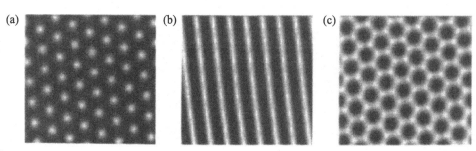

Figure 27.8 Examples of the change in the stable pattern observed for $\theta = -1$ as I is varied at a fixed value of $C = 4.4$: (a) H$^+$ for $I = 2.9$, (b) rolls for $I = 3.3$ and (c) H$^-$ for $I = 4.5$ (from [328]).

instability is related to the same tilted-wave mechanism as described in connection with the free-running laser and the optical parametric oscillator. When the input field intensity $I = F_I^2$ is increased above the instability threshold, a hexagonal pattern H$^+$ is realized (Fig. 27.8(a)). Upon increasing I away from the instability threshold, one finds a stable branch of rolls (Fig. 27.8(b)) and, further on, a honeycomb pattern H$^-$ (i.e. a hexagonal pattern with a phase shift of π among the three sets of tilted standing waves, Fig. 27.8(c)).

It is necessary to observe that in some cases the instability domain starts at a value of β for which $k_\perp = 0$. It may happen, on the other hand, that the instability domain terminates at a value β_c at which $k_c \neq 0$. Hence the instability domain is found at values of the control parameter smaller than the critical value β_c, such that the pattern-forming instability is met when the control parameter is decreased instead of increased. We shall see an example of this configuration in Fig. 30.8(b) later.

As the last point in the discussion of the stationary instability, we observe that, when we linearize Eq. (27.10) around the homogeneous stationary solution and introduce the ansatz (27.21), the diffraction term gives a contribution that can be combined with the cavity detuning θ, which is thus replaced by $\theta + \bar{a}k_\perp^2$ (compare this with Eq. (27.26)). Therefore the expression for the coefficient c_0 in Eq. (27.23) is given by the plane-wave expression in Eq. (C.7) with θ replaced by $\theta + \bar{a}k_\perp^2$ and, if we fix all parameters except X that we select as parameter β, the equation $c_0 = 0$ defines the boundary of the stationary instability in the $(X, k_\perp^2)$ plane.

Let us now come to the case $\omega \neq 0$ illustrated in Fig. 27.1(b), in which, as we said, one has the formation of a dynamical pattern when $k_c \neq 0$ and the solution immediately above the critical point develops with continuity from the unstable homogeneous state. The simplest dynamical intensity pattern is a drifting roll, which is obtained by adding ωt to the argument of the cosine in Eq. (27.28). Solutions of this kind have been numerically predicted [340] and experimentally observed in χ^3 media [341, 342], by using an input field tilted with respect to the mirror(s). When the tilt angle exceeds a certain minimum [343], one generates a drifting instability of this sort.

Beyond this simple case, one can find numerically a rich variety of dynamical patterns, but it is possible to appreciate them adequately only with the help of an animation; hence we will not give examples here (see e.g. [344]).

We close this section with an important remark. The mechanisms for optical pattern formation that we illustrated are identical to those which govern these phenomena in all the other fields. Also in hydrodynamics, for example, pattern selection is determined by appropriate combinations of wave vectors on a critical circle, dictated by the nonlinearity. In nonlinear optics, however, the nonlinearity is associated with the simultaneous absorption and emission of a number of photons. This circumstance creates correlations of quantum nature, and is the very origin of the quantum aspects of optical patterns and of the field of quantum imaging [67].

27.3 Pattern formation in optical parametric oscillators

We have discussed this topic shortly in the previous section. In this section we will provide a more formal demonstration of the formation of a roll pattern. We start from Eqs. (27.16) and (27.17) with F_1 independent of the transverse variables x and y, and set

$$\delta B = F_2 - F_{2s},\tag{27.29}$$

where F_{2s} is the stationary value of F_2 in the trivial homogeneous stationary solution, which is given by

$$F_{1s} = 0,\qquad F_{2s} = \frac{F_1}{1 + i\theta_2},\tag{27.30}$$

in agreement with Eqs. (13.54). Note that the variables F_1 and F_2 in this chapter correspond to x_1 and x_2 in Eqs. (13.51) and (13.52). Equations (27.16) and (27.17) can be written in the forms

$$\frac{\partial F_1}{\partial t} = \kappa_1 \left[-(1 + i\theta_1)F_1 + F_1^* F_{2s} + i\bar{a}_1 \nabla_\perp^2 F_1 \right] + \kappa_1 F_1^* \,\delta B,\tag{27.31}$$

$$\frac{\partial \delta B}{\partial t} = \kappa_2 \left[-(1 + i\theta_2)\delta B + i\bar{a}_2 \nabla_\perp^2 \delta B \right] - \kappa_2 F_1^2.\tag{27.32}$$

If we linearize Eq. (27.31) and its complex conjugate by dropping the last term, we obtain a self-contained set of equations,

$$\frac{\partial F_1}{\partial t} = -\kappa_1 \left[(1 + i\theta_1)F_1 - \frac{F_1}{1 + i\theta_2} F_1^* - i\bar{a}_1 \nabla_\perp^2 F_1 \right],\tag{27.33}$$

$$\frac{\partial F_1^*}{\partial t} = -\kappa_1 \left[(1 - i\theta_1)F_1^* - \frac{F_1}{1 - i\theta_2} F_1 + i\bar{a}_1 \nabla_\perp^2 F_1^* \right],\tag{27.34}$$

where we have taken F_1 real and we have written F_1 instead of δF_1 because $F_{1s} = 0$. If we now introduce the ansatz

$$\begin{pmatrix} F_1(x,t) \\ F_1^*(x,t) \end{pmatrix} = e^{\lambda t + i\mathbf{k}\cdot\mathbf{x}} \begin{pmatrix} F_{10} \\ F_{10}^* \end{pmatrix}\tag{27.35}$$

(for notational simplicity we use $\mathbf{k}$ instead of $\mathbf{k}_\perp$) we obtain a set of two homogeneous algebraic equations, which leads to the characteristic equation

$$\lambda^2 + 2\kappa_1 \lambda + \kappa_1^2 \left[1 + (\theta_1 + \bar{a}_1 k^2)^2 - \frac{F_1^2}{1 + \theta_2^2} \right] = 0.\tag{27.36}$$

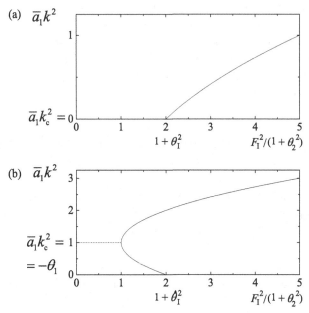

Figure 27.9 The instability domain lies to the right of the curve. (a) For $\theta_1 \geq 0$, $k_c = 0$ and no pattern is formed. (b) For $\theta_1 < 0$, $k_c \neq 0$ and there is the formation of a pattern (see Section 27.2). The value of θ_1 is -1.

The instability boundary is obtained by equating the constant coefficient in Eq. (27.36) to zero, which gives

$$\bar{a}_1 k^2_{\substack{max \\ min}} = \pm \sqrt{\frac{F_I^2}{1 + \theta_2^2} - 1} - \theta_1. \qquad (27.37)$$

The curves (27.37) exist for $F_I^2 > 1 + \theta_2^2$. For $\theta_1 \geq 0$ there is only the curve k^2_{max}, which has the form shown in Fig. 27.9(a). In this case $k_c = 0$ and no pattern is formed. The instability threshold is $F_I^2 = (1 + \theta_1^2)(1 + \theta_2^2)$ and corresponds to the homogeneous stationary solution of the plane-wave theory (compare this with Section 13.4). On the other hand, for $\theta_1 < 0$ the curves $\bar{a}_1 k^2_{max}$ and $\bar{a}_1 k^2_{min}$ are shown in Fig. 27.9(b). In this case $\bar{a}_1 k_c^2 = -\theta_1$ and the threshold for signal field emission is $F_I^2 = 1 + \theta_2^2$, which is smaller than the threshold of the plane-wave theory. As we explained in the previous section, this corresponds to the tilted-wave emission mechanism and the instability gives rise to the formation of a roll pattern [330]. An exact form of the roll solution is not available. However, careful application of perturbation expansions (nonlinear analysis [334]) allows one to find close approximations, whose range of validity extends to values of the input intensity far higher than the instability threshold. For simplicity we analyze the case of equal decay rates $\kappa_1 = \kappa_2$, but the calculations can easily be extended to the general case. Moreover, the axes of the transverse plane have been chosen parallel and perpendicular to the rolls, which are assumed to lie in the horizontal direction. It is important to stress that the following analysis is two-dimensional in space. Equations (27.31) and (27.32) have an optimal form for perturbation expansions close to threshold.

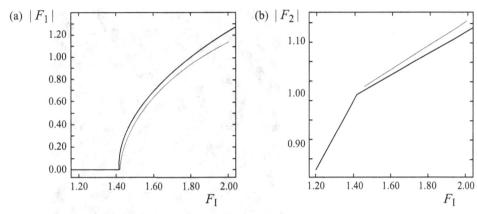

Figure 27.10   A comparison between simulations (grey curves) and the analytical form (27.38)–(27.42) (black curves) for the maximum amplitude of the roll pattern versus the input amplitude F_I for (a) the signal and (b) the pump field. The parameters are $\kappa_1 = \kappa_2$, $\theta_1 = -1$, $\theta_2 = 1$ and $F_I = 1.5$. Reprinted figure with permission from [330]. Copyright (1994) by the American Physical Society.

The detailed calculations are shown in Appendix G. By solving the linearized equations, i.e. Eqs. (27.33) and (27.34), in the stationary state one finds that there is no contribution at first order for the field δB, while the solution for the field F_1 is given by

$$F_1 = \epsilon \left[\theta_2 + i \left(1 - \sqrt{1 + \theta_2^2} \right) \right] \cos(k_c x), \tag{27.38}$$

where the smallness parameter ϵ remains to be evaluated by the solvability condition which fixes the Fredholm alternative, as shown in Appendix G [334]. To find ϵ we have to evaluate both second and third orders of the expansion in powers of ϵ since the solvability condition vanishes at lower orders. The first-order term (27.38) to the signal field yields a smaller correction on the pump field, which can be evaluated at the next order in the perturbation expansion

$$\delta B = \epsilon^2 [\alpha + \beta \cos^2(k_c x)], \tag{27.39}$$

where

$$\alpha = i \frac{2\bar{a}_2 k_c^2}{1 + i\theta_2} \beta, \tag{27.40}$$

$$\beta = \frac{2 \left(1 - \sqrt{1 + \theta_2^2} \right) \left[1 - \theta_2^2 - 4\theta_2 \bar{a}_2 k_c^2 - 2i(\theta_2 + 2\bar{a}_2 k_c^2) \right]}{1 + (\theta_2 + 4\bar{a}_2 k_c^2)^2}. \tag{27.41}$$

The expansion is then closed by evaluating the smallness parameter ϵ by use of the solvability condition at the third order, giving

$$\epsilon = \pm \left[\frac{F_I - \sqrt{1 + \theta_2^2}}{(2\bar{a}_2 k_c^2 + \frac{3}{4}\theta_2) \operatorname{Im} \beta - \frac{3}{4} \operatorname{Re} \beta} \right]^{1/2}. \tag{27.42}$$

One can verify that the denominator is always positive, such that the pattern arises for values of F_I larger than the threshold value $\sqrt{1 + \theta_2^2}$. A comparison between the previous

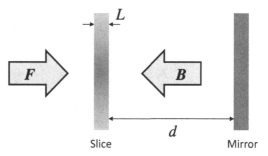

Figure 27.12 A schematic diagram of the model. A thin slice of Kerr material, of thickness L, is illuminated from the left. The mirror, at distance d, reflects this field back through the slice, thus closing the feedback loop.

analytical calculations and the numerical simulations is shown in Fig. 27.10 both for the signal F_1 and for the pump F_2. The agreement is excellent even for input amplitudes F_1 far from the instability threshold.

When the control parameter F_1 is increased much further beyond the instability threshold further instabilities develop. The first leads to the formation of the zig–zag pattern as shown in Fig. 27.11, and the subsequent instabilities produce dynamical patterns [330].

27.4 Systems with a single feedback mirror

In the framework of models with translational symmetry, systems with a single feedback mirror have played a relevant role. As indicated by Fig. 27.12, a thin slice of Kerr medium is irradiated on one side by a stationary plane-wave field, which freely propagates beyond the slice to a mirror that generates a counterpropagating beam in the Kerr slice. If we indicate by φ the nonlinear phase shift the radiation field undergoes upon crossing the slice, the dynamics inside the slice is governed by the time-evolution equation [310]

$$\tau \frac{\partial \varphi}{\partial t} = l^2 \nabla_\perp^2 \varphi + \xi(I_0 + I_1), \tag{27.43}$$

where τ is the response time and l the diffusion length of the Kerr excitation,

$$F_0 = \sqrt{I_0} e^{i\psi_0} \tag{27.44}$$

is the input field and I_1 is the intensity of the counterpropagating field [310]. The parameter ξ is positive (negative) for a self-focussing (self-defocussing) medium. It is assumed that diffusion washes out the grating formed by the two counterpropagating waves. Another hypothesis is that the medium is sluggish, i.e. cannot respond on the roundtrip time scale. Therefore the free propagation of the field from the slice to the mirror and back is described by the equation

$$\frac{\partial F}{\partial z} = \pm \frac{1}{2ik_0} \nabla_\perp^2 F, \tag{27.45}$$

without any time-dependent term (this is an adiabatic elimination of the field); the minus (plus) sign corresponds to propagation to the right (left). The quantity I_1 in Eq. (27.43) is obtained by calculating the field $F_1 = \sqrt{I_1} e^{i\psi_1}$ after the roundtrip with the initial condition

$$F(z = 0) = F_0 e^{i(\varphi_0 + \varphi)}, \tag{27.46}$$

where φ_0 is the linear phase shift induced by the medium. The value of $F_1(x, y)$ obtained in this way turns out to be a nonlinear functional of $\varphi(x, y)$, given by a Fresnel formula. This model was introduced by Firth [310] and extensively analyzed by Firth and D'Alessandro [311]. It was inspired on the one hand by the pioneering experiment of Giusfredi *et al.* [345] and on the other by the previous experience on the problem of counterpropagating waves in Kerr media [339]. In that case, the time-evolution equations include both nonlinearity and diffraction, which leads to computational complexity. In the single-mirror case, instead, the two elements are physically and mathematically separated; because of the sample thinness, diffraction is neglected in the Kerr medium (Eq. (27.43)) and is confined to the case of free propagation (Eq. (27.45)).

The mechanism responsible for pattern formation is simply explained as follows. Assume that a fluctuation creates a small spatial phase modulation in the sample, i.e.

$$F(x, y) = F_0(1 + i\epsilon \cos(kx)), \tag{27.47}$$

where the i (imaginary unit) encapsulates the $\pi/2$ phase shift between carrier and signal characteristic of phase modulation. Now, as one can see from Eq. (27.45), propagation generates a phase slippage relative to the carrier, at a rate $k^2/(2k_0)$ per unit distance. This introduces an element of amplitude modulation. Next, the amplitude modulation gives rise to phase modulation in the Kerr medium (see Eq. (27.43)), and in this way one has the feedback loop needed for pattern formation.

Patterns arise for both self-focussing and self-defocussing media, and may be static or oscillatory, with a period of two roundtrip times. Hexagonal patterns are commonplace here (see the buildup in time from noise shown in Fig. 27.13). On increasing the input field, the hexagons melt and give rise to a turbulence-like behavior.

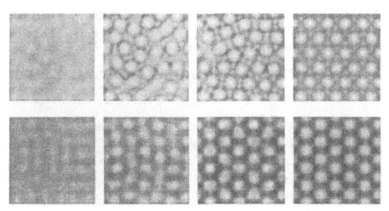

Figure 27.13 The backward field intensity in a focussing (top) and defocussing (bottom) medium. A gray scale, from white (high intensity) to black (low intensity), is used. Time increases from left to right. Reprinted figure with permission from [311]. Copyright (1992) by the American Physical Society.

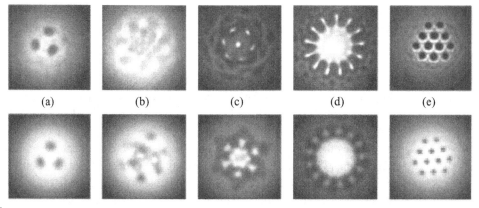

Figure 27.14 Experimental (top) and corresponding simulated (bottom) patterns for increasing input intensities (a)–(d). (e) A well-developed pattern illustrating H^- hexagons. The frames have a size of 2.5 mm $\times$ 2.5 mm. Reprinted figure with permission from [347]. Copyright (1995) by the American Physical Society.

In experiments, however, the input beam has a Gaussian profile, and this introduces interesting boundary effects, which have been studied in [346]. The main result is that the system can realize patterns that are not possible with a plane input, such as, for example, pentagonal patterns; fivefold-symmetric patterns cannot form infinitely periodic lattices in a two-dimensional space.

This kind of system has allowed several experimental realizations. Figure 27.14 shows some experimental observations of Lange, Ackemann and collaborators with sodium vapor in a buffer gas from [347], which were obtained by gradually increasing the input intensity and compared with the numerical simulations from an appropriate microscopic model of the medium. Above the instability threshold, a structure in the form of three dark holes appears (a); the selection of a structure with dihedral symmetry in a situation in which the hexagonal structure is hindered by the boundary conditions agrees with the results of [346]. Experiments with a larger aspect ratio show that a hexagonal lattice of dark holes (honeycomb hexagons) is indeed found (e). In the smaller-aspect-ratio case, instead, increasing the pump power leads to the formation of a pattern with five holes in a pentagonal arrangement, which gives way to the blurred, nonstationary pattern in (b). From this bright spots emerge; the number of spots (ranging from three to seven) increases with power; (c) shows the hexagonal pattern. Further on, flowerlike patterns with a plateau in the beam center are observed (d). The number of petals starts from 7 and increases to 14; in some cases these patterns rotate.

27.5 The analogy with hydrodynamics. Vortices and other defects

The existence of general analogies between optical and hydrodynamical patterns has led to the use of such terminologies as *laser hydrodynamics* [348] and *dry hydrodynamics* [312, 315].

Let us illustrate from a formal viewpoint the analogy between the laser equations and the hydrodynamical ones, as formulated in [348, 349]. We start from Eq. (27.14) with $\theta = 0$ and $F_I = 0$, in the laser case, i.e. with $-2C$ replaced by A,

$$\frac{\partial F}{\partial t''} = -\left(1 - i\overline{\nabla}_\perp^2\right)F + \frac{AF}{1 + |F|^2}, \tag{27.48}$$

where we have used the notations of (27.19) and $\overline{\nabla}_\perp^2 = \partial^2/\partial\bar{x}^2 + \partial^2/\partial\bar{y}^2$. In order to reformulate Eq. (27.48) in the form of *hydrodynamical equations* we set

$$F(\bar{x}, \bar{y}, t'') = |F(\bar{x}, \bar{y}, t'')|\exp[i\varphi(\bar{x}, \bar{y}, t'')/2]. \tag{27.49}$$

Then we introduce a *mass density*,

$$\sigma(\bar{x}, \bar{y}, t'') = |F(\bar{x}, \bar{y}, t'')|^2, \tag{27.50}$$

and a *velocity*,

$$\mathbf{w}(\bar{x}, \bar{y}, t'') = \overline{\nabla}_\perp\varphi(\bar{x}, \bar{y}, t''), \tag{27.51}$$

where $\overline{\nabla}_\perp$ denotes the gradient in the transverse plane. With some algebraic manipulations, Eq. (27.48) and its complex conjugate can be cast in the forms

$$\frac{\partial\sigma}{\partial t''} + \overline{\nabla}_\perp \cdot (\sigma\mathbf{w}) = 2\sigma\left(\frac{A}{1 + \sigma} - 1\right), \tag{27.52}$$

$$\frac{\partial\varphi}{\partial t''} = -\frac{1}{2}|\overline{\nabla}_\perp\varphi|^2 + 2\frac{\overline{\nabla}_\perp^2\sqrt{\sigma}}{\sqrt{\sigma}}. \tag{27.53}$$

We can compare Eqs. (27.52) and (27.53) with the fundamental equations of hydrodynamics in two spatial dimensions, namely the equation of continuity for the mass density σ,

$$\frac{\partial\rho}{\partial t} + \nabla_\perp \cdot (\rho\mathbf{v}) = 0, \tag{27.54}$$

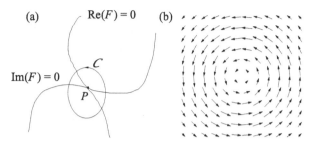

Figure 27.15 (a) A phase singularity is a point P in the transverse plane where both the real part and the imaginary part of the electric field vanish and the circulation of the electric field's phase over a loop C enclosing P is equal to a multiple of 2π. (b) The behavior of the transverse component of the Poynting vector around the phase singularity in a doughnut mode.

and the Bernoulli equation for the velocity potential Φ (such that $\mathbf{v} = \overline{\nabla}_\perp \Phi$),

$$\frac{\partial \Phi}{\partial t} = -\frac{1}{2}|\nabla \Phi|^2 - p, \tag{27.55}$$

which involves the pressure term p. From the comparison between Eqs. (27.52) and (27.53) and Eqs. (27.54) and (27.55) we can conclude that in the optical case the total *mass* is not conserved because of the presence of the two dissipative terms that appear on the r.h.s. of Eq. (27.52), while the analogy with the Bernoulli equation is complete if we define a *pressure*

$$p = -2\frac{\overline{\nabla}_\perp^2 \sqrt{\sigma}}{\sqrt{\sigma}}. \tag{27.56}$$

In this way a formal analogy between lasers and hydrodynamics is established, in the sense that all the relevant physical quantities which describe a fluid, i.e. the mass density, velocity and pressure, can be derived from the modulus and the phase of the electric field emitted by a laser.

One of the main consequences of this analogy is the occurrence of phase singularities, which play the same role as vortices in hydrodynamics and therefore are often called *optical vortices*. These vortices are centered at isolated points where both the real part and the imaginary part of the electric field vanish (Fig. 27.15(a)). At these points the intensity of the radiation is exactly zero and they appear as dark spots.[1] Since F is a single-valued complex function, the variation of its phase $\Phi = \varphi/2$ (see Eq. (27.49)) along a circuit C surrounding one of these points P (Fig. 27.15(a)) is equal to an integer multiple of 2π,

$$\Delta \Phi = \oint_C \nabla \Phi \cdot d\mathbf{l} = 2m\pi. \tag{27.57}$$

The value of the integral does not change if C is shrunk to a loop of infinitesimal length enclosing P. Therefore the gradient of Φ diverges at P and this point is a phase singularity

---

[1] Note that the hydrodynamical analogy holds only outside the vortices because in those singular points the velocity field cannot be written as the gradient of a potential and the pressure defined by Eq. (27.56) diverges.

for the electric field. The smoothness of F implies that a phase singularity can exist only where $F = 0$. The integer m is called the *topological charge* of the phase singularity.

It is interesting to analyze the behavior of the Poynting vector around a phase singularity. First of all we must remind the reader that, even in the paraxial approximation, the electric and magnetic fields associated with the laser radiation possess a longitudinal component. If the electric field is polarized along x and the magnetic field along y, then, from Eq. (3.22), we have that the slowly varying envelope of the electric field is given by

$$\mathbf{E}_0 \propto \left(F, \, 0, \, \frac{i}{k_0} \frac{\partial F}{\partial x} \right). \tag{27.58}$$

Similarly, one obtains that the slowly varying envelope of the magnetic field is given by[2]

$$\mathbf{H}_0 \propto \left(0, \, F, \, \frac{i}{k_0} \frac{\partial F}{\partial y} \right). \tag{27.59}$$

The Poynting vector $\mathbf{S}$ is defined as

$$\mathbf{S} = \mathbf{E} \times \mathbf{H} = \frac{1}{2} \operatorname{Re}(\mathbf{E}_0 \times \mathbf{H}_0^*), \tag{27.60}$$

and represents the time average of the irradiated power per unit area. Taking into account Eqs. (27.58) and (27.59), one has

$$\mathbf{S} \propto |F|^2 \left(\frac{\nabla_\perp \Phi}{k_0}, \, 1 \right). \tag{27.61}$$

The longitudinal components of the electric and magnetic fields produce the transverse component of the Poynting vector, which is proportional to the intensity of the electric field and directed as the gradient of its phase. Figure 27.15(b) shows the behavior of

---

[2] The slowly varying envelopes of the electric and magnetic fields are given by Eq. (3.9) and by

$$\mathbf{H}(x, y, z, t) = \frac{1}{2} \mathbf{H}_0(x, y, z, t) e^{-i(\omega_0 t - k_0 z)} + \text{c.c.}$$

We can also set (see Eq. (3.14))

$$\mathbf{E}(x, y, z, t) = \frac{1}{2} [\hat{\mathbf{e}}_x E_{0x}(x, y, z, t) + \hat{\mathbf{e}}_z E_{0z}(x, y, z, t)] e^{-i(\omega_0 t - k_0 z)} + \text{c.c.},$$

$$\mathbf{H}(x, y, z, t) = \frac{1}{2} [\hat{\mathbf{e}}_y H_{0y}(x, y, z, t) + \hat{\mathbf{e}}_z H_{0z}(x, y, z, t)] e^{-i(\omega_0 t - k_0 z)} + \text{c.c.},$$

where $\hat{\mathbf{e}}_x$, $\hat{\mathbf{e}}_y$ and $\hat{\mathbf{e}}_z$ are the unit vectors of the x, y and z axes, respectively. In the paraxial and slowly varying approximations, E_x and H_y are the dominant contributions. They are linked by a relation that is obtained by inserting the above two equations into the Maxwell equation (3.1). Keeping only the dominant terms, which are obtained by applying the space and time derivatives to the exponential factor, one obtains

$$H_{0y} = \sqrt{\frac{\epsilon_0}{\mu_0}} n_B E_{0x} \propto F.$$

The longitudinal component E_{0z} is given by Eq. (3.22) with $\hat{\mathbf{e}} = \hat{\mathbf{e}}_x$ and $E_{0L} = E_{0z}$ and, similarly, the longitudinal component H_{0z} is obtained from the Maxwell equation (3.2):

$$H_{0z} = \frac{i}{k_0} \nabla_\perp \cdot (\hat{\mathbf{e}}_y H_{0y}) = \frac{i}{k_0} \frac{\partial H_{0y}}{\partial y} \propto \frac{i}{k_0} \frac{\partial F}{\partial y}.$$

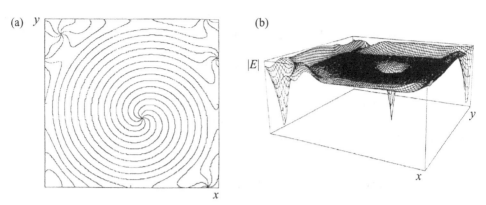

Figure 27.16   (a) Equiphase lines of the electric field. Four vortices are clearly visible. The phase difference between lines is $\pi/4$. (b) The corresponding field amplitude. The minima show the existence of optical vortices. Reprinted from [353, p. 403], with kind permission from Elsevier Science.

the transverse component of **S** for a doughnut mode with positive helicity. The presence of the phase singularity is associated with an angular momentum of the radiation, and the electromagnetic field can exchange angular momentum with matter [350]. The vortex structure is evident in Fig. 27.15.

Defects in the radiation field were first discussed by Berry and collaborators [351, 352]. They considered the linear wave equation and identified appropriate solutions that display patterns with rich and complex structures, sometimes on different spatial scales simultaneously, including subwavelength structures. Vortices in nonlinear optical systems were first predicted by Coullet and collaborators [353] on the basis of the model (27.10)–(27.12) with $\chi(x, y) = 1$; Fig. 27.16 shows an example of the results obtained in their numerical calculations. They display a main vortex in the center with the form of a spiral. In this model, the vortex arises from the presence of two distinct stationary solutions, i.e. the trivial solution $F = 0$ and the homogeneous nontrivial solution above threshold. Solutions that display vortices have an intensity distribution that connects these two basic configurations, in the sense that it vanishes at the center of the vortex and presents a narrow core that rises steeply to the homogeneous nontrivial solution of which the vortex is a defect.

Phase singularities are commonplace in class-A lasers with Fresnel number of order unity, operating with one or a few families of Gauss–Laguerre modes [354], as will be shown in Chapter 29, and in photorefractive materials [355].

The equation (27.15) with F_I independent of x and y was proposed in 1987 [307] with the purpose of formulating a model in the field of optics, that would be capable of producing a symmetry-breaking instability leading to the spontaneous formation of a two-dimensional stationary spatial pattern in an initially uniform system. Such instabilities had previously attracted considerable interest in chemistry and in developmental biology [91, 234, 235, 356], where they are commonly known as Turing instabilities [331]. In such fields they arise generally from the coupling between nonlinear chemical reactions and diffusion. Equation (27.15) provides a simple optical model in which analogous phenomena arise from the coupling between optical nonlinearities and diffraction in an optical cavity. This close analogy was made possible by the use of the low-transmission and single-longitudinal-mode limits, for the first time in a framework that included transverse effects.[1]

Previous models [358] included also propagation and therefore involved all of the variables x, y, z and t in an exceedingly complex framework to identify an analogy with Turing instabilities in reaction–diffusion systems. The paradigmatic simplicity of Eq. (27.15), which arises from the presence of the simple Kerr nonlinearity, played a decisive role in promoting the field of optical pattern formation. Equation (27.15) has even been dubbed "the hydrogen atom" of nonlinear cavities [359], and is usually called the Lugiato–Lefever (LL) equation after the authors of [307].

Another basic feature that emerged from the introduction of Eq. (27.15) is the following. In plane-wave models, stationary instabilities arise for parametric values that are easily accessible experimentally (for example the laser threshold or the emergence of optical bistability), whereas dynamical instabilities of the kind described, for example, in Chapter 20, present rather restrictive parametric conditions (for instance, high instability thresholds or bad-cavity conditions). It turns out that stationary spatial pattern-forming instabilities arise under parametric conditions that are as easily accessible as those of plane-wave stationary instabilities.

In this chapter we discuss the phenomena of spatial pattern formation that one finds in the framework of the model (27.15). In Section 28.1.1 we perform the modulational stability analysis and calculate the stationary instability boundary. The stationary and dynamical hexagonal patterns which arise from the instability in two transverse dimensions are described in Section 28.1.2, whereas Section 28.1.3 is devoted to the subject of stationary and oscillatory Kerr cavity solitons. The topic of cavity solitons is discussed more extensively in Chapter 30, but here we anticipate the results which concern the LL equation.

---

[1] A discussion of the relations between diffractive optical patterns and diffusive chemical patterns in connection with the Turing instability can be found in [357].

On the other hand, Section 28.2 deals with a topic that, strictly speaking, concerns temporal rather than spatial instabilities, but is intimately connected with the LL model. The topic is that of the temporal analogue of the LL model introduced in 1992 by Haelterman, Trillo and Wabnitz [360]. The latter model is equivalent to the LL model in one transverse dimension, and is based on the key role of chromatic dispersion, which replaces the diffraction term of the LL equation. The inclusion of dispersion is the element which distinguishes the temporal instability of Haelterman *et al.* from the temporal instabilities discussed in Part II, and makes it analogous, instead, to the spatial instabilities discussed in Part III.

In Section 28.2.1 we describe briefly the derivation of the temporal analogue equation and emphasize its advantages in connection with the experimental realization of the model, which is very accessible using cavities formed by standard silica fibers, which present a perfect Kerr nonlinearity. The remarkable length of these cavities represents an additional advantage. Subsequently, we describe the pattern-forming modulational instability in the temporal model and its experimental observation.

In Section 28.2.2 we discuss the recently attained experimental observation of Kerr temporal cavity solitons and the promising features of the experimental results with respect to future applications, also concerning the realization of broadband frequency combs.

28.1 Modulational instability and the patterns arising from it

28.1.1 Modulational instability boundary

The homogeneous stationary solution of the model (27.15) with F_I independent of x and y is given by

$$F_I^2 = |F_s|^2 \left[1 + (\theta - |F_s|^2)^2\right],\tag{28.1}$$

where we assumed that F_I is real and we have taken into account that $\eta = \pm 1$. With different symbols, Eq. (28.1) coincides with the cubic equations (11.22) and (13.34). The stationary curve is S-shaped, i.e. it exhibits bistability, for $\theta > \sqrt{3}$ as shown by Fig. 11.6. Next, we linearize Eq. (27.15) around a homogeneous stationary solution by setting $F(x, y, t) = F_s + \delta F(x, y, t)$ and obtain the linearized equations

$$\frac{\partial \delta F}{\partial t''} = -\delta F + i\eta(2|F_s|^2 - \theta)\delta F + i\eta F_s^2 \delta F^* + i\bar{a}\,\tilde{\nabla}_\perp^2 \delta F,\tag{28.2}$$

$$\frac{\partial \delta F^*}{\partial t''} = -\delta F^* - i\eta(2|F_s|^2 - \theta)\delta F^* - i\eta F_s^{*2} \delta F - i\bar{a}\,\tilde{\nabla}_\perp^2 \delta F^*,\tag{28.3}$$

where we have set $t'' = \kappa t$ as usual. By introducing the ansatz (see Eqs. (27.21) and (27.22))

$$\begin{pmatrix} \delta F(x, y, t'') \\ \delta F^*(x, y, t'') \end{pmatrix} = e^{\lambda t''} e^{i\mathbf{k}_\perp \cdot \mathbf{x}} \begin{pmatrix} \delta F_k \\ \delta F_k^* \end{pmatrix}\tag{28.4}$$

we obtain the two homogeneous algebraic equations, where λ is normalized with respect to κ,

$$[\lambda + 1 - i\eta(2|F_s|^2 - \theta) + i\bar{a}k_\perp^2]\delta F_k - i\eta F_s^2 \delta F_k^* = 0, \tag{28.5}$$

$$i\eta(F_s^*)^2 \delta F_k + [\lambda + 1 + i\eta(2|F_s|^2 - \theta) - i\bar{a}k_\perp^2]\delta F_k^* = 0, \tag{28.6}$$

which lead to the characteristic equation

$$\lambda^2 + c_1\lambda + c_0 = 0, \qquad c_1 = 2, \qquad c_0 = \frac{dF_I^2}{d|F_s|^2} + \bar{a}k_\perp^2 \left[\bar{a}k_\perp^2 - 2\eta\left(2|F_s|^2 - \theta\right)\right], \tag{28.7}$$

with

$$\frac{dF_I^2}{d|F_s|^2} = 1 + (|F_s|^2 - \theta)(3|F_s|^2 - \theta). \tag{28.8}$$

Since c_1 is always different from zero there is no oscillatory instability (see the final part of Section 18.1). On the other hand, the stationary instability domain in the $(|F_s|^2, \bar{a}k_\perp^2)$ plane is given by the equation $c_0 < 0$, i.e.

$$a_{\min}(|F_s|^2) \leq \bar{a}k_\perp^2 \leq a_{\max}(|F_s|^2), \tag{28.9}$$

with

$$a_{\max \atop \min}(|F_s|^2) = \eta(2|F_s|^2 - \theta) \pm \sqrt{|F_s|^4 - 1}. \tag{28.10}$$

Let us consider first the self-focussing case $\eta = +1$. In the monostable case $\theta < \sqrt{3}$, the critical point is

$$|F_s|_c^2 = 1, \qquad \bar{a}k_{\perp,c}^2 = 2 - \theta. \tag{28.11}$$

Figure 28.1(a) shows the stationary curve for $\theta = 0$ and its part that is modulationally unstable. Figure 28.1(b) shows the instability domain (28.9) with the critical point.

For $\sqrt{3} \leq \theta < 2$ the stationary curve is S-shaped; a small segment of the lower-transmission branch close to the up-switching threshold is modulationally unstable, as is all of the remaining part of the stationary curve. On the other hand, for $\theta > 2$ the lower-transmission branch is stable, whereas the remaining part of the stationary curve is modulationally unstable and $k_c = 0$ (Fig. 28.1(c) and (d)).

In the self-defocussing case $\eta = -1$ the critical point is

$$|F_s|_c^2 = 1, \qquad \bar{a}k_{\perp,c}^2 = \theta - 2, \tag{28.12}$$

and lies on the lower-transmission branch, as shown in Fig. 28.1(e), whereas Fig. 28.1(f) exhibits the instability domain, which is finite, in contrast with Figs. 28.1(b) and (d). Only a segment of the lower branch is modulationally unstable.

A simple procedure to obtain the instability domain is the following. Let us consider first the self-focussing case $\eta = 1$. By considering Eq. (28.8) and setting $dF_I^2/d|F_s|^2 = 0$, we can plot θ as a function of $|F_s|^2$. If we take into account that the instability boundary is obtained by replacing θ with $\theta + \bar{a}k_\perp^2$ in the equation $dF_I^2/d|F_s|^2 = 0$ (see the final part of Section 27.2), we can conclude that the instability boundary in the $(|F_s|^2, \bar{a}k_\perp^2)$ plane for

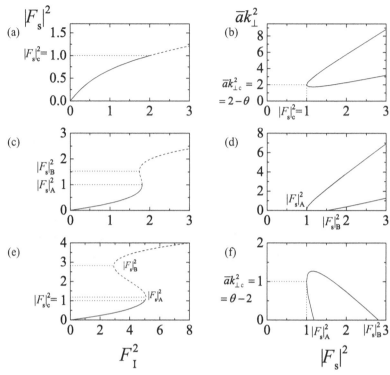

Figure 28.1 Left column: the stationary curve with the unstable part shown as a dashed line. Right column: the unstable domain in the plane $(|F_s|^2, \bar{a}k_\perp^2)$ defined by Eq. (28.9). Parts (a) and (b) refer to the self-focussing case, $\theta = 0$; parts (c) and (d) to the self-focussing case, $\theta = 3$; and parts (e) and (f) to the self-defocussing case, $\theta = 3$.

$\theta = 0$ is immediately obtained from the previous plot by simply replacing the axis θ by the axis $\bar{a}k_\perp^2$. This is precisely the graph in Fig. 28.1(b). As for the instability boundary for $\theta \neq 0$, from Eq. (28.10) with $\eta = 1$ we see that it is simply obtained by shifting downwards by θ the boundary for $\theta = 0$, in this way one obtains the graph in Fig. 28.1(d). This is also evident from the comparison of Figs. 27.9(a) and (b): in the first θ_1 is positive and in the second it is negative. The same arguments hold for the stationary instability domain of Eqs. (27.10)–(27.12) and of Eq. (27.14).

The procedure is different in the self-defocussing case due to the presence of the parameter η, which in this case is equal to -1. On setting $\eta = -1$ in Eq. (27.15) one sees that the instability boundary $c_0 = 0$ is obtained from the equation $dF_I^2/d|F_s|^2 = 0$ by replacing θ by $\theta - \bar{a}k_\perp^2$ instead of $\theta + \bar{a}k_\perp^2$. Hence, reasoning as before, we have that the instability boundary in the $(|F_s|^2, \bar{a}k_\perp^2)$ plane for $\theta = 0$ is immediately obtained from the plot of θ as a function of $|F_s|^2$, obtained from Eq. (28.8) as described before, by reversing the direction of the axis θ and by replacing it by the axis $\bar{a}k_\perp^2$. The graph of the instability boundary for $\theta \neq 0$ is obtained from the graph for $\theta = 0$ by shifting it upwards by θ (see Eq. (28.10) with $\eta = -1$). In this way one obtains Fig. 28.1(f).

A cubic model including the transverse Laplacian, which is similar to the LL equation but for a free-running laser near threshold, has been analyzed in [361].

Transverse patterns in the laser with injected signal have been described in [362, 363].

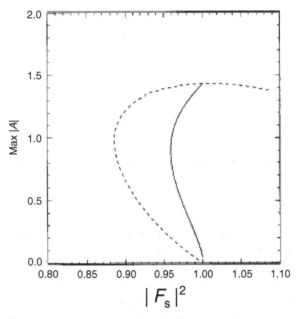

Figure 28.2 The maximum amplitude of patterned stationary solutions versus $|F_s|^2$ for $\theta = 1$. The solutions shown are the hexagonal pattern (dashed line) and the Kerr cavity soliton (solid curve). The slight irregularity in the hexagon curve is a numerical artifact associated with the existence of a continuum of hexagonal patterns with different lattice spacings. Reprinted figure with permission from [365]. Copyright (2002) by the Optical Society of America.

28.1.2 Stationary and dynamical patterns

The original paper [307] considered the case of one transverse dimension, which can be realized in a waveguide geometry and, on the basis of a nonlinear analysis of the kind utilized in Appendix G [329], demonstrated the formation of a stable modulated pattern in the self-focussing case. We will illustrate the case of one-dimensional patterns further in Section 28.2, where we discuss the temporal analogue of the LL model. In this and in the following subsection we discuss two-dimensional patterns arising in the framework of Eq. (27.15) in the self-focussing configuration $\eta = 1$.

The stationary modulated pattern arising in one dimension corresponds, in two dimensions, to a roll solution, which, however, turns out to be unstable and decays into a hexagonal pattern [364]. This is in accord with the discussion of pattern formation in the presence of a Kerr nonlinearity given in Section 27.2. The bifurcation of the hexagonal branch from the homogeneous state is described in Fig. 28.2 (dashed line) [365]. The quantity which is plotted is not the maximum of $|F|$, but the maximum of $|A|$, where $A(x, y) = F(x, y)/F_s - 1$. In Fig. 28.2 the value of θ ($\theta = 1$) is such that the stationary curve of the homogeneous stationary solution is single-valued, as in Fig. 28.1(a). An outstanding feature is that the hexagon branch bifurcates for values of $|F_s|^2$ smaller than the critical value $|F_s|^2 = 1$ (see Eq. (28.11)), i.e. in a range of $|F_s|^2$ where the homogeneous stationary solution is stable. This kind of bifurcation is called *subcritical*. As a general rule, the branch which bifurcates at a subcritical bifurcation is unstable, just as a branch that bifurcates at a supercritical

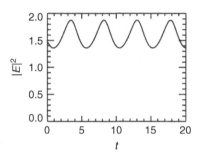

Figure 28.3 The time evolution of the maximum of the hexagonal pattern peak. The wavenumber of the initial unstable hexagon is $\bar{a}k_\perp^2 = 1.1$ and $|F_s|^2 = 1.04$. Reprinted figure with permission from [366]. Copyright (2007) by the American Physical Society.

bifurcation is stable. However, as we see from Fig. 28.2 the unstable bifurcated hexagonal branch at a certain point bends, assuming a positive slope, and this positive-slope branch is stable. Therefore, in a scenario of subcritical bifurcation the modulated pattern appears with a finite amplitude when $|F_s|^2$ overcomes the critical value $|F_s|^2 = 1$ and exhibits hysteretic behavior. This is completely different from the case of a supercritical bifurcation, where the modulated pattern (which bifurcates when the critical point is surpassed) is stable, as happens in the case of a steady-state bifurcation (Fig. 18.2), in which the bifurcated solution emerges in a continuous way at the instability threshold.

The analysis of [366] shows that for each value of $|F_s|^2$ larger than that corresponding to the turning point of the hexagonal branch in Fig. 28.2 there is a whole band of possible stable hexagonal patterns with different wavenumbers $k_\perp$. When the input intensity, and therefore the normalized transmitted intensity $|F_s|^2$, is increased, at a certain point the hexagonal pattern becomes unstable because of a Hopf instability and the pattern exhibits oscillatory behavior of different kinds, of which we shall see now some examples. The simplest configuration is that the peaks of the pattern oscillate synchronously in the same way. Figure 28.3 exhibits the temporal oscillations of the peaks in this regime. Another possibility is that the peaks do not oscillate synchronously, but nearby peaks are dephased.

The last scenario that we consider is seen in Fig. 28.4. In the transient (second and third columns) a new shifted hexagonal lattice appears in the near field. Finally the system ends up in a spatio-temporal chaotic regime that one can call *optical turbulence*. This regime is characterized by the formation of peaks at random positions that, after oscillating several times, vanish, allowing other peaks to appear at nearby locations.

A complete description of the various scenarios one meets in the case $\theta = 1$ can be found in [366]. Already, from the few examples that we have mentioned, one can get an idea of the extraordinary complexity one finds even in this "simple" model.

28.1.3 Kerr cavity solitons

The subject of cavity solitons will be discussed extensively in Chapter 30. However, in this chapter we discuss results concerning cavity solitons that arise in the framework of the model (27.15) which is characterized by a Kerr nonlinearity, and therefore we will call them Kerr cavity solitons.

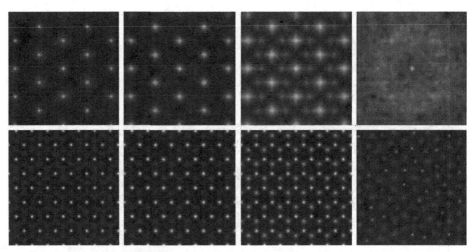

Figure 28.4 The time evolution of the near (bottom) and far (corresponding top) field of an unstable hexagonal pattern. The wavenumber of the initial unstable hexagons is $\bar{a}k_\perp^2 = 1$ and $|F_s|^2 = 1.12$. The time increases from left to right. The top pictures show the central part of the far field. At the end one has a situation of optical turbulence. Reprinted figure with permission from [366]. Copyright (2007) by the American Physical Society.

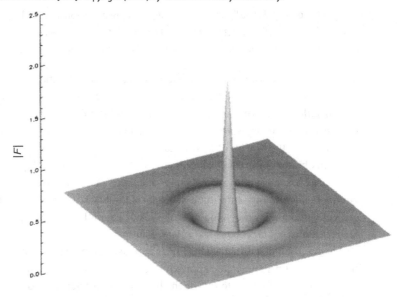

Figure 28.5 A typical Kerr cavity soliton, showing a bright peak on a darker homogeneous background with a few weak diffraction rings. The modulus of the intracavity field F is plotted as a function of the transverse coordinates x and y. The parameters are $\theta = 1.2$ and $|F_s|^2 = 0.9$. Reprinted figure with permission from [365]. Copyright (2002) by the Optical Society of America.

Cavity solitons are isolated peaks of radiation in the transverse structure of the radiation field, which sit on the pedestal of a homogeneous stationary solution. The typical shape of a Kerr cavity soliton is shown in Fig. 28.5. Cavity solitons are generated by shining an address pulse (writing pulse) in the longitudinal direction of the cavity (see Fig. 30.5). While the driving field (holding beam) is broad (ideally homogeneous in the transverse

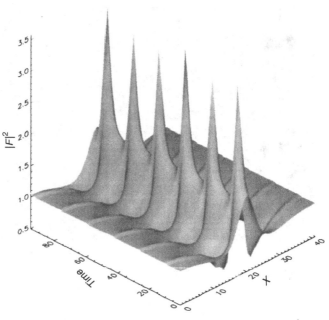

Figure 28.6 The dynamics of an oscillating Kerr cavity soliton beyond the Hopf bifurcation ($\theta = 1.3$, $|F_s|^2 = 0.9$). The cross section $|F(x, y = 0, t)|^2$ is plotted relative to x and time. The oscillation preserves cylindrical symmetry. Reprinted figure with permission from [365]. Copyright (2002) by the Optical Society of America.

section), the addressing beam is narrow and has a short duration. The cavity soliton persists after the passage of the address pulse through the cavity.

A detailed study of Kerr cavity solitons and their stability when the curve of the homogeneous stationary solution is single-valued can be found in [365]. Figure 28.2 shows, in addition to the branch of hexagonal patterns, also the cavity soliton branch (solid line). Also this branch bifurcates subcritically at the critical value $|F_s|^2 = 1$ and the part of the bifurcated branch with positive slope is stable. Therefore also cavity solitons appear discontinuously as large-amplitude stationary solutions when $|F_s|^2$ is increased beyond the critical value and, if the value of $|F_s|^2$ is subsequently decreased, hysteresis is found with respect to the homogeneous stationary solution.

An important fact is that when $|F_s|^2$ is increased enough the cavity-soliton stationary solution becomes unstable. In the approximate range $1 < \theta < 1.22$ the soliton loses its cylindrical symmetry and grows to invade the homogeneous background. In the approximate range $1.22 < \theta < 1.4$ the soliton keeps its cylindrical symmetry but starts oscillating because of a Hopf instability. The dynamics of an oscillating cavity soliton is shown in Fig. 28.6.

28.2 The temporal version of the LL model and its application perspectives

In 1992 Haelterman, Trillo and Wabnitz [360] formulated an ingenious "temporal" analogue of the LL model that, instead of using a broad-area-driven nonlinear cavity, utilizes a driven

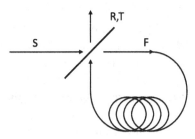

Figure 28.7 The scheme of the driven nonlinear dispersive fiber loop.

nonlinear fiber loop with an input/output mirror as illustrated in Fig. 28.7. In the practical realization the mirror is replaced by input and output fiber coupler(s).

In order to best appreciate the meaning of the step introduced by Haelterman *et al.*, it is convenient to return to Sections 7.2 and 7.3. In Section 7.2 we described the topic of temporal Kerr solitons on the basis of Eq. (7.19), which includes a term with the derivative with respect to z and a term with the second derivative with respect to the retarded time $t - (z/v_g)$, where v_g is the group velocity; this term describes the group velocity dispersion. On the other hand, in Section 7.3 we considered an equation of the same form, with the same derivative with respect to z and the same Kerr nonlinear term, but with the dispersion term replaced by the diffraction term with the transverse Laplacian, i.e. the sum of the second-order derivatives with respect to the transverse spatial variables x and y. If one considers only one transverse variable, the dynamical equation of Section 7.3 becomes identical to that of Section 7.2, and this feature has allowed us to write at once the solution corresponding to the spatial Kerr soliton from that which describes the temporal Kerr soliton.

The step introduced in [360] is basically the same as that involved in passing from Section 7.2 to Section 7.3, but in reverse order and in a dissipative instead of a conservative framework, due to the presence of a cavity. As will be shown in the following, their equation is equivalent to Eq. (27.15) with the diffraction term (second-order derivatives with respect to the transverse spatial variables) replaced by the dispersion term (the second-order derivative with respect to the retarded time). Hence, instead of depending on the temporal variable t and on the two spatial variables x and y, the model of Haelterman *et al.* depends on two variables: the "slow" time variable t, which describes phenomena that occur on the scale of the cavity decay time and is the same as the LL model; and the "fast" time variable τ, which describes phenomena that occur on a temporal scale smaller than the cavity roundtrip time. Therefore the model of Haelterman *et al.* is equivalent to an LL model restricted to one transverse dimension. Correspondingly, the spatial frequencies $k_\perp$ of the spatial model are replaced by temporal frequencies (or, more precisely, by the difference between the temporal frequencies and the reference frequency ω_0).

As we said, the interest aroused by the LL model arises from its simplicity. However, it is difficult to realize experimentally a system that fits such a model. For example, cavity solitons have been experimentally observed, but in semiconductor microresonators, which are described by a model that is more complex than the LL model, as will be discussed in Chapter 30. In addition, the presence of spatial defects in the semiconductor microresonator limits the possibility of controlling cavity solitons. All these problems disappear in the case of the model of Haelterman *et al.*, because standard silica fibers provide a perfect instantaneous Kerr nonlinearity like that assumed in the LL model and the cavity can easily be built using off-the-shelf components and fibers. The temporal case of [360] provides an essentially perfect (time) translational symmetry, which is essential e.g. in the case of Kerr temporal cavity solitons, which will be described in Section 28.2.2 [367].

28.2.1 The temporal modulational instability and its experimental observation

Let us sketch the main steps necessary to derive the temporal analogue of the LL model. One starts from Eq. (7.19) and combines it with the boundary condition of the cavity. Then one applies the low-transmission approximation and, after a number of passages, one arrives at the equation

$$\bar{\kappa}^{-1}\frac{\partial E_0(t,\tau)}{\partial t} = E_\mathrm{I}(\tau) - (1+i\theta)E_0(t,\tau) + i\frac{\tilde{\xi}\mathcal{L}}{\bar{T}}|E_0(t,\tau)|^2 E_0(t,\tau) - i\frac{k''\mathcal{L}}{2\bar{T}}\frac{\partial^2 E_0(t,\tau)}{\partial\tau^2},$$
$$(28.13)$$

where $\mathcal{L}$ is the cavity length and

$$\bar{\kappa} = \frac{\bar{T}}{t_\mathrm{R}} = \frac{\bar{T}v_\mathrm{g}}{\mathcal{L}},$$
$$(28.14)$$

with $\bar{T}$ being an effective transmission coefficient that takes into account also the presence of a linear absorption coefficient α_B along the fiber, as in Eq. (5.51). To be precise, $\bar{T}$ is equal to one half the total cavity losses expressed in terms of the percentage of power lost per cavity roundtrip. t_R is the cavity roundtrip time. The input field E_I is either constant in the case of CW input, or depends on τ when the input consists in a periodic sequence of short pulses with period t_R (synchronous pumping). Because of the fiber loop, the field $E_0(t,\tau)$ obeys a periodic boundary condition with respect to the fast time τ, with period t_R. The slow time t is associated with the temporal scale $\bar{\kappa}^{-1}$, which in the low-transmission limit is much longer than t_R whereas, as we said, the fast time τ is associated with a temporal scale smaller than t_R. Let us now introduce the scaled quantities

$$\bar{t} = \bar{\kappa}t = \frac{t\bar{T}}{t_\mathrm{R}}, \quad \bar{\tau} = \sqrt{\frac{2\bar{T}}{|k''|\mathcal{L}}}\tau, \quad F(t,\tau) = \sqrt{\frac{\tilde{\xi}\mathcal{L}}{\bar{T}}}E_0(t,\tau), \quad F_\mathrm{I}(\tau) = \sqrt{\frac{\tilde{\xi}\mathcal{L}}{\bar{T}}}E_\mathrm{I}(\tau),$$
$$(28.15)$$

so that Eq. (28.13) takes the form

$$\frac{\partial F(\bar{t},\bar{\tau})}{\partial\bar{t}} = F_\mathrm{I}(\bar{\tau}) - F(\bar{t},\bar{\tau}) + i(|F(\bar{t},\bar{\tau})|^2 - \theta)F(\bar{t},\bar{\tau}) - i\bar{\eta}\frac{\partial^2 F(\bar{t},\bar{\tau})}{\partial\bar{\tau}^2}, \qquad (28.16)$$

where $\bar{\eta} = 1$ in the case of normal dispersion ($k'' > 0$) and $\bar{\eta} = -1$ in the case of anomalous dispersion ($k'' < 0$). In connection with the normalization of τ, we assume that

$$\sqrt{\frac{|k''|\mathcal{L}}{2\bar{T}}} < t_\mathrm{R} \quad \Rightarrow \quad |k''| < \frac{t_\mathrm{R}^2}{\mathcal{L}}2\bar{T}. \qquad (28.17)$$

Let us now compare Eq. (28.16) with the LL equation (27.15) written in terms of the variables (27.19). In the case of anomalous dispersion $\bar{\eta} = -1$, Eq. (28.16) coincides with Eq. (27.15) in the self-focussing case $\eta = 1$ with the replacement of t'' by $\bar{t}$ and of the

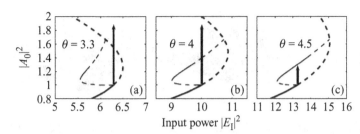

Figure 28.8 Bifurcation diagrams for different values of θ, in the case of normal dispersion. The dashed curves indicate unstable segments and the arrows depict the evolution of the system on reaching the critical point $|A_0|^2 = |F_s|^2 = 1$. Reprinted figure with permission from [368]. Copyright (1999) by IOP Publishing.

transverse Laplacian by the dispersion term. On the other hand, in the case of normal dispersion $\bar{\eta} = 1$, the complex conjugate of Eq. (28.16) reads

$$\frac{\partial F^*(\bar{t}, \bar{\tau})}{\partial \bar{t}} = F_{\mathrm{I}}(\bar{\tau}) - F^*(\bar{t}, \bar{\tau}) - i(|F^*(\bar{t}, \bar{\tau})|^2 - \theta)F^*(\bar{t}, \bar{\tau}) + i\frac{\partial^2 F^*(\bar{t}, \bar{\tau})}{\partial \bar{\tau}^2} \qquad (28.18)$$

and coincides with Eq. (27.15) with the additional replacement of F^* by F. Since this replacement is immaterial from the viewpoint of the intensity, we can say that Eq. (27.15) is equivalent to the LL equation in one dimension.[2]

If we assume now that F_{I} is independent of τ (CW input), the τ-independent solution of Eq. (28.16) coincides with that of Eq. (27.15), and the same holds for its linear-stability analysis. With a nonlinear analysis, similar to that shown in Appendix G for the case of the optical parametric oscillator but with one transverse dimension, in [329] it was shown that the modulated solution bifurcates supercritically as a stable solution for $\theta < 41/30$, and bifurcates subcritically as an unstable solution for $\theta > 41/30$, both in the self-focussing case and in the self-defocussing case. The same results hold for the temporal Kerr model.

Following the historical order of the investigations carried out by Haelterman, Coen and collaborators, let us consider first the case of normal dispersion, which corresponds to the self-defocussing configuration of the spatial case, so that the stability of the stationary curve is described by Figs. 28.1(e) and (f). Since the modulational instability arises for $\theta > 2$ (see Eq. (28.12)), the modulated pattern bifurcates subcritically. The analysis of [368–370] describes the stationary (with respect to the slow time t) bifurcated pattern in terms of a three-mode truncation

$$F(\tau) = A_0 + U\left(e^{i\Omega\tau} + e^{-i\Omega\tau}\right) \qquad (28.19)$$

calculating the quantities A_0, U and Ω and analyzing the stability of the solution.

In Fig. 28.8 one sees the stationary S-shaped curve (compare this with Fig. 28.1(e)) and the bifurcated pattern branch with indication of stable and unstable segments. One sees that, for values of θ smaller than 4, when F_{I}^2 overcomes the critical point in the lower-transmission branch where $|F_s|^2 = 1$, the system simply switches to the higher-transmission stationary state, which is stable (Fig. 28.8(a) and (b)). However, when θ becomes somewhat larger than 4 the system switches discontinuously to a stable modulated pattern (Fig. 28.8(c)) and, if

---

[2] We are grateful to Stephane Coen for this argument.

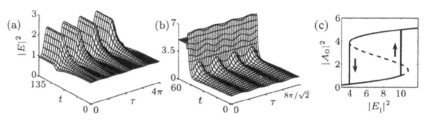

Figure 28.9 The evolution of an initially weakly perturbed stationary state at the critical power $|F_s|^2 = 1$: (a) $\theta = 6$ and (b) $\theta = 4$. Part (c) shows the corresponding truncated hysteresis cycle. Reprinted figure with permission from [368]. Copyright (1999) by IOP Publishing.

one decreases F_1^2, one has hysteresis. Figure 28.9 exhibits the slow time evolution towards the stationary pattern for $\theta = 6$ (Fig. 28.9(a)) and the switch to the higher-transmission stationary state for $\theta = 4$ (Fig. 28.9(b)).

In the normal dispersion configuration, the experiment performed in [368, 369] was done not with a CW input, but with an input $F_1(\tau)$ consisting of a sequence of short pulses with period t_R, so that it has been possible to observe only transients of the time evolution towards the stationary pattern, i.e. the exponential growth of the sidebands of the modulational instability, corresponding to the two terms proportional to U in Eq. (28.19).

In the case of anomalous dispersion, which corresponds to the self-focussing case of the spatial model, it has been possible to operate the system with a CW input (constant F_1) [370]. From an application viewpoint, the modulational instability is regarded as a generator of high-repetition-rate pulse trains, and the system is accordingly called *modulational instability laser* [371].

28.2.2 Temporal Kerr cavity solitons and their prospective applications

The temporal analogue of a spatial cavity soliton is a narrow pulse that circulates indefinitely without deformation (apart from fluctuations) along the fiber cavity, with a period equal to the cavity roundtrip time. It sits over the pedestal of a stable stationary solution and is excited by injecting into the cavity a single address pulse that adds to the driving field. The cavity soliton persists after the passage of the address pulse. A recent article by Leo *et al.* [372] (see also [367]) reports on the first experimental observation of temporal Kerr cavity solitons in a 380-m-long cavity (standard single-mode silica optical fiber), driven by a CW field of wavelength 1551 nm and linewidth on the order of 1 kHz. The cavity roundtrip amounts to 1.85 μs, and the observed cavity solitons have a width on the order of 4 ps.

The system is in the regime of anomalous dispersion and the parameter θ is chosen in such a way that the stationary curve is S-shaped. Figure 28.10 shows numerical results obtained from Eq. (28.16): One sees two stationary curves, one for $\theta = 3.3$ and the other for $\theta = 3.8$, because the experiments were done for these two values. For both cases, the figure shows also the cavity-soliton branch, which bifurcates subcritically at the right turning point of the stationary curve. As in the spatial case (Fig. 28.2), the stable part of the bifurcated branch is that with positive slope. On the same graph, one finds also the temporal width of the soliton (FWHM). The experiments were operated for the two values of $|F_s|^2$ indicated by the crosses, and the black dots indicate the corresponding position in the cavity-soliton branch.

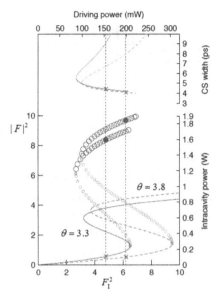

Figure 28.10 The existence and stability chart of temporal Kerr cavity solitons for $\theta = 3.3$ (full line) and $\theta = 3.8$ (broken line) as a function of the driving power F_I^2. The circles indicate the peak power of the solitons, while the top curves show their corresponding temporal duration (FWHM). The lighter segments correspond to unstable states. At the bottom one sees the stationary bistable response of the cavity. The filled circles and the crosses highlight the two operating conditions of the experiment. Note that the left and bottom axes are in normalized units, whereas the top and right axes are in real units. Figure taken from [372] with kind permission from Nature Publishing Group.

In order to generate cavity solitons, one starts with the system lying in the stable lower branch of the stationary curve, for the value of the input power indicated by the cross. One injects into the cavity an addressing pulse, which comes from a mode-locked fiber laser that emits 4-ps pulses with a repetition rate of 10 MHz. An acousto-optic modulator is placed along the path of the addressing beam and allows one to select either a single pulse or to encode a binary pattern in the addressing beam. In the first case one generates a single cavity soliton, which, apart from some fluctuations, circulates steadily in the cavity, producing a sequence of nearly identical pulses with a period equal to the roundtrip cavity time. The temporal duration of the soliton has been measured and found to be in remarkable agreement with the numerical predictions. Since the addressing laser has a frequency quite different from the driving field frequency, the writing process is incoherent.

It has also been verified that temporal cavity solitons are independently addressable and that several solitons can coexist in the cavity. The acousto-optic modulator was used to encode a binary-coded data stream onto the addressing beam and in this way some data were stored in the cavity in the form of a sequence of solitons. The input data stream is sent to the cavity just once. Figure 28.11 shows the cavity output after having written a 15-bit sequence encoding the abbreviation of the institute of the authors of the experiment (ULB, Université Libre de Bruxelles). Each letter is represented by its ordinal position in the alphabet (U $=21$, L $=12$, B $=2$) coded on five bits. The presence of a soliton encodes a binary 1, and its absence in the corresponding time slot encodes a 0. The stored data stream

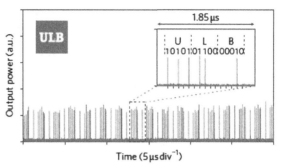

Figure 28.11 The abbreviation ULB of the Université Libre de Bruxelles stored all-optically with Kerr temporal cavity solitons. The abbreviation is encoded as a 15-bit data stream in the passive fiber cavity. Each letter is represented with five bits by its ordinal position in the alphabet (U = 21, L = 12, B = 2). Figure taken from [372] with kind permission from Nature Publishing Group.

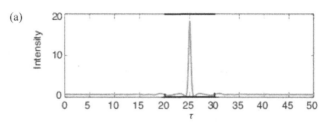

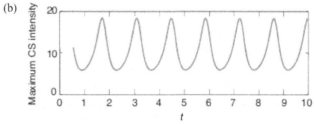

Figure 28.12 A numerical simulation of oscillatory Kerr temporal cavity solitons corresponding to the experiment. Part (a) shows the temporal intensity profile of the soliton circulating inside the cavity when it reaches its maximum peak power, over the fast time scale τ. The slow time evolution of the soliton peak is depicted in (b) over the time scale t; the period of the oscillations corresponds to 10.6 cavity roundtrip times. The values of the parameters are $F_1^2 = 8.5$ and $\theta = 4.1$.

repeats itself over and over again for a storage time exceeding one second, demonstrating that the fiber cavity operates as an all-optical memory.

The low repetition rate of 10 MHz of the addressing laser prevented the authors of [372] from writing more than 18 bits per cavity roundtrip time, but by using appropriate techniques they were able to write much longer streams of cavity solitons, such that thousands of well-separated solitons could be written at quasi-random locations. This allows them to claim a potential 45-kbit memory at 25 Gbits/s, which is within the operating range of realistic telecommunication systems. They have been able also to generate pairs of cavity solitons, the temporal separation of which could be continuously varied between 20 and

60 ps, demonstrating, in particular, that for separations lower than 25 ps the two solitons fuse into a single soliton, whereas above a separation of 40 ps the solitons become basically independent of each other. Numerical simulations show also that addressing pulses somewhat stronger than writing pulses can perform an XOR operation, i.e. they can write a new cavity soliton where there is none, but can also erase a pre-existing soliton.

A later article [373] reports on the experimental observation of breathing cavity solitons, i.e. solitons that oscillate periodically over the slow temporal scale t. This phenomenon, which arises from a Hopf bifurcation of the soliton branch, is similar to that described numerically in the spatial case, see Section 28.1.3 and Fig. 28.6. Figure 28.12 shows a numerical simulation of the oscillations of the temporal cavity-soliton peak.

A very recent evolution of these lines of research connects them with the topic of broadband frequency combs. As a matter of fact, in the spectral domain the periodic trains of cavity solitons correspond to a periodic comb of frequencies. This remark immediately draws a parallel with the generation of broadband frequency combs in high-Q passive monolithic microresonators with a Kerr nonlinearity driven by CW light [374]. The spatial size of the resonators and the driving power levels are completely different from the configuration described in this section, but the physics is strongly related. Microresonator Kerr frequency combs, as they are called, have been predicted to have an important impact as a compact technology for the generation of ultra-stable frequency combs. A recent report has shown that some experimentally reported Kerr combs can be described and simulated efficiently by an LL equation generalized to include higher-order dispersion derivatives [375].

Recent articles by Leo and collaborators have demonstrated the experimental generation of pulse trains of repetition rate 1.6 THz in a passive optical-fiber resonator. The cavity is made of two different fibers, but the observations are still described by the generalized LL equation in which the dispersion coefficient is simply the length-weighted average dispersion of the cavity [376].

29 Spatial patterns in cavities with spherical mirrors

In Chapters 27 and 28 we discussed the topic of optical spatial pattern formation in planar cavities. In this chapter, instead, we consider the case of ring cavities with spherical mirrors, on the basis of the discussion of transverse modes in Chapter 26.

In the first section we derive, first of all, a set of modal equations to be coupled with the atomic equations, with a configuration similar to that of the modal equations of Chapter 12, with the difference that now transverse modes are included in addition to longitudinal modes. This provides an extremely complex context, and therefore in the following we focus on the case of quasi-planar cavities, in which it is possible, by introducing the low-transmission and single-longitudinal-mode approximations, to derive a single-longitudinal-mode model that represents, in the framework of cavities with spherical mirrors, the counterpart of the single-longitudinal-mode model for planar cavities discussed in Chapter 27.

In Section 29.2 we consider the simplest configuration, in which only the fundamental Gaussian transverse mode is assumed to have negligible losses, so that it is possible to consider a single-mode model that includes only the Gaussian mode. This is precisely the model considered in Section 24.3.1 for the comparison with experimental data concerning the single-mode instability of optical bistability. We calculate the stationary curve, and in the laser case we demonstrate the validity of the mode-pulling formula also in this configuration.

In the following section we consider, instead, the general case in which several transverse modes of a laser are involved and demonstrate that the modal equations admit multimodal stationary intensity states, a configuration that is not possible in the case of longitudinal modes, at least in the low-transmission approximation. In this phenomenon, which we call *cooperative frequency locking*, the modes agree a common oscillation frequency that is given by the average of the (mode-pulled) modal frequencies, with weights given by the modal intensities. Numerical and experimental evidence of cooperative frequency locking is provided.

In Section 29.4 we discuss the case in which the atomic line excites only a frequency-degenerate family of transverse modes. We focus on the case of class-A lasers (see Section 22.1) and show that the stationary solutions are governed by a variational principle identical to that illustrated in Section 18.4.5. We discuss the various stationary solutions which arise from the competition of the transverse modes of the frequency-degenerate family, and the phenomenon of the *spontaneous breaking of the cylindrical symmetry*. The stationary solutions display the presence of phase singularities (vortices) and of *phase-singularity crystals*.

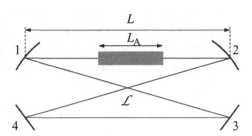

Figure 29.1 The scheme of the ring resonator.

29.1 Modal equations and the single-longitudinal-mode model

Let us consider the ring cavity with spherical mirrors shown in Fig. 29.1. Mirrors 1 and 2 have radius of curvature R and transmissivity T, while the two planar mirrors 3 and 4 are perfectly reflecting. The total length of the cavity is $\mathcal{L}$, while L is the distance between the two spherical mirrors. The active medium is assumed to have the form of a cylinder of length L_A. The dynamics of the system is governed by the Maxwell-Bloch equations (27.1)–(27.3); in this chapter we assume $n_B = 1$. In order to formulate the boundary conditions, it is necessary to expand the field envelope F on the basis of the Gauss–Laguerre modes

$$F(r, \varphi, t) = \sum_{p,l} \psi_{pl}(r, \varphi, z) f_{pl}(z, t), \tag{29.1}$$

where the functions ψ_{pl} are defined by Eq. (26.46) with $\tilde{r}$ and η replaced by $r = w_0 \tilde{r}$ and $z = z_0 \eta$, respectively. If we insert this equation into Eq. (27.1) and take into account that the modes ψ_{pl} obey Eq. (26.1), we obtain the following equation for the modal amplitudes $f_{pl}(z, t)$ [377]

$$\frac{\partial f_{pl}}{\partial z} + \frac{1}{c}\frac{\partial f_{pl}}{\partial t} = g \int_0^\infty dr\, r \int_0^{2\pi} d\varphi\, \psi_{pl}^*(r, \varphi, z) P(r, \varphi, z, t), \tag{29.2}$$

where we have taken into account that the functions ψ_{pl} constitute an orthonormal set for any value of z. On generalizing Eqs. (8.36) and (8.37), the amplitudes f_{pl} are subject to the boundary condition [377]

$$f_{pl}\left(-\frac{L_A}{2}, t\right) = T F_I \delta_{p,0}\delta_{l,0} + R e^{-i(\delta_0 + \delta_{pl})} f_{pl}\left(\frac{L_A}{2}, t - \frac{\mathcal{L} - L_A}{c}\right), \tag{29.3}$$

where for simplicity we have assumed that the input field, if any, has the transverse profile of the fundamental Gaussian mode $p = 0, l = 0$ and we define

$$\delta_{pl} = \frac{\omega_{\bar{j}pl} - \omega_{\bar{j}00}}{c/\mathcal{L}} = \frac{\omega_{\bar{j}pl} - \omega_c}{c/\mathcal{L}} \quad \Rightarrow \quad \delta_0 + \delta_{pl} = \frac{\omega_{\bar{j}pl} - \omega_0}{c/\mathcal{L}}. \tag{29.4}$$

The frequency $\omega_{\bar{j}pl}$ is given by Eq. (26.83) with q replaced by $2p + |l|$, and the value $\bar{j}$ of the index j is identified by the circumstance that $\omega_{\bar{j}00}$ is the longitudinal frequency closest to the reference frequency ω_0, and one has by definition that $\delta_{00} = 0$.

The set of modal equations (29.2) with the boundary conditions (29.3) must be coupled with Eqs. (27.2) and (27.3) with F given by Eq. (29.1) and $\chi(x, y)$ replaced by $\chi(r)$, since we assume that the pumped region is cylindrically symmetrical.

A simplification arises if we assume that the medium is much shorter than the Rayleigh length z_0,

$$L_A \ll z_0, \tag{29.5}$$

because the beam width becomes constant along the atomic sample and the modal functions ψ_{pl} become z-independent:

$$\psi_{pl}(r, \varphi) = \sqrt{\frac{2}{\pi}} \frac{1}{w_0} \sqrt{\frac{p!}{(p + |l|)!}} \left[\frac{2r^2}{w_0^2}\right]^{|l|/2} L_p^{|l|}\left[\frac{2r^2}{w_0^2}\right] e^{-r^2/w_0^2} e^{il\varphi}. \tag{29.6}$$

By performing transformations of the same kind of Eqs. (12.1), (12.8) and (12.9), one can convert the boundary conditions (29.3) into periodic boundary conditions [377] and therefore one can expand the field amplitudes f_{pl} also in terms of longitudinal modes. Thus, the dynamics involves longitudinal and transverse modes and is exceedingly complex in general.

A very partial study was conducted in the laser case, assuming (i) $\omega_a = \omega_c$, (ii) the low-transmission limit, (iii) $\kappa \ll \gamma_\perp$, $\gamma_\parallel$ and (iv) that the frequency spacing $\delta_{pl}c/\mathcal{L}$ of the transverse mode $\bar{j}pl$ from the Gaussian mode $\bar{j}00$ is of the same order of magnitude as the free spectral range. It turned out that the single-mode Gaussian stationary solution can become unstable with an instability threshold substantially smaller than the instability threshold in the Risken–Nummedal–Graham–Haken multimode instability [378].

In this chapter, instead, we focus on parametric conditions that lead to a single-longitudinal-mode model analogous to that considered in Chapter 27 in the case of planar mirrors. To be precise, we assume the low-transmission limit (27.8) and the single-mode limit (27.9). We note that the last condition of Eq. (27.8) in the case of a cavity with spherical mirrors amounts to assuming that the frequency spacing $\delta_{pl}c/\mathcal{L}$ of the relevant transverse modes $\bar{j}pl$ from the Gaussian mode ω_c is much smaller than the free spectral range, i.e.

$$\delta_{pl} = O(T). \tag{29.7}$$

Condition (29.7) can be satisfied in the limit of quasi-planar cavities such that $R \gg \mathcal{L}$, in which the transverse modes become closely spaced as shown in Fig. 26.7(a). From Eqs. (26.83) and (29.4) we have

$$\delta_{pl} = 2(2p + |l|)\left[\arctan\left(\frac{L}{2z_0}\right) + \arctan\left(\frac{\mathcal{L} - L}{2z_1}\right)\right]. \tag{29.8}$$

For $R \gg \mathcal{L}$, by using Eqs. (26.68) and (26.69) we have $z_0 \simeq z_1 \simeq \sqrt{R\mathcal{L}}/2$ and we obtain

$$\delta_{pl} \simeq 2(2p + |l|)\sqrt{\frac{\mathcal{L}}{R}}. \tag{29.9}$$

Hence we can set

$$\sqrt{\frac{L}{R}} = O(T), \quad b = \sqrt{\frac{L}{R}\frac{2}{T}} \quad \Rightarrow \quad \frac{\delta_{pl}}{T} = b(2p + |l|) = \frac{\omega_{\tilde{j}pl} - \omega_c}{\kappa}. \tag{29.10}$$

Before writing the equations obtained in the limits that we defined, we specify that in the following we use dimensionless quantities, i.e. the normalized radial variable $\tilde{r} = r/w_0$, dimensionless modal amplitudes and modal functions. Hence we write

$$F(\tilde{r}, \varphi, t) = \sum_{pl} \tilde{\psi}_{pl}(\tilde{r}, \varphi)\tilde{f}_{pl}(t), \tag{29.11}$$

where

$$\tilde{\psi}_{pl}(\tilde{r}, \varphi) = \sqrt{\frac{2}{\pi}}\sqrt{\frac{p!}{(p + |l|)!}}\left(2\tilde{r}^2\right)^{|l|/2} L_p^{|l|}\left(2\tilde{r}^2\right) e^{-\tilde{r}^2} e^{il\varphi}.$$

The amplitudes $\tilde{f}_{pl}(t)$ obey the equations [228, 377]

$$\frac{d\tilde{f}_{pl}}{dt} = -\kappa\left\{\left[1 + i(\theta + b(2p + |l|))\right]\tilde{f}_{pl} - F_{\mathrm{I}}\delta_{p,0}\delta_{l,0}\right.$$
$$\left. - A\int_0^\infty d\tilde{r}\,\tilde{r}\int_0^{2\pi} d\varphi\,\tilde{\psi}_{pl}^*(\tilde{r}, \varphi)P(\tilde{r}, \varphi, t)\right\}, \tag{29.12}$$

where, as usual, $\theta = \delta_0/T = (\omega_c - \omega_0)/\kappa$. Note that

$$\theta + b(2p + |l|) = \frac{\omega_{\tilde{j}pl} - \omega_0}{\kappa}. \tag{29.13}$$

The modal equations (29.12) must be coupled with Eqs. (27.2) and (27.3) with $\chi(x, y)$ replaced by $\chi(\tilde{r})$.

This set of equations can be directly numerically solved; alternatively, one can expand also P and D over the Gauss–Laguerre mode basis, whereby one obtains a set of ordinary differential equations [377].

Let us now show that the set of modal equations (29.12) can be recast in the form of a partial differential equation for the field F [228, 377]

$$\frac{\partial F}{\partial t} = -\kappa\left[(1 + i\theta)F - F_{\mathrm{I}}\psi_{00}(\tilde{r}, \varphi) - AP - ib\left(1 - \tilde{r}^2 + \frac{1}{4}\tilde{\nabla}_\perp^2\right)F\right], \tag{29.14}$$

with

$$\tilde{\nabla}_\perp^2 = \frac{\partial^2}{\partial\tilde{r}^2} + \frac{1}{\tilde{r}}\frac{\partial}{\partial\tilde{r}} + \frac{1}{\tilde{r}^2}\frac{\partial^2}{\partial\varphi^2}. \tag{29.15}$$

To check this, the simplest procedure is to start from Eq. (29.14), insert the expansion (29.11), and use the identity[1]

$$\left[\frac{1}{4}\tilde{\nabla}_{\perp}^{2}-\tilde{r}^{2}\right]\tilde{\psi}_{pl}(\tilde{r},\varphi)=-(2p+|l|+1)\tilde{\psi}_{pl}(\tilde{r},\varphi) \qquad (29.16)$$

and the orthonormality of the functions ψ_{pl}; in this way one recovers Eqs. (29.12).

The partial differential equation (29.14) represents the counterpart for a cavity with spherical mirrors of Eq. (27.10), which is valid for a cavity with planar mirrors. It can be noted that Eqs. (29.14), (27.2) and (27.3) (with $\chi(x,y)$ replaced by $\chi(\tilde{r})$) can be used also for a quasi-confocal or quasi-concentric cavity, as is demonstrated in [377], provided that the sample length L_A is much smaller than the Rayleigh range (thin medium, see Eq. (29.5)), and the sample is located at cavity center. In the quasi-concentric case the parameter b is negative. In the quasi-confocal case b can be positive or negative, but one must include the additional constraint that F must be even with respect to the parity transformation $\mathbf{x} \to -\mathbf{x}$.

In the case of optical bistability a cubic approximation of Eqs. (29.12), (27.2) and (27.3), which represents the counterpart of Eq. (27.15) for a cavity with spherical mirrors, has been analyzed in [379].

29.2 The single-mode Gaussian model

In this section we assume that all the transverse modes of order higher than the fundamental Gaussian mode $p = l = 0$ suffer from large losses, for example because of the presence of an aperture in the cavity. Therefore we can drop all the modal equations (29.12) with the exception of that for $p = l = 0$, which, using the definitions

$$\tilde{f}_0(t)=\sqrt{\frac{2}{\pi}}\tilde{f}_{00}, \quad \tilde{F}_{\mathrm{I}}=\sqrt{\frac{2}{\pi}}F_{\mathrm{I}}, \qquad (29.17)$$

---

[1] By using Eqs. (29.11) and (29.15) one gets

$$\left[\frac{1}{4}\tilde{\nabla}_{\perp}^{2}-\tilde{r}^{2}\right]\tilde{\psi}_{pl}(\tilde{r},\varphi)=C_{pl}\left[\frac{1}{4}\frac{d^2}{d\tilde{r}^2}L_p^{|l|}(2\tilde{r}^2)+\left(\frac{|l|+\frac{1}{2}}{2\tilde{r}}-\tilde{r}\right)\frac{d}{d\tilde{r}}L_p^{|l|}(2\tilde{r}^2)\right.$$
$$\left. -(|l|+1)L_p^{|l|}(2\tilde{r}^2)\right](2\tilde{r}^2)^{|l|/2}e^{-\tilde{r}^2}e^{il\varphi},$$

with $C_{pl}=\sqrt{2/\pi}\sqrt{p!/(p+|l|)!}$. On performing the transformation $u=2\tilde{r}^2$ the expression on the r.h.s. becomes

$$C_{pl}\left[2u\frac{d^2}{du^2}L_p^{|l|}(u)+2(|l|+1-u)\frac{d}{du}L_p^{|l|}(u)-(|l|+1)L_p^{|l|}(u)\right]u^{|l|/2}e^{-u/2}e^{il\varphi},$$

and, taking into account that the function $L_p^{|l|}(u)$ obeys Eq. (26.42), the expression reduces to

$$-C_{pl}(2p+|l|+1)(2\tilde{r}^2)^{|l|/2}L_p^{|l|}(2\tilde{r}^2)e^{-\tilde{r}^2}e^{il\varphi}=-(2p+|l|+1)\tilde{\psi}_{pl}(\tilde{r},\varphi),$$

which verifies Eq. (29.16).

reads

$$\frac{d\tilde{f}_0}{dt} = -\kappa \left[(1 + i\theta) \tilde{f}_0 - \tilde{F}_I - A \int_0^\infty d\tilde{r}\, 4\tilde{r} e^{-\tilde{r}^2} P(\tilde{r}, t) \right], \tag{29.18}$$

while Eqs. (27.2) and (27.3) read

$$\frac{\partial P(\tilde{r}, t)}{\partial t} = \gamma_\perp \left[D(\tilde{r}, t) \tilde{f}_0(t) e^{-\tilde{r}^2} - (1 + i\Delta) P(\tilde{r}, t) \right], \tag{29.19}$$

$$\frac{\partial D(\tilde{r}, t)}{\partial t} = -\gamma_\parallel \left[\frac{1}{2} (\tilde{f}_0(t) P^*(\tilde{r}, t) + \text{c.c.}) e^{-\tilde{r}^2} + D(\tilde{r}, t) - \chi(\tilde{r}) \right]. \tag{29.20}$$

Equations (29.18)–(29.20) constitute a single-mode [380, 381] model that has the same kind of structure as the single-mode model for the inhomogeneously broadened laser (Eqs. (15.7)–(15.9)) or for the Fabry–Perot laser (Eqs. (14.44), (14.40) and (14.41)). In the case of optical bistability, the model (29.18)–(29.20) is just the Gaussian model which is mentioned in Section 24.3.1 and provides good agreement with the experimental results.

Let us now focus, instead, on the free-running-laser case $F_I = 0$ and on the nontrivial stationary solution of Eqs. (29.18)–(29.20). From Eqs. (29.19) and (29.20) we obtain the following expressions for P and D in the stationary state:

$$P(\tilde{r}) = \frac{1 - i\Delta}{1 + \Delta^2 + |\tilde{f}_0|^2 e^{-2\tilde{r}^2}} \tilde{f}_0 e^{-\tilde{r}^2} \chi(\tilde{r}), \tag{29.21}$$

$$D(\tilde{r}) = \frac{1 + \Delta^2}{1 + \Delta^2 + |\tilde{f}_0|^2 e^{-2\tilde{r}^2}} \chi(\tilde{r}). \tag{29.22}$$

By inserting Eq. (29.21) into Eq. (29.18) in the stationary state we get

$$(1 + i\theta) \tilde{f}_0 = 4A(1 - i\Delta) \int_0^\infty d\tilde{r}\, \tilde{r} \frac{e^{-2\tilde{r}^2} \chi(\tilde{r})}{1 + \Delta^2 + |\tilde{f}_0|^2 e^{-2\tilde{r}^2}} \tilde{f}_0. \tag{29.23}$$

In order to obtain the nontrivial stationary solution, we divide Eq. (29.23) by $\tilde{f}_0$ and equate real and imaginary parts, which gives on the one hand the equation [228]

$$1 = 4A \int_0^{d/w_0} d\tilde{r}\, \tilde{r} \frac{e^{-2\tilde{r}^2}}{1 + \Delta^2 + |\tilde{f}_0|^2 e^{-2\tilde{r}^2}} = A \int_{e^{-2(d^2/w_0^2)}}^1 \frac{ds}{1 + \Delta^2 + |\tilde{f}_0|^2 s}$$

$$= \frac{A}{|\tilde{f}_0|^2} \ln \left(\frac{1 + \Delta^2 + |\tilde{f}_0|^2}{1 + \Delta^2 + |\tilde{f}_0|^2 e^{-2(d^2/w_0^2)}} \right), \tag{29.24}$$

where we have assumed that the function $\chi(\tilde{r})$ is equal to 1 for $\tilde{r} < d$ and equal to 0 for $\tilde{r} > d$, and on the other hand Eq. (12.57), which leads to the mode-pulling formula (12.58). Thus the stationary equation (29.24) is conveniently written in the form

$$A = |\tilde{f}_0|^2 \left[\ln \left(\frac{1 + \Delta^2 + |\tilde{f}_0|^2}{1 + \Delta^2 + |\tilde{f}_0|^2 e^{-2(d^2/w_0^2)}} \right) \right]^{-1} \quad \Rightarrow \quad A_{thr} = \frac{1 + \Delta^2}{1 - e^{-2(d^2/w_0^2)}}, \tag{29.25}$$

where the threshold value of A is obtained in the limit $\tilde{f}_0 \to 0$. The stationary curve of $|\tilde{f}_0|^2$ as a function of A can be obtained by exchanging the axes in the plot of A vs. $|\tilde{f}_0|^2$ given by Eq. (29.25). In Fig. 29.2 we see the graph of $|\tilde{f}_0|^2$ as a function of A for various values of the parameters Δ and d/w_0.

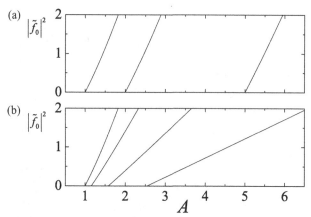

Figure 29.2 Graphs of the stationary values of $|\tilde{f}_0|^2$ as a function of the pump parameter A: (a) $d \gg w_0$ and $\Delta = 0$, 1 and 2 from left to right; (b) $\Delta = 0$ and $2(d/w_0)^2 \gg 1$, $2(d/w_0)^2 = 2$, 1, 0.5 from left to right.

When d/w_0 becomes small, only the flat part of the Gaussian profile contributes to the integral in $\tilde{r}$. It is easy to verify that in the double limit $d/w_0 \to 0$, $A \to \infty$, with $A' = A \times 2(d/w_0)^2$ constant, one recovers the stationary equation of the plane-wave theory $|\tilde{f}_0|^2 = A' - 1 - \Delta^2$.

We observe that the single-mode Gaussian model we have considered in this section holds also when condition (29.7) is not satisfied, provided that the low-transmission and single-longitudinal-mode conditions are satisfied and that all transverse modes except the fundamental mode $p = l = 0$ are lossy.

29.3 The multimodal transverse regime and cooperative frequency locking in the laser

Let us now focus on the laser case ($F_I = 0$, $\theta = -\Delta$), where we take as reference frequency ω_0 the mode-pulled frequency (12.58), and return to the single-longitudinal-mode model defined by Eqs. (29.14), (27.2) and (27.3) with $\chi = \chi(\tilde{r})$. This model includes only one longitudinal mode and a variety of transverse modes.

In Parts I and II of this book we have extensively analyzed the plane-wave multimode model in the low-transmission approximation of Eqs. (29.12), (27.2) and (27.3), which represents the counterpart of Eq. (27.15) for a cavity with spherical mirrors, and we have seen that the mode–mode competition can only lead to two distinct regimes of operation for the laser:

(a) the single-mode regime, where one mode suppresses all the others (we have seen that all the stationary-intensity solutions of the plane-wave laser belong to this class); and

(b) the standard multimode regime, in which a number of modes coexist and, because they have different frequencies, they interfere with one another and produce an output intensity that pulsates in time.

The inclusion of transverse modes introduces a third kind of operation by way of the so-called cooperative frequency locking [228, 309]. In this multimodal regime, at least two transverse modes coexist and contribute significantly to the output intensity. Unlike the standard multimode behavior, the active modes, under cooperative-frequency-locking conditions, do not maintain their frequencies but lock, instead, to a common frequency value and oscillate in a synchronized way.

Let us demonstrate this [309, 348] by showing that the model (29.14), (27.2) and (27.3) admits multimodal stationary-intensity solutions in which the variables have the form

$$F(\tilde{r}, \varphi, t) = e^{-i\delta t} F_s(\tilde{r}, \varphi), \quad P(\tilde{r}, \varphi, t) = e^{-i\delta t} P_s(\tilde{r}, \varphi), \quad D(\tilde{r}, \varphi, t) = D_s(\tilde{r}, \varphi).$$
$$(29.26)$$

If we insert Eq. (29.26) into Eqs. (27.2) and (27.3) we obtain

$$P_s = \frac{1 - i(\Delta - \delta/\gamma_\perp)}{1 + (\Delta - \delta/\gamma_\perp)^2 + |F_s|^2} F_s \chi(\tilde{r}), \quad D_s = \frac{1 + (\Delta - \delta/\gamma_\perp)^2}{1 + (\Delta - \delta/\gamma_\perp)^2 + |F_s|^2} \chi(\tilde{r}), \quad (29.27)$$

so that, by inserting the expression of P_s into Eq. (29.14) and multiplying the resulting equation by F_s^*, we get

$$i\frac{\delta}{\kappa}|F_s|^2 = (1 - i\Delta)|F_s|^2 - ibF_s^* \left(\frac{1}{4}\tilde{\nabla}_\perp^2 - \tilde{r}^2 + 1\right) F_s$$
$$- A\frac{[1 - i(\Delta - \delta/\gamma_\perp)]|F_s|^2}{1 + (\Delta - \delta/\gamma_\perp)^2 + |F_s|^2} \chi(\tilde{r}). \quad (29.28)$$

Next, we add and subtract Eq. (29.28) and its complex conjugate and arrive at the two equations

$$\frac{A|F_s|^2 \chi(\tilde{r})}{1 + (\Delta - \delta/\gamma_\perp)^2 + |F_s|^2} = |F_s|^2 - i\frac{b}{2}\left[F_s^* \left(\frac{1}{4}\tilde{\nabla}_\perp^2 - \tilde{r}^2 + 1\right) F_s - \text{c.c.}\right], \quad (29.29)$$

$$\frac{\delta}{\kappa}|F_s|^2 = -\Delta|F_s|^2 + \left(\Delta - \frac{\delta}{\gamma_\perp}\right)\frac{A|F_s|^2 \chi(\tilde{r})}{1 + (\Delta - \delta/\gamma_\perp)^2 + |F_s|^2}$$
$$- \frac{b}{2}\left[F_s^* \left(\frac{1}{4}\tilde{\nabla}_\perp^2 - \tilde{r}^2 + 1\right) F_s + \text{c.c.}\right]. \quad (29.30)$$

Then, by inserting one into the other, we arrive at

$$\left(\frac{\delta}{\kappa} + \frac{\delta}{\gamma_\perp}\right)|F_s|^2 = -i\left(\Delta - \frac{\delta}{\gamma_\perp}\right)\frac{b}{2}\left[F_s^* \left(\frac{1}{4}\tilde{\nabla}_\perp^2 - \tilde{r}^2 + 1\right) F_s - \text{c.c.}\right]$$
$$- \frac{b}{2}\left[F_s^* \left(\frac{1}{4}\tilde{\nabla}_\perp^2 - \tilde{r}^2 + 1\right) F_s + \text{c.c.}\right]. \quad (29.31)$$

Now we integrate Eq. (29.31) over the transverse plane and take into account the equations

$$\int_0^\infty d\tilde{r}\,\tilde{r} \int_0^{2\pi} d\varphi |F_s|^2 = \sum_{p,l} |\tilde{f}_{pl,s}|^2,$$

$$\int_0^\infty d\tilde{r}\,\tilde{r} \int_0^{2\pi} d\varphi\, F_s^* \left(\frac{1}{4}\tilde{\nabla}_\perp^2 - \tilde{r}^2 + 1 \right) F_s = -\sum_{p,l} |\tilde{f}_{pl,s}|^2 (2p + |l|), \quad (29.32)$$

which follow from Eqs. (29.11) and (29.16) and the orthonormality of the functions $\tilde{\psi}_{pl}$. We arrive at the result

$$\delta = \frac{\kappa\gamma_\perp}{\kappa + \gamma_\perp} \frac{b\sum_{p,l} |\tilde{f}_{pl,s}|^2 (2p + |l|)}{\sum_{p,l} |\tilde{f}_{pl,s}|^2} = \frac{\gamma_\perp}{\kappa + \gamma_\perp} \frac{\sum_{p,l} |\tilde{f}_{pl,s}|^2 (\omega_{\bar{j}pl} - \omega_c)}{\sum_{p,l} |\tilde{f}_{pl,s}|^2}, \quad (29.33)$$

where we used Eq. (29.10). Hence the complete frequency $\omega_s = \omega_0 + \delta$ of the stationary-intensity state is given by [309, 348]

$$\omega_s = \omega_0 + \delta = \frac{\kappa\omega_a + \gamma_\perp \omega_c}{\kappa + \gamma_\perp} + \frac{\gamma_\perp}{\kappa + \gamma_\perp} \frac{\sum_{p,l} |\tilde{f}_{pl,s}|^2 (\omega_{\bar{j}pl} - \omega_c)}{\sum_{p,l} |\tilde{f}_{pl,s}|^2} = \frac{\sum_{p,l} \bar{\omega}_{\bar{j}pl} |\tilde{f}_{pl,s}|^2}{\sum_{p,l} |\tilde{f}_{pl,s}|^2},$$
$$(29.34)$$

where $\bar{\omega}_{\bar{j}pl}$ is the mode-pulled frequency of mode p, l

$$\bar{\omega}_{\bar{j}pl} = \frac{\kappa\omega_a + \gamma_\perp \omega_{\bar{j}pl}}{\kappa + \gamma_\perp}. \quad (29.35)$$

In conclusion, the common oscillation frequency ω_s of the transverse modes in the cooperatively frequency-locked state is the weighted average of the mode-pulled frequencies of the transverse modes, with the weights given by the distribution $|\tilde{f}_{pl}|^2$ of the intensities of the various modes. This is the rule which governs the phenomenon of cooperative frequency locking, in which the transverse modes give rise to a sort of stationary "supermode".

When in a nonlinear system there is competition between oscillators with different frequencies, different scenarios can arise. The simplest is that one frequency prevails, as happens in single-mode operation and in the frequency locking that occurs in a laser with an injected signal as we saw in Section 13.1. A second possibility is that the various frequencies coexist and, in the case of optical systems, in the field intensity one finds the beat notes among the frequencies in play. This happens in the standard multimode operation of longitudinal modes and in a laser with an injected signal when the driving field is weak and cannot impose the frequency locking.

The last scenario we consider arises when the different frequencies come to a sort of agreement and select a common intermediate oscillation frequency, corresponding to a weighted average of the competing frequencies, such that the intensity is again stationary. This happens in a laser when the atomic frequency does not coincide with a cavity frequency, and the behavior is governed by the mode-pulling formula (12.58). This happens also in the phenomenon of cooperative frequency locking, in which the weights in the average are provided by the modal intensities instead of the relaxation times.

The set of equations (29.12), (27.2) and (27.3), with $\chi = 1$ for all values of $\tilde{r}$, was analyzed numerically in [309, 382, 383]. These calculations assume that the laser field

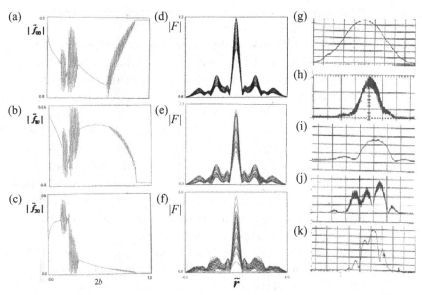

Figure 29.3 Scans of the modal amplitudes of the first three cylindrically symmetric modes, (a) $\tilde{f}_{00}$, (b) $\tilde{f}_{10}$ and (c) $\tilde{f}_{20}$, as a function of the modal spacing b. The numerical spatio-temporal evolution of the modulus of the total field for (d) $b = 0.17$, (e) $b = 0.16$ and (f) $b = 0.125$. The other parameters are selected as stated in the text. Experimental beam transverse profiles for decreasing separation between transverse modes from (g) to (k). Reprinted figure with permission from [383]. Copyright (1989) by the American Physical Society.

remains cylindrically symmetrical around the laser axis and therefore take into account only the modes with $l = 0$. In addition they include diffraction losses by replacing the loss term $-\kappa \tilde{f}_{p0}$ with $-\kappa_p \tilde{f}_{p0}$, where the effective loss parameter κ_p is given by

$$\kappa_p = \kappa(1 + \beta p^4), \tag{29.36}$$

with β constant. Figure 29.3 shows the result of a numerical investigation in which the frequency spacing (normalized with respect to κ) b was swept very slowly from large to smaller values over the indicated range. The values of the other parameters are fixed as follows: $\kappa/\gamma_\perp = 1$, $\gamma_\parallel/\gamma_\perp = 0.05$, $A = 24$, $\Delta = 0.18$ and $\beta = 0.005$. With this choice of β only the three modes with $p = 0$, 1 and 2 have non-negligible intensity. Figures 29.3(a)–(c) show the behavior of the moduli $|\tilde{f}_{p0}|$ (for $p = 0$, 1 and 2) as the parameter b slowly decreases. We note the alternation of regions of stationary behavior with two domains of oscillation. In the direction of decreasing b, the first stationary regime is practically of the single-mode type, because $|\tilde{f}_{10}|$ and $|\tilde{f}_{20}|$ are very small with respect to $|\tilde{f}_{00}|$; the second stationary intensity state is a cooperative frequency-locked state of the two modes with $p = 0$ and 1; and the last stationary interval shows cooperative frequency locking of the three major modal components.

The first oscillatory region exhibits periodic oscillations of the output intensity with a fundamental frequency equal to the mode-pulled frequency separation between modes $p = 0$ and 1. Note, however, that these oscillations are not just the result of a beating effect between these two modes, because, as shown in Fig. 29.3, the moduli of the modal amplitudes also oscillate. The second region of oscillations shows periodic, quasi-periodic or chaotic oscillations according to the value of b.

Figures 29.3(d)–(f) show three superpositions of snapshots of the transverse profile of the modulus of the total electric field obtained for three different fixed values of the frequency spacing b corresponding to periodic (d), quasi-periodic (e) and chaotic (f) oscillations. For a value of the ratio $\kappa/\gamma_\perp = 0.3$, which is closer to the typical values found in CO_2 lasers, the picture remains qualitatively unchanged, apart from the disappearance of chaotic behavior.

These results have been compared with those from experiments [383] conducted by Tredicce and collaborators with a CO_2 laser. The mirrors of the laser cavity are plane and parallel, but the insertion of two lenses inside the resonator makes the configuration equivalent to that of a cavity with spherical mirrors having an effective radius of curvature that is controlled by the distance between the lenses and can be varied at will. Figures 29.3(g)–(k) show a sequence of observed transverse profiles; each picture was obtained from the superposition of individual traces similar to those of Figs. 29.3(d)–(f). Figures 29.3(g)–(k) are in order of decreasing value of the frequency spacing b and display the same type of alternation between stationary and oscillatory regimes as predicted by the theory. In agreement with the numerical results, (g) the first stationary stage is single-mode (Gaussian), (h) the first oscillatory regime is periodic, (i) the second stationary configuration is the result of frequency locking of the first two modes, (j) the second oscillatory regime exhibits periodic and quasi-periodic behaviors, and (k) the third stationary state is of the multimode type. In the last two cases we note a breaking of the cylindrical symmetry that was not predicted by the theory because the numerical computations included only cylindrically symmetrical modes.

Thus the experiment performed in [383] provides a good confirmation of the general trends predicted theoretically. An additional experimental observation was described in [384].

29.4 Laser patterns from frequency-degenerate families of modes. Spontaneous breaking of the cylindrical symmetry. Phase-singularity crystals

Let us now consider again the case of a free-running laser (that is, $F_I = 0$), and assume that the gain line is so narrow that it excites only one frequency-degenerate family of modes, such that the sum in Eq. (29.11) can be restricted to the modes for which $2p + |l| = q$, where q is fixed and identifies the family of modes. In addition, we assume for simplicity that the atomic frequency coincides with that of the frequency-degenerate family and we take as the reference frequency ω_0 just this common frequency, so that $\Delta = 0$. In this way we see that Eqs. (29.12) take the form [228]

$$\frac{d\tilde{f}_{pl}}{dt} = -\kappa \left[\tilde{f}_{pl} - A \int_0^\infty d\tilde{r}\, \tilde{r} \int_0^{2\pi} d\varphi\, \tilde{\psi}_{pl}^*(\tilde{r}, \varphi) P(\tilde{r}, \varphi, t) \right], \tag{29.37}$$

since the expression (29.13) vanishes, and the values of p and l are limited to those such that $2p + |l| = q$. In addition, Eqs. (27.2) and (27.3) read

$$\frac{\partial P}{\partial t} = \gamma_\perp [FD - P], \tag{29.38}$$

$$\frac{\partial D}{\partial t} = -\gamma_\parallel \left[\frac{1}{2}(FP^* + F^*P) + D - e^{-2\tilde{r}^2/\psi^2} \right], \tag{29.39}$$

where we have taken [228]

$$\chi(r) = e^{-r^2/2r_p^2}, \qquad \psi = 2\frac{r_p}{w_0}, \tag{29.40}$$

with r_p being the radius of the pumped region. Note that the model given by (29.37)–(29.39) holds also when condition (29.7) is not satisfied, provided that the low-transmission and single-longitudinal-mode conditions are satisfied and that the atomic line excites only the modes of the frequency-degenerate family q.

A detailed discussion of this problem can be found in [354, 385]. When $\kappa \ll \gamma_\perp, \gamma_\parallel$ (class-A lasers), the atomic variables can be adiabatically eliminated by setting $\partial P/\partial t = \partial D/\partial t = 0$ and, by using Eqs. (29.37)–(29.39) and (29.11), we obtain the set of equations

$$\frac{d\tilde{f}_{pl}}{dt''} = -\tilde{f}_{pl} + A \int_0^\infty \tilde{r}\, d\tilde{r}\, \chi(\tilde{r}) \int_0^{2\pi} d\varphi\, \frac{\tilde{\psi}_{pl}^*(\tilde{r}, \varphi) \sum'_{p'l'} \tilde{\psi}_{p'l'}(\tilde{r}, \varphi) \tilde{f}_{p'l'}}{1 + \left| \sum'_{p'l'} \tilde{\psi}_{p'l'}(\tilde{r}, \varphi) \tilde{f}_{p'l'} \right|^2}, \tag{29.41}$$

where $t'' = \kappa t$ and the symbol $'$ means that the sum is extended only to the modes of the frequency-degenerate family. Therefore we introduce the function

$$V(\{\tilde{f}_{pl}\}, \{\tilde{f}_{pl}^*\}) = \int_0^\infty \tilde{r}\, d\tilde{r} \int_0^{2\pi} d\varphi\, \left[|F(\tilde{r}, \varphi)|^2 - A\chi(\tilde{r}) \ln(1 + |F(\tilde{r}, \varphi)|^2) \right], \tag{29.42}$$

with F given by Eq. (29.11) so that Eqs. (29.41) can be cast in the compact form

$$\frac{d\tilde{f}_{pl}}{dt''} = -\frac{\partial V}{\partial \tilde{f}_{pl}^*}, \tag{29.43}$$

which shows that the stationary states coincide with the stationary points of the "potential" function V, which plays the role of a generalized free energy. From Eq. (29.43) we have also that

$$\frac{dV}{dt''} = \sum_{p,l} \left(\frac{\partial V}{\partial \tilde{f}_{pl}} \frac{d\tilde{f}_{pl}}{dt''} + \frac{\partial V}{\partial \tilde{f}_{pl}^*} \frac{d\tilde{f}_{pl}^*}{dt''} \right) = -2\sum_{p,l} \left| \frac{\partial V}{\partial \tilde{f}_{pl}} \right|^2 \le 0, \tag{29.44}$$

exactly as in the case of Eq. (18.16), so the long-time solutions, i.e. the stable stationary solutions, correspond to the local minima of V in the space of the amplitudes $\tilde{f}_{pl}$ and $\tilde{f}_{pl}^*$. In order to find the stable stationary states, it is essential to allow the amplitude $\tilde{f}_{pl}$ to be complex. Even if the patterns which arise from the interaction and competition of the modes of the family emerge through a spatial hole-burning (gain) mechanism, the process is strongly affected by the phases of the modal amplitudes. We must also keep in mind that, as a consequence of the cylindrical symmetry of the system, any stationary pattern generates an infinity of other possible patterns by rotation.

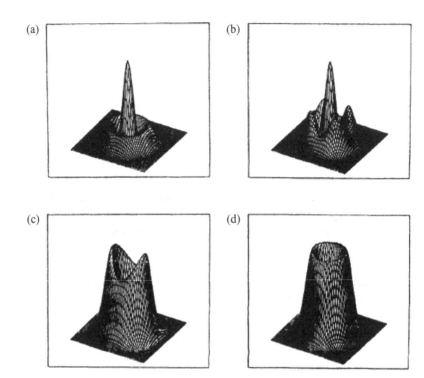

Figure 29.4 Transverse intensity distributions for the stable patterns in the case $2p + |l| = 2$: (a) the Gauss–Laguerre mode $p = 1, l = 0$; (b) the "leopard" configuration; (c) the "oval" pattern; and (d) the doughnut configuration. Reprinted figure with permission from [354]. Copyright (1991) by the American Physical Society.

As an example, let us consider the case of the family $2p + |l| = 2$, in which the family consists of the three modes $(p = 1, l = 0)$ and $(p = 0, l = \pm 2)$. In this case there are three single-mode stationary solutions corresponding to the three modes, and there are also multimode stationary solutions that correspond to combinations of the modes. References [354, 385] include a systematic search for the stationary solutions in the space of the parameters A and ψ and the linear-stability analysis of the single-mode stationary solutions, which is based on the same functions T_1 and T_2 as are considered in Appendix D. The minima of V correspond to four distinct kinds of stationary solutions, which are shown in Fig. 29.4. The structure shown in Fig. 29.4(a) corresponds to the single-mode solution $p = 1, l = 0$, and the doughnut pattern in Fig. 29.4(d) corresponds to the single-mode stationary solutions $p = 0, l = \pm 2$. The configurations shown in Figs. 29.4(b) and (c) are instead of the multimode type. In the structure of Fig. 29.4(b), which we identify as the "leopard", there are five peaks in the transverse intensity configuration and we have $|\tilde{f}_{1,0}| > |\tilde{f}_{0,2}| = |\tilde{f}_{0,-2}|$, while in the pattern of Fig. 29.4(c), the "oval", we have $|\tilde{f}_{1,0}| \gg |\tilde{f}_{0,2}| \gg |\tilde{f}_{0,-2}|$ or $|\tilde{f}_{1,0}| \gg |\tilde{f}_{0,-2}| \gg |\tilde{f}_{0,2}|$. The insides of the doughnut and oval configurations are shown in Fig. 29.5.

Figure 29.6 displays the domains of stability of the four patterns in the parameter plane of the variables ψ and A. This shows that, if we move from one region to another, the laser undergoes a transition from one pattern to the other. For example, if we go from region S

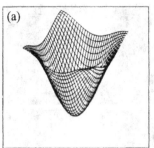
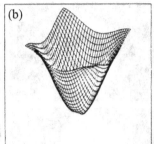
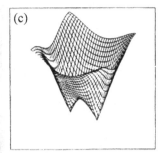

Figure 29.5 Transverse intensity distributions for the inner part of the doughnut in part (a) and of the oval in parts (b) and (c). The sequence of patterns is obtained by increasing the pump parameter A from one case to the next in such a way that one starts from region DL in Fig. 29.6 in part (a) and moves to region OL in parts (b) and (c). Reprinted figure with permission from [354]. Copyright (1991) by the American Physical Society.

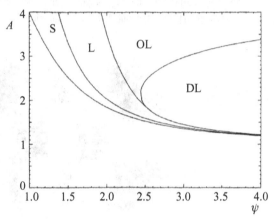

Figure 29.6 The phase-space diagram in the plane of the control parameters A and ψ for the family $2p + |l| = 2$. The letters S, L, O and D denote, respectively, "symmetrical", "leopard", "oval" and "doughnut" and correspond to the patterns shown in Figs. 29.4(a)–(d), respectively. The letters in each region indicate the stable patterns. Reprinted figure with permission from [354]. Copyright (1991) by the American Physical Society.

to region L we observe a continuous transition from the cylindrically symmetric state to the leopard, and this corresponds to a process of spontaneous breaking of the cylindrical symmetry [386]. If we move, instead, from region DL to region OL starting from the doughnut configuration, the pattern deforms itself gradually into an oval, as shown by the sequence in Fig. 29.5. From the viewpoint of the intensity, this is again a phenomenon of spontaneous symmetry breaking. Furthermore, when we cross over to region L, the oval configuration becomes unstable and decays into a leopard.

An experiment with a sodium dimer laser performed by Tamm and Weiss [354] shows exactly these configurations, as shown by Fig. 29.7. The mapping of the states in the (ψ, A) plane shows qualitatively good agreement between experiment and theory. In particular, two phenomena involving spontaneous breaking of the rotational symmetry were observed: (1) the continuous transition from (a) to (b) in Figs. 29.4 and 29.7 with formation of four phase singularities, and (2) the continuous transition from (d) to (c), with the splitting of a

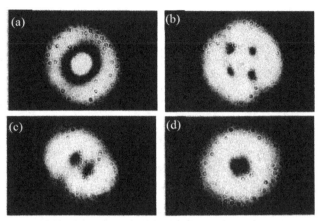

Figure 29.7 Output intensity patterns observed in a sodium dimer laser. The four configurations (a)–(d) correspond to those of Fig. 29.4. The small-scale structures in the photographs are diffraction patterns caused by dust particles. Reprinted figure with permission from [354]. Copyright (1991) by the American Physical Society.

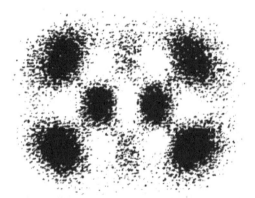

Figure 29.8 Crystal of seven optical vortices formed by the frequency-degenerate family $2p + |l| = 3$. Reprinted figure with permission from [354]. Copyright (1991) by the American Physical Society.

phase singularity of charge 2 into two phase singularities of charge 1. As shown in [387], if one adds stochastic noise terms to the modal equations, similarly to what is done in Eqs. (18.18), the rotational symmetry can be restored by the slow random rotations of the pattern around the axis of the system. For families of order higher than 2, one finds an increasing number of phase singularities, arranged in regular arrays; an example of these "phase-singularity crystals" [354, 385] is shown in Fig. 29.8.

Figures 29.4 and 29.5 provide the simplest example of *spatial multistability*, i.e. the coexistence of distinct stable patterns in the phase space of the system for the same values of the parameters. Spatial multistability is quite different from standard optical multistability, because the difference among the coexisting states lies not so much in the spatial output intensity but, rather, in the spatial configuration of the field.

Discussions of dynamical patterns, starting from the simplest example of a rotating pattern [388, 389], can be found e.g. in [318, 385].

With respect to the goal of encoding information in transverse spatial structures, patterns arising from combinations of a few modes present an intrinsic limitation, because their various parts are strongly correlated with one another. Therefore any local modification induced by an external control in order to encode information affects also other parts of the pattern, or, alternatively, the correlation leads to spontaneous elimination of the modification itself. For example, one might consider a lattice of intensity peaks that arise in the near field from the superposition of a few plane waves generated by a spatial instability as described in Section 27.2 (see Fig. 30.1(a), in which we consider a square lattice). If the peaks were independent of one another, they could be addressed individually as pixels and one could encode information in binary form (Fig. 30.1(b)). However, this is made impossible by the strong correlation between the points in the lattice. In the opposite case of many modes, one opens the door to the world of complexity. However, one must avoid having the behavior of the system become irregular, hence it is necessary to introduce control. In order to ensure the possibility of a reasonably practical implementation, the control procedure must be simple.

One finds that by generating solitons in a nonlinear cavity (*cavity solitons*) one can realize a set of intensity peaks that can be addressed (i.e. written and erased) individually and pinned down to precise locations, so that one is indeed dealing with an array of independent pixels as in Fig. 30.1(b). A set of $N \times N$ solitons constitutes a memory (or, in general, an optical processor) with $2^{N \times N}$ distinct states. This approach was pioneered by the work of Rosanov (see [390–392] and references cited therein) on "diffractive autosolitons" and of Moloney and collaborators [393]. In [390–392] autosolitons arise from the interaction of switching waves connecting two homogeneous stationary states.

In the cases which we analyze in the following the cavity solitons are instead connected with the presence of a modulational, pattern-forming instability and with the so-called localized structures [394, 400] which arise from the coexistence (bistability) of a stable homogeneous stationary solution and a stable patterned stationary solution. They are related to the results obtained in the one-dimensional case in [316, 401]. Later investigations have demonstrated the possibility of realizing these arrays in two dimensions, in particular using semiconductor broad-area microcavities, which is interesting for applications. The control of cavity solitons is reasonably simple and robust, and they behave as a self-organizing set capable of accommodating errors in the addressing of the writing/erasing beam. Thus, cavity solitons materialize the idea of considering the transverse planes orthogonal to the propagation direction of the beam as a blackboard on which light spots can be written and erased in any desired location and in a controlled way. A detailed description will be provided in the following sections.

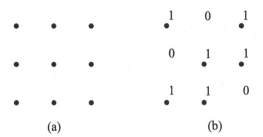

Figure 30.1
(a) A square array of points. (b) If the points are independent of one another, they behave as binary pixels and one can encode a number in binary units.

The topic of Kerr cavity solitons has been discussed in Chapter 28.

The prototype of standard spatial solitons has been described in Section 7.3, and arises in the propagation of light in a Kerr medium. Cavity solitons arise inside an optical resonator, where the escape of photons from the cavity (and, in addition, the occurrence of absorption and amplification) makes the dynamics dissipative [402–405].

In Section 30.1 we discuss the topic of localized structures, which is relevant for the general field of spatial pattern formation, and show that cavity solitons can be considered as elementary localized structures. In Section 30.2 we show how cavity solitons can be written and erased, and how they can be set in controlled motion, typically by introducing phase or amplitude gradients in the holding beam.

The remaining sections of this chapter are devoted to the case of cavity solitons in semiconductor microresonators, namely large-area VCSELs, which indeed provide the ideal framework to realize the concept of a broad-area cavity with planar mirrors which optimally satisfies the conditions of the low-transmission approximation and single-longitudinal-mode approximation specified in Section 27.1. In addition, the small dimensions of this device and its fast relaxation times make it interesting in view of practical applications. In Section 30.3 we illustrate the predictions of an appropriate theoretical model, the experimental observation of cavity solitons in driven VCSELs slightly below threshold (which corresponds to the configuration of an amplifier with an injected signal) and the comparison of the experimental results with the theoretical predictions.

Section 30.4 illustrates the so-called cavity-soliton laser, which provides a more compact system to generate cavity solitons, because it does not require the use of an external coherent driving field, and is capable of emitting a controlled number of narrow beams, corresponding to the cavity solitons.

Reviews of the subject of cavity solitons can be found in [318, 406–411].

30.1 Localized structures in optics

In the case of systems with translational symmetry, one often meets the phenomenon of coexistence (bistability) between a homogeneous stationary state and a spatially modulated stationary solution, in the sense that within a certain interval of the control parameter β (for example, the input field F_{I}) the two solutions are both stable and,

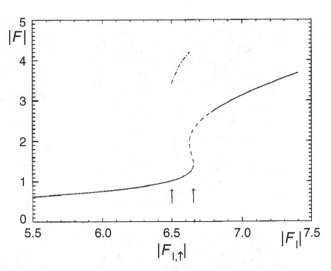

Figure 30.2 A stationary curve for $C = 5.4$ and $\theta = -1$. The broken line plots the part where the modulational instability develops. The dash–dotted line plots the maximum of the modulated pattern. Arrows indicate the region of coexistence between the homogeneous solution and the hexagonal solution. The instability region broadens with increasing C, with larger coexistence domains. Reprinted from [412], with kind permission from IOP Publishing.

depending on the initial conditions, the system evolves to one or other of the two. Although the presence of plane-wave bistability (i.e. bistability between two homogeneous stationary solutions) favors the possibility of localized structures, plane-wave bistability is not necessary.

An example of this localized structure is found in the absorptive model (27.14) [328]. Figure 30.2 shows a case of coexistence between a hexagonal stationary pattern and a homogeneous stationary state [412]. We meet here another phenomenon stemming from this coexistence, which links the field of optics to hydrodynamics and nonlinear chemical reactions, namely the existence of the so-called "localized structures" which have been described e.g. in [394,395] and experimentally observed in fluids, granulates and chemical systems [396–400].

In the case of optics, a localized structure is a solution for the field intensity where the two coexisting solutions appear simultaneously in different regions of the transverse plane. For example one observes a stable, stationary island containing a portion of a hexagonal pattern embedded in a homogeneous background corresponding to the homogeneous stationary solution. The existence of this kind of localized structure in nonlinear optical systems was predicted by Tlidi, Mandel and Lefever [413].

The location, shape and size of localized structures constitute, so to say, free parameters spanning an infinitely degenerate manifold of stable solutions for the global profile of the emitted field. In this section we will illustrate how control over such qualifying parameters can be gained; for the sake of simplicity, we will consider the model for the saturable absorber (27.14). The same techniques have been applied successfully to other models, as will be specified in the following.

Control over the localized structure's location can be achieved by properly acting on the driving field, F_1. The idea is to maintain a homogeneous part whose amplitude locates

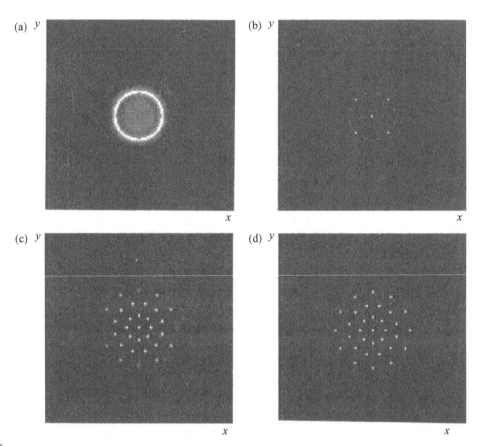

Figure 30.3 The evolution of the output field intensity profile when a broad Gaussian is injected. A stationary localized structure is visible for the regime in panel (d). Reprinted from [318], with kind permission from Elsevier.

the system's operating point somewhere in the region of coexistence with the stable branch of the hexagonal pattern and to start from an initial condition corresponding to the homogeneous stationary state. Then we superimpose an input Gaussian pulse of duration t_p so that we can represent the total input field as [328, 412]

$$F_I(x, y, t) = \begin{cases} \tilde{F}_I + \xi e^{-[(x-x_0)^2 + (y-y_0)^2]/\sigma^2} e^{i\varphi} & t \leq t_p, \\ \tilde{F}_I & t > t_p, \end{cases} \qquad (30.1)$$

where $\tilde{F}_I$ lies within the coexistence region of Fig. 30.2. Assuming that $\varphi = 0$, we thus locally increase F_I by an amount ξ in an area centered around (x_0, y_0), whose size is measured by σ, and is in general much smaller than the medium section. If $\xi > F_{I,\uparrow} - \tilde{F}_I$ the action of this perturbation is to locally bring the system into the modulationally unstable range, so that in that region the only available state is the modulated state and the field realizes a portion of the hexagonal lattice.

When the pulse is switched off, assuming that the pulse duration t_p is large enough to overcome transients, the system persists on the modulated branch since its operating point lies within the coexistence region. A temporal sequence of the transverse intensity profile is shown in Fig. 30.3, where the process of local pattern formation can be followed, including

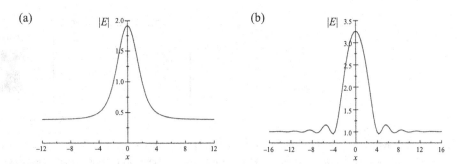

Figure 30.4 A diametral section of the cavity soliton transverse intensity profile. (a) The self-focussing active case: $\mu = 0.9$, $\theta = -2$, $\alpha = 5$ and $d = 0.052$. (b) The self-defocussing passive case: $\theta = -0.716$, $\Delta = 10$, $\mu = -1$ and $d = 0.052$. The parameters refer to Eqs. (30.5) and (30.6) and $\bar{a} = 19.32\ \mu\text{m}^2$. Reprinted from [318], with kind permission from Elsevier.

a further boundary-effect-dominated dynamics that shapes the contour of the localized structure.

It is intuitive that the pulse width σ determines the size of the localized structure while the pulse location coincides with the location where the localized structure will approximatively be centered. The interesting aspect is that on reducing σ the localized structure shrinks, until the portion of the hexagonal structure is reduced and eventually coincides with a single intensity peak, when the pulse width is (approximately) smaller than twice the critical wavelength associated with the modulational instability. This single peak corresponds to a cavity soliton, which is therefore a sort of elementary localized structure, in which the single peak sits on the pedestal of the homogeneous stationary state. Figure 30.4 shows two typical examples for the section of the cavity soliton. When the nonlinearity is dominated by absorption/gain the cavity soliton has a typical "solitonic" structure (Fig. 30.4(a)), wheareas when the refractive nonlinearity dominates the cavity soliton develops rings around its center (Fig. 30.4(b)).

In the case of the absorptive model (27.14), the laser pulse locally creates a bleached area, which allows the local intensity to rise. Thus, the cavity solitons are in this case "optical bullet holes" [414] generated by the laser pulse.

In contrast with propagative solitons (in the nonlinear Schrödinger equation, see Section 7.3) no exact analytical expression for the cavity soliton solution is available, but powerful methods have been devised to calculate them with arbitrary accuracy without solving the time-evolution equations [414], and to perform their linear-stability analysis [410,415]. Since a cavity soliton is a structure related to the modulational instability, the size of cavity solitons is on the order of the inverse of the critical wave vector, although it is not strictly a subelement of the hexagonal lattice, due to the boundary conditions imposing its connection to the homogeneous background, causing a certain deformation in its phase and intensity profile with respect to the single peak of the hexagonal lattice [414]. Thus, the cavity-soliton size scales as $\sqrt{\lambda\mathcal{L}/T}$ (compare this with the parameter $\bar{a}$ in Eq. (27.13), which has the dimensions of an area since ak_c^2 is dimensionless), with the same spatial scale as the spatial patterns. Localized structures in parametric down-conversion are discussed e.g. in [416,417].

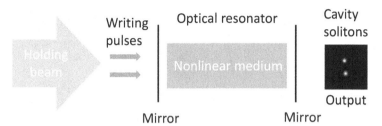

Figure 30.5 A coherent, stationary, quasi-plane-wave holding field drives an optical cavity containing a nonlinear medium. The injection of narrow laser pulses creates persistent localized intensity peaks in the output (cavity solitons).

30.2 Generation and control of cavity solitons

30.2.1 Generation and interaction of cavity solitons

Figure 30.5 illustrates the standard procedure employed to generate cavity solitons by means of optical resonators containing nonlinear materials [408]. The energy is provided to the system by a broad-area, coherent and stationary holding beam that is injected into the cavity and, in the case of semiconductor amplifiers, also by an electrical current. The intensity of the holding beam corresponds to the regime of coexistence between the patterned state and the homogeneous state, and the system lies initially in the homogeneous state. In order to create a cavity soliton, one adds a short and narrow "writing" pulse to the holding beam, of the form described by Eq. (30.1), aimed at a certain transverse point (x, y). Provided that the pulse is in phase with the holding beam $\tilde{F}_\mathrm{I}$, its intensity locally increases the field and finally provides the system with the energy necessary to access the patterned branch. In the output intensity profile, one readily observes the formation of a bright peak. Since the cavity soliton is a subset of a stable solution, when the pulse dies off, the peak indefinitely persists where it has been excited, as if the pulse drove along the resonator axis a channel that became self-sustained by the mirrors' feedback action. Therefore the cavity soliton remains in the memory of the system. It is, of course, straightforward to shoot more pulses at different locations of the transverse section of the system and turn on as many cavity solitons as we like, provided that the resulting distances between them will remain above the interaction range which we shall discuss later.

In order to switch a single cavity soliton off, with no consequences for the other cavity solitons, it suffices to locally bring the system to a regime where the patterned state can no longer persist. This is achieved by shooting, at the location where a cavity soliton lies, an "erasing" pulse similar to the "writing" one but with opposite phase with respect to the holding beam. The local intensity thus decreases and the system precipitates to the homogeneous state, thus restoring homogeneity where the cavity soliton was [412]. Of course, there exists a minimum value of the duration t_p of the erasing pulse (see Eq. (30.1)) which causes the erasure for a given value of the pulse amplitude ξ and, as a rule of thumb, one observes that the product ξt_p is approximately constant. The advantage of this procedure is its robustness relative to the choices of t_p and the pulse width σ and also with respect to the phase φ: There exists a broad set of values for φ for which the cancelation takes place [412]. Furthermore, some error in the superposition of the erasing pulse with respect to the cavity soliton can be accommodated by the system.

The question of the maximum density of independent cavity solitons that can be generated, and what happens when the solitons interact, now arises. It is thus essential to evaluate the minimum distance from an existing cavity soliton at which a second one can be created, without interacting [408, 412].

In the light of the previous comments on cavity-soliton characteristics, it is intuitive to expect that the two subelements of the global lattice will interact when their distance will be on the order of or smaller than the transverse wavelength of the hexagonal lattice. Though this is a simplification, the idea is fundamentally substantiated by results.

Simulations in which the second cavity soliton is excited at locations progressively closer to the first one, by using the same form (30.1) with $\varphi = 0$, indicate that two critical distances can be defined: D_c and d_c (with $D_c > d_c$). Two cavity solitons do not interact provided that their distance is larger than a minimal distance D_c. Below this distance, there is the possibility that the two cavity solitons spontaneously reach an equilibrium distance determined by the locking of their tails. Actually, there can be more than one equilibrium distance, and there is the possibility of the formation of clusters of solitons [418]. Finally, when the distance is smaller than a critical minimum d_c, the two cavity solitons fuse into a single cavity soliton that has the same characteristics (width, height, etc.) as the original cavity solitons. Solitons with properties similar to those of cavity solitons can be generated in single-feedback-mirror systems (see [419] and references cited therein).

30.2.2 Motion of cavity solitons

Under conditions of translational invariance in the transverse plane, cavity solitons can move in the presence of noise, performing a slow random walk, which, of course, is undesired for applications. Another cause for the motion of cavity solitons is the presence of amplitude or phase gradients in the holding beam. Under the influence of an amplitude gradient, cavity solitons tend to move to the nearest local maximum of intensity with a velocity proportional to the gradient, at least for a constant gradient [406, 415]. For example, if the holding beam has a Gaussian profile, all of the solitons move to the top and merge so that, finally, only one soliton remains. This effect is also undesired for applications. On the other hand, phase modulation in the holding beam can be very useful for applications and, in particular, can be utilized to neutralize the negative effects arising from the amplitude gradients and from noise.

In general, let us assume that the driving field F_I in the field equation (see e.g. Eqs. (27.10), (27.14) and (27.15)) has the form

$$F_I(x, y) = F_{I,0} e^{i\Phi(x,y)}, \tag{30.2}$$

where $\Phi(x, y)$ is an arbitrary phase modulation. If we now set $F(x, y, t) = F'(x, y, t)e^{i\Phi(x,y)}$ we obtain for $F'(x, y, t)$ an equation with a convective derivative [414],

$$\left(\frac{\partial}{\partial t} + \mathbf{v} \cdot \nabla\right) F'(x, y, t) = -\kappa\left[(1 + i\theta)F'(x, y, t) - F_{I,0} - i\bar{a}\,\nabla_\perp^2 F'(x, y, t) + \cdots\right], \tag{30.3}$$

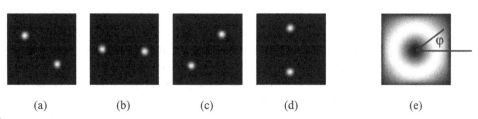

Figure 30.6 In parts (a)–(d) two cavity solitons perform a uniform rotatory motion along the crater of a doughnut-shaped holding beam (e), under the action of the angular gradient (numerical simulation). Reprinted from S. Barland *et al. Europhysics News* **34**, 136 (2003) with kind permission from EDP Sciences.

with the velocity **v** given by

$$\mathbf{v} = 2\kappa\bar{a}\,\nabla\Phi. \tag{30.4}$$

In writing Eq. (30.4) we have neglected a term with second-order derivatives of Φ, assuming that the phase modulation is much weaker than the spatial variation of F. Hence any cavity soliton drifts in the transverse plane with a velocity proportional to the gradient of the phase profile, and the local maxima of the phase profile represent equilibrium positions for the soliton, where the motion stops.

By introducing phase profiles in the holding beam it is possible to set a cavity soliton in controlled motion, and this opportunity confers on cavity solitons a special property of plasticity. Typical velocities of the solitons are on the order of 1 μm/ns = 1 km/s.

Figure 30.6 shows the case of a holding beam that has the shape of a doughnut mode (Fig. 30.6(e), see also the end of Section 28.2). Two cavity solitons exhibit a circular motion along the ring-shaped region where the intensity of the doughnut mode (which is shaped like the crater of a volcano) is maximum. The motion is induced by the fact that the phase of the doughnut mode varies as $\mathrm{e}^{\pm i\varphi}$, where φ is the angle (see Fig. 30.6(e)) which creates a constant phase gradient along the crater, and the direction of motion is determined by the $\pm$ sign.

If, instead, one tailors the holding beam in such a way that it displays a periodic phase modulation (Fig. 30.7(b)), this constitutes the immaterial support for an array of cavity-soliton pixels (Figs. 30.7(a)–(d)), which can be individually set on and off by shining laser pulses. By varying the phase landscape, one can reconfigure the array, and by suitably introducing further gradients cavity solitons can be brought into controlled interactions. On the other hand, Figs. 30.7(e) and (f) show the experimental realization of arrays of cavity solitons in semiconductor microcavities [420], see Section 30.3. It is interesting to compare such arrays of cavity solitons with etched arrays of micropixels in semiconductors [421]; in the latter case the pixels are rigid and the array is not reconfigurable.

The possibility of controlling the motion of cavity solitons can be exploited for practical applications. In addition to reconfigurable optical memories, examples of possible applications are serial-to-parallel converters, signal amplification, realization of cellular automata, pattern recognition and optical tweezers.

From an experimental viewpoint, the results of [419] in sodium vapor with a single feedback mirror show some of the main functionalities of cavity solitons.

It is interesting to mention that a further cause of motion for cavity solitons is thermal effects [422, 423]. This circumstance emerges if one includes temperature in the models

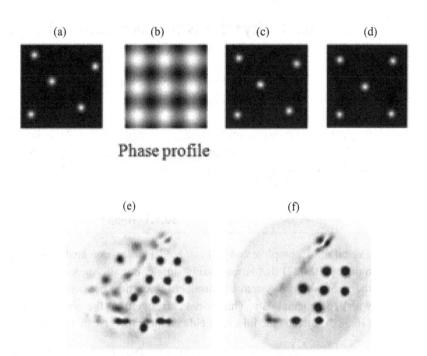

Phase profile

Figure 30.7 Frame (a) shows the initial position of five cavity solitons. Frame (b) shows the phase profile of the holding beam, where the white spots correspond to the local maxima. The cavity solitons move towards the nearest local maximum of the phase profile, (c) and (d). Parts (e) and (f) show experimental realizations of arrays of cavity solitons. Reprinted with permission from [420]. Copyright [2006], Air Publishing LLC.

which describe optical pattern formation. In the case of semiconductor materials, there is a thermal shift of the cavity frequency that introduces a linear dependence on temperature of the cavity detuning θ [422]. By adding to the time-evolution equations, which will be discussed in the following section, an equation that governs the time evolution of the temperature, one obtains a closed set of dynamical equations. The analysis of such a model leads to the prediction [422, 423] of an instability that produces an intrinsic motion of cavity solitons due to temperature, under appropriate parametric conditions. In the presence of translational and rotational invariance, this motion occurs along a straight line in a random direction, in the absence of any gradient in the holding beam. In other words, cavity solitons move spontaneously in a self-propelled fashion. This spontaneous drift, which occurs on the slow time scale (microseconds) which characterizes thermal effects, can be controlled, however, by using holding beams with phase or amplitude modulations. A peculiar phenomenon has been numerically shown [424] in the case of a two-dimensional phase modulation in the holding beam of the kind of Fig. 30.7(b): For suitable values of the modulation amplitude, the combination of the thermally induced motion and the attracting action of the maxima of the phase profile gives rise to a random walk in which the cavity soliton visits in sequence various maxima, moving from one maximum to an adjacent maximum in a random manner, on a slow time scale.

30.3 Cavity solitons in semiconductor microresonators

In this section we focus on the especially relevant case of cavity solitons in semiconductor devices. To be precise, we consider a broad-area VCSEL capable of accommodating a significant number of cavity solitons in its section, with a value of the injected current J larger than the transparency value J_0, so that it amplifies the radiation, but is also slightly (10%–20%) below the threshold value for the free-running laser. In addition, there is a coherent field that is injected into the cavity, so that the device corresponds to an amplifier with injected signal.

30.3.1 Theory

Even if more complete and realistic models have been analyzed [425], we consider here a model [426, 427] that is reasonably simple, but allows one to grasp the basic features of the physics of the system and produces theoretical predictions that are in good agreement with experimental data. This model corresponds to the generalization of Eqs. (16.57) and (16.58) to include the injected field and the transverse effects, and reads

$$\frac{\partial \mathcal{F}(\bar{x}, \bar{y}, t'')}{\partial t''} = \mathcal{F}_\mathrm{I} - \left[1 + i\theta - (1 - i\alpha)\tilde{D}(\bar{x}, \bar{y}, t'')\right]\mathcal{F}(\bar{x}, \bar{y}, t'') + i\,\bar{\nabla}_\perp^2 \mathcal{F}(\bar{x}, \bar{y}, t''),$$

(30.5)

$$\frac{\partial \tilde{D}(\bar{x}, \bar{y}, t'')}{\partial t} = \gamma' \left[\mu\chi(\bar{x}, \bar{y}) - \tilde{D}(\bar{x}, \bar{y}, t'')(1 + |\mathcal{F}(\bar{x}, \bar{y}, t'')|^2) + d\,\bar{\nabla}_\perp^2 \tilde{D}(\bar{x}, \bar{y}, t'')\right],$$

(30.6)

where $\mathcal{F}_\mathrm{I}$ denotes the injected field (holding beam plus addressing beam, see Eq. (30.1)) and θ is the usual cavity detuning parameter $(\omega_\mathrm{c} - \omega_0)/\kappa$, with ω_0 being the frequency of the input field and ω_c the nearest cavity frequency. We use the normalized time $t'' = \kappa t$ and

$$\gamma' = \gamma_\mathrm{nr}/\kappa,$$

(30.7)

where γ_nr is the nonradiative recombination rate of the carriers. The scaled transverse variables $\bar{x}$ and $\bar{y}$ correspond to x and y divided by $\sqrt{\bar{a}}$, respectively, where the constant $\bar{a}$ multiplies the transverse Laplacian in Eqs. (27.10) etc., and $\bar{\nabla}_\perp^2 = \partial^2/\partial\bar{x}^2 + \partial^2/\partial\bar{y}^2$. The dimensionless parameter d is defined as

$$d = l_\mathrm{d}^2/\bar{a},$$

(30.8)

where l_d is the carrier diffusion length. Remember that the variable $\tilde{D}$ is linked to the carrier density N by Eq. (16.54). The function $\chi(\bar{x}, \bar{y})$ describes the shape of the pumped region.

The transverse effects arise from the terms with the transverse Laplacian, i.e. from light diffraction (Eq. (30.5)) and carrier diffusion (Eq. (30.6)). The carrier diffusion term has a real coefficient, whereas the diffraction term has an imaginary coefficient; typically the parameter d is small, which reduces the importance of the diffusion term.

If one drops the terms with the transverse Laplacian and sets $\chi = 1$, Eqs. (30.5) and (30.6) reduce to Eqs. (16.57) and (16.58) if we consider the case of the free-running laser,

Table 30.1 Typical parameters of a VCSEL		
Wavelength in vacuum	$\lambda_0 = 2\pi c/\omega_0$	850 nm
Background refraction index	n_B	3.5
Wavelength in medium	$\lambda = \lambda_0/n_B$	242.86 nm
Mirror transmissivity	T	4×10^{-3}
Cavity length	L	2 µm
Cavity damping rate	$\kappa = cT/(2n_BL)$	$8.57 \times 10^{10}\text{s}^{-1}$
Time unit	κ^{-1}	11.7 ps
Carrier nonradiative damping time	γ_{nr}^{-1}	1 ns
Scaled carrier damping rate	$\gamma = \gamma_{nr}/\kappa$	1.17×10^{-2}
Diffraction coefficient	$\bar{a}$	19.32 µm^2
Spatial unit	$\sqrt{\bar{a}}$	4.41 µm
Diffusion length	l_d	1 µm
Diffusion coefficient	$d = l_d^2/\bar{a}$	0.052

in which $\mathcal{F}_I = 0$ and $\theta = -\alpha$. As noted in Section 16.5, Eqs. (16.55) and (16.56) (and therefore also Eqs. (16.57) and (16.58)) have the same form as the dynamical equations for class-B two-level lasers. In the case of two-level free-running lasers $\theta = -\Delta$ (see Eq. (12.57)), and the parameter α for the semiconductor corresponds to Δ for the two-level laser.

Indeed, it is easy to check that, within the adiabatic elimination of the atomic polarization, the two-level model (27.10)–(27.12) is equivalent to the set of equations (30.5) and (30.6) with $d = 0$. To verify this, one sets $\partial P/\partial t = 0$, $\tilde{F} = F/\sqrt{1+\Delta^2}$, $\tilde{F}_I = F_I/\sqrt{1+\Delta^2}$, $\mu = A/(1+\Delta^2)$, $\tilde{D} = \mu D$, introduces the scaled coordinates (27.19) and arrives at Eqs. (30.5) and (30.6) with $\mathcal{F}$ replaced by $\tilde{F}$, α replaced by Δ, γ' defined as $\gamma_\parallel/\kappa$ and $d = 0$.

Note that Eqs. (30.5) and (30.6) describe not only the active case $\mu > 0$, but also the passive configuration $\mu < 0$, in the latter case $\tilde{D}$ is negative, in contrast with the variable D of the two-level model, which is always positive. In the passive case, the model (30.5)–(30.6) describes fairly well the configuration of an excitonic nonlinearity, which is similar to the two-level Lorenzian resonance [160].

Example values for the parameters of the system in a GaAs VCSEL cavity are shown in Table 30.1. A broad-area VCSEL is a disk with a diameter on the order of 200 µm. A typical order of magnitude for the transparency carrier density is $N_0 = 10^{18}$ cm^{-3}.

Let us now consider the homogeneous stationary solutions of Eqs. (30.5) and (30.6) when $\mathcal{F}_I$ is independent of $\tilde{x}$ and $\tilde{y}$, and $\chi = 1$ everywhere. The stationary value of $\tilde{D}$ is linked to the stationary value of $|\mathcal{F}_s|^2$ by the relation

$$\tilde{D}_s = \frac{\mu}{1+|\mathcal{F}_s|^2}, \tag{30.9}$$

while the stationary normalized intensity $|\mathcal{F}_s|^2$ is linked to the normalized input intensity $|\mathcal{F}_I|^2$ by the stationary equation

$$|\mathcal{F}_I|^2 = |\mathcal{F}_s|^2 \left[\left(1 - \frac{\mu}{1+|\mathcal{F}_s|^2} \right)^2 + \left(\theta + \frac{\mu\alpha}{1+|\mathcal{F}_s|^2} \right)^2 \right], \tag{30.10}$$

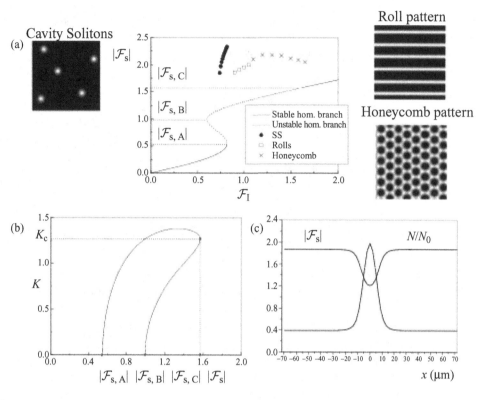

Figure 30.8 (a) The stationary curve of intracavity field vs. input field for the homogeneous stationary solutions. In appropriate ranges of the input field one has the formation of a honeycomb pattern (crosses), a roll pattern (squares) or cavity solitons (circles; the position of the circle indicates the value of the peak height of the soliton). The three structures are shown. The parameters are $\mu = 0.9, \theta = -2$ and $\alpha = 5$. (b) The boundary of the stationary modulational instability for $d = 0.052$. Here $k_\perp = K$. (c) The profile of the field amplitude and of the carrier density (normalized with respect to its transparency value N_0) for the cavity soliton obtained with $\mathcal{F}_I = 0.75, \bar{a} = 19.32\ \mu m^2$ and $J/J_0 - 1 = 1$. Reprinted figure with permission from [427]. Copyright (1998) by the American Physical Society.

In the case of the free-running laser ($|\mathcal{F}_I|^2 = 0$) the nontrivial stationary solution is $|\mathcal{F}_s|^2 = \mu - 1$, so that the threshold corresponds to $\mu = 1$, and $\theta = -\alpha$.

Figure 30.8(a) shows the stationary curve of transmitted vs. incident intensity for a value of the parameter μ such that the free-running laser is operating slightly below threshold. The stationary curve is S-shaped and, since the free-running laser is below threshold, in the plane-wave theory (i.e. neglecting the transverse Laplacian) all the parts of the curve with positive slope are stable, as in the case of Fig. 13.2(a)). However, if we include the transverse Laplacian we can find a stationary instability that leads to the formation of stationary patterns.

A straightforward way to calculate the boundary of the stationary spatial instability in the plane of the variables $(|\mathcal{F}_s|, k_\perp)^1$ when all the other parameters are fixed is the following.

---

[1] As in Chapter 20, we use $|\mathcal{F}_s|$ as a parameter, instead of $|\mathcal{F}_I|$.

For reasons that will become clear in a moment, let us write Eq. (30.6) with $\chi = 1$ as follows:

$$\frac{\partial \tilde{D}(\bar{x}, \bar{y}, t'')}{\partial t''} = \gamma' \left[\mu - \tilde{D}(\bar{x}, \bar{y}, t'')(\eta + |\mathcal{F}(\bar{x}, \bar{y}, t'')|^2) + d \, \bar{\nabla}_\perp^2 \tilde{D}(\bar{x}, \bar{y}, t'') \right], \quad (30.11)$$

where η is a fictitious parameter, which, in our case, has value $\eta = 1$. Correspondingly, the stationary equation (30.10) becomes

$$|\mathcal{F}_I|^2 = |\mathcal{F}_s|^2 \left[\left(1 - \frac{\mu}{\eta + |\mathcal{F}_s|^2} \right)^2 + \left(\theta + \frac{\mu\alpha}{\eta + |\mathcal{F}_s|^2} \right)^2 \right]. \quad (30.12)$$

If we linearize Eqs. (30.5) and (30.11) around a homogeneous stationary solution and introduce the ansätze (27.21) and (27.22), we arrive at the characteristic equation (27.23). The boundary of the stationary instability ($\omega = 0$) in the plane of the variables ($|\mathcal{F}_s|, k_\perp$), when all the other parameters are fixed, corresponds to the equation $c_0(|\mathcal{F}_s|, k_\perp) = 0$, as discussed in Section 27.2.

In the final part of Section 27.2, we argued that the effect of the term with the transverse Laplacian in the linearized field equation is to replace the cavity detuning parameter θ by $\theta + \bar{a}k_\perp^2$. This happens in Eq. (30.5) (with $\bar{a}$ replaced by 1). Similarly, in Eq. (30.11), once linearized, the effect of the term with the transverse Laplacian is to replace the parameter η by $\eta + dk_\perp^2$.

Another key point is that in the plane-wave theory, as we show at the end of Appendix C, c_0 is proportional to the derivative $d|\mathcal{F}_I|^2/d|\mathcal{F}_s|^2$, which, using Eq. (30.12), is given by

$$\frac{d|\mathcal{F}_I|^2}{d|\mathcal{F}_s|^2} = 1 + \theta^2 + \frac{2\eta\mu(\alpha\theta - 1)}{(\eta + |\mathcal{F}_s|^2)^2} + \frac{\mu^2(1 + \alpha^2)\left(|\mathcal{F}_s|^2 - \eta\right)}{(\eta + |\mathcal{F}_s|^2)^3}. \quad (30.13)$$

Therefore the instability boundary is obtained by equating to 0 the r.h.s. of Eq. (30.13) with θ replaced by $\theta + k_\perp^2$ and η replaced by $1 + dk_\perp^2$. In this way, one obtains the following equation for $k_\perp^2$

$$\left[1 + \left(\theta + k_\perp^2 \right)^2 \right] \left(1 + dk_\perp^2 + |\mathcal{F}_s|^2 \right)^3$$
$$+ 2\mu \left[\alpha \left(\theta + k_\perp^2 \right) - 1 \right] \left(1 + dk_\perp^2 \right) \left(1 + dk_\perp^2 + |\mathcal{F}_s|^2 \right)$$
$$+ \mu^2 \left(1 + \alpha^2 \right) \left(|\mathcal{F}_s|^2 - 1 - dk_\perp^2 \right) = 0, \quad (30.14)$$

the solutions of which provide the instability boundary in the plane $((|\mathcal{F}_s|, k_\perp))$. Since the coefficient d is usually small, the solutions are given, to a good approximation, by those obtained in the limit $d = 0$, which read

$$k_\perp^2 = -\theta - \frac{\mu}{(\eta + |\mathcal{F}_s|^2)^2} \pm \frac{\sqrt{\mu^2\alpha^2|\mathcal{F}_s|^4 - \left[\mu - \left(\eta + |\mathcal{F}_s|^2 \right)^2 \right]^2}}{(\eta + |\mathcal{F}_s|^2)^2}. \quad (30.15)$$

Figure 30.8(b) shows this boundary corresponding to Fig. 30.8(a) for $d = 0.052$. In Fig. 30.8(a) one finds also the peak height of the patterns (honeycomb, roll, cavity soliton) which emerge from the instability. On the other hand, Fig. 30.8(c) shows the profile of the cavity soliton for $|\mathcal{F}_I| = 0.75$, and the corresponding profile of the carrier density normalized with respect to the transparency value, i.e. N/N_0. The cavity-soliton profile has a width

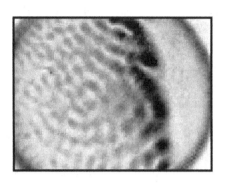 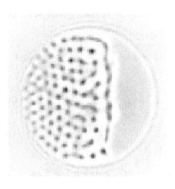

(a)   (b)

Figure 30.9   The time-averaged intensity profile of a VCSEL driven by a coherent holding beam: (a) experiment and (b) numerical simulation. A patterned region appears on the left, a homogeneous domain on the right. Reprinted from S. Barland *et al. Europhysics News* **34**, 136 (2003) with kind permission from EDP Sciences.

that is typically on the order of 10 μm. The carrier density exhibits a dip corresponding to the cavity-soliton peak, and is obtained from Eq. (16.54) with $\mu = 0.9$ and $J/J_0 - 1 = 1$. An important point is that, since the cavity solitons arise from the interaction of light with the semiconductor, it is a composite structure that appears not only in the electric field, but also in the carrier density, as shown in Fig. 30.8(c). This composite structure of cavity solitons occurs not only in semiconductors, but in any system that generates cavity solitons. Usually this composite structure is overlooked because one emphasizes the field aspects only.

The case of cavity solitons in driven VCSELs for values of parameters such that the free-running laser is operating above threshold is discussed theoretically and experimentally in [428].

30.3.2 Experiments

The goal of the experimental demonstration of cavity solitons in semiconductor microresonators was reached by the group of Tredicce, Barland and collaborators [429] using broad-area VCSELs (bottom-emitter, diameter 150 μm), operating slightly below threshold, constructed at the University of Ulm. The VCSEL is injected by a coherent field generated by a high-power edge-emitting laser, whose wavelength is tunable in the range 960–980 nm. A typical time-averaged transverse intensity profile of the VCSEL driven by the holding beam is shown in Fig. 30.9(a). One observes a homogeneous area on the right-hand side of the sample, and a patterned region on the left. The numerical simulation shown in Fig. 30.9(b) indicates that this behavior can be explained by taking into account the gradient of cavity length along the sample due to the standard epitaxial growth technique. This corresponds to a linear variation of the cavity detuning parameter θ in the horizontal direction in Fig. 30.9. The numerical simulations were done using Eqs. (30.5) and (30.6) with a function $\chi(\bar{x}, \bar{y})$ that simulates the spatial profile of the injected current.

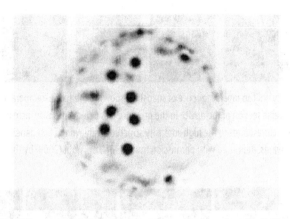

Figure 30.10   The intensity distribution of the output field showing the presence of seven cavity solitons. The power of the holding beam is 25 mW, the injected current is 290 mA. Reprinted from S. Barland *et al. Europhysics News* **34**, 136 (2003) with kind permission from EDP Sciences.

As suggested by the numerical simulations, the most appropriate region to generate cavity solitons lies immediately to the right of the line (orthogonal to the gradient) which separates the homogeneous area and the patterned area. Starting with no spot, the writing beam (power about 50 μW, compared with 8 mW for the holding beam) is capable of generating a high-intensity spot when it is in phase with the holding beam, as predicted by theory. When the writing beam is removed, the bright spot remains on indefinitely. After writing one spot, the writing beam is displaced and a new pulse is injected, causing a second cavity soliton to turn on and persist after the writing beam is extinguished. The two spots are subsequently erased by aiming the writing beam at the existing cavity solitons and changing its phase by π.

Figure 30.10 exhibits an example with seven cavity solitons. This was obtained thanks to the introduction of a misalignment of the holding beam parallel to the frequency gradient: The spatial region where the parameter values allow the existence of stable cavity solitons is consequently enlarged. The solitons in Fig. 30.10 can be controlled independently; however, due to technical reasons concerning the alignment precision of the writing beam, the erasing procedure works with only four of the solitons.

The presence of the cavity length gradient implies that a cavity soliton is subjected to a force that tends to shift it towards the patterned region. The fact that the cavity solitons instead sit stably at given positions on the sample is interpreted as an effect of small local defects in the sample that are strong enough to compensate for the overall gradient and pin the cavity soliton position. The picture is similar to that of a rock (the cavity soliton) rolling downhill (the cavity length gradient) and getting stopped by small potholes and bumps (the sample roughness). This interpretation is confirmed by numerical simulations that include the roughness of the layers of the Bragg reflectors.

A defect is also capable of generating a cavity soliton spontaneously. This circumstance gives rise to an interesting phenomenon when the holding beam presents a phase gradient. The gradient tends to set the cavity soliton in motion, and therefore to detach it from the defect, which happens provided that the gradient is not too weak. While the cavity

Figure 30.11 A cavity-soliton tap. A sequence of snapshots showing the spatio-temporal dynamics of drifting solitons in the transverse section of the device in the presence of a defect. Intensity increases from black to white. The position of the defect corresponds to the high-intensity structure visible in the first panel, and the phase gradient is directed rightwards. Reprinted with permission from [430]. Copyright (2009) by the American Physical Society.

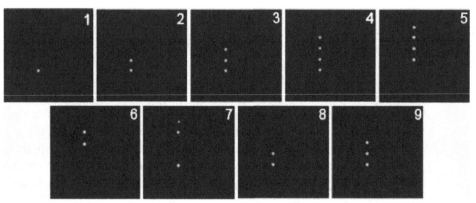

Figure 30.12 A sequence of snapshots illustrating different stages of the continuous creation and inhibition of cavity solitons. The defect position corresponds to the soliton in the first frame, and the phase gradient is directed upwards. The injection of the writing beam starts just before frame 5 and stops after frame 6. From left to right, the frames were taken every 12 ns. Reprinted with permission from [430]. Copyright (2009) by the American Physical Society.

soliton moves away from the defect, after a while the defect spontaneously generates another soliton, which is then in turn detached and moves away. In this manner, the system generates spontaneously a sequence of cavity solitons that travel, in an ordered line and at regular intervals of time, in the direction of the phase gradient. A numerical simulation of this effect, which has been also observed experimentally, is shown in Fig. 30.11 [430]. We have called this phenomenon a *cavity-soliton tap*, because it is similar to the case of a tap that is not completely closed, so that a droplet of water forms and hangs from it, until the gravity force makes it fall, then another droplet forms and subsequently falls, and so on again and again.

In this case, if one injects the writing beam into the defect, the flux of cavity solitons stops. Therefore, by repeatedly injecting the writing beam into the defect, one can set in motion a certain number of cavity solitons, then stop the flux, set some other solitons into motion at will, then stop, and so on. In such a controlled way one can generate a sequence of 1s and 0s that travel in the transverse plane with a velocity on the order of one micrometer per nanosecond, i.e. of one kilometer per second. This principle is illustrated in Fig. 30.12.

Cavity solitons localized in all three spatial dimensions are called cavity light bullets, and have been theoretically analyzed e.g. in [431–433].

(a)   (b)

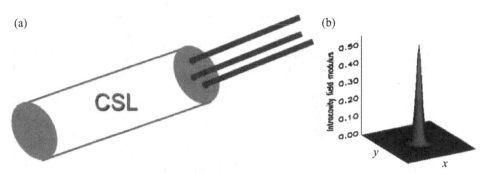

Figure 30.13 (a) A cavity-soliton laser (CSL) emits a set of narrow beams (cavity solitons), the number and position of which can be controlled. Part (b) shows the shape of a cavity soliton in a cavity-soliton laser.

30.4 The "cavity-soliton laser"

A remarkable simplification in the configuration of systems capable of generating cavity solitons (see Fig. 30.5) is the realization of a *cavity-soliton laser*, a device capable of emitting cavity solitons even in the absence of an external holding beam, so that the cavity solitons emitted can be seen as a set of self-assembled microlasers (see Fig. 30.13). In this case the contrast between the cavity soliton and the background is maximum because the background consists in pure spontaneous-emission noise. Three types of cavity-soliton laser have recently been realized experimentally, and have a VCSEL as the basic element: a VCSEL with frequency-selective feedback [434], two coupled VCSELs in a face-to-face configuration [435], and a monolithic VCSEL with a saturable absorber integrated in the cavity [436].

The possibility of realizing a cavity-soliton laser using a laser with a saturable absorber was first theoretically predicted by Rosanov and collaborators [437, 438]. Subsequently a model adequate to describe the behavior of a monolithic cavity-soliton laser of the kind realized in [436] was introduced [439]. The set of equations which is used generalizes the set (30.5) and (30.6) to include a third equation that governs the time evolution of the absorber. A term that describes the radiative carrier recombination which is quadratic in the carrier density N and therefore in $\tilde{D}$ (see Eq. (16.54)) can be included in the material equations [440, 441]. Carrier diffusion is neglected in the model. The homogeneous stationary solution has the typical form of an S-shaped curve. The linear-stability analysis of the homogeneous stationary solution shows that the nontrivial homogeneous stationary solution is unstable everywhere; Fig. 30.14 shows the height of the cavity-soliton peak as a function of the pump parameter μ.

In the case of cavity solitons generated in systems driven by a holding beam, cavity solitons can be switched on and off (i.e. written and erased) by using a coherent writing beam, in phase or in opposition of phase with respect to the coherent holding beam as described in Section 30.2.1. This is not possible in the case of a cavity-soliton laser, because

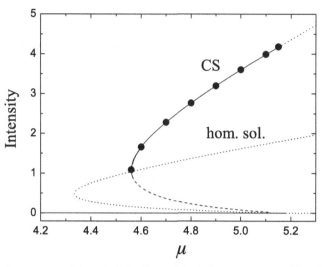

Figure 30.14 The intensity of the homogeneous stationary solution (hom. sol.) and of the cavity-soliton (CS) peaks as a function of the pump parameter μ. The only stable solutions are the laser-off solution up to laser threshold and the cavity solitons in the interval represented by the solid line. Reprinted figure with permission from H. Vahed *et al. Phys. Rev. A* **84**, 063814 (2011). Copyright (2011) by the American Physical Society.

of the absence of a holding beam. It is possible, however, to switch cavity solitons on and off by using incoherent or semi-coherent procedures, as described in details in [439, 442].

Another difference from the case of cavity solitons in driven systems is that cavity solitons cannot be set in motion by introducing phase or amplitude gradients in the holding beam. This is possible, however, by introducing gradients in the injected current J, i.e. in the pump parameter μ. In addition, it is very remarkable that, under appropriate parametric conditions, solitons in cavity-soliton lasers with saturable absorbers become self-propelled, similarly to the case of the spontaneous motion in the absence of any gradient induced by thermal effects which was discussed in Section 30.2.2. The motion is caused by a dynamical instability and, in the absence of defects in the material, is along a straight line in a random direction and with constant velocity [443, 444].

The trajectory of the self-propelled soliton bends in the presence of pump boundaries, as shown by numerical simulations, thus, if we consider a square pump profile, we realize a cavity-soliton billiard. In [445] we show that the soliton is reflected when it impinges the boundary and the sum of incidence and reflection angles is 90°, which is quite different from the standard reflection law. In this regime, the cavity soliton may cover two trajectories, one clockwise and the other counterclockwise, as shown in Fig. 30.15. The resulting scenario is similar to that exhibited by walking droplets in a vibrated liquid bath [446, 447].

On the other hand, in the presence of defects, which act as scatterers for the soliton, the trajectory may became open and may cover ergodically the entire available square section [445]. If one has just one scatterer located at an intersection of the two closed orbits, it is not enough to obtain an open trajectory, because the trajectory leaves the forbidden square orbits but ends on another stable orbit that lies along one diagonal of the square. An open orbit can be realized only by setting a second scatterer at the center of the square, in

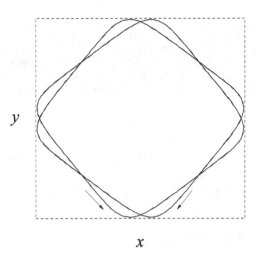

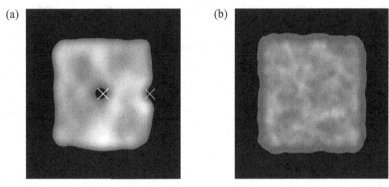

such a way that the two diagonals are also forbidden. The averaged output in the presence of two scatterers is shown in Fig. 30.16(a). Another method to make closed orbits unstable consists in switching on at least three solitons in the device. The average intensity profile over long times produced by the three moving solitons is substantially uniform as shown in Fig. 30.16(b).

In the case of Fig. 30.16(a), the existence of defects (scatterers) in the material is revealed by the average motion, such that the device works as a sort of *soliton-force microscope* that is based on the spontaneous motion of solitons. A soliton-force microscope has been realized experimentally not in a cavity-soliton laser, but in a driven system. The soliton explores the section of the device under the action of gradients in the holding-beam profile, revealing the presence of defects [448]. The use of a cavity-soliton laser, with the result shown in Fig. 30.16(a), would correspond to a simpler and more compact soliton-force microscope.

Appendix A The Routh–Hurwitz stability criterion

We describe the criterion without proof. With reference to the characteristic equation (18.8) of degree n, a stationary solution of the system of nonlinear equations (18.1) is stable if and only if the following conditions are satisfied:

(i) $c_0/c_n > 0, c_1/c_n > 0, \ldots, c_{n-1}/c_n > 0$; and

(ii) the determinants

$$D_1 = c_{n-1}, \tag{A.1}$$

$$D_2 = \begin{vmatrix} c_{n-1} & c_n \\ c_{n-3} & c_{n-2} \end{vmatrix}, \tag{A.2}$$

$$D_3 = \begin{vmatrix} c_{n-1} & c_n & 0 \\ c_{n-3} & c_{n-2} & c_{n-1} \\ c_{n-5} & c_{n-4} & c_{n-3} \end{vmatrix}, \tag{A.3}$$

$$\vdots$$

$$D_{n-1} = \begin{vmatrix} c_{n-1} & c_n & \cdots & 0 & 0 \\ c_{n-3} & c_{n-2} & \cdots & \cdots & \cdots \\ \vdots & \vdots & \ddots & \vdots & \vdots \\ \cdots & \cdots & \cdots & c_2 & c_3 \\ 0 & 0 & \cdots & c_0 & c_1 \end{vmatrix}, \tag{A.4}$$

are positive. The coefficients c_k must be replaced by 0 when $k < 0$.

Appendix B Calculation of the oscillatory instability boundary

In this appendix we illustrate the method that allows one to obtain the equation for the boundary of the instability domain for oscillatory instabilities, as well as the frequency of the oscillation at the boundary.

We start from Eqs. (18.10) for the two polynomials $P_1(\omega, \{\beta\})$ and $P_2(\omega, \{\beta\})$. If the order of the characteristic equation is n, one polynomial contains the powers $n - 2j$ and the other the powers $n - 1 - 2j$ of ω with $j = 1, 2 \ldots$. Since we are looking for oscillatory instabilities with $\omega \neq 0$, we can divide by ω the one of the two polynomials containing only odd powers. In this way we obtain two equations containing only even powers of ω. To be precise, both equations contain all the even powers from 0 to $n - 1$ if n is odd, whereas one equation contains all the even powers from 0 to n and the other all the even powers from 0 to $n - 2$ if n is even.

By writing appropriate linear combinations of the two resulting equations, we can progressively reduce the order until we arrive at two linear equations in ω^2. They provide two different expressions for ω^2, which must be equated, giving the equation for the instability boundary in the form $h_n(\{\beta\}) = 0$, where h_n is a function of the coefficients of the polynomials. The equation can be solved by fixing all the parameters $\beta_i, i = 1, \ldots, m$ but one, and the solution is the critical value of that parameter for which the Hopf bifurcation occurs. If the critical value is plugged into one of the two equations for ω^2 we also get the oscillation frequency at the bifurcation point. Let us illustrate the method for characteristic equations of increasing order.

B.1 The cubic case

For $n = 3$ Eqs. (18.10) are

$$c_2\omega^2 - c_0 = 0, \tag{B.1}$$

$$\omega^2 - c_1 = 0, \tag{B.2}$$

which are already two linear equations in ω^2. They yield (assuming $c_2 \neq 0$)

$$\omega^2 = c_1 = \frac{c_0}{c_2} \tag{B.3}$$

for the oscillation frequency and

$$h_3(\{\beta\}) = c_1 c_2 - c_0 = 0 \tag{B.4}$$

for the boundary in the parametric space of oscillatory instability.

B.2 The quartic case

For $n = 4$ the two equations read

$$\omega^4 - c_2\omega^2 + c_0 = 0, \tag{B.5}$$

$$c_3\omega^2 - c_1 = 0. \tag{B.6}$$

We multiply Eq. (B.5) by c_1 and Eq. (B.6) by c_0, sum the two equations and divide by ω^2, obtaining

$$c_1\omega^2 - c_1c_2 + c_0c_3 = 0. \tag{B.7}$$

From this equation and Eq. (B.6), assuming $c_3 \neq 0$ and $c_1 \neq 0$, we get the two expressions for ω^2,

$$\omega^2 = \frac{c_1}{c_3} = \frac{c_1c_2 - c_0c_3}{c_1}, \tag{B.8}$$

and the equation for the boundary is

$$h_4(\{\beta\}) = c_1^2 - c_3(c_1c_2 - c_0c_3) = 0. \tag{B.9}$$

B.3 The quintic case

For $n = 5$ the two equations are

$$c_4\omega^4 - c_2\omega^2 + c_0 = 0, \tag{B.10}$$

$$-\omega^4 + c_3\omega^2 - c_1 = 0. \tag{B.11}$$

We multiply Eq. (B.11) by c_4 and sum, obtaining

$$(c_3c_4 - c_2)\omega^2 = c_1c_4 - c_0. \tag{B.12}$$

Next, we multiply Eq. (B.10) by c_1 and Eq. (B.11) by c_0, sum and divide by ω^2, obtaining

$$(c_1c_4 - c_0)\omega^2 = c_1c_2 - c_0c_3. \tag{B.13}$$

Note that in this way, by starting from two quadratic equations in ω^2, we derived two linear equations in ω^2. Assuming $c_3c_4 \neq c_2$ and $c_1c_4 \neq c_0$, we obtain from Eqs. (B.12) and (B.13) two expressions for ω^2,

$$\omega^2 = \frac{c_1c_4 - c_0}{c_3c_4 - c_2} = \frac{c_1c_2 - c_0c_3}{c_1c_4 - c_0}, \tag{B.14}$$

from which it follows that the equation defining the boundary is

$$h_5(\{\beta\}) = (c_1c_4 - c_0)^2 - (c_3c_4 - c_2)(c_1c_2 - c_0c_3) = 0. \tag{B.15}$$

The procedure can be straightforwardly generalized to any value of n. Of course, the complexity of the final expressions increases very quickly with the order n of the characteristic equation.

Appendix C Coefficients of the characteristic equation (20.20)

The coefficients of Eq. (20.20) are given by

$$c_5^{(n)} = 1, \tag{C.1}$$

$$c_4^{(n)} = 2 + \tilde{\gamma} + 2\tilde{\kappa} - 3i\tilde{\alpha}_n, \tag{C.2}$$

$$c_3^{(n)} = 1 + \Delta^2 + \tilde{\gamma}(2 + X) + 2\tilde{\kappa}\left[2 - (1 + \Delta^2)AB\right] + 2\tilde{\kappa}\tilde{\gamma} + \tilde{\kappa}^2(1 + \theta^2)$$
$$- 2i\tilde{\alpha}_n(2 + \tilde{\gamma} + 3\tilde{\kappa}) - 3\tilde{\alpha}_n^2, \tag{C.3}$$

$$c_2^{(n)} = \tilde{\gamma}/B + 2\tilde{\kappa}(1 + \Delta^2)(1 - AB) + 2\tilde{\kappa}^2\left[1 + \theta^2 - AB(1 + \Delta^2)\right]$$
$$+ \tilde{\kappa}\tilde{\gamma}\left[4 + X(2 + AB) - 2AB(1 + \Delta^2)\right] + \tilde{\kappa}^2\tilde{\gamma}(1 + \theta^2)$$
$$- i\tilde{\alpha}_n\left[1 + \Delta^2 + 8\tilde{\kappa} - 4\tilde{\kappa}AB(1 + \Delta^2) + 3\tilde{\kappa}^2(1 + \theta^2) + \tilde{\gamma}(2 + 4\tilde{\kappa} + X)\right]$$
$$- \tilde{\alpha}_n^2(2 + \tilde{\gamma} + 6\tilde{\kappa}) + i\tilde{\alpha}_n^3, \tag{C.4}$$

$$c_1^{(n)} = 2\left(\tilde{\kappa}\tilde{\gamma}/B\right)\left[1 - (1 + \Delta^2)AB^2\right] + \tilde{\kappa}^2(1 + \Delta^2)\left[(1 - AB)^2 + (\theta + \Delta AB)^2\right]$$
$$+ \tilde{\kappa}^2\tilde{\gamma}\left[(2 + X)(1 + \theta^2) - 2(1 + \Delta^2)AB + X(1 + \Delta\theta)AB\right]$$
$$- i\tilde{\kappa}\tilde{\alpha}_n\left[2(1 + \Delta^2)(1 - AB) + 4\tilde{\gamma} + \tilde{\gamma}X(2 + AB) + 4\tilde{\kappa}(1 + \theta^2)\right.$$
$$\left. - 2(2\tilde{\kappa} + \tilde{\gamma})(1 + \Delta^2)AB + 2\tilde{\kappa}\tilde{\gamma}(1 + \theta^2)\right]$$
$$+ \tilde{\kappa}\tilde{\alpha}_n^2\left[2AB(1 + \Delta^2) - 4 - 2\tilde{\gamma} - 3\tilde{\kappa}(1 + \theta^2)\right] + 2i\tilde{\kappa}\tilde{\alpha}_n^3, \tag{C.5}$$

$$c_0^{(n)} = \left(\tilde{\kappa}^2\tilde{\gamma}/B\right)\left[(1 - AB)^2 + (\theta + \Delta AB)^2 + 2\left(1 - AB - \Delta^2 AB - \Delta\theta\right)AB^2X\right]$$
$$- i\tilde{\kappa}^2\tilde{\alpha}_n\left\{A^2B^2(1 + \Delta^2)^2 + AB\left[2(1 + \Delta^2)(\Delta\theta - 1 - \tilde{\gamma}) + \tilde{\gamma}X(\Delta\theta + 1)\right]\right.$$
$$\left. + \left[1 + \Delta^2 + \tilde{\gamma}(2 + X)\right](1 + \theta^2)\right\}$$
$$+ \tilde{\kappa}^2\tilde{\alpha}_n^2\left[2AB(1 + \Delta^2) - (2 + \tilde{\gamma})(1 + \theta^2)\right] + i\tilde{\kappa}^2\tilde{\alpha}_n^3(1 + \theta^2), \tag{C.6}$$

where $\tilde{\kappa} = \kappa/\gamma_\perp$, $\tilde{\gamma} = \gamma_\parallel/\gamma_\perp$, $\tilde{\alpha}_n = \alpha_n/\gamma_\perp$ and $X = |f_{0,s}|^2$. Moreover, we have set $B = (1 + \Delta^2 + X)^{-1}$ so that the equations for the stationary state are $AB = 1$ and $\theta = -\Delta$ for a laser, $Y = X[(1 - AB)^2 + (\theta + \Delta AB)^2]$ for a laser with an injected signal and $Y = X[(1 + 2CB)^2 + (\theta - 2C\Delta B)^2]$ for optical bistability. The coefficients $c_k^{(0)}$ ($k = 0, 1, \ldots, 5$) for the characteristic equation associated with the block of single-mode linearized equations (20.15)–(20.19) are obtained by setting $\tilde{\alpha}_n = 0$ in Eqs. (C.1)–(C.6). In particular,

$$c_0^{(0)} = \frac{\tilde{\kappa}^2\tilde{\gamma}}{B}\left[(1 - AB)^2 + (\theta + \Delta AB)^2 + 2\left(1 - AB - \Delta^2 AB - \Delta\theta\right)AB^2X\right], \tag{C.7}$$

and $c_0^{(0)} = 0$ for a laser, whereas $c_0^{(0)} = \left(\tilde{\kappa}^2\tilde{\gamma}/B\right) dY/dX$ for a laser with an injected signal and for optical bistability.

Appendix D Derivation of equations (20.27) and (20.28)

The purpose of this appendix is to sketch the derivation of Eqs. (20.27) and (20.28) and to define all relevant symbols. We consider Eqs. (20.12)–(20.14), with their l.h.s. set equal to zero. From Eq. (20.14) we obtain

$$\delta \tilde{d}_n = \Gamma x^* \left[\delta \tilde{p}_n + B(1 + i\Delta)\delta \tilde{f}_n \right] + \Gamma x \left[\delta \tilde{p}_{-n}^* + B(1 - i\Delta)\delta \tilde{f}_{-n}^* \right], \tag{D.1}$$

$$\Gamma = \frac{1}{2} \frac{\tilde{\gamma}}{i\tilde{\alpha}_n - \tilde{\gamma}}, \tag{D.2}$$

$$B = \frac{1}{1 + \Delta^2 + X}. \tag{D.3}$$

By plugging the expression for $\delta \tilde{d}_n$ into Eqs. (20.12) and (20.13) we obtain a set of algebraic equations for $\delta \tilde{p}_n$ and $\delta \tilde{p}_{-n}^*$,

$$S_1(\Delta)\delta \tilde{p}_n - \Gamma x^2 \, \delta \tilde{p}_{-n}^* = S_2(\Delta)\delta \tilde{f}_n + S_3(\Delta)x^2 \, \delta \tilde{f}_{-n}^*, \tag{D.4}$$

$$-\Gamma x^{*2} \, \delta \tilde{p}_n + S_1(-\Delta)\delta \tilde{p}_{-n}^* = S_3(-\Delta)x^{*2} \, \delta \tilde{f}_n + S_2(-\Delta)\delta \tilde{f}_{-n}^*, \tag{D.5}$$

with

$$S_1(\Delta) = 1 - \Gamma X + i(\Delta - \tilde{\alpha}_n), \tag{D.6}$$

$$S_2(\Delta) = B(1 + \Delta^2) + B\Gamma X(1 + i\Delta), \tag{D.7}$$

$$S_3(\Delta) = B\Gamma(1 - i\Delta). \tag{D.8}$$

The solutions of Eqs. (D.4) and (D.5) can be put into the form

$$\delta \tilde{p}_n = T_1(\Delta)\delta \tilde{f}_n + T_2(\Delta)x^2 \, \delta \tilde{f}_{-n}^*, \tag{D.9}$$

$$\delta \tilde{p}_{-n}^* = T_2(-\Delta)x^{*2} \, \delta \tilde{f}_n + T_1(-\Delta)\delta \tilde{f}_{-n}^*, \tag{D.10}$$

where T_1 and T_2 are defined by

$$T_1(\Delta) = \frac{(1 + \Delta^2)(1 - i\tilde{\alpha}_n - i\Delta) - i\Gamma\tilde{\alpha}_n(1 + i\Delta)X}{(1 - i\tilde{\alpha}_n)(1 - i\tilde{\alpha}_n - 2\Gamma X) + \Delta^2} B, \tag{D.11}$$

$$T_2(\Delta) = \frac{\Gamma(1 - i\Delta)(2 - i\tilde{\alpha}_n)}{(1 - i\tilde{\alpha}_n)(1 - i\tilde{\alpha}_n - 2\Gamma X) + \Delta^2} B. \tag{D.12}$$

Appendix E Coefficients of equations (20.60) and (20.61)

E

The coefficients $a_k^{(n)}$ which appear in Eq. (20.60) are

$$a_3^{(n)} = 1, \tag{E.1}$$

$$a_2^{(n)} = 1 + \tilde{\kappa} + \tilde{\gamma} - 2i\tilde{\alpha}_n, \tag{E.2}$$

$$a_1^{(n)} = \tilde{\kappa}(1 - AB) + \tilde{\gamma}(1 + X) + \tilde{\kappa}\tilde{\gamma} - i\tilde{\alpha}_n(1 + 2\tilde{\kappa} + \tilde{\gamma}) - \tilde{\alpha}_n^2, \tag{E.3}$$

$$a_0^{(n)} = \frac{\tilde{\kappa}\tilde{\gamma}}{B}\left(1 - AB + 2AB^2 X\right) - i\tilde{\kappa}\tilde{\alpha}_n(1 + \tilde{\gamma} - AB) - \tilde{\kappa}\tilde{\alpha}_n^2, \tag{E.4}$$

while the coefficients $b_k^{(n)}$ which appear in Eq. (20.61) are

$$b_2^{(n)} = 1, \tag{E.5}$$

$$b_1^{(n)} = 1 + \tilde{\kappa} - i\tilde{\alpha}_n, \tag{E.6}$$

$$b_0^{(n)} = \tilde{\kappa}(1 - AB) - i\tilde{\kappa}\tilde{\alpha}_n. \tag{E.7}$$

In these equations $B = (1 + X)^{-1}$.

Appendix F The exact boundary of the Risken–Nummedal–Graham–Haken instability

Equation (22.48) can be cast in the form of a cubic equation for λ,

$$\lambda^3 + \lambda^2(\kappa + \gamma_\perp + \gamma_\| + i\alpha_n) + \lambda[\gamma_\|(\gamma_\perp A + \kappa) + i\alpha_n(\gamma_\perp + \gamma_\|)]$$
$$+ 2\kappa\gamma_\perp\gamma_\|(A - 1) + i\alpha_n\gamma_\perp\gamma_\| A = 0. \tag{F.1}$$

We note that for $n = 0$ this equation reduces to the single-mode eigenvalue equation (18.8) with $n = 3$ and the coefficients defined by Eq. (22.34).

By following Risken and Nummedal [124], let us calculate the stability boundary where λ is purely imaginary, i.e. $\lambda = -i\beta$ with β real. By inserting this definition into Eq. (F.1) and separating real and imaginary parts we obtain the two equations

$$\beta^3 - \alpha\beta^2 - \gamma_\|(\gamma_\perp A + \kappa)\beta + \alpha\gamma_\perp\gamma_\| A = 0, \tag{F.2}$$

$$-(\kappa + \gamma_\perp + \gamma_\|)\beta^2 + \alpha(\gamma_\perp + \gamma_\|)\beta + 2\kappa\gamma_\perp\gamma_\|(A - 1) = 0, \tag{F.3}$$

where we have written α instead of α_n. From the first equation we obtain the relation

$$\frac{\alpha}{\beta} = \frac{\beta^2 - \gamma_\|(\gamma_\perp A + \kappa)}{\beta^2 - \gamma_\perp\gamma_\| A}, \tag{F.4}$$

which links α and β. By inserting Eq. (F.4) into Eq. (F.3) we obtain a quadratic equation for β^2 whose solutions are

$$\beta_\pm^2 = \frac{\gamma_\|}{2}\left\{3\gamma_\perp(A - 1) - \gamma_\| \pm \left[\gamma_\perp^2(A - 1)(A - 9) - 6\gamma_\|\gamma_\perp(A - 1) + \gamma_\|^2\right]^{1/2}\right\}. \tag{F.5}$$

The crucial point is that when $\kappa \ll \gamma_\perp$ we have from Eq. (F.4) that $\beta = \alpha$, so expression (F.5) holds also for α^2, i.e. for α_n^2. Finally, if we take into account that $\tilde{\alpha}_n = \alpha_n/\gamma_\perp$, $\tilde{\gamma} = \gamma_\|/\gamma_\perp$ and $X = A - 1$, we see that the expressions for α_+ and α_- coincide with the expressions for $\tilde{\alpha}_{\max}$ and $\tilde{\alpha}_{\min}$ given by Eq. (20.39).

Appendix G Nonlinear analysis of the roll solution

In this appendix we derive an approximated analytical solution for the roll pattern in the optical parametric oscillator that emerges above the instability boundary studied in Section 27.3. We start by rewriting the second of Eqs. (27.30) as

$$F_{2s} = \frac{F_1}{1 + i\theta_2} = \frac{\eta + \sqrt{1 + \theta_2^2}}{1 + i\theta_2}, \tag{G.1}$$

where we have introduced the parameter η which represents the difference between the input field F_1 and its value at the pattern-forming threshold $\sqrt{1 + \theta_2^2}$.

If we now insert this expression into Eqs. (27.31) and (27.32) where we also set for simplicity $\kappa_1 = \kappa_2 = \kappa$, we obtain

$$\kappa^{-1} \frac{\partial F_1}{\partial t} = -(1 + i\theta_1)F_1 + (1 - i\theta_2)\frac{\eta + \sqrt{1 + \theta_2^2}}{1 + \theta_2^2}F_1^* + i\bar{a}_1 \nabla_\perp^2 F_1 + F_1^* \delta B, \tag{G.2}$$

$$\kappa^{-1} \frac{\partial \delta B}{\partial t} = -(1 + i\theta_2)\delta B + i\bar{a}_2 \nabla_\perp^2 \delta B - F_1^2. \tag{G.3}$$

We now want to study the form of a stationary solution above threshold, which will exhibit a spatial modulation due to the pattern-formation mechanism. To this end we set to zero the time derivatives and introduce the real and imaginary parts for the fields by setting $F_1 = X + iY$ and $\delta B = Z + iW$. We can thus rewrite the stationary equations in the form

$$L_0 V + \eta M V + N = 0, \tag{G.4}$$

where the four-dimensional vector V represents the solution

$$V = \begin{pmatrix} X \\ Y \\ Z \\ W \end{pmatrix}, \tag{G.5}$$

L_0 and M are linear operators defined as

$$L_0 = \begin{pmatrix} -1 + \frac{1}{\sqrt{1 + \theta_2^2}} & \theta_1 - \frac{\theta_2}{\sqrt{1 + \theta_2^2}} - \bar{a}_1 \nabla_\perp^2 & 0 & 0 \\ -\theta_1 - \frac{\theta_2}{\sqrt{1 + \theta_2^2}} + \bar{a}_1 \nabla_\perp^2 & -1 - \frac{1}{\sqrt{1 + \theta_2^2}} & 0 & 0 \\ 0 & 0 & -1 & \theta_2 - \bar{a}_2 \nabla_\perp^2 \\ 0 & 0 & -\theta_2 + \bar{a}_2 \nabla_\perp^2 & -1 \end{pmatrix}, \tag{G.6}$$

$$M = \frac{1}{1+\theta_2^2}\begin{pmatrix} 1 & -\theta_2 & 0 & 0 \\ -\theta_2 & -1 & 0 & 0 \\ 0 & 0 & 0 & 0 \\ 0 & 0 & 0 & 0 \end{pmatrix}, \tag{G.7}$$

and the nonlinear terms are isolated in the vector

$$N = \begin{pmatrix} XZ + YW \\ XW - YZ \\ -X^2 + Y^2 \\ -2XY \end{pmatrix}. \tag{G.8}$$

We now assume that, as the input field is slightly increased above threshold, a periodically modulated solution will emerge continuously from the homogeneous steady state (the null state, with our previous choices). From the considerations developed in Section 27.3 the ansatz can be made that the spatial dependence of the solution just above the threshold will be of the type $\cos(k_c x)$.

We will thus expand the bifurcation parameter η and the solution as follows:

$$\eta = \epsilon \eta_1 + \epsilon^2 \eta_2 + \cdots,$$
$$V = \epsilon V_1 \cos(k_c x) + \epsilon^2 V_2(x) + \cdots, \tag{G.9}$$
$$N = \epsilon^2 N_2 + \epsilon^3 N_3 + \cdots,$$

where ϵ is a smallness parameter that will be determined, and the quantities V_1 and η_i and the function $V_2(x)$ will be determined. In writing Eqs. (G.9) we assume that the rolls are in the horizontal direction. Hence the Laplacian appearing in L_0 reduces to d^2/dx^2. The vector components will be denoted by

$$V_i = \begin{pmatrix} X_i \\ Y_i \\ Z_i \\ W_i \end{pmatrix}, i = 1, 2, \ldots \rightarrow \begin{pmatrix} X \\ Y \\ Z \\ W \end{pmatrix} = \epsilon \begin{pmatrix} X_1 \\ Y_1 \\ Z_1 \\ W_1 \end{pmatrix} \cos(k_c x) + \epsilon^2 \begin{pmatrix} X_2(x) \\ Y_2(x) \\ Z_2(x) \\ W_2(x) \end{pmatrix} + \cdots. \tag{G.10}$$

Owing to the block-diagonal structure of the linear operators, it is easy to substitute the above expansion into Eq. (G.4) and solve at first order in ϵ to find

$$L_0 V_1 \cos(k_c x) = 0, \tag{G.11}$$

Since $\bar{a}_1 k_c^2 = -\theta_1$, Eq. (G.11) amounts to the explicit linear system

$$\left(\frac{1}{\sqrt{1+\theta_2^2}} - 1\right) X_1 - \frac{\theta_2}{\sqrt{1+\theta_2^2}} Y_1 = 0, \tag{G.12}$$

$$\frac{\theta_2}{\sqrt{1+\theta_2^2}} X_1 + \left(\frac{1}{\sqrt{1+\theta_2^2}} + 1\right) Y_1 = 0, \tag{G.13}$$

$$-Z_1 + (\theta_2 + \bar{a}_2 k_c^2) W_1 = 0, \tag{G.14}$$
$$-(\theta_2 + \bar{a}_2 k_c^2) Z_1 - W_1 = 0. \tag{G.15}$$

The solutions of the system can be written as

$$X_1 = \xi\theta_2, \tag{G.16}$$

$$Y_1 = \xi\left(1 - \sqrt{1 + \theta_2^2}\right), \tag{G.17}$$

$$Z_1 = 0, \tag{G.18}$$

$$W_1 = 0, \tag{G.19}$$

where ξ is an arbitrary constant, hence

$$V_1 = \begin{pmatrix} X_1 \\ Y_1 \\ 0 \\ 0 \end{pmatrix}, \tag{G.20}$$

where ξ will be chosen equal to 1, so that we achieve the result of Eq. (27.38).

For the remainder of the analysis we will need the solution of the adjoint equation $L_0^+ \tilde{V}_1 = 0$, where L_0^+ is the adjoint operator of L_0. It is straightforward to verify that $\tilde{V}_1$ coincides with V_1.

At second order, the equation to be solved is

$$L_0 V_2(x) = -\eta_1 M V_1 \cos(k_c x) - N_2, \tag{G.21}$$

which, by virtue of the Fredholm alternative [334], has a nontrivial solution if and only if the r.h.s. of Eq. (G.21) is orthogonal to the nucleus of L_0^+. We must thus require that

$$\left\langle \tilde{V}_1 \cos(k_c x) \middle| -\eta_1 M V_1 \cos(k_c x) - N_2 \right\rangle = 0, \tag{G.22}$$

where $\langle V(x)|W(x)\rangle = \lambda_c^{-1} \int_0^{\lambda_c} dx\, V(x) \cdot W(x)$, $\lambda_c = 2\pi/k_c$ and $V \cdot W$ is the scalar product of the two vectors. This amounts to solving the following equation:[1]

$$\left\langle \tilde{V}_1 \cos(k_c x) \middle| -\eta_1 M V_1 \cos(k_c x) \right\rangle$$
$$= \frac{\eta_1}{1+\theta_2^2} \frac{1}{\lambda_c} \int_0^{\lambda_c} dx \left(Y_1^2 - X_1^2 + 2\theta_2 X_1 Y_1\right) \cos^2(k_c x)$$
$$= \eta_1\left(1 - \sqrt{1 + \theta_2^2}\right) = 0, \tag{G.23}$$

where we used the first-order solutions (G.16) and (G.17). Hence we must have $\eta_1 = 0$. We can then proceed to solve (G.21), which now reads

$$L_0 V_2(x) = -N_2 = \begin{pmatrix} 0 \\ 0 \\ X_1^2 - Y_1^2 \\ 2X_1 Y_1 \end{pmatrix} \cos^2(k_c x). \tag{G.24}$$

---

[1] Note that the term involving N_2 yields a similar integral, but it depends on the odd function $\cos^3(k_c x)$ and thus vanishes.

The first two equations of the above system read

$$\left(\frac{1}{\sqrt{1+\theta_2^2}} - 1\right) X_2 + \left(\theta_1 - \frac{\theta_2}{\sqrt{1+\theta_2^2}}\right) Y_2 - \bar{a}_1 \frac{d^2 Y_2}{dx^2} = 0, \qquad \text{(G.25)}$$

$$-\left(\theta_1 + \frac{\theta_2}{\sqrt{1+\theta_2^2}}\right) X_2 + \bar{a}_1 \frac{d^2 X_2}{dx^2} - \left(\frac{1}{\sqrt{1+\theta_2^2}} + 1\right) Y_2 = 0. \qquad \text{(G.26)}$$

We note that $X_2 = \xi' X_1 \cos(k_c x)$ and $Y_2 = \xi' Y_1 \cos(k_c x)$, with ξ' being an arbitrary constant and X_1 and Y_1 given by Eqs. (G.16) and (G.17), are solutions of Eqs. (G.25) and (G.26). We choose $\xi' = 0$ so that

$$V_1 = \begin{pmatrix} 0 \\ 0 \\ Z_2(x) \\ W_2(x) \end{pmatrix}. \qquad \text{(G.27)}$$

The third and fourth equations of (G.24) read

$$-Z_2 + \theta_2 W_2 - \bar{a}_2 \frac{d^2 W_2}{dx^2} = -2\left(1 - \sqrt{1+\theta_2^2}\right) \cos^2(k_c x), \qquad \text{(G.28)}$$

$$-\theta_2 Z_2 + \bar{a}_2 \frac{d^2 Z_2}{dx^2} - W_2 = 2\theta_2 \left(1 - \sqrt{1+\theta_2^2}\right) \cos^2(k_c x), \qquad \text{(G.29)}$$

which will be solved by introducing the following assumptions:

$$Z_2(x) = \delta + \gamma \cos^2(k_c x), \qquad \text{(G.30)}$$

$$W_2(x) = \sigma + \tau \cos^2(k_c x). \qquad \text{(G.31)}$$

We can now substitute Eqs. (G.30) and (G.31) into Eqs. (G.28) and (G.29) and separate the constant terms from those containing $\cos^2(k_c x)$. Then the four resulting algebraic equations can be solved to obtain

$$\delta = \frac{2\bar{a}_2 k_c^2}{1 + \theta_2^2} (\theta_2 \gamma - \tau), \qquad \text{(G.32)}$$

$$\gamma = 2\left(1 - \sqrt{1+\theta_2^2}\right) \frac{1 - \theta_2^2 - 4\theta_2 \bar{a}_2 k_c^2}{1 + \left(\theta_2 + 4\bar{a}_2 k_c^2\right)^2}, \qquad \text{(G.33)}$$

$$\sigma = \frac{2\bar{a}_2 k_c^2}{1 + \theta_2^2} (\gamma + \theta_2 \tau), \qquad \text{(G.34)}$$

$$\tau = -4\left(1 - \sqrt{1+\theta_2^2}\right) \frac{\theta_2 + 2\bar{a}_2 k_c^2}{1 + \left(\theta_2 + 4\bar{a}_2 k_c^2\right)^2}. \qquad \text{(G.35)}$$

Owing to Eq. (G.27), the second-order correction affects only the pump field, which reads

$$\delta B = \epsilon^2 (Z_2 + i W_2) = \epsilon^2 \left[\alpha + \beta \cos^2(k_c x)\right], \qquad \text{(G.36)}$$

where, taking into account Eqs. (G.30) and (G.31),

$$\alpha = \delta + i\sigma, \qquad \beta = \gamma + i\tau, \tag{G.37}$$

which, using Eqs. (G.32)–(G.35), yields Eqs. (27.40) and (27.41).

The final step is to solve the Fredholm alternative for the equation at third order, which reads

$$L_0 V_3(x) = R = -\eta_2 M V_1 \cos(k_c x) - N_3, \tag{G.38}$$

and the r.h.s., in vector form, reads

$$R = -\frac{\eta_2}{1 + \theta_2^2} \begin{pmatrix} X_1 - \theta_2 Y_1 \\ -\theta_2 X_1 - Y_1 \\ 0 \\ 0 \end{pmatrix} \cos(k_c x) - \begin{pmatrix} X_1 Z_2 + Y_1 W_2 \\ X_1 W_2 - Y_1 Z_2 \\ 0 \\ 0 \end{pmatrix} \cos(k_c x). \tag{G.39}$$

The requirement is again

$$\langle \tilde{V}_1 \cos(k_c x) | R \rangle = 0, \tag{G.40}$$

which leads to the equation

$$\frac{\eta_2}{1 + \theta_2^2} \frac{1}{\lambda_c} \int_0^{\lambda_c} dx (-X_1^2 + 2\theta_2 X_1 Y_1 + Y_1^2) \cos^2(k_c x)$$
$$= \frac{1}{\lambda_c} \int_0^{\lambda_c} dx \left[X_1^2 Z_2(x) + X_1 Y_1 W_2(x) + Y_1 X_1 W_2(x) - Y_1^2 Z_2(x) \right] \cos^2(k_c x). \tag{G.41}$$

The l.h.s. of this equation contains the same integral as that appearing in Eq. (G.23), which is equal to $\lambda_c(1 + \theta_2^2)(1 - \sqrt{1 + \theta_2^2})$. Using this result and the solutions at second order (G.30) and (G.31), the equation reduces to

$$\eta_2 \left(1 - \sqrt{1 + \theta_2^2} \right) = \frac{1}{\lambda_c} \int_0^{\lambda_c} dx \left[(X_1^2 - Y_1^2)(\delta + \gamma \cos^2(k_c x)) \right.$$
$$\left. + 2 X_1 Y_1 \left(\sigma + \tau \cos^2(k_c x) \right) \right] \cos^2(k_c x). \tag{G.42}$$

By using again Eqs. (G.16) and (G.17) with $\xi = 1$ and performing the integration of the cosine polynomial we find

$$\eta_2 = (\theta_2 \sigma - \delta) + \frac{3}{4}(\theta_2 \tau - \gamma) = 2\bar{a}_2 k_c^2 \tau + \frac{3}{4}(\theta_2 \tau - \gamma), \tag{G.43}$$

where in the last step we used Eqs. (G.32) and (G.35). Since by definition (remember that $\eta_1 = 0$) we assumed $\eta = F_{\rm I} - \sqrt{1 + \theta_2^2} = \epsilon^2 \eta_2$, we can finally write

$$\epsilon = \pm \left[\frac{F_{\rm I} - \sqrt{1 + \theta_2^2}}{\left(2\bar{a}_2 k_c^2 + \frac{3}{4}\theta_2 \right) \operatorname{Im} \beta - \frac{3}{4} \operatorname{Re} \beta} \right]^{\frac{1}{2}}, \tag{G.44}$$

which coincides with Eq. (27.42). It is worth noting that, above threshold, ϵ^2 is always positive if $\theta_2 \operatorname{Im} \beta - \operatorname{Re} \beta = \theta_2 \tau - \gamma > 0$. Using Eqs. (G.32) and (G.35) this quantity can be written as

$$\theta_2 \tau - \gamma = \frac{2\left(\sqrt{1 + \theta_2^2} - 1\right)\left(1 + \theta_2^2\right)}{1 + \left(\theta_2 + 4\bar{a}_2 k_{\mathrm{c}}^2\right)^2}, \tag{G.45}$$

and, since $\theta_2 \neq 0$, the condition is verified.

References

[1] A. Einstein, Zur Quantentheorie der Strahlung, *Phys. Z.* **18**, 121 (1917)

[2] M. Planck, Über eine Verbesserung der Wienschen Spektralgleichung, *Verhandl. Dtsch. Phys. Ges.* **2**, 202 (1900); Zur Theorie des Gesetzes der Energieverteilung im Normalspekrum, *Verhandl. Dtsch. Phys. Ges.* **2**, 237 (1900)

[3] C. Cohen-Tannoudji, B. Diu, and F. Laloë, *Quantum Mechanics*, Vols. 1 and 2 (Paris: Wiley and Hermann, 1977)

[4] C. Cohen-Tannoudji, J. Dupont-Roc, and G. Grynberg, *Photons and Atoms, Introduction to Quantum Electrodynamics*, (New York: Wiley, 1989)

[5] C. Cohen-Tannoudji, J. Dupont-Roc, and G. Grynberg, *Atom–Photon Interactions: Basic Processes and Applications* (New York: Wiley, 1992)

[6] O. Svelto, *Principles of Lasers* (New York: Plenum, 1989)

[7] H. Haken, *Laser Theory* in S. Flugge and L. Genzel (eds.), *Handbuch der Physik (Encyclopedia of Physics)* vol. XXV/2C (Berlin: Springer, 1970)

[8] M. Sargent III, M. O. Scully, and W. Lamb, Jr., *Laser Physics* (Reading, MA: Addison-Wesley, 1974)

[9] A. Yariv, *Quantum Electronics* (New York: Wiley, 1975)

[10] A. Yariv, *Optical Electronics* (New York: Wiley, 1977)

[11] H. Haken, *Light*, Vols. 1 and 2 (Amsterdam: North-Holland, 1981)

[12] M. Bertolotti, *Masers and Lasers* (Bristol: Adam Hilger, 1983)

[13] P. L. Knight *Quantum Electronics and Electro-optics* (New York: Wiley, 1983)

[14] A. E. Siegman, *Lasers* (Mill Valley, CA: University Science Books, 1986)

[15] A. N. Oraevski, *Research on Laser Theory* (Commack, NY: Nova, 1988)

[16] P. M. Milonni and J. H. Eberly, *Lasers* (New York: Wiley, 1986)

[17] K. Shimoda, *Introduction to Laser Physics* (Berlin: Springer, 1981)

[18] G. Grynberg, A. Aspect, and C. Fabre, *Introduction to Quantum Optics* (Cambridge: Cambridge University Press, 2010)

[19] D. Meschede, *Light and Lasers* (Weinheim: Wiley-VCH, 2004)

[20] B. E. A. Saleh and N. C. Teich, *Fundamentals of Photonics*, 2nd edn. (New York: Wiley, 2007)

[21] V. Degiorgio and I. Cristiani, *Photonics. A Short Course* (Springer, Berlin, 2014)

[22] M. O. Scully and M. S. Zubairy, *Quantum Optics* (Cambridge: Cambridge University Press, 2001)

[23] L. Allen and J. H. Eberly, *Optical Resonance & Two-Level Atoms* (New York: Wiley, 1975), reprinted in 1987 (New York: Dover)

[24] L. Mandel and E. Wolf, *Optical Coherence and Quantum Optics* (Cambridge: Cambridge University Press, 1995)

[25] W. P. Schleich, *Quantum Optics in Phase Space* (Wiley-VCH, Berlin, 2005)

[26] J. von Neumann, *Mathematical Foundations of Quantum Mechanics* (Princeton, MA: Princeton University Press, 1955)

[27] I. I. Rabi, Space quantization in a gyrating magnetic field, *Phys. Rev.* **51**, 652 (1937)

[28] F. Bloch, Nuclear induction, *Phys. Rev.* **60**, 460 (1946)

[29] P. Mandel, *Nonlinear Optics, An Analytical Approach* (Weinheim: VCH Verlagsgesellschaft, 2010)

[30] M. Lax, W. H. Louisell, and W. B. McKnight, From Maxwell to paraxial wave optics, *Phys. Rev.* A **11**, 1365 (1975)

[31] O. Hess, *Spatio-Temporal Dynamics of Semiconductor Lasers* (Berlin: Wissenschaft und Technik Verlag, 1993)

[32] F. T. Arecchi and R. Bonifacio, Theory of optical maser amplifiers, *IEEE J. Quantum Electron.* **1**, 169 (1965)

[33] A. Icsevgi and W. E. Lamb, Jr., Propagation of light pulses in a laser amplifier, *Phys. Rev.* **195**, 517 (1969)

[34] S. L. McCall and E. L. Hahn, Self-induced transparency, *Phys. Rev.* **183**, 457 (1969)

[35] R. H. Dicke, Coherence in spontaneous radiation processes, *Phys. Rev.* **93**, 99 (1954)

[36] N. E. Rehler and J. H. Eberly, Superradiance, *Phys. Rev.* A **3**, 1735 (1971)

[37] R. Bonifacio, P. Schwendimann, and F. Haake, Quantum statistical theory of superradiance. Parts I and II, *Phys. Rev.* A **4**, 302, 854 (1971)

[38] F. T. Arecchi and E. Courtens, Cooperative phenomena in resonant electromagnetic propagation, *Phys. Rev.* A **2**, 1730 (1970)

[39] H. M. Gibbs, Q. H. F. Vrehen, and H. M. J. Hikspoors, Single-pulse superfluorescence in cesium, *Phys. Rev. Lett.* **39**, 547 (1977)

[40] R. Bonifacio and L. A. Lugiato, Cooperative radiation processes in two-level systems: Superfluorescence, *Phys. Rev.* A **11**, 1507 (1975)

[41] N. Skribanowitz, I. P. Herman, J. C. MacGillivray, and M. S. Feld, Observation of Dicke superradiance in optically pumped HF gas, *Phys. Rev. Lett.* **30**, 309 (1973)

[42] R. Bonifacio and G. Preparata, Coherent spontaneous emission, *Phys. Rev.* A **2**, 336 (1970)

[43] M. Tavis and F. W. Cummings, Exact solution for an N-molecule-radiation-field Hamiltonian, *Phys. Rev.* **170**, 379 (1968)

[44] E. T. Jaynes and F. W. Cummings, Comparison of quantum and semiclassical radiation theories with application to the beam maser, *Proc. IEEE* **51**, 89 (1963)

[45] P. Goy, J. M. Raimond, M. Gross, and S. Haroche, Observation of cavity-enhanced single-atom spontaneous emission, *Phys. Rev. Lett.* **50**, 1903 (1983)

[46] D. Meschede, H. Walther, and G. Müller, One-atom maser, *Phys. Rev. Lett.* **54**, 551 (1985)

[47] B. R. Mollow, Power spectrum of light scattered by two-level systems, *Phys. Rev.* **188**, 1969 (1969)

[48] E. V. Goldstein and P. Meystre, Dipole–dipole interaction in optical cavities, *Phys. Rev.* A **56**, 5135 (1997)

[49] P. Meystre and M. Sargent III, *Elements of Quantum Optics* (Berlin: Springer, 1990)

[50] J. D. Jackson, *Classical Electrodynamics*, 3rd edn. (New York: Wiley, 1998)

[51] P. A. Franken, A. E. Hill, C. W. Peters, and G. Weinreich, Generation of optical harmonics, *Phys. Rev. Lett.* **7**, 118 (1961)

[52] J. A. Armstrong, N. Bloembergen, J. Ducuing, and P. S. Persham, Interactions between light waves in a nonlinear dielectric, *Phys. Rev.* **127**, 1918 (1962)

[53] N. Bloembergen and P. S. Pershan, Light waves at the boundary of nonlinear media, *Phys. Rev.* **128**, 606 (1962)

[54] N. Bloembergen and Y. R. Shen, Quantum-theoretical comparison of nonlinear susceptibilities in parametric media, lasers, and Raman lasers, *Phys. Rev.* A **133**, 37 (1964)

[55] A. C. Newell and J. V. Moloney, *Nonlinear Optics* (Reading, MA: Addison-Wesley, 1992)

[56] N. Bloembergen, *Nonlinear Optics* (Singapore: World Scientific, 1996)

[57] R. Menzel, *Photonics – Linear and Nonlinear Interactions of Laser Light and Matter* (Berlin: Springer, 2001)

[58] Y. R. Shen, *The Principles of Nonlinear Optics* (New York: Wiley, 2002)

[59] R. W. Boyd, *Nonlinear Optics* (Amsterdam: Elsevier, 2008)

[60] G. I. Stegeman and R. A. Stegeman, *Nonlinear Optics: Phenomena, Materials and Devices* (New York: Wiley, 2012)

[61] V. G. Dmitriev, G. G. Gurzadyan, and N. Nikogosyan, *Handbook of Nonlinear Optical Crystals* (Berlin: Springer, 1991)

[62] R. Loudon, *The Quantum Theory of Light* (Oxford: Oxford University Press, 2000)

[63] G. Milburn and D. F. Walls, *Quantum Optics*, 2nd edn. (Berlin: Springer, 2008)

[64] C. C. Gerry and P. L. Knight, *Introductory Quantum Optics* (Cambridge: Cambridge University Press, 2005)

[65] G. Grynberg, A. Aspect, and C. Fabre, *Introduction to Lasers and Quantum Optics* (Cambridge: Cambridge University Press, 2010)

[66] G. S. Agarwal, *Quantum Optics* (Cambridge: Cambridge University Press, 2012)

[67] A. Gatti, E. Brambilla and L. A. Lugiato, Quantum imaging, in E. Wolf (ed.) *Progress in Optics* vol. LI (Amsterdam: Elsevier, 2008)

[68] V. E. Zacharov and A. B. Shabat, Exact theory of two-dimensional self-focusing and one-dimensional self-modulation of waves in nonlinear media, *Sov. Phys. JETP* **34**, 62 (1972)

[69] L. F. Mollenauer, R. H. Stolen, and J. P. Gordon, Experimental observation of picosecond pulse narrowing and solitons in optical fibers, *Phys. Rev. Lett.* **45**, 1095 (1980)

[70] A. A. Barthelemy, S. Maneuf, and C. Froelich, Soliton propagation and self confinement of laser beams by Kerr optical nonlinearity, *Opt. Commun.* **55**, 201 (1985)

[71] J. S. Aitchison, Y. Silberberg, A. M. Weiner *et al.*, Spatial optical solitons in planar glass waveguides, *J. Opt. Soc. Am.* B **8**, 1290 (1991)

[72] J. E. Bjorkholm and A. Ashkin, CW self-focusing and self-trapping of light in sodium vapor, *Phys. Rev. Lett.* **32**, 129 (1974)

[73] B. Crosignani, M. Segev, D. Engin *et al.* Self-trapping of optical beams in photore-fractive media, *J. Opt. Soc. Am.* B **10**, 446 (1993)

[74] G. L. Lamb, *Elements of Soliton Theory* (New York: Wiley, 1980)

[75] G. P. Agrawal and R. W. Boyd, *Contemporary Nonlinear Optics* (New York: Academic Press, 1992)

[76] J. R. Taylor, *Optical Solitons, Theory and Experiment* (Cambridge: Cambridge University Press, 1992)

[77] N. N. Akhmediev and A. Ankiewicz, *Solitons: Nonlinear Pulses and Beams* (London: Chapman and Hall, 1997)

[78] A. V. Buryak, P. Di Trapani, D. V. Skryabin, and S. Trillo, Optical Solitons due to quadratic nonlinearities: from basic physics to futuristic applications, *Phys. Rep.* **370**, 63 (2002)

[79] Y. S. Kivshar and G. P. Agrawal, *Optical Solitons: From Fiber to Photonic Crystals* (Amsterdam: Academic–Elsevier Science, 2003)

[80] F. Abelès, Sur la propagation des ondes dans les milieux stratifiés, *Nuovo Cim.* **9**, Issue 3 Supplement, 214 (1952)

[81] K. Ishiguro and T. Kato, The reflection and transmission of a multi-layer film, *J. Phys. Soc. Japan* **8**, 77 (1953)

[82] V. Degiorgio, Phase shift between the transmitted and the reflected optical fields of a semireflecting lossless mirror is $\pi/2$, *Am. J. Phys.* **48**, 81 (1980)

[83] A. Zeilinger, General properties of lossless beam splitters in interferometry, *Am. J. Phys.* **49**, 882 (1981)

[84] Yu. V. Troitskii, *The energy conservation law for optical two-port devices*, *Opt. Spectrosc.* **92**, 555 (2002)

[85] R. Bonifacio and L. A. Lugiato, Bistable absorption in a ring cavity, *Lett. Nuovo Cim.* **21**, 505 (1978)

[86] L. A. Lugiato, Theory of optical bistability, in E. Wolf (ed.) *Progress in Optics* vol. XXI (Amsterdam: North-Holland, 1984), p. 69

[87] L. A. Lugiato, L. M. Narducci, E. V. Eschenazi, D. K. Bandy, and N. B. Abraham, Multimode instabilities in a homogeneously broadened ring laser, *Phys. Rev.* A **32**, 1563 (1985)

[88] L. M. Narducci, J. R. Tredicce, L. A. Lugiato, N. B. Abraham, and D. K. Bandy, Mode–mode competition and unstable behavior in a homogeneously broadened ring laser, *Phys. Rev.* A **33**, 1842 (1986)

[89] V. Degiorgio and M. O. Scully, Analogy between the laser threshold region and a second-order phase transition, *Phys. Rev.* A **2**, 1170 (1970)

[90] R. Graham and H. Haken, Laserlight – First example of a second order phase transition far from thermal equilibrium, *Z. Phys.* **237**, 31 (1970)

[91] H. Haken, *Synergetics – An Introduction* (Heidelberg: Springer, 1983)

[92] H. Haken, *Advanced Synergetics* (Heidelberg: Springer, 1983)

[93] G.-L. Oppo and A. Politi, Center-manifold Reduction for Laser Equation with Detuning, *Phys. Rev.* A **40**, 1422 (1989)

[94] T. H. Maiman, Stimulated optical radiation in ruby, *Nature* **187**, 493 (1960)

[95] A. Szöke, V. Daneu, J. Goldhar, and N. A. Kurnit, Bistable optical element and its applications, *Appl. Phys. Lett.* **15**, 376 (1969)

[96] R. Bonifacio and L. A. Lugiato, Cooperative effects and bistability for resonance fluorescence, *Opt. Commun.* **19**, 172 (1976)

[97] R. Bonifacio and L. A. Lugiato, Optical bistability and cooperative effects in resonance fluorescence, *Phys. Rev.* A **18**, 1129 (1978)

[98] R. Bonifacio, M. Gronchi, and L. A. Lugiato, Photon statistics of a bistable absorber, *Phys. Rev.* A **18**, 2266 (1978)

[99] L. A. Lugiato, Optical bistability, *Contemp. Phys.* **24**, 333 (1983)

[100] F. S. Felber and J. H. Marburger, Theory of nonresonant multistable optical devices, *Appl. Phys. Lett.* **28**, 731 (1976)

[101] H. M. Gibbs, S. L. McCall, and T. N. C. Venkatesan, Differential gain and bistability using a sodium-filled Fabry–Perot interferometer, *Phys. Rev. Lett.* **36**, 1135 (1976)

[102] R. Roy and M. S. Zubairy, Beyond the mean-field theory of dispersive optical bistability, *Phys. Rev.* A **21**, 274 (1980)

[103] R. Bonifacio, M. Gronchi, and L. A. Lugiato, Dispersive bistability in homogeneously broadened systems, *Nuovo Cim.* B **53**, 311 (1979)

[104] R. Bonifacio and L. A. Lugiato, Mean field model for absorptive and dispersive bistability with inhomogeneous broadening, *Lett. Nuovo Cim.* **21**, 517 (1978)

[105] S. S. Hassan, P. D. Drummond, and D. F. Walls, Dispersive optical bistability in a ring cavity, *Opt. Commun.* **27**, 480 (1978)

[106] G. P. Agrawal and H. J. Carmichael, Optical bistability through nonlinear dispersion and absorption, *Phys. Rev.* A **19**, 2074 (1979)

[107] W. J. Sandle and A. Gallagher, Optical bistability by an atomic vapor in a focusing Fabry–Perot cavity, *Phys. Rev.* A **24**, 2017 (1981)

[108] E. Arimondo, A. Gozzini, L. Lovich, and E. Pistelli, Microwave dispersive bistability in a confocal Fabry–Perot microwave cavity, in C. M. Bowden, M. Ciftan, and H. R. Robl (eds.) *Optical Bistability, Proc. Int. Conf. on Optical Bistability*, Asheville (New York: Plenum, 1980)

[109] H. M. Gibbs, *Optical Bistability: Controlling Light by Light* (New York: Academic Press, 1985)

[110] A. Joshi and Min Xiao, *Controlling Steady–state and Dynamical Properties of Atomic Optical Bistability* (Singapore: World Scienific, 2012)

[111] S. L. McCall, Instabilities in continuous-wave light propagation in absorbing media, *Phys. Rev.* A **9**, 1515 (1974)

[112] P. Meystre, On the use of the mean-field theory in optical bistability, *Opt. Commun.* **26**, 277 (1978)

[113] E. Abraham, R. K. Bullough, and S. S. Hassan, Space and time-dependent effects in optical bistability, *Opt. Commun.* **29**, 109 (1979)

[114] E. Abraham, S. S. Hassan, and R. K. Bullough, Dispersive optical bistability in a Fabry–Perot cavity, *Opt. Commun.* **33**, 93 (1980)

[115] E. Abraham and S. S. Hassan, Effects of inhomogeneous broadening on optical bistability in a Fabry–Perot cavity, *Opt. Commun.* **35**, 291 (1980)

[116] R. Roy and M. S. Zubairy, Analytic solutions of the optical bistability equations for a standing wave cavity, *Opt. Commun.* **32**, 163 (1980)

[117] H. J. Carmichael, The mean-field approximation and validity of a truncated Bloch hierarchy in absorptive bistability, *Opt. Acta* **27**, 147 (1980)

[118] J. A. Hermann, Spatial effects in optical bistability, *Opt. Acta* **27**, 159 (1980)

[119] H. J. Carmichael and J. A. Hermann, Analytic description of optical bistability including spatial effects, *Z. Phys.* B **38**, 365 (1980)

[120] K. G. Weyer, H. Wiedenmann, M. Rateike *et al.*, Observation of absorptive optical bistability in a Fabry–Perot cavity containing multiple atomic beams, *Opt. Commun.* **37**, 426 (1981)

[121] V. Benza and L. A. Lugiato, Dressed mode description of optical bistability, *Z. Phys.* B **35**, 383 (1979)

[122] L. A. Lugiato, Many-mode quantum statistical theory of optical bistability, *Z. Phys.* B **41**, 85 (1981)

[123] L. M. Narducci, J. R. Tredicce, L. A. Lugiato, N. B. Abraham, and D. K. Bandy, Multimode laser with an injected signal: Steady-state and linear stability analysis, *Phys. Rev.* A **32**, 1588 (1985)

[124] H. Risken and R. Nummedal, Self-pulsing in lasers, *J. Appl. Phys.* **39**, 4662 (1968)

[125] M. B. Spencer and W. E. Lamb, Jr., Laser with a transmitting window, *Phys. Rev.* A **5**, 884 (1972)

[126] L. A. Lugiato, Instabilities in the laser with injected signal and laser-phase transition analogy, *Lett. Nuovo Cim.* **23**, 609 (1978)

[127] R. Salomaa and S. Stenholm, Gas laser with saturable absorber. I. Single-mode characteristics, *Phys. Rev.* A **8**, 2695 (1973); Gas laser with saturable Absorber. II. Single-mode stability, *Phys. Rev.* A **8**, 2711 (1973)

[128] L. A. Lugiato, P. Mandel, S. T. Dembinski, and A. Kossakowsi, Semiclassical and quantum theories of bistability in lasers containing saturable absorbers, *Phys. Rev.* A **18**, 238 (1978)

[129] S. T. Dembinski, A. Kossakowski, P. Pepłowski, L. A. Lugiato, and P. Mandel, Laser instability below threshold, *Phys. Lett.* A **68**, 20 (1978)

[130] N. B. Abraham, P. Mandel, and L. M. Narducci, Dynamical instabilities and pulsations in lasers in E. Wolf (ed.) *Progress in Optics* vol. XXV (Amsterdam: North-Holland, 1988), p. 1

[131] P. H. Lee, P. B. Schaefer, and W. B. Barker, Single-mode power from 6328 Å laser incorporating neon absorption, *Appl. Phys. Lett.* **13**, 373 (1968)

[132] V. N. Lisitsyn and V. P. Chebotaev, Hysteresis and hard excitation in a gas laser, *JETP Lett.* **7**, 1 (1968)

[133] S. Ruschin and S. H. Bauer, Bistability hysteresis and critical behavior of CO_2 laser, with SF_6 intracavity as a saturable absorber, *Chem. Phys. Lett.* **66**, 100 (1979)

[134] S. Ruschin and S. H. Bauer, Bistability of a CO_2 laser with SF_6 intracavity as an absorber: Transient effects, *Appl. Phys.* **24**, 45 (1981)

[135] E. Arimondo, F. Casagrande, L. A. Lugiato, and P. Glorieux, Repetitive passive Q-switching and bistability in lasers with saturable absorber, *Appl. Phys.* B **30**, 57 (1983)

[136] M. Brambilla, F. Castelli, L. A. Lugiato, F. Prati, and G. Strini, Nondegenerate four-wave mixing in a cavity: instabilities and quantum noise reduction, *Opt. Commun.* **83**, 367 (1991)

[137] P. D. Drummond, K. J. McNeil, and D. F. Walls, Non-equilibrium transitions in sub/second harmonic generation, I. Semiclassical theory, *Opt. Acta* **27**, 321 (1980)

[138] P. D. Drummond, K. J. McNeil, and D. F. Walls, Non-equilibrium transitions in sub/second harmonic generation II. Quantum theory, *Opt. Acta* **28**, 211 (1981)

[139] L. A. Lugiato, C. Oldano, C. Fabre, E. Giacobino, and R. J. Horowicz, Bistability, self-pulsing and chaos in optical parametric oscillators, *Nuovo Cim.* D **10**, 959 (1988)

[140] C. Richy, K. I. Petsas, E. Giacobino, C. Fabre, and L. A. Lugiato, Observation of bistability and delayed bifurcation in a triply resonant optical parametric oscillator, *J. Opt. Soc. Am.* B **12**, 456 (1994)

[141] L. A. Lugiato and L. M. Narducci, Nonlinear dynamics in a Fabry–Perot cavity, *Z. Phys.* **71**, 129 (1988)

[142] W. E. Lamb, Theory of an optical maser, *Phys. Rev.* A **134**, 1429 (1964)

[143] C. L. Tang, H. Statz, and G. DeMars, Spectral output and spiking behaviour of solid-state lasers, *J. Appl. Phys.* **34**, 2289 (1963)

[144] H. J. Carmichael, Multimode instability for a standing wave cavity containing a saturable absorber, *Opt. Commun.* **53**, 122 (1985)

[145] K. Ikeda, Multiple-valued stationary state and its instability of the transmitted light by a ring cavity system, *Opt. Commun.* **30**, 257 (1979)

[146] K. Ikeda, H. Daido, and O. Akimoto. Optical turbulence: Chaotic behavior of transmitted light from a ring cavity, *Phys. Rev. Lett.* **45**, 709 (1980)

[147] K. Ikeda and M. Mizuno, Modeling of nonlinear Fabry–Perot resonators by difference-differential equations, *IEEE J. Quantum Electron.* **21**, 1429 (1985)

[148] L. A. Lugiato and F. Prati, Difference differential equations for a resonator with a very thin nonlinear medium, *Phys. Rev. Lett.* **104**, 233902 (2010)

[149] F. Prati and L. A. Lugiato, Instabilities for a coherently driven Fabry–Perot cavity with a very thin absorber, *Eur. Phys. J. Special Topics* **203**, 117 (2012)

[150] J. Mulet and S. Balle, Mode-locking dynamics in electrically driven vertical-external-cavity surface-emitting lasers, *IEEE J. Quantum Electron.* **41**, 1148 (2005)

[151] N. B. Abraham, L. A. Lugiato, P. Mandel, L. M. Narducci, and D. K. Bandy, Steady-state and unstable behaviour of a single-mode inhomogeneously broadened laser, *J. Opt. Soc. Am.* B **2**, 35 (1985)

[152] P. Mandel, Properties of a Lorentz-broadened single-mode unidirectional ring laser, *J. Opt. Soc. Am.* B **2**, 112 (1985)

[153] G. H. B. Thompson, *Physics of Semiconductor Laser Devices* (New York: Wiley, 1980)

[154] J. E. Carrol, *Rate Equations in Semiconductor Electronics* (Cambridge: Cambridge University Press, 1985)

[155] P. A. Markovich, *The Stationary Semiconductor Device* (Berlin: Springer, 1986)

[156] G. P. Agrawal, *Long Wavelength Semiconductor Lasers* (New York: Van Nostrand Reinhold, 1986)

[157] B. Mroziewicz, M. Bugajski, and W. Nakwaski, *Physics of Semiconductor Lasers* (Amsterdam: North-Holland, 991)

[158] J. Buus, *Single Frequency Semiconductor Lasers* (Bellingham, WA: SPIE, 1991)

[159] M. Ohtsu, *Highly Coherent Semiconductor Lasers* (Boston, MA: Artech House, 1992)

[160] W. W. Chow, S. W. Koch, and M. Sargent III, *Semiconductor-Laser Physics* (Berlin: Springer, 1994)

[161] C. F. Klingshirn, *Semiconductor Optics* (Berlin: Springer, 1995)

[162] W. W. Chow and S. W. Koch, *Semiconductor Laser Fundamentals* (Berlin: Springer, 1999)

[163] J. Ohtsubo, *Semiconductor Lasers – Stability, Instability and Chaos* (Berlin: Springer, 2005)

[164] C. H. Henry, Theory of the linewidth of semiconductor lasers, *IEEE J. Quantum Electron.* **18**, 259 (1982)

[165] C. Wilmsen, H. Temkin, and L. A. Coldren (eds.), *Vertical-Cavity Surface-Emitting Lasers* (Cambridge: Cambridge University Press, 1999)

[166] M. San Miguel, Q. Feng, and J. V. Moloney, Light-polarization dynamics in surface-emitting semiconductor lasers, *Phys. Rev.* A, **52**, 1728 (1995)

[167] K. Panajotov and F. Prati, Polarization dynamics of VCSELs, in R. Michalzik (ed.) *VCSELs* (Berlin: Springer-Verlag, 2012) p. 181

[168] J. Mompart and R. Corbalan, Lasing without inversion, *J. Opt.* B **2**, R7 (2000)

[169] E. Arimondo, Coherent population trapping in laser spectroscopy, in E. Wolf (ed.) *Progress in Optics* vol. XXXV (Amsterdam: North-Holland, 1996), p. 257

[170] G. Alzetta, A. Gozzini, L. Moi, and G. Orriols, An experimental method for the observation of RF transitions and laser beat resonances in oriented Na vapour, *Nuovo Cim.* B **36**, 5 (1976)

[171] E. Arimondo and G. Orriols, Nonabsorbing atomic coherences by coherent two-photon transitions in a three-level optical pumping, *Lett. Nuovo Cim.* **17**, 333 (1976)

[172] R. M. Whitley and C. R. Stroud, Jr., Double optical resonance, *Phys. Rev.* A **14**, 1498 (1976)

[173] H. R. Gray, R. M. Whitley and C. R. Stroud, Jr., Coherent trapping of atomic populations, *Opt. Lett.* **3**, 218 (1978)

[174] G. Alzetta, L. Moi, and G. Orriols, Nonabsorption hyperfine resonances in a sodium vapour irradiated by a multimode dye-laser, *Nuovo Cim.* B **52**, 209 (1979)

[175] J. D. Stettler, C. M. Bowden, N. M. Witriol, and J. H. Eberly, Population trapping during laser induced molecular excitation and dissociation, *Phys. Lett.* A **73**, 171 (1979)

[176] R. J. Dalton and P. L. Knight, Population trapping and ultranarrow Raman lineshapes induced by phase-fluctuating fields, *Opt. Commun.* **42**, 411 (1982)

[177] S. Swain, Conditions for population trapping in a three-level system, *J. Phys.* B **15**, 3405 (1982)

[178] P. M. Radmore and P. L. Knight, Population trapping and dispersion in a three-level system, *J. Phys.* B **15**, 561 (1981); Two-photon ionisation: Interference and population trapping, *Phys. Lett.* A **102**, 180 (1984)

[179] G. S. Agarwal and N. Nayak, Effects of long-lived incoherences on coherent population trapping, *J. Phys.* B **19**, 3375 (1986)

[180] K. Zaheer and M. S. Zubairy, Phase sensitivity in atom–field interaction via coherent superposition, *Phys. Rev.* A **39**, 2000 (1989)

[181] S. E. Harris, J. E. Field, and A. Imamoğlu, Nonlinear optical processes using electromagnetically induced transparency, *Phys. Rev. Lett.* **64**, 1107 (1990)

[182] K. H. Hahn, D. A. King, and S. E. Harris, Nonlinear generation of 104.8-nm radiation within an absorption window in zinc, *Phys. Rev. Lett.* **65**, 2777 (1990)

[183] K. Hakuta, L. Marmet, and B. Stoicheff, Electric-field-induced 2nd-harmonic generation with reduced absorption in atomic-hydrogen, *Phys. Rev. Lett.* **66**, 596 (1991)

[184] K.-J. Boller, A. Imamoğlu, and S. E. Harris, Observation of electromagnetically induced transparency, *Phys. Rev. Lett.* **66**, 2593 (1991)

[185] J. E. Field, K. H. Hahn, and S. E. Harris, Observation of electromagnetically induced transparency in collisionally broadened lead vapor, *Phys. Rev. Lett.* **67**, 3062 (1991)

[186] A. Javan, Theory of a three-level maser, *Phys. Rev.* **107**, 1579 (1956)

[187] T. W. Hänsch and P. E. Toschek, Theory of a three-level gas laser amplifier, *Z. Phys.* **236**, 213 (1970)

[188] V. Arkhipkin and Yu. Heller, Radiation amplification without population inversion at transitions to autoionizing states, *Phys. Lett.* A **98**, 12 (1983)

[189] O. Kocharovskaya and Ya. I. Khanin, Coherent amplification of an ultrashort pulse in a 3-level medium without a population-inversion, *JETP Lett.* **48**, 630 (1988)

[190] S. E. Harris, Lasers without inversion – Interference of lifetime-broadened resonances, *Phys. Rev. Lett.* **62**, 1033 (1989)

[191] M. O. Scully, S.-Y. Zhu, and A. Gavrielides, Degenerate quantum-beat laser – Lasing without inversion and inversion without lasing, *Phys. Rev. Lett.* **62**, 2813 (1989)

[192] S. E. Harris and J. H. Macklin, Lasers without inversion – Single-atom transient-response, *Phys. Rev.* A **40**, 4135 (1989)

[193] A. Imamoğlu, Interference of radiatively broadened resonances, *Phys. Rev.* A **40**, 2835 (1989)

[194] A. Lyras, X. Tang, P. Lambropoulos, and J. Zhang, Radiation amplification through auto-ionizing resonances without population-inversion, *Phys. Rev.* A **40**, 4131 (1989)

[195] O. Kocharovskaya and P. Mandel, Amplification without inversion – The double-lambda scheme, *Phys. Rev.* A **42**, 523 (1990)

[196] E. E. Fill, M. O. Scully, and S.-Y. Zhu, Lasing without inversion via the lambda-quantum-beat laser in the collision-dominated regime, *Opt. Commun.* **77**, 36 (1990)

[197] O. Kocharovskaya, R.-D. Li, and P. Mandel, Lasing without inversion – The double-lambda scheme, *Opt. Commun.* **77**, 215 (1990)

[198] V. R. Blok and G. M. Krochik, Theory of lasers without inversion, *Phys. Rev.* A **41**, 1517 (1990)

[199] G. S. Agarwal, S. Ravi, and J. Cooper, DC-field-coupled autoionizing states for laser action without population-inversion, *Phys. Rev.* A **41**, 4721 (1990); Lasers without inversion – Raman transitions using autoionizing resonances, *Phys. Rev.* A **41**, 4727 (1990)

[200] S. Basile and P. Lambropoulos, Radiation amplification without population-inversion in discrete 3-level systems, *Opt. Commun.* **78**, 163 (1990)

[201] A. Imamoğlu, J. E. Field, and S. E. Harris, Lasers without inversion – A closed lifetime broadened system, *Phys. Rev. Lett.* **66**, 1154 (1991)

[202] G. S. Agarwal, Origin of gain in systems without inversion in bare or dressed states, *Phys. Rev. A* **44**, R28 (1991)

[203] L. M. Narducci, H. M. Doss, P. Ru, M. O. Scully, and C. Keitel, A simple model of a laser without inversion, *Opt. Commun.* **81**, 379 (1991)

[204] J. A. Bergou and P. Bogár, Quantum theory of a noninversion laser with injected atomic coherence, *Phys. Rev. A* **43**, 4889 (1991)

[205] O. Kocharovskaya, Amplification and lasing without inversion *Phys. Rep.* **219**, 175 (1992)

[206] M. O. Scully, From lasers and masers to phaseonium and phasers *Phys. Rep.* **219**, 191 (1992)

[207] M. O. Scully, S.-Y. Zhu, and H. Fearn, Lasing without inversion. I. Initial atomic coherence, *Z. Phys. D* **22**, 471 (1992); Lasing without inversion. II. Raman process created atomic coherence, *Z. Phys. D* **22**, 483 (1992)

[208] M. Fleischhauer, C. H. Keitel, L. M. Narducci *et al.*, Lasing without inversion – Interference of radiatively broadened resonances in dressed atomic systems, *Opt. Commun.* **94**, 599 (1992)

[209] M. O. Scully, Resolving conundrums in lasing without inversion via exact solutions to simple models, *Quantum Optics* **6**, 203 (1994)

[210] O. Kocharovskaya and P. Mandel, Basic models of lasing without inversion – General form of amplification condition and problem of self-consistency, *Quantum Optics* **6**, 217 (1994)

[211] N. B. Abraham, L. A. Lugiato, and L. M. Narducci (eds.), Feature Issue on Instabilities in Active Optical Media, *J. Opt. Soc. Am.* B **2** (January 1985)

[212] D. K. Bandy, A. N. Oraevsky, and J. R. Tredicce (eds.), Feature Issue on Nonlinear Dynamics of Lasers, *J. Opt. Soc. Am.* B **5** (May 1988)

[213] R. W. Boyd, M. G. Raymer, and L. M. Narducci (eds.), *Optical Instabilities* (Cambridge: Cambridge University Press, 1986)

[214] J. Chrostowsky and N. B. Abraham (eds.), *Optical Chaos* (Bellingham, MA: SPIE, 1986)

[215] E. R. Pike and S. Sarkar (eds.), *Frontiers of Quantum Optics* (Bristol: Hilger, 1986)

[216] F. T. Arecchi and R. G. Harrison (eds.), *Instabilities and Chaos in Quantum Optics* (Berlin: Springer, 1987)

[217] E. R. Pike and L. A. Lugiato (eds.), *Chaos, Noise and Fractals* (Bristol: Hilger, 1987)

[218] N. B. Abraham, F. T. Arecchi, and L. A. Lugiato (eds.), *Instabilities and Chaos in Quantum Optics II* (New York: Plenum Press, 1988)

[219] E. J. Quel, J. R. Tredicce, and L. M. Narducci (eds.), *Laser Physics and Quantum Optics* (Singapore: World Scientific, 1990)

[220] P. W. Milonni, J. Akerhalt, and M.-L. Shih, *Chaos in Laser–Matter Interactions* (Singapore: World Scientific, 1987)

[221] L. M. Narducci and N. B. Abraham, *Laser Physics and Laser Instabilities* (Singapore: World Scientific, 1988)

[222] C. O. Weiss and R. Vilaseca, *Dynamics of Lasers* (Weinheim: VCH, 1991)

[223] R. G. Harrison and D. J. Biswas, Pulsating instabilities and chaos in lasers, *Prog. Quantum Electron.* **10**, 147 (1985)

[224] J. R. Ackerhalt, P. W. Milonni, and M.-L. Shih, Chaos in quantum optics *Phys. Rep. (Phys. Lett. C)* **128**, 205 (1985)

[225] C. O. Weiss, Chaotic laser dynamics, *Opt. Quantum Electron.* **20**, 1 (1988)

[226] J. C. Englund, R. R. Snapp, and W. C. Schieve, Fluctuations, instabilities and chaos in the laser-driven nonlinear ring cavity in E. Wolf (ed.) *Progress in Optics* vol. XXI (Amsterdam: North-Holland, 1984), p. 355

[227] R. Vilaseca and R. Corbalan, *Nonlinear Dynamics and Quantum Phenomena in Optical Systems* (Berlin: Springer, 1991)

[228] L. A. Lugiato and L. M. Narducci, Multistability, chaos and spatio-temporal dynamics, in J. Dalibard, J. M. Raymond, and J. Zinn-Justin (eds.) *Fundamental Systems in Quantum Optics*, Les Houches, Session LIII, 1990 (Amsterdam: Elsevier, 1992)

[229] R. G. Harrison and J. S. Uppal (Eds), *Nonlinear Dynamics and Spatial Complexity in Optical Systems* (Bristol and Philadelphia: Scottish University Summer School in Physics and Institute of Physics Publishing, 1993)

[230] Ya. I. Khanin, *Principles of Laser Dynamics* (Amsterdam: North-Holland, 1995)

[231] K. Otsuka, *Nonlinear Dynamics in Optical Complex Systems* (Dordrecht: Kluwer Academic, 1999)

[232] P. Mandel, *Theoretical Problems in Cavity Nonlinear Optics* (Cambridge: Cambridge University Press, 2005)

[233] T. Erneux and P. Glorieux, *Laser Dynamics* (Cambridge: Cambridge University Press, 2010)

[234] G. Nicolis and I. Prigogine, *Self-organization in Non-Equilibrium Systems*, (New York: Wiley, 1974)

[235] G. Nicolis, *Introduction to Nonlinear Science* (Cambridge: Cambridge University Press, 1995)

[236] M. Marsden, *The Geometry of the Zeros of a Polynomial in a Complex Variable* (Providence, RI: American Mathematical Society, 1949)

[237] J. D. Farmer, Chaotic attractors of an infinite-dimensional dynamical system, *Physica D* **4**, 366 (1982)

[238] P. Grassberger and I. Procaccia, Characterization of strange attractors, *Phys. Rev. Lett.* **50**, 346 (1983)

[239] E. Lorenz, Deterministic nonperiodic flow, *J. Atmos. Sci.* **20**, 130 (1963)

[240] H. Haken, Analogy between higher instabilities in fluids and lasers, *Phys. Lett.* A **53**, 77 (1975)

[241] J. P. Eckmann, Roads to turbulence in dissipative dynamical systems, *Rev. Mod. Phys.* **53**, 643 (1981)

[242] R. Thom, *Stabilité structurelle et morphogenèse* (Paris: Interéditions, 1972)

[243] G. P. Puccioni, F. T. Arecchi, G.-L. Lippi, and J. R. Tredicce, Deterministic chaos in laser with injected signal, *Opt. Commun.* **51**, 308 (1984)

[244] G.-L. Oppo and A. Politi, Toda potential in laser equations, *Z. Phys.* B **59**, 111 (1985)

[245] L. W. Casperson, Stability criteria for high-intensity lasers, *Phys. Rev.* A **21**, 911 (1980)

[246] S. T. Hendow and M. Sargent III, The role of population pulsation in single-mode laser instabilities, *Opt. Commun.* **40**, 385 (1982)

[247] L. W. Hillman, R. W. Boyd, and C. R. Stroud, Jr., Natural modes for the analysis of optical bistability and laser instability, *Opt. Lett.* **7**, 426 (1982)

[248] Y. Silberberg and I. Bar-Joseph, The mechanism of instabilities in an optical cavity, *Opt. Commun.* **48**, 53 (1983)

[249] W. J. Firth, E. M. Wright, and E. Cummins, Connection between Ikeda instabilities and phase conjugation in C. R. Bowden, H. M. Gibbs, and S. L. McCall (eds.) *Optical Bistability 2* (New York: Plenum Press, 1984), p. 111

[250] L. A. Lugiato and L. M. Narducci, Single-mode and multimode instabilities in lasers and related optical systems, *Phys. Rev.* A **32**, 1576 (1985)

[251] B. R. Mollow, Stimulated emission and absorption near resonance for driven systems, *Phys. Rev.* A **5**, 2217 (1972)

[252] A. M. Bonch-Bruevich, V. A. Khodovoi, and N. A. Chigir, Changes in the absorption spectrum and of dispersion of a two-level system in a rotating monochromatic radiation field, *Sov. Phys. JETP* **40**, 1027 (1975)

[253] F. Y. Wu, S. Ezekiel, M. Ducloy, and B. R. Mollow, Observation of amplification in a strongly driven two-level atomic system at optical frequencies, *Phys. Rev. Lett.* **38**, 1077 (1977)

[254] M. J. Feigenbaum, Quantitative universality for a class of non-linear transformations, *J. Statist. Phys.* **19**, 25 (1978); The universal metric properties of nonlinear transformations, *J. Statist. Phys.* **21**, 669 (1979)

[255] C. O. Weiss and J. Brock, Evidence for Lorenz-type chaos in a laser, *Phys. Rev. Lett.* **57**, 2804 (1986)

[256] J. Pujol, F. Laguarta, R. Vilaseca, and R. Corbalan, Influence of pump coherence on the dynamic behavior of a laser, *J. Opt. Soc. Am.* B **5**, 1004 (1988)

[257] R. Graham and H. Haken, Quantum theory of light propagation in a fluctuating laser-active medium, *Z. Phys.* **213**, 420 (1968)

[258] E. M. Pessina, G. Bonfrate, L. A. Lugiato, and F. Fontana, Experimental observation of the Risken–Nummedal–Graham–Haken multimode laser instability, *Phys. Rev.* A **56**, 4086 (1997)

[259] E. M. Pessina, F. Prati, J. Redondo, E. Roldán, and G. J. de Valcárcel, Multimode instability in a ring fiber laser, *Phys. Rev.* A **60**, 2517 (1999)

[260] T. Voigt, M. O. Lenz, and F. Mitschke, Risken–Nummedal–Graham–Haken instability finally confirmed experimentally, *Proc. SPIE* **4429**, 112 (2001)

[261] E. Roldán, G. J. de Valcárcel, F. Prati, F. Mitschke, and T. Voigt, Multilongitudinal mode emission in ring cavity class B lasers, in O. G. Calderon and J. M. Guerra (eds.) *Trends in Spatiotemporal Dynamics in Lasers, Instabilities, Polarization Dynamics, and Spatial Structures* (Kerala: Research Signpost, 2005), p. 1

[262] P. Gerber and M. Buttiker, Stability domain of coherent laser waves, *Z. Phys.* B **33**, 219 (1979)

[263] S. T. Hendow and M. Sargent III, Theory of single-mode laser instabilities, *J. Opt. Soc. Am.* B **2**, 84 (1985)

[264] J. R. Tredicce, L. M. Narducci, D. K. Bandy, L. A. Lugiato, and N. B. Abraham, Experimental evidence of mode competition leading to optical bistability in homogeneously broadened lasers *Opt. Commun.* **56**, 435 (1986)

[265] A. Gordon, C. Y. Wang, L. Diehl *et al.*, Multimode regimes in quantum cascade lasers: From coherent instabilities to spatial hole burning, *Phys. Rev.* A 77, 053804 (2008)

[266] D. K. Bandy, L. M. Narducci, L. A. Lugiato, and N. B. Abraham, Time-dependent behaviour of a unidirectional ring laser with inhomogeneous broadening, *J. Opt. Soc. Am.* B **2**, 56 (1985)

[267] E. Roldán, G. J. de Valcárcel, F. Silva, and F. Prati, Multimode emission in inhomogeneously broadened lasers, *J. Opt. Soc. Am.* B **18**, 1601 (2001)

[268] J. Y. Zhang, H. Haken, and H. Ohno, Self–pulsing instability in inhomogeneously broadened traveling-wave lasers, *J. Opt. Soc. Am.* B **2**, 141 (1985)

[269] L. W. Casperson, Spontaneous coherent pulsations in laser oscillators, *IEEE J. Quantum Electron.* **14**, 756 (1978)

[270] J. Bentley and N. B. Abraham, Mode-pulling, mode-splitting and pulsing in a high gain He–Xe laser, *Opt. Commun.* **41**, 52 (1982)

[271] R. Bonifacio and P. Meystre, Critical slowing down in optical bistability, *Opt. Commun.* **27**, 147 (1979)

[272] V. Benza and L. A. Lugiato, Analytical treatment of the transient in absorptive optical bistability, *Lett. Nuovo Cim.* **26**, 405 (1979)

[273] S. Barbarino, A. Gozzini, I. Longo, F. Maccarrone, and R. Stampacchia, Critical slowing-down in microwave absorptive bistability, *Nuovo Cim.* B **71**, 183 (1982)

[274] E. Garmire, J. H. Marburger, S. D. Allen, and H. G. Winful, Transient response of hybrid bistable optical devices, *Appl. Phys. Lett.* **34**, 374 (1979)

[275] F. Mitsche, R. Deserno, J. Mlynek, and W. Lange, Transients in all-optical bistability using transverse optical pumping: Observation of critical slowing down, *Opt. Commun.* **46**, 135–140 (1983)

[276] G. Broggi and L. A. Lugiato, Transient noise-induced optical bistability, *Phys. Rev.* A **29**, 2949 (1984)

[277] F. Mitschke, R. Deserno, J. Mlynek, and W. Lange, Transients in optical bistability: Experiments with external noise, *IEEE J. Quantum Electron.* **21**, 1435–1440 (1985)

[278] R. Bonifacio and L. A. Lugiato, Instabilities for a coherently driven absorber in a ring cavity, *Lett. Nuovo Cim.* **21**, 510 (1978)

[279] R. Bonifacio, M. Gronchi, and L. A. Lugiato, Self-pulsing in bistable absorption, *Opt. Commun.* **30**, 129 (1979)

[280] L. A. Lugiato, Self pulsing in dispersive optical bistability, *Opt. Commun.* **33**, 108 (1980)

[281] B. Segard and B. Macke, Self-pulsing in intrinsic optical bistability with two-level molecules, *Phys. Rev. Lett.* **69**, 412 (1988)

[282] B. Segard, B. Macke, L. A. Lugiato, F. Prati, and M. Brambilla, The multimode instability in optical bistability, *Phys. Rev.* A **39**, 703 (1989)

[283] A. J. van Wonderen and L. G. Suttorp, Instabilities for absorptive optical bistability in a nonideal Fabry–Perot cavity, *Phys. Rev.* A **40**, 7104 (1989)

[284] M. Le Berre, E. Ressayre, and A. Tallet, Physics in counterpropagating light-beam devices: Phase-conjugation and gain concepts in multiwave mixing, *Phys. Rev.* A **44**, 5958 (1991)

[285] H. M. Gibbs, F. A. Hopf, D. L. Kaplan, and R. L. Shoemaker, Observation of chaos in optical bistability, *Phys. Rev. Lett.* **46**, 474 (1981)

[286] L. A. Lugiato, M. L. Asquini, and L. M. Narducci, The relation between the Bonifacio–Lugiato and the Ikeda instabilities in optical bistability, *Opt. Commun.* **41**, 450 (1982)

[287] R. R. Snapp, H. J. Carmichael, and W. C. Schieve, Period doubling and chaos in the optical bistability, *Opt. Commun.* **40**, 68 (1981)

[288] H. J. Carmichael, R. R. Snapp, and W. C. Schieve, Oscillatory instabilities leading to optical turbulence in a bistable ring cavity, *Phys. Rev.* A **26**, 3408 (1982)

[289] L. A. Lugiato, L. M. Narducci, and M. F. Squicciarini, Exact linear stability analysis of the plane-wave Maxwell–Bloch equations for a ring laser, *Phys. Rev.* A **34**, 3101 (1986)

[290] K. Ikeda and O. Akimoto, Instability leading to periodic and chaotic self-pulsations in a bistable optical cavity, *Phys. Rev. Lett.* **48**, 617 (1982)

[291] L. A. Lugiato, L. M. Narducci, D. K. Bandy, and C. A. Pennise, Self-pulsing and chaos in a mean field model of optical bistability, *Opt. Commun.* **43**, 281 (1982)

[292] L. A. Orozco, A. T. Rosenberger, and H. J. Kimble, Intrinsic dynamical instability in optical bistability with two-level atoms, *Phys. Rev. Lett.* **53**, 2547 (1984)

[293] L. A. Orozco, H. J. Kimble, A. T. Rosenberger *et al.*, Single-mode instability in optical bistability, *Phys. Rev.* A **39**, 1235 (1989)

[294] L. A. Lugiato, L. M. Narducci, D. K. Bandy, and C. A. Pennise, Breathing, spiking, and chaos in a laser with injected signal, *Opt. Commun.* **46**, 64 (1983)

[295] J. R. Tredicce, F. T. Arecchi, G.-L. Lippi, and G. P. Puccioni, Instabilities in lasers with an injected signal, *J. Opt. Soc. Am.* B **2**, 173 (1985)

[296] D. K. Bandy, L. M. Narducci, and L. A. Lugiato, Coexisting attractors in a laser with an injected signal, *J. Opt. Soc. Am.* B **2**, 148 (1985)

[297] J. C. Boulnois, P. Cottin, A. Van Lenberghe, F. T. Arecchi, and G. P. Puccioni, Self-pulsing in a CO_2 ring laser with an injected signal, *Opt. Commun.* **58**, 124 (1986)

[298] E. Brun, B. Derighetti, D. Meier, R. Holzner, and M. Ravani, Observation of order and chaos in a nuclear spin-flip laser, *J. Opt. Soc. Am.* B **2**, 156 (1985)

[299] J. C. Antoranz, L. L. Bonilla, J. Gea, and M. G. Velarde, Bistable limit cycles in a model for a laser with a saturable absorber, *Phys. Rev. Lett.* **49**, 35 (1982)

[300] P. Mandel and T. Erneux, Stationary, harmonic, and pulsed operations of an optically bistable laser with saturable absorber, *Phys. Rev.* A **30**, 1893 (1984)

[301] H. T. Powell and G. J. Wolga, Repetitive passive Q-switching of single-frequency lasers, *IEEE J. Quantum Electron.* **7**, 213 (1971)

[302] O. R. Wood and S. E. Schwartz, Passive Q-switching of a CO_2 laser, *Appl. Phys. Lett.* **11**, 88 (1967)

[303] F. de Tomasi, D. Hennequin, B. Zambon, and E. Arimondo, Instabilities and chaos in an infrared laser with saturable absorber. Experiments and vibro-rotational model, *J. Opt. Soc. Am.* B **6**, 45 (1989)

[304] F. L. Hong, M. Tachikawa, T. Oda, and T. Shimizu, Chaotic passive Q-switching pulsation in N_2O laser with a saturable absorber, *J. Opt. Soc. Am.* B **6**, 1378 (1989)

[305] B. Zambon, Theoretical investigations of models for the laser with saturable absorber. A case of homoclinic tangency to a periodic orbit, *Phys. Rev.* A **44**, 688 (1991)

[306] L. A. Lugiato, Transverse nonlinear optics: Introduction and review, *Chaos, Solitons and Fractals* **4**, 1251 (1994)

[307] L. A. Lugiato and R. Lefever, Spatial dissipative structures in passive optical systems, *Phys. Rev. Lett.* **58**, 2209 (1987)

[308] L. A. Lugiato and C. Oldano, Stationary spatial patterns in passive optical systems: Two level atoms, *Phys. Rev.* A **37**, 3896 (1988)

[309] L. A. Lugiato, C. Oldano, and L. M. Narducci, Cooperative frequency locking and stationary spatial structures in lasers, *J. Opt. Soc. Am.* B **5**, 879 (1988)

[310] W. J. Firth, Spatial instabilities in a Kerr medium with with a single feedback mirror, *J. Mod. Opt.* **37**, 151 (1990)

[311] G. P. D'Alessandro and W. J. Firth, Hexagonal spatial patterns for a Kerr slice with a feedback mirror, *Phys. Rev.* A **46**, 537 (1992)

[312] F. T. Arecchi, Space–time complexity in nonlinear optics, *Physica* D **51**, 450 (1991)

[313] L. A. Lugiato, Spatio-temporal structures. Part I, *Phys. Rep.* **219**, 293 (1992)

[314] C. O. Weiss, Spatio-temporal structures. Part II. Vortices and defects in lasers. *Phys. Rep.* **219**, 311 (1992)

[315] F. T. Arecchi, Optical morphogenesis: Pattern formation and competition in nonlinear optics, *Nuovo Cim.* A **107**, 1111 (1994)

[316] W. J. Firth, Pattern formation in passive nonlinear optical systems, in M. Vorontsov and W. B. Miller (eds.) *Self-organization in Optical Systems and Application to Informaton Technology* (Berlin: Springer, 1995)

[317] N. N. Rosanov, Transverse patterns in wide-aperture nonlinear optical systems, in E. Wolf (ed.), *Progress in Optics* vol. XXXV (Amsterdam: North-Holland, 1996), p. 1

[318] L. A. Lugiato, M. Brambilla, and A. Gatti, Optical pattern formation, in B. Bederson and H. Walther (eds.), *Adv. Mol. Opt. Phys.* **40**, 229 (1999)

[319] F. T. Arecchi, S. Boccaletti, and P.-L. Ramazza, Pattern formation and competition in nonlinear optics, *Phys. Rep.* **318**, 1 (1999)

[320] N. N. Rosanov, *Spatial Hysteresis and Optical Patterns* (Berlin: Springer, 2002)

[321] K. Staliunas and V. J. Sanchez-Morcillo, *Transverse Patterns in Nonlinear Optical Resonators* (Berlin: Springer, 2003)

[322] C. Denz, M. Schwob, and C. Weilnau, *Transverse Pattern Formation in Photore-fractive Optics* (Berlin: Springer-Verlag, 2003)

[323] P. Mandel and M. Tlidi, Transverse dynamics in cavity nonlinear optics (2000–2003), *J. Opt.* B **6**, R60 (2004)

[324] N. B. Abraham and W. J. Firth (eds.), Feature issues on transverse effects in nonlinear optics and transverse effects in nonlinear optical systems, *J. Opt. Soc. Am.* B **7**(6, 7) (1990)

[325] M. Saffman and Y. Wang, Collective focussing and modulational instability of light and cold atoms, in N. Akhmediev and A. Ankiewicz (eds.) *Dissipative Solitons: From Optics to Biology and Medicine* (Berlin: Springer, 2008)

[326] H. A. Haus, *Waves and Fields in Optoelectronics* (Englewood Cliffs, NJ: Prentice-Hall, 1984)

[327] P. Ru, L. M. Narducci, J. R. Tredicce, D. K. Bandy, and L. A. Lugiato, The Gauss–Laguerre modes of a ring resonator, *Opt. Commun.* **63**, 310 (1987)

[328] W. J. Firth and A. J. Scroggie, Spontaneous pattern formation in an absorptive system, *Europhys. Lett.* **26**, 521 (1994)

[329] L. A. Lugiato and R. Lefever, Diffractive stationary patterns in passive optical systems, in *Interaction of Radiation with Matter*, a volume in honour of Adriano Gozzini (Pisa: Quaderni della Scuola Normale Superiore, 1987), p. 311

[330] G.-L. Oppo, M. Brambilla, and L. A. Lugiato, Formation and evolution of roll patterns in optical parametric oscillators, *Phys. Rev.* A **49**, 2028 (1994)

[331] A. M. Turing, The chemical basis of morphogenesis, *Phil. Trans. R. Soc. London* B **237**, 37 (1952)

[332] G. Grynberg, E. Le Bihan, P. Verkerk *et al.*, Observation of instabilities due to mirrorless four-wave mixing oscillation in sodium, *Opt. Commun.* **67**, 363 (1988)

[333] W. H. Press, S. A. Teukolsky, W. T. Vetterling, and B. P. Flannery, *Numerical Recipes* (Cambridge: Cambridge University Press, 2002)

[334] P. Manneville, *Dissipative Structures and Weak Turbolence* (San Diego, CA: Academic Press, 1990)

[335] P. K. Jakobsen, J. V. Moloney, A. C. Newell, and R. Indik, Space-time dynamics of wide-gain-section lasers, *Phys. Rev.* A **45**, 8129 (1992)

[336] T. P. Dawling, M. O. Scully, and F. De Martini, Radiative patterns of a classical dipole in a cavity, *Opt. Commun.* **82**, 415 (1991)

[337] P. K. Jakobsen, J. Lega, Q. Feng *et al.*, Nonlinear transverse modes of large aspect ratio, homogeneously broadened laser I. Analysis and numerical simulation, *Phys. Rev.* A **49**, 4189 (1994)

[338] J. Lega, P. K. Jakobsen, J. V. Moloney, and A. C. Newell, Nonlinear transverse modes of large aspect ratio homogeneously broadened lasers II. Pattern analysis near and beyond threshold, *Phys. Rev.* A **49**, 4201 (1994)

[339] W. J. Firth and C. Paré, Transverse modulational instabilities for counterpropagating beams in Kerr media, *Opt. Lett.* **13**, 1096 (1988)

[340] M. Haelterman and G. Vitrant, Drift instability and spatiotemporal dissipative structures in a nonlinear Fabry–Perot resonator under oblique incidence, *J. Opt. Soc. Am.* B **9**, 1563 (1992)

[341] P. La Penna and G. Giusfredi, Spatiotemporal instabilities in a Fabry–Perot resonator filled with sodium vapor, *Phys. Rev.* A **48**, 2299 (1993)

[342] A. Petrossian, L. Dambly, and G. Grynberg, Drift instability for a laser beam transmitted through a rubidium cell with feedback mirror, *Europhys. Lett.* **29**, 209 (1995)

[343] J. P. Seipenbusch, T. Ackemann, B. Schapers, B. Berge, and W. Lange, Drift instability and locking behaviour of optical patterns, *Phys. Rev.* A **56**, R4401 (1997)

[344] Yu. A. Logvin, B. A. Samson, A. A. Afanasév, A. M. Samson, and N. A. Loiko, Triadic Hopf-static structures in two-dimensional optical pattern formation, *Phys. Rev.* A **54**, R4548 (1996)

[345] G. Giusfredi, J. F. Valley, R. Pon, G. Khitrova, and H. M. Gibbs, Optical instabilities in sodium vapor, *J. Opt. Soc. Am.* B **5**, 1181 (1988)

[346] F. Papoff, G. D'Alessandro, G.-L. Oppo and W. J. Firth, Local and global effects of boundaries on optical-pattern formation in Kerr media, *Phys. Rev.* A **48**, 634 (1993)

[347] T. Ackemann, Yu. A. Logvin, A. Heuer, and W. Lange, Transition between positive and negative hexagons in optical pattern formation, *Phys. Rev. Lett.* **75**, 3450 (1995)

[348] M. Brambilla, L. A. Lugiato, V. Penna *et al.*, Transverse laser patterns II. Variational principle for pattern selection, spatial multistability and laser hydrodynamics, *Phys. Rev.* A **43**, 5114 (1991)

[349] S. A. Akhmanov, R. V. Khoklov, and A. P. Suchkorukov, Self-focusing, self-defocusing and self-modulation of laser beams, in F. T. Arecchi and E. O. Schultz-DuBois (eds.), *Laser Handbook* (Amsterdam: North-Holland, 1972), p. 1151

[350] L. Allen, M. W. Beijenbergen, R. J. Spreeuw and J. P. Woerdman, Orbital angular momentum of light and the trasformation of Laguerre–Gaussian laser modes, *Phys. Rev.* A **45**, 8185 (1992)

[351] M. V. Berry, Singularities in waves and rays, in R. Balian, M. Kléman, and J.-P. Poirier (eds.), *Physics of Defects*, Les Houches, Session XXXV (North-Holland, 1981), p. 453 and references cited therein

[352] M. V. Berry, J. F. Nye, and F. J. Wright, The elliptic umbilic diffraction catastrophe, *Phil. Trans. Roy. Soc. Lond.* **291**, 453 (1979)

[353] P. Coullet, L. Gil, and F. Rocca, Optical vortices, *Opt. Commun.* **73**, 403 (1989)

[354] M. Brambilla, F. Battipede, L. A. Lugiato *et al.*, Transverse laser patterns. I. Phase singularity crystals, *Phys. Rev.* A **43**, 5090 (1991)

[355] F. T. Arecchi, G. Giacomelli, P.-L. Ramazza, and S. Residori, Vortices and defect statistics in optical chaos, *Phys. Rev. Lett.* **67**, 3749 (1991)

[356] S. Mertens, G. Dewel, P. Borckmans, and R. Engelhardt, Pattern selection in bistable systems, *Europhys. Lett.* **37**, 109 (1997)

[357] G.-L. Oppo, Formation and control of Turing patterns and phase fronts in photonics and chemistry, *J. Math. Chem.* **45**, 95 (2009)

[358] J. V. Moloney in F. T. Arecchi and R. G. Harrison (eds.) *Instabilities and Chaos* in Quantum Optics (Berlin: Springer, 1987), p. 139 and references cited therein

[359] W. J. Firth and C. O. Weiss, Cavity and feedback solitons, *Optics Photonics News* **13**, 54 (2002)

[360] H. Haelterman, S. Trillo, and S. Wabnitz, Dissipative modulation instability in a nonlinear dispersive ring cavity, *Opt. Commun.* **91**, 401 (1992)

[361] R. Lefever, L. A. Lugiato, Wang Kaige, and N. B. Abraham, Phase dynamics of transverse diffraction patterns in the laser, *Phys. Lett.* A **135**, 254 (1989); Wang Kaige, N. B. Abraham, and L. A. Lugiato, Leading role of the optical phase instabilities in the formation of certain laser transverse patterns, *Phys. Rev.* A **47**, 1263 (1993)

[362] S. Longhi, Transverse patterns in a laser with an injected signal, *Phys. Rev.* A **56**, 2397 (1997)

[363] S. Longhi and A. Geraci, Roll–hexagon transition in an active optical system, *Phys. Rev.* A **57**, R2281 (1998)

[364] W. J. Firth, A. J. Scroggie, G. S. McDonald, and L. A. Lugiato, Hexagonal patterns in optical bistability, *Phys. Rev.* A **46**, 3609 (1992); A. J. Scroggie, W. J. Firth, G. S. McDonald *et al.*, Pattern formation in a passive Kerr cavity, *Chaos, Solitons and Fractals* **4**, 1323 (1994)

[365] W. J. Firth, G. K. Harkness, A. Lord *et al.*, Dynamical properties of two-dimensional Kerr cavity solitons, *J. Opt. Soc. Am.* B **19**, 747 (2002)

[366] D. Gomila and P. Colet, Dynamics of hexagonal patterns in a self-focussing Kerr cavity, *Phys. Rev.* E **76**, 016217 (2007)

[367] W. J. Firth, Temporal cavity solitons – Buffering optical data, *Nature Photon.* **4**, 415 (2010)

[368] S. Coen, M. Haelterman, Ph. Emplit *et al.*, Bistable switching induced by modulational instability in a normally dispersive all-fibre optical cavity, *J. Opt.* B **1**, 36 (1999)

[369] S. Coen and M. Haelterman, Competition between modulational instability and switching in optical bistability, *Opt. Lett.* **24**, 80 (1999)

[370] S. Coen and M. Haelterman, Continuous-wave ultrahigh-repetition-rate pulse-train generation through modulational instability in a passive fiber cavity, *Opt. Lett.* **26**, 39 (2001)

[371] S. Coen, *Passive Nonlinear Optical Fiber Resonators – Fundamentals and Applications*, Thèse de doctorat, Université Libre de Bruxelles (2000)

[372] F. Leo, S. Coen, P. Kockaert *et al.*, Temporal cavity solitons in one-dimensional media as bits in an all-optical buffer, *Nature Photon.* **4**, 471 (2010)

[373] F. Leo, L. Gelens, Ph. Emplit, M. Haelterman, and S. Coen, Dynamics of one-dimensional Kerr cavity solitons, *Opt. Express* **21**, 9180 (2013)

[374] P. Del'Haye, A. Schliesser, O. Arcizet *et al.*, Optical frequency comb generation from a monolithic microresonator, *Nature* **450**, 1214 (2007)

[375] S. Coen, H. G. Randle, Th. Sylvestre, and M. Erkintalo, Modeling of octave-spanning Kerr frequency combs using a generalized Lugiato–Lefever model, *Opt. Lett.* **38**, 37 (2013)

[376] F. Leo, P. Kockaert, Ph. Emplit *et al.*, Experimental generation of 1.6-THz repetition-rate pulse-trains in a passive optical fiber resonator, in *Proceedings of Lasers and Electro-Optics, 2009* and *2009 Conference on Quantum Electronics and Laser Science.* CLEO/QELS (2009); doi: 10.1364/CLEO. 2009.JTuD111

[377] L. A. Lugiato, G.-L.Oppo, J. R. Tredicce, L. M. Narducci, and M. A. Pernigo, Instabilities and spatial complexity in a laser, *J. Opt. Soc. Am.* B **7**, 1019 (1990)

[378] L. A. Lugiato, F. Prati, L. M. Narducci *et al.*, Role of transverse effects in laser instabilities, *Phys. Rev.* A **37**, 3847 (1988)

[379] L. A. Lugiato, Wang Kaige, and N. B. Abraham, Spatial pattern formation in resonators with nonlinear dispersive media, *Phys. Rev.* A **49**, 2049 (1994)

[380] P. D. Drummond, Optical bistability in a radially varying mode, *IEEE J. Quantum Electron.* **17**, 301–306 (1981)

[381] L. A. Lugiato and M. Milani, Effects of Gaussian averaging on laser instabilities, *J. Opt. Soc. Am.* B **2**, 15 (1985)

[382] L. A. Lugiato, G.-L.Oppo, M. A. Pernigo *et al.*, Spontaneous spatial pattern-formation in lasers and cooperative frequency locking, *Opt. Commun.* **68**, 63 (1988)

[383] J. R. Tredicce, E. J. Quel, A. M. Ghazzawi *et al.*, Spatial and temporal instabilities in a CO_2 laser, *Phys. Rev. Lett.* **62**, 1274 (1989)

[384] C. Tamm, Frequency locking of two transverse optical modes of a laser, *Phys. Rev.* A **38**, 5960 (1988)

[385] F. Prati, M. Brambilla, and L. A. Lugiato, Pattern formation in lasers, *Riv. Nuovo Cim.* **17**, 1 (1994)

[386] L. A. Lugiato, F. Prati, L. M. Narducci, and G.-L. Oppo, Spontaneous breaking of the cylindrical symmetry in lasers, *Opt. Commun.* **69**, 387 (1989)

[387] P. Colet, M. San Miguel, M. Brambilla, and L. A. Lugiato, Fluctuations in transverse laser patterns, *Phys. Rev.* A **43**, 3862 (1991)

[388] A. B. Coates, C. O. Weiss, C. Green *et al.*, Dynamical transverse laser patterns, II. Experiments, *Phys. Rev.* A **49**, 1452 (1994)

[389] I. Boscolo, A. Bramati, M. Malvezzi, and F. Prati, Three-mode rotating pattern in a CO_2 laser with high cylindrical symmetry, *Phys. Rev.* A **55**, 738 (1997)

[390] N. N. Rosanov and G. V. Khodova, Autosolitons in bistable interferometers, *Opt. Spectrosc.* **65**, 449 (1988)

[391] N. N. Rosanov and G. V. Khodova, Diffractive autosolitons in nonlinear interferometers, *J. Opt. Soc. Am.* B **7**, 1057 (1990)

[392] N. N. Rosanov, V. A. Smirnov, and N. V. Vyssotina, Numerical simulations of interaction of bright spatial solitons in medium with saturable nonlinearity, *Chaos, Solitons and Fractals* **4**, 1767 (1994)

[393] D. W. McLaughlin, J. V. Moloney, and A. C. Newell, New class of instabilities in passive optical cavities, *Phys. Rev. Lett.* **51**, 75 (1983)

[394] S. Fauve and O. Thual, Localised structures generated by subcritical instability, *J. Physique.* **49**, 1829 (1988)

[395] L. Y. Glebsky and L. M. Lerman, On small stationary localized solutions for the generalized Swift–Hohenberg equation, *Chaos* **5**, 424 (1995)

[396] P. B. Umbanhowar, F. Melo, and H. L. Swinney, Localised excitations in a vertical vibrated granular layer, *Nature* **382**, 793 (1986)

[397] K. A. Gorshkov, L. N. Korzinov, M. I. Rabinovich, and L. S. Tsimring, Random pinning of localized states and the birth of deterministic disorder within gradient models, *J. Statist. Phys.* **74**, 1033 (1994)

[398] O. Lioubashevski, H. Arbell, and J. Fineberg, Dissipative solitary states in driven surface waves, *Phys. Rev. Lett.* **76**, 3959 (1996)

[399] L. S. Tsimring and I. Aranson, Cellular and localized structures in a vibrated granular layer, *Phys. Rev. Lett.* **79**, 213 (1997)

[400] J. Dewel, P. Borckmans, A. De Wit *et al.*, Pattern selection and localized structures in reaction–diffusion systems, *Physica* A **213**, 181 (1995)

[401] G. S. McDonald and W. J. Firth, Spatial solitary wave optical memory, *J. Opt. Soc. Am.* B **7**, 1328 (1990)

[402] C. I. Christov and M. G. Velarde, Dissipative solitons *Physica* D **86**, 323 (1995)

[403] N. Akhmediev and A. Ankiewicz (eds.) *Dissipative Solitons* (Berlin: Springer, 2005)

[404] N. Akhmediev and A. Ankiewicz (eds.) *Dissipative Solitons: From Optics to Biology and Medicine* (Berlin: Springer, 2008)

[405] E. A. Ultanir, G. J. Stegeman, D. Michaelis, C. H. Lange, and F. Lederer, Stable dissipative solitons in semiconductor optical amplifiers, in N. Akhmediev and A. Ankiewicz (eds.) *Dissipative Solitons: From Optics to Biology and Medicine* (Berlin: Springer, 2008) p. 37

[406] W. J. Firth and G. K. Harkness, Cavity solitons, *Asian J. Phys.* **7**, 665 (1998)

[407] W. J. Firth, Theory of cavity solitons, in A. D. Boardman and A. P. Sukhorukov (eds.), *Soliton-driven Photonics* (London: Kluwer, 2001), p. 459

[408] L. A. Lugiato, Introduction to the feature section on cavity solitons: An overview, *IEEE J. Quantum Electron.* **39**, 193 (2003)

[409] L. A. Lugiato, F. Prati, G. Tissoni *et al.*, Cavity solitons in semiconductor devices, in N. Akhmediev and A. Ankiewicz (eds.) *Dissipative Solitons: From Optics to Biology and Medicine* (Berlin: Springer, 2008), p. 978

[410] Th. Ackemann, W. J. Firth, and G.-L. Oppo, Fundamentals and applications of spatial dissipative solitons in photonic devices, in P. R. Berman, E. Arimondo, and Chun C. Lin (eds.) *Advances in Atomic, Molecular and Optical Physics* vol. 57 (Amsterdam: Elsevier, 2009), p. 323

[411] S. Barbay, R. Kuszelewicz, and J. R. Tredicce, Cavity solitons in VCSEL devices, *Adv. Opt. Technol.* 628761 (2011)

[412] M. Brambilla, L. A. Lugiato, and M. Stefani, Interaction and control of optical localised structures, *Europhys. Lett.* **34**, 109 (1996)

[413] M. Tlidi, P. Mandel, and R. Lefever, Localised structures and localised patterns in optical bistability, *Phys. Rev. Lett.* **73**, 640 (1994)

[414] W. J. Firth and A. J. Scroggie, Optical bullet holes: Robust controllable localised states of a nonlinear cavity, *Phys. Rev. Lett.* **76**, 1623 (1996)

[415] T. Maggipinto, M. Brambilla, G. K. Harkness, and W. J. Firth, Cavity solitons in semiconductor microresonators: Existence, stability, and dynamical properties, *Phys. Rev.* E **62**, 8726 (2000)

[416] G.-L. Oppo, A. J. Scroggie, and W. J. Firth, From domain walls to localized structures in degenerate optical parametric oscillators, *J. Opt.* B **1**, 133 (1999)

[417] C. Etrich, D. Michaelis, and F. Lederer, Bifurcation, stability and multistability of cavity solitons in parametric downconversion, *J. Opt. Soc. Am.* B **19**, 792 (2002)

[418] A. G. Vladimirov, J. M. McSloy, D. V. Skryabin, and W. J. Firth, Two-dimensional clusters of solitary structures in driven optical cavities, *Phys. Rev.* E **65**, 0046606 (2002)

[419] B. Schaepers, Th. Ackemann, and W. Lange, Properties of feedback solitons in a single-mirror experiment, *IEEE J. Quantum Electron.* **39**, 227 (2003)

[420] F. Pedaci, P. Genevet, S. Barland, M. Giudici, and J. R. Tredicce, Positioning cavity solitons with a phase mask, *Appl. Phys. Lett.* **89**, 221111 (2006)

[421] J. L. Oudar, T. Rivera, R. Kuszelewicz, and F. Ladan, Etched arrays of quantum well optical bistable microresonators, *J. Physique* III **4**, 2361 (1994)

[422] L. Spinelli, G. Tissoni, L. A. Lugiato, and M. Brambilla, Thermal effects and transverse structures in semiconductor microcavities with population inversion, *Phys. Rev.* A **66**, 023817 (2002)

[423] A. J. Scroggie, J. M. McSloy, and W. J. Firth, Self-propelled cavity solitons in semiconductor microcavities, *Phys. Rev.* E **66**, 036607 (2002)

[424] R. Kheradmand, L. A. Lugiato, G. Tissoni, M. Brambilla, and H. Tajalli, Cavity soliton mobility in semiconductor microresonators, *Math. Computing Simulations* **69**, 346 (2005)

[425] G. Tissoni, L. Spinelli, M. Brambilla *et al.*, Cavity solitons in bulk semiconductor microcavities: microscopic model and modulational instabilities, *J. Opt. Soc. Am.* B **16**, 2083 (1999); Cavity solitons in bulk semiconductor microcavities: dynamical properties and control, *J. Opt. Soc. Am.* B **16**, 2095 (1999)

[426] M. Brambilla, L. A. Lugiato, F. Prati, L. Spinelli, and W. J. Firth, Spatial soliton pixels in semiconductor devices, *Phys. Rev. Lett.* **79**, 2042 (1997)

[427] L. Spinelli, G. Tissoni, M. Brambilla, F. Prati, and L. A. Lugiato, Spatial solitons in semiconductor microcavities, *Phys. Rev.* A **58**, 2542 (1998)

[428] X. Hachair, F. Pedaci, E. Caboche *et al.*, Cavity solitons in a driven VCSEL above threshold, *IEEE J. Sel. Top. Quantum Electron.* **12**, 339 (2006)

[429] S. Barland, J. R. Tredicce, M. Brambilla *et al.*, Cavity solitons as pixels in semiconductors, *Nature* **419**, 699 (2002)

[430] E. Caboche, S. Barland, M. Giudici *et al.*, Cavity soliton motion in presence of device defects, *Phys. Rev.* A **80**, 053814 (2009)

[431] M. Brambilla, L. Columbo, T. Maggipinto, and G. Patera, 3D cavity light bullets in a nonlinear optical resonator, *Phys. Rev. Lett.* **93**, 203901 (2004)

[432] M. Tlidi and P. Mandel, Three-dimensional optical crystals and localized structures in cavity second harmonic generation, *Phys. Rev. Lett.* **83**, 4995 (1999)

[433] S. D. Jenkins, F. Prati, L. A. Lugiato, L. Columbo, and M. Brambilla, Cavity light bullets in a dispersive Kerr medium, *Phys. Rev.* A **80**, 033832 (2009)

[434] Y. Tanguy, Th. Ackemann, W. J. Firth, and R. Jaeger, Realization of a semiconductor-based cavity soliton laser, *Phys. Rev. Lett.* **100**, 013907 (2008)

[435] P. Genevet, S. Barland, M. Giudici, and J. R. Tredicce, Cavity soliton laser based on mutually coupled semiconductor microresonators, *Phys. Rev. Lett.* **101**, 123905 (2008)

[436] T. Elsass, K. Gauthron, G. Beaudoin *et al.*, Fast manipulation of laser localized structures in a monolithic vertical cavity with saturable absorber, *Appl. Phys.* B **98**, 307 (2010)

[437] N. N. Rozanov and S. V. Fedorov, Diffraction switching waves and autosolitons in a laser with saturable absorption, *Opt. Spectrosc.* **72**, 782 (1992)

[438] S. V. Fedorov, A. G. Vladimirov, G. V. Khodova, and N. N. Rosanov, Effect of frequency detunings and finite relaxation rates on laser localized structures, *Phys. Rev. E* **61**, 5814 (2000)

[439] M. Bache, F. Prati, G. Tissoni *et al.*, Cavity soliton laser based on VCSEL with saturable absorber, *Appl. Phys. B* **81**, 913 (2005)

[440] F. Prati, P. Caccia, G. Tissoni *et al.*, Effects of carrier radiative recombination on a VCSEL-based cavity soliton laser, *Appl. Phys. B* **88**, 405 (2007)

[441] G. Tissoni, K. M. Aghdami, F. Prati, M. Brambilla, and L. A. Lugiato, Cavity soliton laser based on a VCSEL with saturable absorber, in O. Descalzi, M. Clerc, S. Residori, and G. Assanto (eds.) *Localized States in Physics: Solitons and Patterns* (Berlin: Springer, 2011), p. 187

[442] K. M. Aghdami, F. Prati, P. Caccia *et al.*, Comparison of different switching techniques in a cavity soliton laser, *Eur. Phys. J. D* **47**, 447 (2008)

[443] S. V. Fedorov, N. N. Rosanov, and N. A. Shatsev, Two-dimensional solitons in B-class lasers with saturable absorption, *Opt. Spectrosc.* **102**, 449 (2007)

[444] F. Prati, G. Tissoni, L. A. Lugiato, K. M. Aghdami, and M. Brambilla, Spontaneously moving solitons in a cavity soliton laser with circular section, *Eur. Phys. J. D* **59**, 73 (2010)

[445] F. Prati, L. A. Lugiato, G. Tissoni, and M. Brambilla, Cavity soliton billiards, *Phys. Rev. A* **84**, 053852 (2011)

[446] Y. Couder, S. Protiere, E. Fort, and A. Boudaoud, Dynamical phenomena: Walking and orbiting droplets, *Nature* **437**, 208 (2005)

[447] A. Eddi, E. Fort, F. Moisy, and Y. Couder, Unpredictable tunneling of a classical wave–particle association, *Phys. Rev. Lett.* **102**, 240401 (2009)

[448] F. Pedaci, G. Tissoni, S. Barland, M. Giudici, and J. R. Tredicce, Mapping local defects of extended media using localized structures, *Appl. Phys. Lett.* **93**, 111104 (2008)

Index

Printed in the United States
By Bookmasters